Flow and Creep in the Solar System: Observations, Modeling and Theory

NATO ASI Series

Advanced Science Institutes Series

A Series presenting the results of activities sponsored by the NATO Science Committee, which aims at the dissemination of advanced scientific and technological knowledge, with a view to strengthening links between scientific communities.

The Series is published by an international board of publishers in conjunction with the NATO Scientific Affairs Division

A Life Sciences	Plenum Publishing Corporation
B Physics	London and New York
C Mathematical	Kluwer Academic Publishers
and Physical Sciences	Dordrecht, Boston and London
D Behavioural and Social Sciences	
E Applied Sciences	
F Computer and Systems Sciences	Springer-Verlag
G Ecological Sciences	Berlin, Heidelberg, New York, London,
H Cell Biology	Paris and Tokyo
I Global Environmental Change	

NATO-PCO-DATA BASE

The electronic index to the NATO ASI Series provides full bibliographical references (with keywords and/or abstracts) to more than 30000 contributions from international scientists published in all sections of the NATO ASI Series.
Access to the NATO-PCO-DATA BASE is possible in two ways:

– via online FILE 128 (NATO-PCO-DATA BASE) hosted by ESRIN,
Via Galileo Galilei, I-00044 Frascati, Italy.

– via CD-ROM "NATO-PCO-DATA BASE" with user-friendly retrieval software in English, French and German (© WTV GmbH and DATAWARE Technologies Inc. 1989).

The CD-ROM can be ordered through any member of the Board of Publishers or through NATO-PCO, Overijse, Belgium.

Series E: Applied Sciences - Vol. 391

Flow and Creep in the Solar System: Observations, Modeling and Theory

edited by

David B. Stone

Geophysical Institute and
Department of Geology and Geophysics,
University of Alaska,
Fairbanks, AK, U.S.A.

and

S. K. Runcorn

University of Alaska,
Fairbanks, AK, U.S.A. and
Physics Department,
Imperial College,
London, U.K.

Kluwer Academic Publishers

Dordrecht / Boston / London

Published in cooperation with NATO Scientific Affairs Division

ASTRONOMY

Proceedings of the NATO Advanced Study Institute on
Dynamic Modeling and Flow in the Earth and Planets
Fairbanks, AK, U.S.A.
June 17–28, 1991

ISBN 0-7923-2148-0

Published by Kluwer Academic Publishers,
P.O. Box 17, 3300 AA Dordrecht, The Netherlands.

Kluwer Academic Publishers incorporates the publishing programmes of
D. Reidel, Martinus Nijhoff, Dr W. Junk and MTP Press.

Sold and distributed in the U.S.A. and Canada
by Kluwer Academic Publishers,
101 Philip Drive, Norwell, MA 02061, U.S.A.

In all other countries, sold and distributed
by Kluwer Academic Publishers Group,
P.O. Box 322, 3300 AH Dordrecht, The Netherlands.

Printed on acid-free paper

TABLE OF CONTENTS

vi

PREFACE

The NATO ASI held in the Geophysical Institute, University of Alaska Fairbanks, June 17-28, 1991 was, we believe, the first attempt to bring together geoscientists from all the disciplines related to the solar system where fluid flow is a fundamental phenomenon. The various aspects of flow discussed at the meeting ranged from the flow of ice in glaciers, through motion of the solar wind, to the effects of flow in the Earth's mantle as seen in surface phenomena. A major connecting theme is the role played by convection. For a previous attempt to review the various ways in which convection plays an important role in natural phenomena one must go back to an early comprehensive study by J. Wasiutynski in "Astrophysica Norvegica" vol. 4, 1946. This work, little known now perhaps, was a pioneering study. In understanding the evolution of bodies of the solar system, from accretion to present-day processes, ranging from interplanetary plasma to fluid cores, the understanding of flow hydrodynamics is essential. From the large scale in planetary atmospheres to geological processes, such as those seen in magma chambers on the Earth, one is dealing with thermal or chemical convection. Count Rumford, the founder of the Royal Institution, studied thermal convection experimentally and realized its practical importance in domestic contexts. It was not however until early in the present century, when the famous experiment of Benard demonstrated the regularity of the cells produced in a layer of fluid heated from below, that a scientific discipline was launched. The study of such instabilities became an important theme in applied mathematics, as pursued by Rayleigh, Jeffreys, Vening-Meinesz and finally Chandrasekhar.[1] These studies and the comprehensive work "Hydrodynamic and Hydromagnetic Stability" of Chandrasekhar provided a mathematical account of convection and the patterns which develop in various geometries at marginal stability. But experiment has been productive too, perhaps starting with Hide's experiments on convection in rotating cylindrical annuli. Many qualitatively different phenomena are observed as the various non-dimensional parameters are altered. In the field of geosciences, thermal convection was first recognized as being of importance in meteorology in the observed regular structures sometimes seen in cloud formations. It was in connection with explaining the rolls seen in clouds that an early laboratory experiment[2] was done, showing that the familiar hexagonal pattern of the Benard cell in a layer of fluid heated from below changed to a pattern of parallel rolls in the presence of a vertical shearing flow. Very much more recently, apparently in ignorance of the earlier one, a similar laboratory experiment was carried out to explain features of the geoid over the Pacific plate which seemed to imply the existence of convecting rolls in the upper mantle with their axes in the direction of the plate motion.

This and other such examples suggested to us that it would be a valuable exercise to bring together geoscientists attacking convection and other flow problems in their own disciplines, some from the theoretical side, and some observing the effects of fluid motions. The sharing of common problems, experience and insights between these disciplines, particularly now that powerful computers are being used to model such problems, could be of great value. This experiment, which could have foundered on the different languages—or jargon in the several disciplines—in fact was judged to be worthwhile. Of course, a study institute of this sort is only as good as its participants, and we would like to thank them all for their enthusiasm and willingness to exchange ideas. A list of the papers presented is given at the end of this Preface.

We wish to express our appreciation of the sterling work Nancy Smoyer did in making preparations for the meeting and ensuring its smooth running. We also express our appreciation for the secretarial assistance given to us by Joan Roderick and Sue Poling, and for the invaluable assistance of Barbara Severin in the editorial work. We wish to thank many students whose help with logistics, field trips, projectors, etc. was most valuable. Informal and educative evening lectures were given by Terrence Cole of the History department on the Development of Fairbanks, by R. G. White on Alaskan wildlife, and by Neal Brown on the Poker Flat Rocket Research Range. We also owe a debt of thanks to NATO for funding the meeting, the National Science Foundation for travel grants and the Geophysical Institute of the University of Alaska for financial support and the use of their facilities. We also gratefully acknowledge the enthusiastic help of the international organizing committee: Syun-Ichi Akasofu, Director of the Geophysical Institute, University of Alaska Fairbanks; Claude Froidevaux, Laboratoire de Geologie De l'Ecole Normale Superieure, Paris; Raymond Hide, Robert Hooke Institute, Clarendon Laboratory, Oxford; and Thomas Royer, Institute of Marine Science, University of Alaska Fairbanks. These proceedings, which give a very good coverage of the variety of topics discussed, are presented to the wider audience of those geoscientists who recognize in the solar system the truth of the ancient philosophy that "all is in change, nothing is stationary," or as Heraclitus put it in the 6th century B.C. "You never step into the same river twice."

1. Chandrasekhar, S. (1961) Hydrodynamic and Hydromagnetic Stability. Oxford: Clarendon Press. 652 pp.
2. Wasiutynski, J. (1946) Studies in Hydrodynamics and Structures of Stars and Planets. Astrophysica Norvegica 4, 497 pp.

D. B. Stone
S. K. Runcorn

PARTICIPANTS

Syun-Ichi Akasofu
Geophysical Institute
University of Alaska
Fairbanks, AK 99775

Don L. Anderson
Seismological Laboratory 252-21
California Institute of Technology
Pasadena, CA 91125

Benoit Barriere
Geological Sciences
2112 Snee Hall
Cornell University
Ithaca, NY 14853

Laurent Bastian
Lamont Doherty Geological Observatory
Columbia University
Palisades, NY 10964

John R. Baumgardner
Mail Stop B-216
Los Alamos National Laboratory
Los Alamos, NM 87545

Phillipe Cardin
Dept. of Earth and Planetary Sciences
Johns Hopkins University
Baltimore, MD 21210

Douglas Christensen
Geophysical Institute
University of Alaska
Fairbanks, AK 99775

Scott Condie
Research School of Earth Sciences
Australian National University
G.P.O. Box 4
Canberra, A.C.T. 2600
AUSTRALIA

Matthew J. Cordery
Dept. Earth, Atmos. & Planetary Sci.
Massachusetts Institute of Technology
Cambridge, MA 02139

Steve Crumley
Geophysical Institute
University of Alaska
Fairbanks, AK 99775

Bertrand Daudre
Department of Earth Sciences
Free University
De Boelelaan 1085
1081 HV Amsterdam
THE NETHERLANDS

Keith Echelmeyer
Geophysical Institute
University of Alaska
Fairbanks, AK 99775

Claude Froidevaux
Laboratoire de Geologie
de l'Ecole Normale Superieur
24 rue Lhomond
75231 Paris
FRANCE

Ina Geyer
Inst. für Geophysik und Meteorologie
Universität zu Köln
Albertus-Magnus-Platz
5000 Köln - 41
GERMANY

* Laurent Goeb
Room 545, International House
2299 Piedmont Ave.
Berkeley, CA 94720

Orhan Gokcol
Istanbul Technical University
Dept. of Space Sciences & Technology
Maslak
80626 Istanbul
TURKEY

David Gubbins
The University of Leeds
Leeds LS2 9JT
UNITED KINGDOM

Susan L. Hautala
School of Oceanography WB-10
University of Washington
Seattle, WA 98195

Raymond Hide
Robert Hooke Institute
Old Observatory
Clarendon Laboratory
Parks Road
Oxford OX2 3PU
UNITED KINGDOM

Gauthier Hulot
Institut de Physique du Globe de Paris
4, Place Jussieu
75252 Paris
FRANCE

Keita Iga
Ocean Research Institute
University of Tokyo
15-1 I-chome Minamidai Nakano-ku
Tokyo
JAPAN

Barclay Kamb
Division of Geol. & Planetary Sci. 170-25
California Institute of Technology
Pasadena, CA 91125

Joseph Kan
Geophysical Institute
University of Alaska
Fairbanks, AK 99775

Koji Kawasaki
Geophysical Institute
University of Alaska
Fairbanks, AK 99775

Sharon Kedar
Seismological Laboratory 252-12
California Institute of Technology
Pasadena, CA 91125

Juergen Kienle
Geophysical Institute
University of Alaska
Fairbanks, AK 99775

James Kirklin
Dept. of Earth & Planetary Sciences
Johns Hopkins University
Baltimore, MD 21210

Laurent Labous
Scripps Institute of Oceanography
UCSD
La Jolla, CA 92037

Tine Larsen
Dept. of Geology & Geophysics
310 Pillsbury Dr.
University of Minnesota
Minneapolis, MN 55455

Lawrence A. Lawver
Institute for Geophysics
8701 N. Mopac Expy.
Austin, TX 78759-8345

* Yves Le Stunff
Room 565, International House
2299 Piedmont Ave.
Berkeley, CA 94720

Craig Lingle
Geophysical Institute
University of Alaska
Fairbanks, AK 99775

Louis A. Lliboutry
3 Avenue de la Foy
38700 Corenc
FRANCE

David Loper
Geophysical Fluid Dynamics Institute
Florida State University
Tallahassee, FL 32306

Yu-Qing Lou
Dept. of Astronomy & Astrophysics
The University of Chicago
5640 South Ellis Ave.
Chicago, IL 60637

Keith MacGregor
High Altitude Observatory-NCAR
P.O. Box 3000
Boulder, CO 80307-3000

Alain Mazaud
Centre des Faibles Radioactivites
Laboratoire Mixte CNRS-CEA
Avenue de la Terrasse
91198 Gif-sur-Yvette Cedex
FRANCE

Dagmar Olbertz
Dept. of Geophysics, Inst. of Earth Science
University of Utrecht
P.O. Box 80.021
3508 TA Utrecht
THE NETHERLANDS

John Olson
Geophysical Institute
University of Alaska
Fairbanks, AK 99775

Carol Peterson
Geophysical Institute
University of Alaska
Fairbanks, AK 99775

Nick Petford
Department of Earth Sciences
University of Liverpool
P.O. Box 147
Liverpool LB9 3BX
UNITED KINGDOM

Alexei Poliakov
Institute of Geology
Uppsala University
Box 555,751 22 Uppsala
SWEDEN

David Rees
Atmospheric Physics Laboratory
University College London
67-73 Riding House Street
London WIP 7PP
UNITED KINGDOM

Alan Rice
Department of Physics
Box 390
University of Colorado
Boulder, CO 80309

Jose A. C. Rocha e Silva
University of East Asia
Faculty of Science and Technology
P.O. Box 3001
Macau via Hong Kong

Juan Roederer
Geophysical Institute
University of Alaska
Fairbanks, AK 99775

Tom Royer
Institute of Marine Sciences
University of Alaska
Fairbanks, AK 99775

Bert Rudels
Institut für Meereskunde
Univ. Hamburg
Troplowitzstrasse 7
D-2000 Hamburg 54
GERMANY

S. Keith Runcorn
Geophysical Institute
University of Alaska
Fairbanks, AK 99775 and
Physics Department
Imperial College
London
UNITED KINGDOM

William Sackinger
Geophysical Institute
University of Alaska
Fairbanks, AK 99775

Mujgan Salk
Dokuz Eylul University
Fak. Jeoloji Bolumu
Bornova 35100
Izmir
TURKEY

Graeme Sarson
Department of Earth Sciences
University of Leeds
Leeds LS2 9JT
UNITED KINGDOM

Jorg Schmalzl
Institute of Earth Science
Dept. of Theoretical Geophysics
Budapestlaan 4
3508 TA Utrecht
THE NETHERLANDS

Cheryl Searcy
Geophysical Institute
University of Alaska
Fairbanks, AK 99775

Craig Searcy
Geophysical Institute
University of Alaska
Fairbanks, AK 99775

Steven T. Siems
Department of Geophysics, AK-50
University of Washington
Seattle, WA 98195

Viatcheslav Solomatov
Seismological Laboratory 252-21
California Institute of Technology
Pasadena, CA 91125

Sharon Spitzak
Geophysical Institute
University of Alaska
Fairbanks, AK 99775

Scott Stihler
Geophysical Institute
University of Alaska
Fairbanks, AK 99775

David B. Stone
Geophysical Institute
University of Alaska
Fairbanks, AK 99775

Paul J. Tackley
Seismological Laboratory 252-21
California Institute of Technology
Pasadena, CA 91125

Hiroshi Tanaka
Institute of Geoscience
University of Tsukuba
Tsukuba, Ibaraki 305
JAPAN

Nathalie Thomas
Institut de Physique du Globe de Paris
4, Place Jussieu
75252 Paris
FRANCE

Ronald van Balen
Department of Earth Sciences
Free University
De Boelelaan 1085
1081 HV Amsterdam
THE NETHERLANDS

Margaret S. Woyski
Dean's Office MH-166
California State University
Fullerton, CA 92634

David Yuen
Minnesota Supercomputer Institute
1200 Washington Avenue South
Minneapolis, MN 55415

Keke Zhang
Department of Mathematics
University of Exeter
Exeter EX4 4QJ
UNITED KINGDOM

Yihong Zhang
Geophysical Institute
University of Alaska
Fairbanks, AK 99775

Dapeng Zhao
Seismological Laboratory 252-21
California Institute of Technology
Pasadena, CA 91125

* Current address unknown

PAPERS PRESENTED:

Akasofu, S.: Reversals of the magnetic field of the sun
Anderson, D.: The style of convection in the earth's mantle
Baumgardner, J.: Convective motion in 3-D shells
Cardin, P.; Olson, P.: Experiments on convection in a rapidly rotating spherical shell
Daudre, B.: Dynamics of salt diapirs
Echelmeyer, K.: Ice flow measurements: Greenland, Alaska and Antarctica
Froidevaux, C.: Geoid and mantle plumes
Geyer, I.: Atmosphere of Io
Gokcol, O.: Interplanetary magnetic field and its possible effects on the ionospheric critical frequencies
Gubbins, D.: Core dynamo models and reversals
Gubbins, D.: Geomagnetism and inferences for core motions
Hautala, S.; Riser, S.: Does east pacific rise buoyancy flux drive a large scale gyre in the deep south pacific ocean?
Hide, R.: Flow in the outer planets
Hide, R.: Motions in rotating fluids, Earth and planetary applications
Iga, K.; Kimura, R.: Bubble convection—Theory and numerical simulation
Kamb, W. B.: Ice, rheology and flow of glaciers
Kan, J.: Nonlinear interactions between the solar wind and the magnetosphere-ionosphere system
Kedar, S.; Anderson, D.; Stevenson, D.: Relationship between hotspots and mantle structure—correlation with whole mantle seismic tomography
Kienle, J.: Volcanic plumes
Lawver, L. : Paleogeographic reconstructions, breakup of Gondwanaland
Lliboutry, L.: Ice-sheet dynamics
Lliboutry, L.: Strategies for modeling climate change
Loper, D.: Structure and process within the Earth's interior
MacGregor, K.: Convection and flow in the sun and stars
Mazaud, A.; Bard, E.; Laj, C.; Arnold, M.; Tric, E.: A geomagnetic calibration of the radiocarbon time-scale
Mazaud, A.; Laj, C.; Fuller, M.: Lateral variations at the core-mantle boundary revealed by geomagnetic reversal patterns
Niebauer, J.: Ocean circulations and ocean-atmosphere coupling
Olbertz, D.; Hansen, U.: Topography at the core-mantle-boundary as a consequence of thermal and thermochemical boundary layer instabilities
Rees, D.: A review of the dynamics of lower and upper thermosphere
Rice, A.: Geophysical fluid dynamics in the crust
Rice, A.: Volcanism: Geometry and magma motion
Rudels, B.: High latitude ocean convection
Runcorn, S. K.: Electromagnetic monitoring of ocean currents
Runcorn, S. K.: Fundamentals—A review of the key role of flow in the solar system
Runcorn, S. K.: Lunar convection and paleomagnetism
Schmalzl, J.; Hansen, U.: The dynamics of subcritical double-diffusive convection in the southern ocean
Siems, S.: A numerical investigation of cloud-top entrainment
Tackley, P.; Stevenson, D.: The production of large off-axis non-hotspot seamounts by Rayleigh-Taylor instabilities at the solidus
Tanaka, H.: Atmospheric convection
van Balen, R.: Flow in sedimentary basins
Yuen, D. A.: Strongly chaotic thermal convection
Zhang, K.: Hydrodynamic waves in the earth's core and in the planetary system

CONVECTION AND FLOWS IN THE SUN AND STARS

K. B. MACGREGOR
High Altitude Observatory
National Center for Atmospheric Research[1]
P.O. Box 3000
Boulder, Colorado 80307, U.S.A.

ABSTRACT. Fluid motions associated with a variety of hydrodynamical and MHD processes occur continuously in the atmosphere and interior of the Sun. The present brief review focuses on one such flow, namely, the internal solar rotation. After summarizing some of what is known about the structure and physics of the Sun's interior, we discuss the properties of solar acoustic oscillations and how analysis of them can reveal information about the internal angular velocity distribution. The results of such analyses are then compared with the predictions of models based on particular angular momentum transport mechanisms.

1. Introduction

Much of the discussion at the present meeting has centered on the observation, measurement, and interpretation of material motions within the Earth's interior. Among the topics receiving attention have been the structure of convection in the mantle and the nature of the fluid motions which sustain magnetic dynamo activity in the core. These diverse phenomena have in common the fact that much of our quantitative understanding of them has been obtained through application of the physical laws governing the dynamics and energetics of rotating, viscous, electrically conducting fluids.

On a broader scale, most of the matter contained in the visible universe appears to have the following physical attributes: (i) it is gaseous; (ii) it has angular momentum; and (iii) it is (for the most part) partially to fully ionized. Hence, to the extent to which collisions between the microscopic constituents of the gas are frequent enough to make a fluid description viable, the general approach described above can be utilized in studying a variety of astrophysical systems. Of all such systems, the Sun, by virtue of its proximity, affords us the unique opportunity to observe hydrodynamical and magnetohydrodynamical processes in an astrophysical context with relatively good spatial/temporal resolution. Some of these processes must be operative in more distant, unresolvable objects; some are the stellar analogues of processes which occur inside the Earth, albeit, in a fluid with considerably different transport properties.

A variety of flows and flow-related phenomena take place throughout the solar interior and atmosphere. These range from the rotationally induced, meridional circulation which encompasses much of the Sun's radiative core to the thermally driven, supersonic expansion of the

---

[1]The National Center for Atmospheric Research is sponsored by the National Science Foundation.

D. B. Stone and S. K. Runcorn (eds.),
Flow and Creep in the Solar System: Observations, Modeling and Theory, 1–11.
© 1993 *Kluwer Academic Publishers. Printed in the Netherlands.*

outer layers of the high-temperature corona to form the solar wind. In the present paper, we confine our attention to a single example from the diversity of fluid motions within the Sun, namely, the internal solar rotation. In recent years, much has been learned about the rotational state of the Sun's interior, largely through the use of techniques which are quite similar to those of terrestrial seismology. After first presenting a brief overview of the structure of the solar interior, we describe how analysis of the acoustic oscillation modes observed at the Sun's surface can reveal information about the depth and latitude dependence of the solar angular velocity. These observational inferences are then discussed in the context of mechanisms for angular momentum redistribution inside the Sun. Here, particular emphasis is given to the interaction between differential rotation and a large-scale, poloidal magnetic field, and how this process might affect angular momentum transport in the solar radiative interior. We conclude by examining the implications of these considerations for the rotational evolution of solar-type stars in general.

2. Structure of the Solar Interior: An Overview

We begin with a summary of the solar vital statistics. As a star, the most readily measured physical properties of the Sun are its mass ($M_\odot$), radius ($R_\odot$), and bolometric luminosity ($L_\odot$; i.e., the radiative flux emitted at all wavelengths, integrated over the entire solar surface). The values of these quantities determined from observations are $M_\odot = 1.989 \times 10^{33}$ g, $R_\odot = 6.96 \times 10^{10}$ cm, and $L_\odot = 3.82 \times 10^{33}$ erg -s^{-1}. The effective temperature T_{eff} of a star is the temperature of a black body having a total radiative flux equal to that emergent from the visible stellar surface; for the Sun, $T_{eff,\odot} = 5770$ K. Because the Sun rotates, it is slightly oblate. Observations indicate that the difference between the Sun's equatorial and polar radii (as a fraction of the polar radius) is $\leq 2 \times 10^{-5}$. Finally, from geophysical and meteoritic evidence, the age of the Sun is estimated to be within the range $t_\odot = 4.5$–4.7×10^9 years.

In contrast to the measurable characteristics enumerated above, almost everything which is known about the solar interior has been inferred from theoretical models. In the absence of complications arising from the inclusion of rapid rotation and strong, internal magnetic fields, such models are constructed by solving the system of equations expressing conservation of mass, momentum, and energy for a spherical, hydrostatic star (see, e.g., Schwarzschild, 1958; Stix, 1989). When supplemented by (i) an expression for the energy generation rate due to nuclear reactions, (ii) an equation of state and radiative opacities, and (iii) appropriate boundary conditions at the center and surface, these equations provide the means for determining the structure of the solar interior, and tracing its evolution in time, starting from an initially chemically homogeneous state prior to the onset of nuclear burning. If the chemical composition of this initial state is chosen such that the fractional abundances (by mass) of hydrogen, helium, and heavier elements are, respectively, X = 0.73, Y = 0.25, and Z = 0.02, then the following picture of the present-day solar interior emerges from model calculations.

The internal structure of the Sun is representative of that of low-mass stars whose radiative output derives principally from the conversion of hydrogen to helium via the set of nuclear reactions which constitute the proton-proton chain (cf., Stix, 1989). In so-called "standard" solar models, the prevailing thermodynamic conditions are sufficient to sustain energy generation by such nuclear processes only within a spherical volume of radius $r \approx 0.3 R_\odot$ about the Sun's center ($r = 0$). At $r = 0$, the gas pressure implied by the constraint of hydrostatic equilibrium is $p \approx 2.5 \times 10^{17}$ dyne -cm^{-2} ($\sim GM_\odot^2/R_\odot^4$), while the mass density ρ and gas temperature T have the values $\rho \approx 150$ g-cm^{-3} ($\sim M_\odot/R_\odot^3$) and $T \approx 1.5 \times 10^7$ K ($\sim m_p p/\rho k$). Note that the gradual transmutation of hydrogen into helium leads, over time, to a decrease in the central value of the hydrogen mass fraction X and an increase in the central value of the mean molecular weight μ. For a model with $X = 0.73$ and $\mu = 0.61$ m_H (m_H is the hydrogen mass) at the start of core hydrogen-burning, $X = 0.34$ and $\mu = 0.85$ m_H at $r = 0$ by the time the solar age is attained.

Hydrogen burning within the innermost portion of the Sun's core liberates an amount of energy slightly in excess of 26 MeV for each helium nucleus produced. Deep inside the Sun, this energy is transported radially outward by radiative processes. Within the energy-producing central region of the Sun (e.g., 95% of $L_\odot$ is generated in the volume $r \le 0.21R_\odot$ which contains a mass $0.38M_\odot$), virtually all chemical species are fully ionized, electron scattering dominates the opacity, and radiative transport is diffusive in character. Denoting the mass absorption coefficient of the stellar material by κ (units: cm^2 - g^{-1}), the radiative flux F depends on local physical conditions according to $F \sim -(T^3/\kappa\rho)(dT/dr)$, and is essentially constant (i.e., radiative equilibrium obtains) for $0.3R_\odot \lesssim r \lesssim 0.74R_\odot$. Throughout this interval, both ρ and T decrease monotonically from their central values; at $r = 0.74R_\odot$, $\rho \approx 0.1$ g -cm^{-3} and $T \approx 2 \times 10^6$ K. Associated with this overall outward decrease in the magnitudes of ρ and T is an increase in the magnitude of the opacity κ. In the outer layers of the solar interior, hydrogen and helium, which by virtue of their abundances are the largest contributors to κ, become partially ionized. Near $r = 0.74R_\odot$, κ becomes sufficiently large (due to the increasing importance of bound-free transitions) that the magnitude of the temperature gradient required to transport the outward energy flux by radiative diffusion exceeds the local adiabatic gradient, and convective instability occurs.

The radius $r = 0.74R_\odot$ marks the base of the solar convection zone. Throughout the region $0.74R_\odot \le r \le R_\odot$ (which contains only 0.02 of the solar mass), the energy flux generated by nuclear reactions in the deep interior is transported outward by the material motions of the convectively unstable fluid. In the context of standard models for the solar interior, convective energy transport is almost always treated using the results of mixing length theory (see, e.g., Stix, 1989; Spruit, 1974). In this approach, it is supposed that energy is carried to the solar surface by buoyant convective elements which have mean temperature excess ΔT relative to the surrounding material. The upward progress of each such gas parcel is presumed to terminate after it has traversed a distance l (the mixing length), whereupon the thermal energy associated with ΔT is deposited into the background through dissolution of the element. In utilizing mixing length theory to construct models of the solar interior, the quantity l is specified; in most such applications, it is given as a constant multiple $\alpha(\ge 1)$ of the local pressure scale height (cf., Spruit, 1974, and references therein).

Although mixing length theory neglects effects such as overshoot at the base of the convection zone and dynamical modifications to the hydrostatic equilibrium near the surface, comparisons with numerical simulations of turbulent compressible convection suggest that it does a reasonable job of describing the average energy transport (Cattaneo et al., 1991). For the standard solar model with $\alpha = 1$, this transport is efficient enough that the temperature gradient throughout most of the convection zone is only slightly superadiabatic. The largest convective velocities predicted by mixing length theory occur about 1 scale height below $r = R_\odot$ and have magnitude ~ 1 km s^{-1}. The top of the convection zone is located just below the visible solar surface, at which point $\rho \approx 3 \times 10^{-7}$ g -cm^{-3} and $T \approx 7000$ K. At higher, photospheric levels of the solar atmosphere, the continuum optical depth at wavelengths near

λ = 5000 Å is less than unity, so that a significant portion of the photon flux can stream freely rather than diffuse. Radiative equilibrium is re-established, and with it, a temperature distribution which is characterized by a sub-adiabatic temperature gradient.

3. Solar Oscillations and Internal Rotation

The rotation of the Sun has been recognized since early in the seventeenth century (if not before) when a number of observers noted the transit of dark blemishes (sunspots) across the solar disk. In 1612, Galileo interpreted the foreshortening of these features as evidence for their being located on or near the surface of the Sun. He subsequently suggested that if the spots were attached to the Sun, then their apparent motion was consistent with an overall rotation of the Sun. On the basis of continued observations, he estimated the solar rotational period to be about one month.

With the passage of time, alternative (e.g., spectroscopic) methods have been developed for measuring the rate of rotation of the solar surface. For the most part, however, modern observers have continued to use variations of the basic technique of measuring the positions of discrete features on the photosphere at different times to establish the surface rotation rate (see, e.g., Snodgrass, 1983). From such observations, it has come to be known that the surface of the Sun rotates differentially with latitude. More specifically, the measured solar equatorial angular velocity of rotation is about 2.9×10^{-6} radians per second; moving toward the poles, the rotation rate decreases with increasing latitude, reaching a value $\approx 2.2 \times 10^{-6}$ radians per second at latitude $\pm 80°$.

A goal of particular interest to solar physicists is the elucidation of the basic properties of the internal solar rotation. Among other things, the magnitude and sense of the depth and latitude variation of the angular velocity $\Omega(r,\theta)$ have profound implications for the operation of the solar dynamo (see, e.g., Parker, 1979). Until recently, observational evidence pertaining to the rotational state of the Sun's interior was largely circumstantial. However, advances in the measurement and analysis of the frequencies of solar oscillations have made possible the determination of the angular velocity distribution throughout a substantial portion of the convection zone and radiative interior.

That the Sun as a compressible fluid body supports numerous acoustic modes of oscillation has been known since the early 1960s. Observations designed to detect solar atmospheric velocity fields by measuring Doppler shifts of spectral lines formed in the photosphere revealed the occurrence of vertical, oscillatory motions in limited areas on the disk of the Sun (Leighton et al., 1962). Such wave-like motions take place with a period of about 300 s, a typical velocity amplitude of about 0.5 km s^{-1}, and are observed to last for several periods in a given location. It was subsequently suggested (Ulrich, 1970; Leibacher and Stein, 1971) that these so-called "five-minute oscillations" are the surface manifestations of standing acoustic waves, trapped within an internal, resonant cavity whose upper boundary is located just below the photosphere. This interpretation was later substantiated when the relation between mode frequency and horizontal wave-number delineated from observations was found to be in good agreement with theoretical predictions (Deubner, 1975). The physical process by means of which the modes are excited is presently not well understood. One hypothesis which has received considerable attention is that the modes are stochastically excited, originating in the acoustic emission which is a by-product of turbulent, compressible convection (Goldreich and Keeley, 1977).

A complete description of the principles and practice of helioseismology is beyond the scope of the present paper; the interested reader is referred to the recent, detailed reviews by Toomre (1986), Libbrecht (1988), and Christensen-Dalsgaard (1990) for additional information. Instead, we herein briefly summarize the results of an asymptotic analysis of the modes. Individual p-modes (i.e., "pressure" modes) are quite linear, with velocity amplitudes ~ 10 cm s^{-1}. They are therefore justifiably treated as small perturbations about the existing solar internal structure. Each such normal mode of oscillation is characterized by a radial order n, an angular degree l, and an azimuthal order $m (-l \le m \le l)$. Consider first the case of a spherical, non-rotating Sun; under these conditions, the absence of a preferred axis implies that the mode frequencies are independent of m. Ray theory or WKB analysis is applicable to high-frequency, short-wavelength modes having vertical wave numbers k_r such that k_r^{-1} is smaller than both r and the scale heights in the background solar structure. An approximate, local representation of such modes can then be given in terms of plane sound waves obeying the dispersion relation $\omega^2 = (k_r^2 + k_h^2)c^2$, where ω is the constant wave frequency, c is the speed of sound, and $k_h = \sqrt{l(l+1)}/r$ is the horizontal component of the wave vector.

Near the surface of the Sun, $(\omega/c)^2 \gg l(l+1)/R_\odot^2$ (for moderate values of l), so that $k_r \gg k_h$ and sound waves propagate nearly radially. At greater depth, both c and k_h increase, implying that k_r must decrease. Downward propagating waves are therefore refracted and turned back toward the surface at an inner turning point $r = r_t$ where $k_r = 0$; from the approximate dispersion relation given above, the location $r = r_t$ is determined by $r_t^2/c^2(r_t) = l(l+1)/\omega^2$. Hence, modes with low l values penetrate almost to the solar center, while those with high angular degree remain confined to the outer portion of the convection zone. Upon returning to $r \approx R_\odot$ from $r = r_t$, waves with frequencies less than the acoustic cutoff frequency ω_{ac} ($\sim g/c$, where g is the solar gravitational acceleration) are reflected downward. This effect is not described by the simple dispersion relation utilized heretofore, and arises because the wavelengths of nearly vertically propagating waves with $\omega < \omega_{ac}$ become greater than the density scale height in the solar surface layers (see, e.g., Gough, 1986 for a detailed discussion).

It can now be seen that the upper and lower boundaries of the acoustic cavity referred to earlier in this section are formed by the reflecting surface at $r \approx R_\odot$ and the turning point at $r = r_t$. Note that the condition for a standing wave to exist within this cavity is that the phase change incurred by a wave traveling between r_t and $R_\odot$ be an integral multiple of π. Application of this constraint using the simple sound wave dispersion relation then yields an approximate expression for the p-mode oscillation eigen-frequencies. For modes of very low angular degree l, $r_t \approx 0$, and one obtains the result $v_{nl} \sim (n + \frac{1}{2}l + \frac{1}{4} + \alpha)\Delta v$, where α accounts for the phase shift arising from the reflection at $r \approx R_\odot$ and $\Delta v = [2 \int_0^{R_\odot}(dr/c)]^{-1}$ is the inverse of twice the time required by a radially propagating sound wave to travel between $r = 0$ and $r = R_\odot$ (Christensen-Dalsgaard et al., 1985). It is apparent from even this rudimentary analysis that the inversion of helioseismic data (i.e., measured oscillation frequencies) makes it possible to acquire information about the physical conditions which prevail in regions which were previously inaccessible to observation. For example, much effort has been expended in determining the sound speed profile $c(r)$ throughout the Sun and, concomitantly, the location of the base of the convection zone (see, e.g., Christensen-Dalsgaard et al., 1991, and references therein).

When the rotation of the Sun is included in the analysis of internal acoustic oscillations, the mode frequencies are no longer degenerate with respect to the azimuthal order m. Note that based on the surface rotation rate, both centrifugally induced departures from sphericity and the effects of Coriolis acceleration on propagating sound waves are likely to be small throughout the Sun. Hence, the principal way in which the Sun's rotation influences the oscillations is by advecting the sound waves of which they are composed, thereby shifting the mode frequencies (see, e.g., Brown, 1986; Gough, 1991). Specifically, when observed from an inertial frame of reference, waves having a propagation vector component in the direction of the solar rotation are increased in frequency, while those propagating in the opposite sense are diminished in frequency. For the particularly simple case in which the internal angular velocity is assumed to be a function of radius only (i.e., $\Omega = \Omega(r)$), the mode frequencies ν_{nlm} are related to the frequencies ν_{nl} of non-rotating modes of the same radial order and angular degree by $\nu_{nlm} = \nu_{nl} - m\bar{\Omega}$ (cf. Brown, 1986). Here, $\bar{\Omega}$ is an average of $\Omega(r)$ over the depth range $r_t \leq r \leq R_\odot$ traversed by a mode with given values of n and l. If Ω depends on both r and θ, the relation between ν_{nlm} and ν_{nl} remains the same, but the average rotation frequency $\bar{\Omega}$ is now computed over the radial *and* angular extent of the volume sampled by the mode. In this regard, we note that although r_t is, to a good approximation, a function only of n and l, a mode of azimuthal order m is further confined to the region external to a cone of semi-angle $\theta \approx \sin^{-1}[m/(l + 1/2)]$, with axis along the rotation axis and apex at $r = 0$ (cf. Gough, 1991).

The internal solar angular velocity is determined by first measuring the frequency shifts (i.e., rotational splittings) of modes having a range of n and l values. The quantities $\bar{\Omega}$ realized from this procedure correspond to averages of $\Omega(r,\theta)$ over different portions of the solar interior. Information about the actual variation of Ω with depth and latitude is then extracted by applying inverse techniques to the observational results. The execution of these steps is described in detail in the previously cited reviews and papers, and will not be considered further herein. Instead, we close this section by briefly summarizing the picture of the internal solar rotation which has emerged from the work of numerous, independent observers (see, e.g., Brown et al., 1989, and references therein). From these results, it appears that the dependence of Ω on latitude observed at the surface persists throughout the convection zone. Hence, within the Sun's convective envelope, Ω is constant on radii rather than on cylinders as has been suggested by some computational results (e.g., Glatzmaier and Gilman, 1982). Just below the convection zone, the angular velocity of the polar regions undergoes a slight increase, while that of equatorial regions decreases. With increasing depth, the polar and equatorial angular velocities tend toward a common value which is intermediate between their respective photospheric values. The radial extent of the transition zone over which this variation takes place is presently unresolved; it could conceivably be quite narrow. Down to a depth $r \approx 0.5R_\odot$ within the solar radiative interior, Ω appears to be nearly independent of both radius and latitude. That is, a large portion of the Sun's radiative core rotates approximately uniformly, with an angular velocity which is about the same as that of the photosphere. At present, inferences drawn about the rotational state of the region interior to $r \approx 0.5R_\odot$ are unreliable. This is due, in large part, to difficulties in observing and measuring the properties of the lowest l-value modes that propagate to the vicinity of the solar center. We close this section by noting that kinematic solar dynamo models require a rather significant inward increase in Ω to account for the equator-ward migration of activity with time over the course of the solar cycle. Hence, the observational inference that $d\Omega/dr \sim 0$ in the outer layers of the solar interior has led some to suggest that dynamo activity is not distributed throughout the convective envelope but is instead concentrated in the transition zone just beneath it (Gilman et al., 1989).

4. Implications for Angular Momentum Transport

The rotational state of the solar interior as inferred from measured p-mode splittings exhibits significant differences from that expected on the basis of theoretical models. Of particular interest is the apparent tendency toward rigid rotation in the outer layers of the Sun's radiative core. Despite its tentative character (see, e.g., Schou et al., 1992), this interpretation of the observations has a number of implications for proposed mechanisms of angular momentum redistribution inside the Sun. Among the most salient of these is that angular momentum transport must be efficient, occurring at all radii within the volume in question over a time interval which is short in comparison to evolutionary time scales.

Numerous hydrodynamical mechanisms for the elimination of non-uniform rotation in stellar radiative interiors have been identified and studied (for a summary see, Endal and Sofia, 1978; Tassoul, 1978; Zahn, 1983; Spruit, 1990). Included among these are angular momentum transport by such processes as large-scale meridional circulation, Ekman flow, waves, and turbulent diffusion arising from any of a variety of shear flow instabilities. Models for the rotational evolution of a 1 $M_\odot$ star which utilize parametric descriptions of transport by these processes are generally characterized by marked differential rotation within the radiative core by the time they attain solar age (Endal and Sofia, 1981; Pinsonneault et al., 1989). Specifically, in models of this kind, the angular velocity of much of the central radiative interior exceeds that of the surface when $t = t_\odot$. This property is a consequence of the fact that the efficiencies of some hydrodynamical modes of transport depend sensitively on the composition of the material within which they operate. In the deep interior, continued nuclear processing causes a gradient in mean molecular weight to develop with age (cf., §2), thereby reducing the rate at which the angular momentum content of the affected region can subsequently be changed.

There are a variety of reasons to suppose that the solar radiative interior contains a large-scale, possibly primordial magnetic field (Mestel and Weiss, 1987). If so, the interaction between such a field and differential rotation provides an alternative mechanism for angular momentum redistribution. Consider a rotating, spherical star having an internal magnetic field **B** which is symmetric about the rotation axis (assumed to be aligned with the z-axis of a Cartesian coordinate system). If the electrically conducting fluid which constitutes the stellar interior rotates non-uniformly, then the poloidal component $\mathbf{B}_p(= [\mathbf{B}_r, \mathbf{B}_\theta]$ in spherical polar coordinates) of **B** is sheared to produce a toroidal magnetic field $\mathbf{B}_\phi$. It is straightforward to show that the current density associated with $\mathbf{B}_\phi$ possesses a component perpendicular to $\mathbf{B}_p$ and that the Lorentz force exerted on the fluid is directed counter to the sense of rotation, acting to eliminate the shear. In general, if an impulsive, transverse shear is locally applied to a poloidal line of force, the resulting disturbance propagates along $\mathbf{B}_p$ at the Alfvén speed. The inclusion of dissipation ensures that such disturbances are ultimately damped, thereby effectively redistributing angular momentum over the length of the field line. For a weakly dissipative fluid, the system tends toward a steady state described by the law of isorotation, in which the angular velocity is constant along each poloidal line of force (Mestel et al., 1988).

To illustrate how magnetic fields might mediate the transfer of angular momentum inside the Sun, we consider the following model calculation (Charbonneau and MacGregor, 1992). We represent the solar "interior" by the volume contained between two concentric spheres, an inner "core" of radius R_c and an outer "surface" of radius R_* $(> R_c)$. The volume is uniformly filled with a fluid having constant values of the magnetic diffusivity η and kinematic viscosity v. In the initial $(t = 0)$ state of the system, the fluid is at rest and is threaded by a specified, axisymmetric poloidal magnetic field $\mathbf{B}_p(r,\theta)$. For $t > 0$, the core is impulsively set into uniform rotation with specified angular velocity Ω_C. The subsequent evolution of the fluid velocity $\mathbf{u}$ and the magnetic field $\mathbf{B}$ is determined by numerical solution of the time-dependent momentum and induction equations, assuming that these quantities can be written as $\mathbf{u} = u_\phi(r,\theta,t)\,\mathbf{e}_\phi$ and $\mathbf{B} = \mathbf{B}_p(r,\theta) + \mathbf{B}_\phi(r,\theta,t)\,\mathbf{e}_\phi$. Analogous problems concerning the spin-up of a viscous, conducting fluid have been studied analytically by Benton and Loper (1969) and Loper (1976a,b).

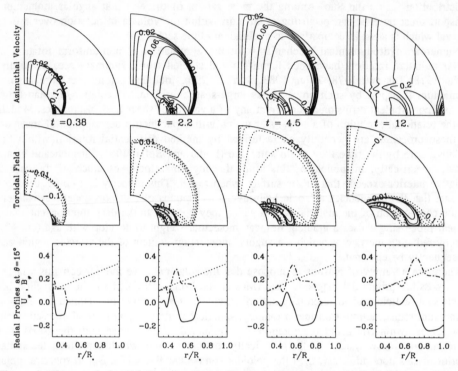

Figure 1. Time evolution of the azimuthal velocity (u_ϕ) and magnetic field (B_ϕ) for the spin-up problem described in the text. Shown are contours of constant u_ϕ (top row), contours of constant B_ϕ (middle row), and profiles of u_ϕ and B_ϕ along a radius located 15° above the equational plane.

A sample of results from the calculation is given in Figure 1, where we have depicted the early time evolution of u_ϕ and B_ϕ for a case in which $\mathbf{B}_p$ is dipolar with all field line foot points attached to the core. For this solution, $R_C/R_* = 0.3$, and the values adopted for the mass density and poloidal magnetic field strength at $r = R_C$ are $\rho_c = 1$ g-cm^{-3} and $B_c = 1G$ respectively. Characteristic values for the Alfvén speed u_A ($\equiv B_c/\sqrt{4\pi\rho_c}$) and Alfvén time t_A ($\equiv R_*/u_A$) are therefore $u_A = 0.28$ cm s^{-1} and $t_A = 7833$ years, for $R_* = R_\odot$. The results shown in the figure are derived from a computation performed in an inertial frame of reference. Their applicability is therefore restricted to systems whose rate of rotation is sufficiently slow that transient MHD phenomena are not significantly modified by the Coriolis force (see Hide, these proceedings, for a complete discussion). In the present case, the angular velocity of the core after spin-up is $\Omega_c = 0.33 t_A^{-1} = 1.34 \times 10^{-12}$ s^{-1} ($\ll \Omega_\odot$), so that waves generated by the impulsive shear at $r = R_c$ are unaffected by Coriolis acceleration.

In the upper row of Figure 1, we show contours of constant azimuthal velocity (measured in units of u_A) in a meridional quadrant, for several different times (measured in units of t_A). For this model, the Reynolds numbers R_m ($\equiv u_A R_*/\eta$) and R_v ($\equiv R_* u_A/\nu$) both have the value $R_m = R_v = 10^4$. The implied values of the diffusivities are sufficiently small that the core spin-up produces a sheared, wave-like disturbance which propagates toward the surface at the local Alfvén speed while maintaining a relatively sharp front. This behavior is also apparent in the second row of Figure 1, where we depict the induced azimuthal field as a function of time. The contours are labelled with the values of the logarithm of (B_ϕ/B_c), with dotted lines indicating negatively directed ($-\mathbf{e}_\phi$) field and solid lines positively directed ($+\mathbf{e}_\phi$) field. The polarity reversal which occurs near the equator at early times and later spreads to higher latitudes is indicative of wave reflection in the plane $\theta = \pi/2$ where symmetry considerations dictate that $B_\phi = 0$. Note that as a consequence of increased path length, waves that emanate from progressively higher latitudes on $r = R_c$ require a longer time to reach the equatorial plane. As a result, large gradients in B_ϕ develop in the direction perpendicular to $\mathbf{B}_p$. The growth of short-wavelength structure across field lines (so-called "phase mixing") leads to the dissipation of B_ϕ in affected areas in a time which is much shorter than either of the estimated diffusion time scales R_*^2/η or R_*^2/ν.

The occurrence and effects of phase mixing can also be seen in the third row of Figure 1. There we depict the time evolution of profiles of (u_ϕ/u_A) and (B_ϕ/B_c) along the radius $R_c \leq r \leq R_*$ at latitude 15°. The dotted line represents the azimuthal velocity profile corresponding to uniform rotation with angular velocity Ω_c. The development in time of oscillations with substantial radial gradients is clearly evident, as is the subsequent dissipative relaxation of the fluid behind the leading wavefront to a state with $B_\phi \approx 0$ and $\Omega \approx \Omega_c$. As is apparent from the last panel in each of the three rows of Figure 1, nearly uniform rotation prevails over much of the volume under consideration by time $t = 12 t_A$. If the calculation is followed still farther in time, approximately rigid rotation is enforced over the entire region by $t \approx 50 t_A$. Hence, the time required to establish solid-body rotation of the volume is longer than the Alfvén transit time across it, but, in the case of the solar interior, considerably shorter than evolutionary time scales. We conclude by noting that for the present problem, a uniformly rotating final state is in fact predicted at the outset by application of the law of isorotation. In general, however, the ultimate rotational state attained by the configuration depends sensitively on the behavior of the prescribed $\mathbf{B}_p$ at the boundaries $r = R_c, R_*$. The interested reader is referred to the paper by Charbonneau and MacGregor (1992) for detailed discussion.

5. Conclusion

Recent inversions of helioseismic data have provided a tantalizing first glimpse of the rotational state of the solar interior. It is anticipated that results from new experiments designed to observe oscillation modes of very low angular degree will become available over the next several years, thereby clarifying the current picture of rotation in the region $r \lesssim 0.5 R_\odot$. If near uniformity proves to be a robust description of rotation throughout the solar radiative core, then preliminary calculations suggest that angular momentum redistribution by a large-scale internal magnetic field is an effective mechanism for enforcing such a state.

Additional constraints on processes affecting the internal solar angular momentum distribution may be obtainable from observations of rotation among solar-type stars. Advances in detector technology have made possible the spectroscopic determination of the surface rotational velocities of numerous solar-mass stars having a broad range of ages (see, e.g., Stauffer, 1991, and references therein). These observations suggest that shortly after the commencement of nuclear burning (age $\sim 4 \times 10^7$ years), many such stars experience rapid rotational deceleration, often from velocities > 50 km s^{-1} to values < 20 km s^{-1} in a time $\sim$ few $\times 10^7$ years. This spin-down is presumably the result of the torque applied to the stellar convective envelope by a magnetically coupled wind, analogous to the solar wind which emanates from the outer layers of the corona. The rapidity of the observed rotational braking of young stars (i.e., the corresponding spin-down time for the Sun is $\sim 7 \times 10^9$ years) may be indicative of the rate at which the processes responsible for angular momentum transfer from the core to the convection zone operate.

Acknowledgment. I am greatly indebted to Paul Charbonneau, both for collaborative assistance and for reviewing the manuscript. I am also grateful to the workshop organizers for financial support.

6. References

Bahcall, J. N. and Cribier, M. (1990) The standard solar model, in Inside the Sun, eds. G. Berthomieu and M. Cribier (Dordrecht: Kluwer), pp. 21-41.

Benton, E. R. and Loper, D. E. (1969) On the spin-up of an electrically conducting fluid. Part I. The unsteady hydromagnetic Ekman-Hartmann boundary layer problem. J. Fluid Mech. 39, 561-586.

Brown, T. M. (1986) Measuring the Sun's internal rotation using solar p-mode oscillations, in Seismology of the Sun and Distant Stars, ed. D. O. Gough (Dordrecht: Reidel), pp. 199-214.

Brown, T. M., Christensen-Dalsgaard, J., Dziembowski, W. A., Goode, P., Gough, D. O. and Morrow, C. A. (1989) Inferring the Sun's internal angular velocity from observed p-mode frequency splittings. Ap. J. 343, 526-546.

Cattaneo, F., Brummell, N. H., Toomre, J., Malagoli, A. and Hurlburt, N. E. (1991) Turbulent compressible convection. Ap. J. 370, 282-294.

Charbonneau, P. and MacGregor, K. B. (1992) Angular momentum transport in magnetized stellar radiative zones. I. Numerical solutions to the core spin-up model problem. Ap. J. 387, 639-661.

Christensen-Dalsgaard, J. (1990) Helioseismic investigation of solar internal structure, in Inside the Sun, eds. G. Berthomieu and M. Cribier (Dordrecht: Kluwer), pp. 305-326.

Christensen-Dalsgaard, J., Duvall, T. L., Gough, D. O., Harvey, J. W. and Rhodes, E. J. (1985) Speed of sound in the solar interior. Nature 315, 378-382.

Christensen-Dalsgaard, J., Gough, D. O. and Thompson, M. J. (1991) The depth of the solar convection zone. Ap. J. 378, 413-437.

Deubner, F. L. (1975) Observations of low wave number nonradial eigenmodes of the Sun. Astr. Ap. 44, 371-375.

Endal, A. S. and Sofia, S. (1978) The evolution of rotating stars. II. Calculations with time-dependent redistribution of angular momentum for 7 and 10 $M_\odot$ Stars. Ap. J. 220, 279-290.

Endal, A. S. and Sofia, S. (1981) Rotation in solar-type stars. I. Evolutionary models for the spin-down of the Sun. Ap. J. 338, 424-452.

Gilman, P. A., Morrow, C. A., and DeLuca, E. E. (1989) Angular momentum transport and dynamo action in the Sun: Implications of recent oscillation measurements. Ap. J. 338, 528-537.

Glatzmaier, G. A. and Gilman, P. A. (1982) Compressible convection in a rotating spherical shell. V. Induced differential rotation and meridional circulation. Ap. J. 256, 316-330.

Goldreich, P. and Keeley, D. A. (1977) Solar seismology. II. The stochastic excitation of the solar p-modes by turbulent convection. Ap. J. 212, 243-251.

Gough, D. O. (1986) Asymptotic sound-speed inversions, in Seismology of the Sun and Distant Stars, ed. D. O. Gough (Dordrecht: Kluwer), pp. 125-140.

Gough, D. O. (1991) Internal solar rotation, in Angular Momentum Evolution of Young Stars, eds. S. Catalano and J. R. Stauffer (Dordrecht: Kluwer), pp. 271-279.

Leibacher, J. and Stein, R. F. (1971) A new description of the solar five-minute oscillations. Astrophys. Lett. 7, L191-L192.

Leighton, R. B., Noyes, R. W. and Simon, G. W. (1962) Velocity fields in the solar atmosphere. I. Preliminary report. Ap. J. 135, 474-499.

Libbrecht, K. G. (1988) Solar and stellar seismology. Space Sci. Reviews 47, 275-301.

Loper, D. E. (1976a) On the spin-up of a stably stratified, electrically conducting fluid. Part 1. Spin-up controlled by the Hartmann layer. Geophys. Fluid Dynamics 7, 133-156.

Loper, D. E. (1976b) On the spin-up of a stably stratified, electrically conducting fluid. Part 2. Spin-up controlled by the MAC layer. Geophys. Fluid Dynamics 7, 175-203.

Mestel, L. and Weiss, N. O. (1987) Magnetic fields and non-uniform rotation in stellar radiative zones. MNRAS 226, 123-135.

Mestel, L., Moss, D. L. and Taylor, R. J. (1988) The mutual interaction of magnetism, rotation, and meridian circulation in stellar radiative zones. MNRAS 231, 873-885.

Parker, E. N. (1979) Cosmical Magnetic Fields (Oxford: Clarendon Press).

Pinsonneault, M. H., Kawaler, S. D., Sofia, S. and Demarque, P. (1989) Evolutionary models of the rotating Sun. Ap. J. 338, 424-452.

Schou, J., Christensen-Dalsgaard, J. and Thompson, M. J. (1992) The resolving power of current helioseismic inversions for the Sun's internal rotation. Ap. J. Lett. 385, L59-L62.

Schwarzschild, M. (1958) Structure and Evolution of the Stars. (New York: Dover).

Snodgrass, H. B. (1983) Magnetic rotation of the solar photosphere. Ap. J. 270, 288-290.

Spruit, H. C. (1974) A model of the solar convection zone. Solar Phys. 34. 277-290.

Spruit, H. C. (1990) Angular momentum transport and magnetic fields in the solar interior, in Inside the Sun, eds. G. Berthomieu and M. Cribier (Dordrecht: Kluwer), pp. 415-423.

Stauffer, J. R. (1991) Rotational velocities of low mass stars in young clusters, in Angular Momentum Evolution of Young Stars, eds. S. Catalano and J. R. Stauffer (Dordrecht: Reidel), pp. 117-134.

Stix, M. (1989) The Sun. (Berlin: Springer-Verlag).

Tassoul, J.-L. (1978) Theory of Rotating Stars. (Princeton: Princeton University Press).

Toomre, J. (1986) Properties of solar oscillations, in Seismology of the Sun and the Distant Stars, ed. D. O. Gough (Dordrecht: Kluwer), pp. 1-22.

Ulrich, R. K. (1970) The five-minute oscillations on the solar surface. Ap. J. 161, 993-1002.

Zahn, J. P. (1983) Instability and mixing processes in upper main sequence stars, in Astrophysical Processes in Upper Main Sequence Stars (Geneva: Observatoire Geneve), pp. 253-329.

A REVIEW OF THE DYNAMICS OF THE LOWER AND UPPER THERMOSPHERE

DAVID REES
Atmospheric Physics Laboratory
Department of Physics and Astronomy
University College London
67-73 Riding House St., London W1P 7PP, UK

ABSTRACT. A Global Thermospheric Model (developed at UCL), and an Ionospheric Model (developed at Sheffield University) have been numerically and computationally merged into a self-consistent coupled thermospheric / ionospheric model. This is an extremely valuable diagnostic tool for examining interactions between the strongly-coupled thermosphere and ionosphere regions. It may be used to investigate the behaviour of the thermosphere and ionosphere under a wide range of solar, geomagnetic activity and seasonal conditions. It may also be used to compare diverse and fragmentary observations of the thermosphere and ionosphere obtained by widely differing techniques, at different locations and times, and under varying solar, seasonal and geomagnetic conditions. The models have been extensively validated by comparing specific simulations with data from ground-based, rocket and satellite investigations. The thermosphere presents many interesting examples of fluid flow, some of which will be discussed in the review.

1. INTRODUCTION

The thermosphere and ionosphere, regions extending from approximately 85 km to 600 km in altitude, are strongly affected by solar radiation in the UV and EUV spectral range, causing both heating and ionisation. In response to solar heating, peak temperatures and ionisation rates are usually created at upper thermospheric levels on the low-latitude dayside. Energy is then conducted downward to the lower thermosphere and mesosphere, where it can be re-radiated to space by radiatively-active molecules such as CO_2 and NO. Photoionisation generates a relatively complex ionospheric response, with several layers (D, E, F_1, F_2) being formed, usually between 80 and about 400 km, by the combination of somewhat diverse ionisation mechanisms, charge exchange, recombination, dissociative recombination and transport due to electric fields and neutral winds. The ionosphere should always be thought of as the ionised component of the thermosphere, which is thus a weakly ionised gas, where the ionised component is always less than 10^{-3} of the neutral gas density.

The thermosphere and ionosphere also respond dramatically to intense localised forcing. For example, during disturbed geomagnetic conditions and under the influence of auroral electron and ion precipitation combined with the large-scale magnetospheric convective electric field, the high latitude regions are often dramatically disturbed. Auroral electrons and ions cause local heating and ionisation of the neutral atmosphere, and dissociation of molecular nitrogen. The direct heating effect of auroral electrons is generally modest, except at high altitudes and in the dayside cusp. However, the additional auroral ionisation greatly

13

D. B. Stone and S. K. Runcorn (eds.),
Flow and Creep in the Solar System: Observations, Modeling and Theory, 13–44.
© 1993 *Kluwer Academic Publishers. Printed in the Netherlands.*

increases the Pedersen conductivity of the ionosphere. The combination of high values of Pedersen conductivity and the magnetospheric convection electric field greatly increases Joule heating, generating the dominant energy source at high latitudes during geomagnetically disturbed periods. Locally, Joule heating may be 2 or 3 orders of magnitude larger than globally-averaged solar UV / EUV heating. When the Pedersen conductivity is enhanced, momentum transfer from convecting ionospheric plasma (mainly the ions) to the neutral gas is similarly enhanced, a process known as ion-drag. This momentum transfer and the effect of pressure gradients induced by enhanced Joule heating together drive intense thermospheric wind systems (400 - 1000 ms^{-1}) observed throughout the high latitude regions during geomagnetically disturbed periods. The dissociation of molecular nitrogen by electron precipitation, through the odd nitrogen chemistry, enhances the number density of nitric oxide, which then plays very important roles in ionospheric chemistry and in the neutral energy balance, due to its infrared radiative properties.

Within the lower thermosphere (85 km and up to approx. 150 km), the effects of disturbances propagating from the lower atmosphere - tidal, gravity and planetary wave, can also be discerned, in addition to the signatures of solar activity and geomagnetic activity. These lower atmosphere sources are variable and somewhat unpredictable, as is the incidence of large magnetospheric disturbances. Thermospheric climatology is predictable, by means of empirical or theoretical models. However, thermospheric 'weather' is very difficult to predict in detail, although it is now possible to predict the overall magnitude, location and occurrence of disturbances induced by magnetospheric disturbances.

2. APPROACHES TO NUMERICAL MODELLING

Numerical simulation of the thermosphere from first principles requires that the most important physical processes occurring within the thermosphere are properly treated (Fuller-Rowell and Rees, 1980, 1983; Dickinson et al., 1981, 1984; Roble et al., 1982). It is normally assumed that most of the energy and momentum sources driving the thermosphere are predetermined, and invariant to the response of the thermosphere. However, major changes of polar thermospheric winds, temperature and composition occur in response to ion convection, auroral particle precipitation and Joule heating within the auroral oval and polar cap. These changes are now well documented by ground-based and spaceborne observation (Hays et al., 1984; Rees et al., 1983, 1985, 1986). Some of these thermospheric responses may change the nature or magnitude of the forcing itself, by modifications of the total electromotive force (EMF) and ionospheric conductivity. Significant dynamical and thermal effects, particularly in the lower thermosphere, also occur as the result of internal forcing from the lower atmosphere, due to the combination of tidal waves, gravity waves and planetary waves. The first can be handled numerically within a thermospheric model by introducing self-consistent changes of wind and temperature corresponding to the amplitudes and phases of specific propagating tidal modes in a 'flexible' lower boundary region. The actual propagation of such tides through the lower and middle atmosphere is not simulated explicitly by such a procedure; however, the amplitudes and phases of several modes can be adjusted, by experiment, until the resulting lower thermosphere wind and temperature fields correspond to observed tidal amplitude and phase variations of temperature and wind (etc) as functions of altitude, season and latitude.

It is found, as would be expected, that the relatively large amplitudes of observed tidal winds in the lower thermosphere (100 - 150 km) can be successfully simulated by introducing such propagating tides. This is not possible if only the tides generated in-situ are considered. Additionally, large non-linear interactions occur between the individual propagating tidal modes themselves, and between them and the tides generated in-situ by solar or geomagnetic forcing. The studies carried out by Fesen et al. (1986) and Parish et al. (1990) are still at the preliminary stage, since the real tidal and meteorological behaviour of the middle atmosphere is still very poorly known. However, they do provide a very useful insight into the complexities of the thermosphere.

The fundamental influences of the magnetosphere on the thermosphere and ionosphere come from magnetospheric plasma convection. This convection drives ionospheric plasma within the polar region into rapid motion, following a twin-cell vortex within the polar cap and auroral oval. These rapid plasma motions accelerate thermospheric winds via the process commonly known as ion drag (Rishbeth, 1972). The structure and variability of polar convective fields were not well understood until the late 1960's and 1970's, with the advent of rocket and satellite measurements of ion drifts and electric fields (Heppner, 1977; Heppner and Maynard, 1987). A statistical model of the convection field, along with associated variations of the auroral electron precipitation as a function of 'auroral activity' are illustrated in Figure 1 (Foster et al., 1986). Detailed information on the patterns of electron precipitation became available in the mid-1980's with the models of Hardy et al. (1985) and Fuller-Rowell and Evans (1987).

The ion-neutral frictional drag resulting from differential ion and neutral gas motions causes direct heating of both ions and neutrals, commonly known as Joule heating (Cole, 1962, 1971). Induced winds may increase or decrease (but generally decrease) the ion drag, and the resulting frictional heating. The induced winds (or more correctly, changed winds, since there is always a complex wind system in existence prior to a given geomagnetic disturbance) may induce a 'back-EMF', opposing the initial magnetospheric convective electric field. This wind system (subtly modified by gas pressure changes due to neutral gas heating) will also induce ion drifts parallel to the local magnetic field. Such 'parallel' ion drifts will also induce a field-aligned electron flow, to maintain charge neutrality (or very nearly). Thus the entire vertical plasma distribution will respond to wind changes, an effect which becomes increasingly important at altitudes above 300 km in the F-region. Such induced changes of ion density distribution will consequently modify the ion drag on the neutrals, the wind acceleration terms, and finally the neutral gas winds.

Diffusive separation of neutral species occurs within the thermosphere. Lighter atomic species tend to become increasingly dominant over heavier molecular species at higher altitudes. Atomic oxygen, formed by dissociation of molecular oxygen, is normally the dominant thermospheric neutral species above 150 km altitude, with molecular nitrogen being the most important 'minor' species. However, large changes of upper thermospheric composition are induced by combined advection and convection, resulting from strong thermospheric heating in the polar regions during geomagnetic disturbances. Enhanced concentrations of molecular nitrogen are particularly pronounced due to persistent upwelling within the summer geomagnetic polar region at F-region heights, particularly during geomagnetic storm periods. Such regions of disturbed chemistry may be advected to lower latitudes by the strong night-time thermospheric winds blowing from the poles toward the equator induced during geomagnetic storm periods. Enhanced concentrations of molecular nitrogen cause significant depletions of F-region plasma densities by greatly increasing the effective recombination coefficient, while the ionisation rates, due to the combination of solar photoionisation and auroral precipitation, are only slightly changed.

At E-region altitudes, induced neutral wind changes, particularly those caused by ion drag, generally oppose the external magnetospheric EMF. This decreases ion drag acceleration (limiting the maximum induced winds), the local electrojet current (at all heights) and the Joule heating, since the total electromotive force i.e. ($\underline{E} + \underline{V}_n \wedge \underline{B}$), as seen in a frame of reference moving with the neutral gas, is decreased. These processes are independent of any plasma density and conductivity modifications, however, the thermosphere will respond to the reduction in both the electromotive force and the Joule heating.

3. THE UCL / SHEFFIELD COUPLED THERMOSPHERE / IONOSPHERE MODEL

The UCL Three Dimensional Thermosphere / Ionosphere Model (also called a General Circulation Model or GCM) simulates the time-dependent structure of the vector wind, temperature, density and composition of the neutral atmosphere, by numerically solving the non-linear equations of momentum, energy and continuity, and a time-dependent mean mass equation. The global atmosphere is divided into a series of elements in geographic latitude, longitude and pressure. Each grid point rotates with the earth to define a non-inertial frame of reference in a spherical polar coordinate system. The latitude resolution is 2°, the longitude resolution is 18°, and each longitude slice sweeps through all local times, with a 1 min. time step. In the vertical direction the atmosphere is divided into 15 levels in log (pressure), each layer is equivalent to one scale height thickness, from a lower boundary of 1 Pascal at 80km height.

The time-dependent variables of southward and eastward neutral wind, total energy density, and mean molecular mass are evaluated at each grid point by an explicit time stepping numerical technique. After each iteration the vertical wind is derived, together with temperature, heights of pressure surfaces, density, and atomic oxygen and molecular nitrogen concentrations. The data can be interpolated to fixed heights for comparison with experimental data, or with empirical models. The momentum equation is non-linear and the solutions fully describe the horizontal and vertical advection, i.e. the transport of momentum.

The neutral atmosphere numerical model is computed using an Eulerian approach. A natural approach to computing the ionospheric code is to use a Lagrangian scheme, following parcels of convecting and co-rotating plasmas. In this, the complex convection patterns imposed by a magnetospheric electric field on plasma movements within the polar regions would be referenced to a fixed Sun - Earth frame, assuming pure $\underline{E}$ x $\underline{B}$ drifts. The electric field is derived by merging a model of magnetospheric convection with the co-rotation potential (induced by the earth's rotation). Parcels of plasma are traced along their convection paths, which are often complex. In practice, however, despite the natural attractions of the Lagrangian scheme, a great deal of computational overhead results from the exchange of data between the thermospheric and ionospheric models, and thus a modified Eulerian system has been adapted recently for both models.

In the ionospheric code, atomic (H^+ and O^+) and molecular ion concentrations are evaluated over the height range from 100 to 1500 km, and used in the thermospheric code within 50° magnetic latitude of the north and south magnetic poles. The use of the self-consistent ionosphere at high- and mid-latitudes and an empirical description at low- and mid-latitudes can result in a discontinuity. The ionospheric code is thus presently being extended to include the self-consistent calculation at low-latitudes.

The initial versions of the global 3-dimensional time-dependent numerical thermospheric models developed by Fuller-Rowell and Rees (1980, 1983), and by Dickinson et al. (1981, 1984) and Roble et al. (1982) used theoretical models or the simple empirical Chiu (1975) global model of the ionosphere to calculate ion drag and Joule heating. However, the lack of any response at high latitudes to geomagnetic processes (precipitation, convection) within the Chiu ionospheric model caused a gross underestimate of the magnitude of ion drag and of Joule heating, particularly at E-region altitudes in the auroral oval. The Chiu model did not under-estimate so seriously real F-region plasma densities. Thus, when the Chiu model was used in the 3-D T-D (or GCM) models, it was possible to simulate F-region winds and temperatures within the upper thermosphere for quiet geomagnetic conditions. However, under disturbed conditions, simulations using the Chiu (1975) ionospheric model generated winds and currents in the auroral E-region, which were always unrealistically low (Rees et al., 1983, 1985, 1986). The underestimate of E-region momentum forcing and heating also caused very serious problems when simulating the geomagnetic response in neutral gas chemistry and in total density, as well as the E-region winds etc (Rees et al., 1985, 1986).

Many of these difficulties have been removed by the numerical coupling of the UCL thermospheric code with the Sheffield ionospheric code. This allows the realistic response of the ionosphere to auroral precipitation and plasma convection, plus the effects of the neutral wind to be computed. The resulting ion drag, Joule heating and winds are thus 'corrected' for the thermospheric and ionospheric interactions. The Sheffield ionospheric code has been described in a series of papers by Quegan et al. (1982, 1986), Watkins (1978) and Allen et al. (1986). The subsequent development of coupled models of the polar ionosphere and thermosphere has been described by Quegan et al. (1982), Fuller-Rowell et al. (1984, 1987, 1988) and by Rees and Fuller-Rowell (1987 a,b, 1989). The latter papers describe a fully coupled model, exchanging ionospheric and thermospheric parameters throughout the region within 50° of the geomagnetic pole. When the physics of the major ionospheric-thermospheric interactions are included within the coupled model, many of the additional 'geomagnetic' energy sources required previously to explain observations became unnecessary (Rees et al., 1988; Rees and Fuller-Rowell, 1989).

Basic Equations:

There are a series of basic equations for the neutral atmosphere. While the equations are relatively simple to write down, their solution is less trivial, and global simulations require considerable super-computer resources. For a pressure-coordinate system, the equations can be written, as:

Hydrostatic Equation:
$$\frac{\partial}{\partial p}(gh) + \frac{1}{\rho} = 0$$

Continuity Equation:
$$\frac{\partial \omega}{\partial p} + \underline{\nabla}_p \underline{V} = 0$$

The Momentum Equation (Navier-Stokes Equation) is:

$$\frac{\partial \underline{V}}{\partial t} + \left(\underline{V} \cdot \underline{\nabla}_p\right)\underline{V} + \omega \frac{\partial}{\partial p} \underline{V} = - \underline{\nabla}_p \Phi - \left(2\Omega + \frac{V\phi}{r\sin\theta}\right) \cdot \underline{k} \wedge \underline{V}$$

$$+ g \frac{\partial}{\partial p}\left[\frac{1}{H} \cdot (\mu_m + \mu_T) \, p \frac{\partial}{\partial_p} \underline{V}\right] + \frac{1}{\rho} \cdot (\mu_m + \mu_T) \nabla_p^2 \underline{V} + \frac{1}{\rho} \left(\underline{J} \wedge \underline{B}\right)$$

The Energy Equation is written as:

$$\frac{\partial \varepsilon}{\partial t} + V \cdot \nabla_p \left(\varepsilon + gh\right) + \omega \frac{\partial}{\partial_p} \left(\varepsilon + gh\right)$$

$$= Q_{EUV} + Q_{IR} + g \frac{\partial}{\partial p}\left[\frac{1}{H} \cdot (K_m + K_T) \, p \frac{\partial T}{\partial p}\right]$$

$$+ \frac{1}{\rho} \left(K_m + K_T\right) \nabla_p^2 T - g \frac{\partial}{\partial p} \frac{K_T}{C_p} \cdot g + \frac{J.E}{\rho} + Q_{VIS}$$

This latter equation includes all the terms appropriate to describe the internal energy of the gas, the various transport, buoyancy and diffusion terms and the radiative sources and sinks. (Sources from the magnetosphere due to particle and Joule (friction) heating are also included).

In a pressure coordinate system, the true vertical gas velocity is the transport across pressure surfaces, plus the vertical motion of the pressure surface, thus the true vertical velocity equation is:

$$V_z = \frac{1}{\rho}\left[\frac{\partial \Phi}{\partial t}_p\right] + \frac{1}{g} \underline{V} \cdot \underline{\nabla}_p \Phi - \frac{\omega}{\rho g}$$

The horizontal current density equation, combining the effects of an external (usually magnetospheric) electric field, and an internal dynamo, due to horizontal neutral gas winds is given by:

$$\underline{J} = \underline{\sigma} \cdot \left(\underline{E} + \underline{V} \wedge \underline{B}\right)$$

Within the thermosphere, diffusive separation between light, atomic species (i.e. atomic oxygen) and the heavier molecular species (N_2, O_2) occurs. In practice, it is possible to neglect photochemistry and recombination for these major species, and the mean molecular mass equation for the resulting two species gas, describing diffusive separation resulting from vertical convection and horizontal advection, is given by:

$$\frac{\partial m}{\partial t} + \underline{V} \cdot \underline{\nabla}_p m + \omega \frac{\partial m}{\partial p} = (D + K) \nabla_p^2 m$$

$$+ \frac{1}{H} p \frac{\partial}{\partial p}\left[\frac{D + K}{H} \cdot p \frac{\partial}{\partial p} m\right] + \frac{1}{H} p \frac{\partial}{\partial p}\left[\frac{D.}{H.m}(m - m_2) \cdot (m - m_1)\right]$$

The notation which is used in these equations is generally conventional for work in Atmospheric Physics, and is explained in detail in the Annex. The equivalent equations for the ionospheric code, and the detailed equations describing the ionospheric production and loss terms will not be described here.

3.1. KEYING MODELS TO SOLAR AND GEOMAGNETIC ACTIVITY

The solar spectrum in the UV, EUV and X-ray spectral regions is very complex and specific emissions vary in a poorly understood way with aspects of solar activity, such as active regions, flares, faculae (Lean, 1988). However, for the purposes of statistical analysis of observations, and for numerical simulations, the variation of the solar spectrum is usually keyed to the solar $F_{10.7}$ cm radio flux, since this is, almost uniquely, available at all times.

As the level of geomagnetic activity increases, the auroral oval generally expands equatorward and broadens, while the auroral precipitation intensifies. These features of auroral precipitation are well shown in statistical surveys and analyses of the energetic electron precipitation (Hardy et al., 1985; Fuller-Rowell and Evans, 1987), as is illustrated in Figure 1.

These statistical surveys, complementing individual ground or space-based observations, show that the signature of increasing geomagnetic activity is that convective electric fields and auroral precipitation intensify, and that there is an equatorward expansion of the regions affected. This also implies that field-aligned currents, Joule heating and ion-drag acceleration of thermospheric winds all increase. The temporal and spatial variability of all of the previously-mentioned terms, broadly speaking - 'geomagnetic forcing' - also increases, particularly at very high activity levels. Disturbances of the lower ionosphere, up to about 120 km, respond directly to 'auroral' ionisation inputs. At higher levels of geomagnetic activity there are strong interactions between the ionosphere and the induced chemical and dynamical changes of the thermosphere, which cause the ionospheric response to become much more complex.

To relate the geomagnetic inputs to the global models as geomagnetic activity levels change, it is necessary to match empirical models of polar electric fields with comparable models of auroral electron precipitation. The NOAA / TIROS precipitation models have been matched to the Millstone Hill I.C.S. Radar convection models for a range of activity levels (Foster et al., 1986, Figure 1), while the Heppner and Maynard (1987) electric fields have been matched to the DMSP electron precipitation models (Rich et al., 1989). Using relationships established between the global indices of Kp, Ap, AE (Auroral Electrojet) and NOAA / TIROS Activity Level, it is then possible to input the matched precipitation and convection patterns with a time-variation as short as approx. 10 min. It is very important to recall, however, that such inputs do not adequately simulate the localised spatial and time-dependent structures of the 'real' geomagnetic inputs, but only modulate the 'global average' inputs, following the statistical patterns. Nevertheless, that approximation is the best which can be currently achieved.

3.2. FEED-BACK MECHANISMS

At times of large geomagnetic disturbances, the Polar E-region winds, driven by magnetospheric electric fields, reach 50% of the $\underline{E} \times \underline{B}$ ion drift velocity (Rees, 1971, 1973; Pereira et al., 1980). Such induced winds reduce the total electromotive force since the electromotive force $(\underline{E} + \underline{V}_n \wedge \underline{B})$ is decreased. This would be expected to decrease the horizontal ionospheric current and thus reduce the Field Aligned Currents connecting the magnetosphere to the auroral ionosphere. However, it not clear under which circumstances the magnetospheric generator acts either as a constant current, or a constant voltage source. Therefore, the simulated results have only a limited validity for such studies. Although some experiments using numerical models to investigate these problems are in progress (e.g. Harel et al., 1981), theoretical and experimental exploration of these problems is still at a very preliminary phase.

Simulations using the fully-coupled model showed that thermospheric winds induced by ion drag and Joule heating caused significant changes in plasma density distribution. E-Region plasma densities within the auroral oval were greatly enhanced compared with those of the early empirical models, as the direct result of precipitation. The resulting induced thermospheric winds and heating were generally greatly increased compared with previous simulations using uncoupled models. These stronger winds were also more effective at creating vertical plasma transport, at both E-Region and F-Region heights.

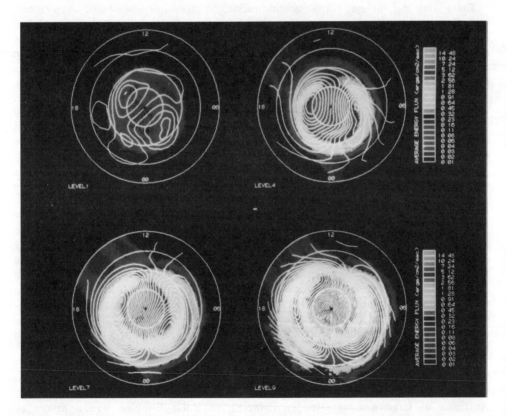

Figure 1: Statistical patterns of auroral particle energy influx and equipotential convection contours for four increasing levels of geomagnetic activity.

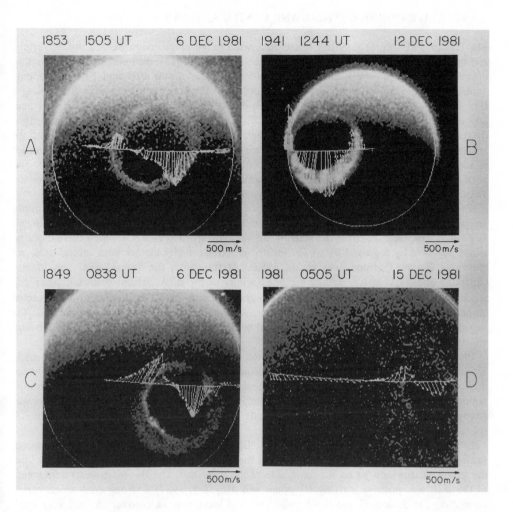

Figure 2: The DE-2 and DE-1 spacecraft simultaneously observed the northern auroral oval on a number of occasions during the winter of 1981 / 1982 (Killeen et al., 1988). These four elements were obtained on Dec. 6, 1981, at 1505 UT; Dec. 12, 1981, at 1244 UT; Dec. 6, 1981 at 0838 UT; and Dec. 15, at 0505 UT, respectively. DE-2 provided the wind vectors, while the large-scale auroral images were obtained from the DE-1 Spin-scan imager.

4. SATELLITE-BORNE OPTICAL AND IN-SITU RESULTS

The Dynamics Explorer satellites observed a wide range of geomagnetic activity conditions and individual events. Figure 2, taken from Killeen et al. (1988), was obtained by combining upper thermospheric wind data (approx 240 km) from the WATS and FPI instruments onboard DE-2, with EUV images of the auroral oval obtained by the Spin-Scan Imager (Frank et al., 1981) on board DE-1. For each of the four intervals depicted, DE-1 was near apogee over the northern polar region while DE-2 was passing underneath, close to its perigee (near 300 km). The results, in each case, show the classical neutral wind signatures of the auroral oval and polar cap: strong sunward winds in the dusk auroral oval, weaker sunward winds in the dawn auroral oval (sometimes only a reduction of the antisunward flow). The polar cap contains a region of strong antisunward flow, while at latitudes lower than the auroral oval there is a general 'global' antisunward flow of around 150 - 200 ms^{-1}.

The four events cover a very wide range of geomagnetic activity. The extremes of activity are depicted by elements B (Dec. 12, 1981) and D (Dec. 15, 1981). Element B was obtained at 1244 UT on Dec. 12, near the start of the very disturbed period for which thermospheric winds observed by a ground-based FPI located at Kiruna, under the northern auroral oval, are shown in Figure 3 (from Rees et al., 1982). At this time, the auroral oval was greatly expanded and very active, generating peak winds in the dusk auroral oval and polar cap reaching nearly 1 kms^{-1}. Element D, in contrast, shows weak emissions and a contracted auroral oval which was only just discernible. Although the auroral emission was weak, the signature in the winds within the dusk auroral oval was still distinct, even if the peak sunward values were less than 200 ms^{-1}.

Element A (Dec. 6, 1505 UT) shows a varied situation: there was a strong (600 - 800 ms^{-1}) antisunward wind flow in the polar cap, confined to the dawn side of the polar cap. On the dusk side, the antisunward wind flow was generally weak, and there was actually a limited region of sunward flow. During this period the Interplanetary Magnetic Field (IMF) field was strong and directed northward. Sunward winds in the dusk auroral oval were less than 300 ms^{-1}, a relatively weak value.

Element C shows data from Dec. 6, 1981, 0838 UT, (southward IMF). On this occasion, sunward winds of more than 600 ms^{-1} within the dusk auroral oval matched the winds within the strong antisunward jet on the dawn side of the polar cap. The B$_Y$ (east-west) component of the IMF was positive, driving (Heppner and Maynard, 1987) a strong convection vortex connecting the dusk oval and dawn polar cap, and leaving weak convection (and very weak sunward winds, for a relatively disturbed period), in the dawn auroral oval. This can be contrasted with Element B, which shows symmetric antisunward flow within the polar cap, and 300 ms^{-1} winds in the dawn auroral oval (30% of the peak dusk oval sunward winds). The B$_Y$ component of the IMF was small during the Dec. 12, 1981 event.

A detailed analysis of the situation displayed in Element A is shown in Figure 4 (also taken from Killeen et al., 1988). Included in this presentation are data from the Ion Drift Meter (IDM, Heelis et al., 1981) and the Langmuir Probe Instrument (LAPI, Krehbiel et al., 1981). The former provided the electric field component equivalent to the ion velocity perpendicular to the satellite track (zonal component), while the LAPI provided thermal plasma density, shown in the section at left. The thermal plasma density was further used to calculate the ion-neutral frictional coupling time constant, shown in the lowest panel of Figure 4.

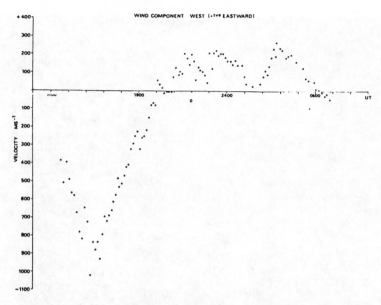

Figure 3: A period of intense westward winds in the dusk sector of the auroral oval, as observed by a ground-based FPI, Dec. 12, 1981, from Kiruna, Sweden.

Figure 5 shows the interval (DE-2 orbit 1853) when B_Z was directed northward. At this Universal Time, the northern polar cap is displaced sunward to almost its maximum extent during the day. Polar cap plasma densities were thus high on the dawn side of the polar cap where, resulting from photoionisation and strong plasma transport, a tongue of high ionisation density has been carried antisunward from the dayside. This feature is discussed by Fuller-Rowell et al. (1988), who show that this plasma tongue is a consistent feature of the modelled polar cap plasma density distribution during the winter period. When the IMF B_Y component is positive, for several hours either side of 18 UT, there is strong direct plasma transport from the sunlit polar cusp region antisunward on the dawn side of the polar cap. On the dusk side of the polar cap, plasma densities are low, as the result of sunward transport from the nightside, combined with the absence of photoionisation and weak precipitation in this part of the polar cap. Heppner and Maynard (1987) describe the polar convection under these conditions (IMF B_Z positive) as a highly distorted two-cell vortex pattern, rather than a multi-cell system. It is of considerable interest and importance that in the region of intense antisunward flow, the plasma densities are also high, so that the neutral-ion time constant is quite low (10 min. or less). The neutral flow on the dawn side of the polar cap thus matches the ion flow quite accurately, whereas in the dusk part of the auroral oval, the dusk part of the polar cap and the dawn part of the auroral oval, the plasma density is much lower, the neutral-ion time constant is around 3 hours, and neutral flows are much weaker than the ion flows. For example, the winds do not turn sunward in the dawn part of the auroral oval, despite sunward ion flows of 600 - 800 ms^{-1}.

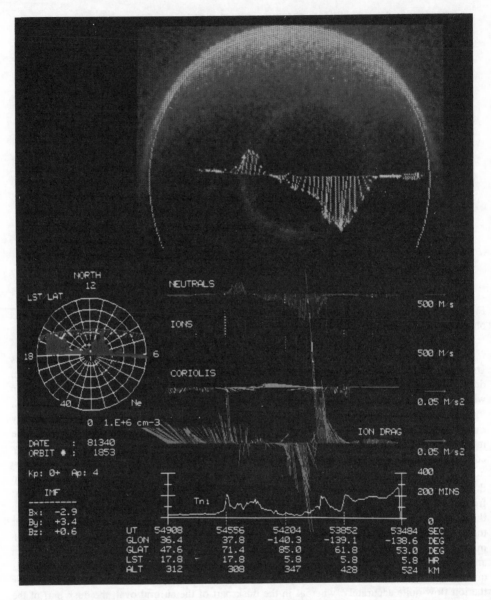

Figure 4: A detailed study of the situation shown in element A of Figure 6 (Killeen et al., 1988). Properties of the thermal plasma (ion density and ion drift) within the polar region are shown, along with a determination of the ion-neutral collisional time constant, and various important forces coupling the high-latitude ionosphere and thermosphere.

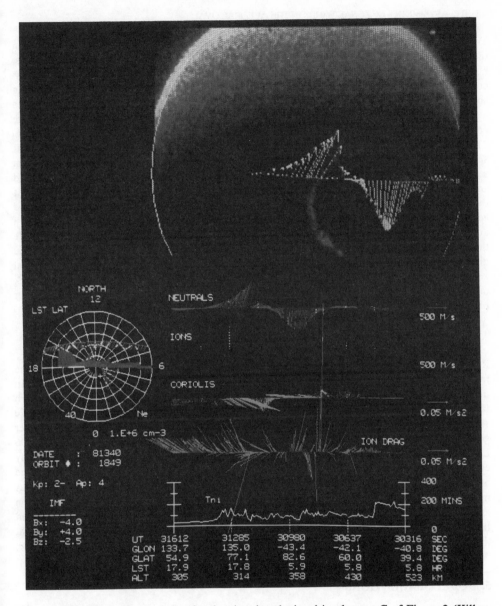

Figure 5: Similar to Figure 4, but for the situation depicted in element C of Figure 2 (Killeen et al., 1988).

4.1. THE ASYMMETRY OF ZONAL WINDS IN THE AURORAL OVAL

The sunward (westward) winds in the dusk auroral oval are almost always much stronger than the sunward (eastward) winds in the dawn auroral oval. The mean ion drag acceleration is, however, equally strong in both regions, taking account of IMF (and particularly the B_Y component) variations. This asymmetry is a natural consequence of fluid flow in a rotating spherical planetary atmosphere. In the dusk auroral oval, the westward ion drag generates Coriolis and advection forces which are oppositely directed. The result is that air parcels are entrained, spending a long period in the region of strong westward acceleration by ion drag. Their paths then follow the 'clockwise' curvature of the auroral oval, and they join in an anti-sunward flow over the polar cap completing a strong clockwise circulation cell. In the dawn auroral oval, sunward or eastward ion drag acting on the gas results in both Coriolis and advection accelerations which are directed equatorward. Gas parcels are literally 'spun out' of the auroral oval, into regions of lower ion drag, and are not entrained in the region of rapid acceleration. This does enhance equatorward thermospheric winds flowing from the dawn auroral oval to mid-latitudes, which has significant consequences for mid-latitude thermospheric composition, and the ionosphere during and following major geomagnetic disturbances. Also, despite the lower zonal winds generated in the dawn auroral oval, the total zonal momentum transferred is probably similar to that in the dusk auroral oval. Due to lack of entrainment, the momentum is transferred into a much larger air mass, with resulting lower peak winds.

4.2. THERMOSPHERIC COMPOSITION IN THE POLAR REGIONS

Rees et al. (1985) discuss in some detail a comparison between two sets of DE-2 data, one from the northern hemisphere obtained during December 1981 and the second obtained in the southern hemisphere in October 1981. They used a number of model simulations which were generated specially to investigate the major terms causing the observed thermospheric structures in neutral thermospheric temperature and wind velocities in the winter and summer geomagnetic polar regions. In December 1981, the DE-2 orbit was in the 18.0 LST to 06 LST time plane. Wind, temperature and composition patterns within the polar regions display complex structures as functions of latitude and Universal time, as shown in Figure 6. The highest wind velocities, and strong enhancements of temperature and of mean molecular mass, closely follow the location (displayed as true anomaly) of the geomagnetic polar region as a function of UT. These statistical studies confirm that there are strong antisunward wind flows within the central polar cap, strong sunward winds in the dusk auroral oval. However, at the location of the dawn auroral oval, there is only a reduction of the global antisunward flow.

5. MODEL SIMULATIONS

The following section presents a series of numerical simulations of the coupled UCL Sheffield Thermosphere / Ionospheric Model. Most of the observational data presented earlier can be understood and interpreted in the context of these model simulations.

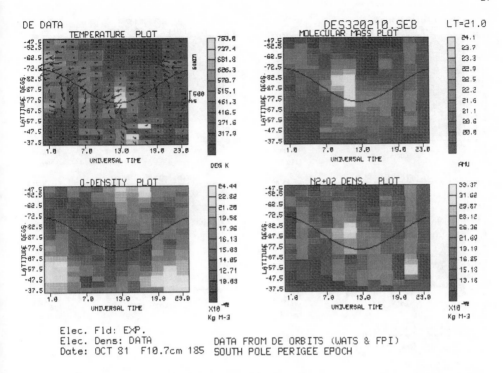

Figure 6: Statistical distributions of neutral composition, winds and temperature for upper thermosphere (300 km) in the south polar region during October 1981 as observed by the Dynamics Explorer-2 satellite.

5.1. GLOBAL TEMPERATURE AND COMPOSITION STRUCTURE

The global distributions of thermospheric temperature and composition have become well established by observation, for example, the recent empirical models of the thermosphere (CIRA, 1986). The model simulations can then be used to examine the causes of the redistribution of thermospheric major and minor species.

Figure 7 illustrates the global distributions of temperature, mean molecular mass, atomic oxygen and molecular nitrogen concentrations in the upper thermosphere as simulated for conditions of moderate solar activity and low geomagnetic activity, at the December solstice. The region from the north to the south geographic pole is illustrated at pressure level 12 close to 300 km altitude, and at 18 UT.

28

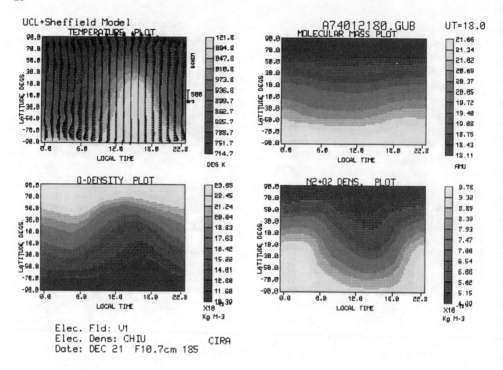

Figure 7: Simulation (3-D coupled model) of neutral composition, winds and temperature in December, for pressure level 12 (approx. 300 km), at 18 UT, at high solar activity and low geomagnetic activity.

5.2. THE THERMOSPHERIC HADLEY CELL

Two features are immediately apparent in Figure 7. Firstly, there is a strong diurnal response of thermospheric temperature, particularly at low and middle latitudes, to the solar forcing, a factor which is unique to the thermosphere. Secondly, there is a strong latitudinal gradient of temperature and composition, with values decreasing from the summer polar regions to the winter polar region. For this situation of low geomagnetic activity, the high-latitude 'geomagnetic' sources are fairly weak and are only apparent in the slightly enhanced wind magnitudes within the polar regions. The global circulation responds to the latitudinal temperature, and related pressure, gradients.

At upper thermospheric altitudes, above 200 km, there is a mean wind circulation, a global Hadley cell, with air flowing from the summer polar regions to the winter polar region. This can occur in the thermosphere, due to the large value of kinetic viscosity. A weak return flow occurs in the lower thermosphere / upper mesosphere (or perhaps lower). This inter-hemispheric horizontal flow is associated with upwelling at summer high latitudes and downwelling at winter high latitudes. The compositional structure is a direct consequence of this global circulation cell. In regions of systematic upwelling, such as the summer polar region, the concentration of heavy molecular species (N_2, O_2) is enhanced relative to light atomic species (O). In contrast in the regions of downwelling such as the winter polar and high latitude region, the heavy molecular species are depleted, relative to light atomic species.

In the simulations, geomagnetic activity is increased by increasing the magnitude of the polar convective electric field and corresponding intensity of particle precipitation at high latitudes. The global structure of the thermosphere and ionosphere is thus modified. Figure 8 shows the same four parameters for the same season and solar activity, but for a level of moderately high geomagnetic activity, equivalent to a situation when Kp would be steady about 3 to 3+. The increased geomagnetic input strengthens the polar winds and further heats the polar regions. The resulting global pressure gradients are modified, as are the wind cells. The circulation from summer to winter pole is stronger, but no longer quite extends to the winter pole. A cell of opposing circulation has developed between the winter pole and high winter mid-latitudes, reversing the prevailing wind. This circulation change also alters the compositional response. At moderate or high geomagnetic activity, the lowest values of temperature and mean molecular mass are now found at high winter mid-latitudes and a second region of high values of mean molecular mass occurs near the winter pole, in response to the upwelling caused by the increased geomagnetic input.

As the level of geomagnetic activity increases further, some of the causes of the extreme neutral wind observations discussed previously become apparent. Figure 9 illustrates the neutral wind and temperature distributions following a period where the Kp would have been around 7 for about six hours, a minor geomagnetic storm. Neutral winds approaching 1 kms^{-1} are generated, with a particularly strong response in the dusk auroral oval (sunward winds) and over the polar cap (antisunward winds). These are particular features seen in the ground-based and satellite observations during very disturbed periods. Even in the dawn auroral oval, sunward winds of nearly 300 ms^{-1} are generated. Such strong eastward winds are rather rare, although a sunward component of 600 - 700 ms^{-1} was observed on Feb. 12, 1982 from Kiruna, N. Sweden, under very extreme conditions.

5.3. SEASONAL AND IMF-By CONTROL OF THE HIGH LATITUDE COMPOSITION AND PLASMA DENSITY

The geomagnetic control of the thermospheric composition at high latitudes was illustrated in Figure 6. Figure 10 shows a model simulation for a similar period, corresponding to October 1981, when solar activity was very high. The average geomagnetic activity was also high. The neutral temperature, mean molecular mass, atomic oxygen density and molecular nitrogen density are presented as a function of Universal Time and co-latitude, as it would be sampled by a satellite in a fixed local-time plane, 09 to 21 LT, at 320 km altitude, essentially an equivalent format to that displayed in Figure 6. Both the average values and the structure of the modelled composition data are in excellent agreement with the FPI, WATS and NACS measurements, and clearly illustrate their high-latitude structures are associated with the geomagnetic polar region. Major disturbances in both composition and temperature response follow closely the location of the geomagnetic polar cap as a function of UT and latitude (the satellite's closest approach to the geomagnetic pole is illustrated by the sinusoidal curve). The general agreement between observation and the simulation is good both quantitatively as well as qualitatively.

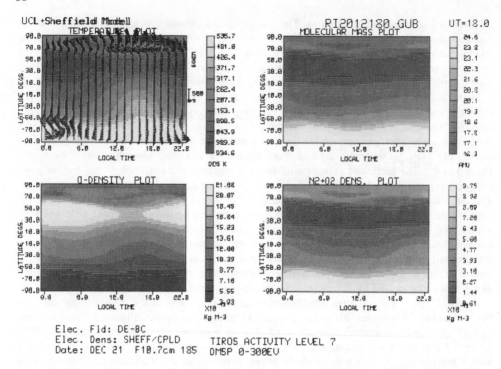

Figure 8: Model simulation of neutral composition, winds and temperature in December, for pressure level 12 (approx. 300 km), at 18 UT, at high solar activity and moderate geomagnetic activity.

To illustrate the thermospheric and ionospheric response at high-latitudes to the orientation of the B_Y component of the IMF in summer and winter we will use data from four simulations of the UCL / Sheffield coupled model. These simulations have been generated for IMF B_Z negative (southward), and for conditions when the IMF B_Y component was either strongly positive, or strongly negative, for a geomagnetic activity level corresponding to approximately Kp = 3 to 4, and for moderately high solar activity ($F_{10.7}$ cm = 185). Two of the simulations are for the December solstice and two for June. The simulations are time-dependent, that is they are UT-dependent, and the results are diurnally reproducible. However, the external solar and geomagnetic inputs are time-independent. These four simulations use an offset dipole representation of the geomagnetic field.

The characteristic UT variations of the summer and winter polar regions are dependent on the offset of the geomagnetic poles from the geographic poles. During the UT day, at all seasons, the geomagnetic polar caps are carried into and out of sunlight. There is, therefore, a large diurnal modulation of the solar photoionisation and UV/EUV heating of the geomagnetic polar regions which also causes large UT variations in plasma density, conductivity, ion drag and Joule and solar heating of the polar thermosphere. There are consequent large UT modulations of the thermospheric and ionospheric response. These characteristic UT variations of thermospheric and ionospheric structures, and the associated thermospheric-ionospheric interactions, are discussed by Fuller-Rowell et al. (1988).

In this paper we will select the 18 UT period to illustrate the response of neutral wind, composition, and plasma density to the orientation of IMF-B_Y, in summer and winter, and to illustrate the coupling between these parameters. Figure 11 is a montage of four polar plots from 50° to 90° geographic latitude for the northern hemisphere. Mean molecular mass is shown at pressure level 12, around 300 km altitude, together with the neutral wind vectors. The winter simulations of B_Y negative and positive are shown in Figure 11 a and b respectively, and summer solstice conditions are shown in Figure 11 c and d for B_Y negative and positive respectively. The corresponding plasma density distributions are illustrated in Figure 12.

At F-region altitudes, only relatively small differences in the wind patterns can be ascribed to seasonal effects. In the winter, the winds in the dawn auroral oval show only a slight tendency to follow the relatively weak sunward ion convection (B_Y positive). In the summer, there is no significant indication of sunward wind acceleration at all in the dawn cell for B_Y positive. Only minor seasonal differences occur in the strong clockwise circulation wind cell which follows the strong sunward ion convection in the dusk auroral oval and antisunward ion flow over the dawn side of the polar cap. On the nightside, at latitudes below the auroral oval, the wind flow is distinctly equatorward in the June simulation, and rather weaker in the December simulation (100 rather than 200 ms^{-1}).

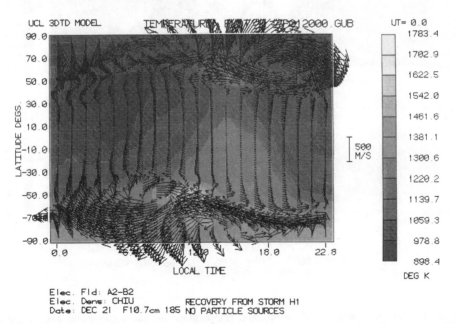

Figure 9: Model simulation of neutral winds and temperature in December, for pressure level 12 (approx. 300 km), at 18 UT, at high solar activity after a period of very high geomagnetic activity.

The By dependence of the neutral wind response, therefore, is clear for either season. The most distinct effect is that the location of the peak antisunward flow moves from the dawn to the dusk side of the polar cap as the IMF-By changes from positive to negative. In the auroral dawn oval there is a weak sunward wind when By is negative, while for By positive, there is only a reduction of the antisunward wind. In the dusk auroral oval again the sunward winds are stronger for the negative By orientation. The polar neutral winds observed by the DE satellite which are presented in the previous Figures agree well with the predictions of the model for the different orientations of By.

The highest values of mean molecular mass which occur poleward of 50° latitude in the winter polar F-region are about equal to the lowest value in the equivalent summer polar region. At high winter mid-latitudes, the mean molecular mass is close to 16, indicating a composition which is nearly pure atomic oxygen. There is then a plateau covering the geomagnetic polar region, due to heating and associated upwelling and outflow, where the mean molecular mass reaches 20.

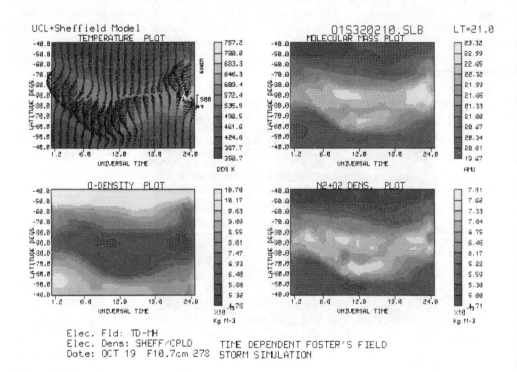

Figure 10: Neutral composition, winds and temperature for the south polar region during October 1981 as simulated by the thermospheric model, presented in the same format as Figure 6.

This plateau value of mean molecular mass in the winter polar region is very dependent on geomagnetic activity. At very low activity levels the mean molecular mass may be as low as 16 - 17, while during extended geomagnetic storm conditions the mean molecular mass may reach values as high as 22.

In the summer polar cap, the lowest values of mean molecular mass above 50° are 20, and the highest values, within the geomagnetic polar cap, reach 24 to 25. At this constant pressure level (12), this implies a 4-fold reduction in the amount of atomic oxygen being transported from high summer mid-latitudes to the pole, and atomic oxygen concentrations which are a factor of 10 lower than those found in the high winter mid-latitude region. The variation of molecular nitrogen density is in direct anti-phase, and compensates for the atomic oxygen changes.

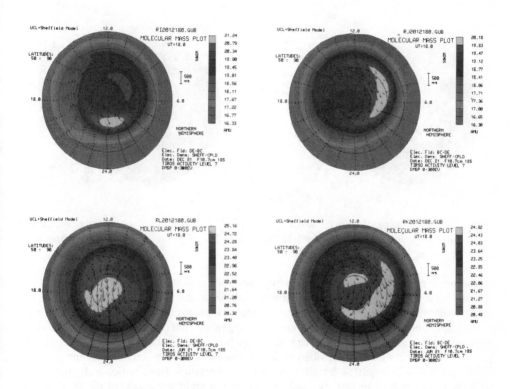

Figure 11: A montage of four polar plots, 50° to 90° N latitude, of mean molecular mass and neutral wind from the model at pressure level 12: a) Winter B$_Y$ negative, b) Winter B$_Y$ positive, c) Summer B$_Y$ negative, d) Summer B$_Y$ positive.

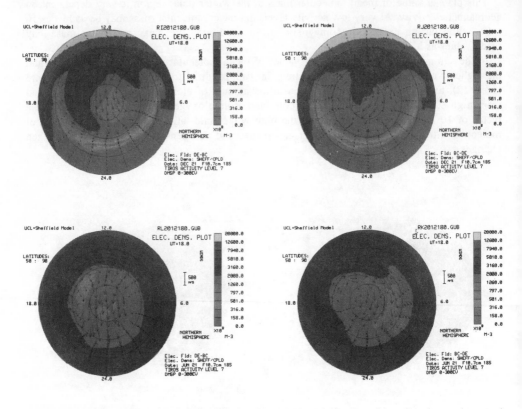

Figure 12: A montage of four polar plots, 50° to 90° N latitude, of ion density and neutral wind from the model at pressure level 12: a) Winter B_Y negative, b) Winter B_Y positive, c) Summer B_Y negative, d) Summer B_Y positive.

The polar plasma density structure shows two types of features, mainly dependent on the season. Most of the middle and upper thermosphere in the winter polar region has a low mean molecular mass, except under very disturbed geomagnetic conditions, reflecting a neutral composition dominated by atomic oxygen. Under such conditions, ionisation recombination time constants are long, and transport of ionisation is very important. The plots of the plasma density distribution in the winter polar cap thus show enhancements, where plasma is transported by convection and co-rotation from regions of high photo-ionisation. The polar cap may be partly filled by the rapid anti-sunward transport of flux tubes containing high density plasma from the dayside at high mid-latitudes, particularly around 18UT. The mid-latitude, or sub-auroral trough, in contrast, is a winter-time feature where flux tubes containing low plasma densities are transported westward from the night-side, where due to a long period out of sunlight or auroral precipitation, the plasma density has decayed to low values. There is also usually a stagnation trough or 'hole' present within the winter polar cap, a region which maintained circulating within the polar cap out of sunlight and away from significant auroral precipitation.

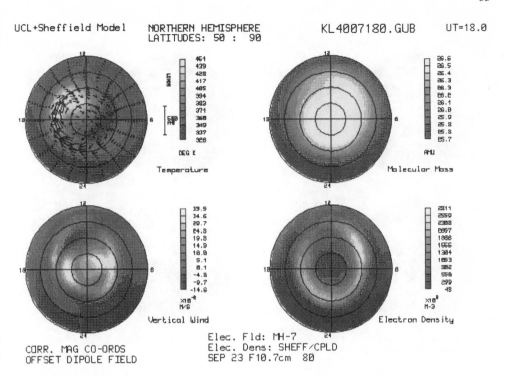

Figure 13: Wind velocity and temperature, mean molecular mass, vertical wind and plasma density distributions taken from the UCL coupled ionosphere / thermosphere GCM simulation for equinox conditions for IMF By zero. The data are presented at pressure level 7 (approx. 120 km), for the northern polar region (poleward of 50° geographic latitude) and at 18 UT.

In the summer polar cap, the mean molecular mass of the middle and upper thermosphere is much higher, due to the combination of convection and advection generated by solar heating and intense Joule heating (particularly during disturbed geomagnetic conditions). The ionospheric chemistry reflects the increased presence of molecular nitrogen, and the typical ionospheric recombination time constants are much lower. Transport is thus less important in structuring the plasma density, and regions of decreased plasma density result from unusually intense heating (with increased upwelling, increased nitrogen concentrations, and further increased ionisation recombination coefficients). Regions of intense plasma convection may thus be associated with particular enhancements of Joule heating and very low plasma densities, although the ionisation rates may be similar throughout the summer polar region. The features within the summer polar cap may thus image the regions of most intense plasma convection as trough-like features or 'holes' in the F-region plasma density distribution.

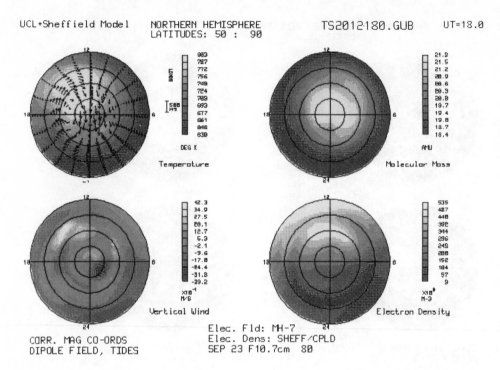

UCL+Sheffield Model   NORTHERN HEMISPHERE   TS2012180.GUB   UT=18.0
                       LATITUDES: 50 : 90

CORR. MAG CO-ORDS   Elec. Fld: MH-7
DIPOLE FIELD, TIDES   Elec. Dens: SHEFF/CPLD
                        SEP 23 F10.7cm 80

Figure 14: Wind velocity and temperature, mean molecular mass, vertical wind and plasma density distributions taken from the UCL coupled ionosphere / thermosphere GCM simulation for equinox conditions for IMF By zero, but including a simulated lower atmosphere tidal source for semi-diurnal tides. The data are presented at pressure level 12 (approx. 280 km), for the northern polar region (poleward of 50° geographic latitude) and at 18 UT.

5.4. LOWER THERMOSPHERE RESPONSE: GEOMAGNETIC FORCING - ION DRAG

The response of the lower thermosphere and ionosphere to magnetospheric forcing is as marked and varied as that of the upper thermosphere. The lower thermosphere has been notoriously difficult to monitor experimentally. Much of the initial exploration was done mainly by rocket experiments and more recently via ground and satellite remote sensing. However, the Incoherent Scatter Radar technique can probe the plasma environment most effectively (Babcock and Evans, 1979), and it is possible to derive winds and neutral gas temperature from the radar measurements. In the lower thermosphere, neutral winds generated by ion drag forcing and by Joule heating are weaker in magnitude than in the upper thermosphere, for example around 120 km altitude, neutral winds are roughly around 50% of those near 300 km. The wind vectors are also rotated in direction, due to the different balances of Coriolis, pressure gradient, and ion-drag forces. Solar heating and tidal forcing from the lower atmosphere are also important, particularly during periods when geomagnetic activity is low.

Figure 13 shows the simulated neutral wind, temperature, mean molecular mass, vertical wind, and ion density at pressure level 7, near 120 km, from 50° to 90°N latitude, for equinox, obtained by using a symmetric electric field pattern, for a moderate level of geomagnetic activity. The thermospheric wind patterns are quite similar to those obtained at upper thermospheric altitudes (Figure 14), although the wind velocities are approximately a factor of 3 lower, similar to the factor that the ion velocity is decreased from ($\underline{E} \wedge \underline{B}$), since the ratio of ion / neutral collision to ion gyro frequency is approx. unity. The ion velocity is also rotated into the $\underline{E}$ direction, as a result of these collisions with the neutrals. It is perhaps initially surprising that the 'steady-state' neutral winds have such similar patterns to those of the upper thermosphere. However, this is the result of the intrinsic properties of a rotating planetary atmosphere, with relatively high viscosity and ion drag acting, and where pressure gradients, Coriolis and advection terms are also in near-equilibrium. Were the same electric fields and particle precipitation imposed for only a short period, the resulting circulation would be quite different. The initial thermospheric response to a suddenly-imposed convection / precipitation pattern is the generation of a convergent clockwise vortex in the dusk auroral oval / dusk polar cap, and a divergent anti-clockwise vortex in the dawn auroral oval / polar cap region. The pressure gradients generated by the initial convergence and divergence act to force the stable wind circulation to the patterns observed in Figure 13.

5.5. LOWER THERMOSPHERE RESPONSE: LOWER ATMOSPHERE TIDAL FORCING

Figure 15 is taken from a simulation for the same equinox conditions as Figure 13, with the exception that tidal forcing at the lower boundary of the model has also been imposed (Parish et al., 1990). The tidal forcing from the middle atmosphere has a modest effect at high-latitudes for this level of geomagnetic activity, since a large-amplitude semi-diurnal variation is induced by ion drag acting in the auroral oval. However, in the mid-latitude regions, the lower-atmosphere tidal influences cause a pronounced semi-diurnal variation in both the temperature and wind distributions. The ion density distribution in the auroral oval is dominated by the distribution of auroral (electron) precipitation, which can be seen to have the elliptical mapping, resulting from the 'distortion' of the IGRF magnetic field model used to map the convection and precipitation onto the polar regions. Typical E-region ion densities of between 2 and $3*10^{11}$ occur in the auroral oval. Outside the auroral oval, solar photo-ionisation generates ion densities a factor 5 lower than those in the auroral oval.

5.6. MAGNETOSPHERIC COUPLING

Interactions between the polar ionosphere and the magnetosphere have not been considered in a self-consistent way in these simulations. The magnetosphere is assumed to be a zero-resistance source of the Field-Aligned-Current (FAC) required to match the Pedersen currents produced by the applied potential. The ionospheric conductivity is produced by the combination of auroral precipitation and solar photoionisation sources. Thermosphere / ionosphere interactions, including the induced neutral wind dynamo (which induces a back-EMF), cause considerable modulation of ionospheric electric currents. Most probably, the real magnetosphere is unable to sustain fully the consequences of ionospheric / thermospheric feedback. The external convection electric field and the FAC should respond to the induced wind dynamo, and to ionospheric current changes as conductivity responds. As these two major features are modulated, it might be expected that the energetic particle populations and perhaps the morphology of the FAC and magnetospheric magnetic field geometry would change significantly.

At the present time it is only possible to consider such changes qualitatively. We can also anticipate that as the magnetosphere reacts to thermosphere and ionosphere feedback, the thermosphere and ionosphere will respond to the new magnetospheric inputs, in a continuous feedback loop. In practice, however, the magnetospheric configuration has also to respond to continuous variations of the solar wind and IMF, so that a condition of full stability throughout the solar wind-magnetosphere-thermosphere and ionosphere can never exist.

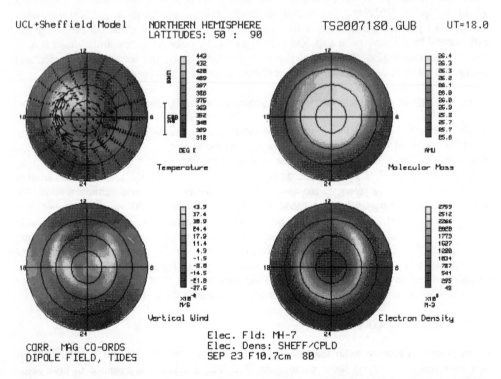

Figure 15: Wind velocity and temperature, mean molecular mass, vertical wind and plasma density distributions taken from the UCL coupled ionosphere / thermosphere GCM simulation for equinox conditions for IMF By zero, but including a simulated lower atmosphere tidal source for semi-diurnal tides. The data are presented at pressure level 7, (120 km), for the northern polar region (poleward of 50° geographic latitude) and at 18 UT.

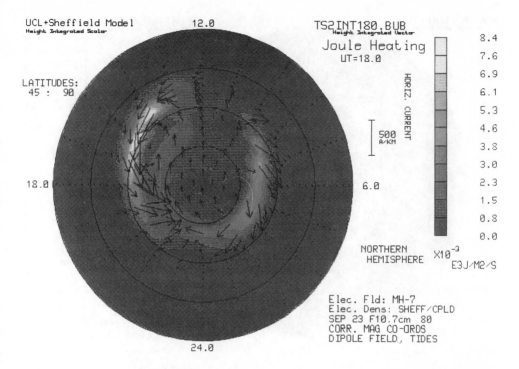

Figure 16: The height-integrated horizontal current density flowing in the high-latitude ionosphere, and the distribution of Joule heating from the same simulation as that used for Figure 14 and 15. The peak value of Joule heating is 8.4 x 10^{-3} Joules M^{-2} s^{-1} or 8.4 mW M^{-2}.

During the development of the coupled ionospheric / thermospheric model, we experimented with the magnetosphere / ionosphere feedback process, and found that it was stable to slowly varying external (magnetospheric) forcing. However, considering the inherent instability of many plasma physics processes, we may anticipate that some ionospheric - magnetospheric interactions may be unstable, non-linear or chaotic. Certainly, many processes involve highly unstable interactions between energetic particle precipitation, the ionosphere, counter-streaming plasma populations and the FAC and magnetic field geometry, with characteristic frequencies in the frequency range from quasi-DC to many MHz. These processes involve physics occurring at a much smaller scale than the grids of present models, and thus have to be parameterised. At the frequency range required for the model simulations, it has to be assumed that the magnetospheric-ionospheric feedback is stable, even when high frequency and sub grid-mesh processes are ignored.

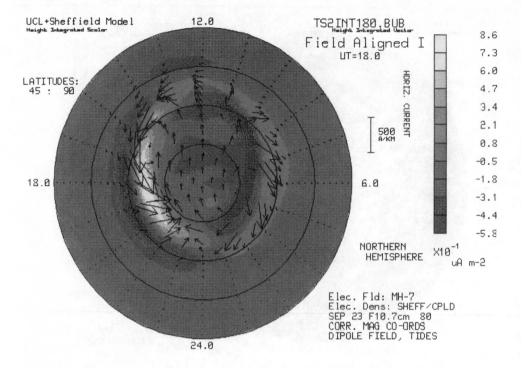

Figure 17: The field-aligned current distribution from the same simulation as that used for Figure 14 and 15. The peak value of field-aligned current is 8.6 x 10⁻¹ μA M⁻² or 0.86 μA M⁻².

The next major task for the development of these coupled ionosphere / thermosphere models will be the experimental process of attempting to couple into the advanced magnetosphere models. Figure 16 shows the height-integrated horizontal current density (vectors) flowing in the northern polar regions, and also the total Joule heating, taken from the same simulation as that used for Figures 13 and 14. Figure 17 shows the corresponding pattern of Field Aligned Current (FAC). The latter is derived from the divergence of the horizontal current system, and indicates the current demand on the magnetosphere. Upward current is denoted as positive (yellow), downward current is negative. The region 1 and region 2 Birkland currents can be clearly seen in both the dusk and dawn sectors of the auroral oval. It is an implicit assumption in this class of numerical modelling that the Magnetosphere / ionosphere system can supply both the electric potential and the charge carriers to support the ionospheric currents, the ion drag, the Joule heating and the Field-Aligned Current. Also, the charge carriers have to include the auroral precipitation particles, required to provide the enhanced Pedersen conductivity and enhanced ion drag. As yet, few of this group of interesting questions have been addressed by observational data sets available from ground-based or from satellite measurements.

Acknowledgments. I would like to acknowledge the assistance of John Harmer in the preparation of a number of the numerical simulations used in the preparation of this review, and to thank Anasuya Aruliah for her assistance in reviewing the manuscript. The work was supported in part by Research Grants from the UK Science and Engineering Research Council.

6. REFERENCES

Allen B.T., G.J. Bailey and R.J. Moffett (1986) Ion distributions in the high-latitude topside ionosphere. Ann. Geophysicae 4 A, 97-106.

Babcock R.R. and J.V. Evans (1979) Effects of geomagnetic disturbances on neutral winds and temperatures in the thermosphere observed over Millstone Hill. J. Geophys. Res. 84, 5349-5354.

Cole K.D. (1962) Joule heating of the upper atmosphere. Aust. J. Phys. 15, 223-235.

Cole K.D. (1971) Electrodynamic heating and movement of the thermosphere. Planet. Space Sci. 19, 59-75.

Chiu Y.T. (1975) An improved phenomenological model of ionospheric density. J. Atmos. Terr. Phys. 37, 1563-1570.

Dickinson R.E., E.C. Ridley and R.G. Roble (1981) A three-dimensional general circulation model of the thermosphere. J. Geophys. Res. 86, 1499-1512.

Dickinson R.E., E.C. Ridley and R.G. Roble (1984) Thermospheric general circulation with coupled dynamics and composition. J. Atmos. Terr. Phys. 41, 205-219.

Fesen C.G., R.E. Dickinson and R.G. Roble (1986) Simulations of the thermospheric tides at equinox with the National Center for Atmospheric Research Thermospheric General Circulation Model. J. Geophys. Res. 91, 4471-4489.

Foster J.C., J.M. Holt, R.G. Musgrove and D.S. Evans (1986) Ionospheric convection associated with discrete levels of particle precipitation. Geophys. Res. Lett. 13, 656-659.

Frank L.A., J.D. Craven, K.L. Ackerson, M.R. English, R.H. Eather and R.L. Carovillano (1981) Global auroral imaging instrumentation for the Dynamics Explorer mission. Space Sci. Instrum. 5, 369-393.

Fuller-Rowell T.J. and D. Rees (1980) A three-dimensional, time-dependent, global model of the thermosphere. J. Atmos. Sci. 37, 2545-2567.

Fuller-Rowell T.J. and D. Rees (1983) Derivation of a conservative equation for mean molecular weight for a two constituent gas within a three-dimensional, time-dependent model of the thermosphere. Planet. Space Sci. 31, 1209-1222.

Fuller-Rowell T.J. and D.S. Evans (1987) Height-integrated Pedersen and Hall conductivity patterns inferred from the TIROS-NOAA satellite data. J. Geophys. Res. 92, 7606-7618.

Fuller-Rowell T.J., D. Rees, S. Quegan, G.J. Bailey and R.J. Moffett (1984) The effect of realistic conductivities on the high-latitude thermospheric circulation. Planet. Space Sci. 32, 469-480.

Fuller-Rowell T.J., S. Quegan, D. Rees, R.J. Moffett and G.J. Bailey (1987) Interactions between neutral thermospheric composition and the polar ionosphere using a coupled ionosphere-thermosphere model. J. Geophys. Res. 92, 7744-7748.

Fuller-Rowell T.J., D. Rees, S. Quegan, R.J. Moffett and G.J. Bailey (1988) Simulations of the seasonal and universal time variations of the thermosphere and ionosphere using a coupled, three-dimensional, global, model. PAGEOPHYS Vol. 127, No.2/3.

Hardy D.A., M.S. Gussenhoven and E. Holeman (1985) A statistical model of auroral electron precipitation. J. Geophys. Res. 90, 4229-4248.

Harel M., R.A. Wolf, P.H. Reiff, R.W. Spiro, W.J. Burke, F.J. Rich and M. Smiddy (1981) Quantitative Simulation of a Magnetospheric Substorm, 1, Model Logic and Overview. J. Geophys. Res. 86, 2217-2241.

Hays P.B., T.L. Killeen and B.C. Kennedy (1981) The Fabry-Perot interferometer on Dynamics Explorer. Space Sci. Inst. 5, 395-416.

Hays P.B., T.L. Killeen, N.W. Spencer, L.E. Wharton, R.G. Roble, B.A. Emery, T.J. Fuller-Rowell, D. Rees, L.A. Frank and J.D. Craven (1984) Observations of the dynamics of the polar thermosphere. J. Geophys. Res. 89, 5547-5612.

Heelis R.A., W.B. Hanson, C.R. Lippincott, D.R. Zuccaro, L.H. Harmon, B.J. Holt, J.E. Doherty and R.A. Power (1981) The ion drift meter for Dynamics Explorer-B. Space Sci. Instrum. 5, 511-521.

Heppner J.P. (1977) Empirical Models of High Latitude Electric Field. J. Geophys. Res. 82, 1115-1125.

Heppner J.P. and N.C. Maynard (1987) Empirical High-Latitude Electric Field Models. J. Geophys. Res. 92, 4467-4490.

Killeen T.L., P.B. Hays, N.W. Spencer and L.E. Wharton (1983) Neutral winds in the polar thermosphere as measured from Dynamics Explorer. Adv. Space Res. 2, No. 10, 133-136.

Killeen T.L., J.D. Craven, L.A. Frank, J.-J. Ponthieu, N.W. Spencer, R.A. Heelis, L.H. Brace, R.G. Roble, P.B. Hays and G.R. Carignan (1988) On the relationship between the dynamics of the polar thermosphere and the morphology of the aurora: Global scale observations from Dynamics Explorers 1 and 2. J. Geophys. Res. 93, 2675-2692.

Krehbiel J.P., L.H. Brace, R.F. Theis, W.H. Pinkus and R.B. Kaplan (1981) The Dynamics Explorer Langmuir probe instrument. Space Sci. Instrum. 5, 493-502.

Lean J. (1988) Solar EUV Irradiances and Indices. Adv. Space Res. 7, 263-292, Pergamon.

Parish H., T.J. Fuller-Rowell, D. Rees. T.S. Virdi and P.J.S. Williams (1990) Numerical simulations of the seasonal response of the thermosphere to propagating tides. Adv. Space Res. Vol. 10, No.6 (6)287-(6)291.

Pereira E., M.C. Kelley, D. Rees, I.S. Mikkelson, T.S. Jorgensen and T.J. Fuller-Rowell (1980) Observations of neutral wind profiles between 115 and 176 km altitude in the dayside auroral oval. J. Geophys. Res. 85, 2935-2940.

Quegan S., G.J. Bailey, R.J. Moffett, R.A. Heelis, T.J. Fuller-Rowell, D. Rees and R.W. Spiro (1982) Theoretical study of the distribution of ionization in the high-latitude ionosphere and the plasmasphere: First results on the mid-latitude trough and the light-ion trough, J. Atmos. Terr. Phys. 44, 619-640.

Quegan S., G.J. Bailey, R.J. Moffett and L.C. Wilkinson (1986) Universal time effects on the plasma convection in the geomagnetic frame. J. Atmos. Terr. Phys. 48, 25-40.

Rees D. (1971) Ionospheric winds in the auroral zone. J. Brit. Interplan. Soc. 24, 233-346.

Rees D. (1973) Neutral wind structure in the thermosphere during quiet and disturbed geomagnetic periods. In Physics and Chemistry of Upper Atmospheres (Edited by B.M. McCormac) 11-23, Reidel, Dortrecht.

Rees D. and T.J. Fuller-Rowell (1987a) Comparison of theoretical models and observations of the thermosphere and ionosphere during extremely disturbed geomagnetic conditions during the last solar cycle. Adv. Space Res. 7, 827-838, Pergamon.

Rees D. and T.J. Fuller-Rowell (1987b) A Theoretical Thermosphere Model for CIRA. Adv. Space Res. 7, 10185-10197, Pergamon.

Rees D. and T.J. Fuller-Rowell (1989) Geomagnetic Response of the Polar Thermosphere and Ionosphere. Electromagnetic Coupling in the Polar Clefs and Caps, p 355-391. Editors P.E. Sandholt and A. Egeland.

Rees D., P.A. Rounce, P. Charleton. T.J. Fuller-Rowell, I. Mcwhirter and K. Smith (1982) Thermospheric Winds during the Energy Budget Campaign: Ground-based Fabry-Perot Observations Supported by Dynamical Simulations with a Three-dimensional, Time-dependent Thermospheric Model. J. Geophys. Res. 50, 202-211.

Rees D., T.J. Fuller-Rowell, R. Gordon, T.L. Killeen, P.B. Hays, L.E. Wharton and N.W. Spencer (1983) A comparison of the wind observations from the Dynamics Explorer satellite with the predictions of a global time-dependent model. Planet. Space Sci. 31, 1299-1314.

Rees D., R. Gordon, T.J. Fuller-Rowell, M.F. Smith, G.R. Carignan, T.L. Killeen, P.B. Hays and N.W. Spencer (1985) The composition, structure, temperature and dynamics of the upper thermosphere in the polar regions during October to December 1981. Planet. Space Sci. 33, 617-666.

Rees D., T.J. Fuller-Rowell, R. Gordon, M.F. Smith, J.P. Heppner, N.C. Maynard, N.W. Spencer, L.E. Wharton, P.B. Hays and T.L. Killeen (1986) A theoretical and empirical study of the response of the high-latitude thermosphere to the sense of the "Y" component of the interplanetary magnetic field. Planet. Space Sci. 34, 1-40.

Rees D., T.J. Fuller-Rowell, S. Quegan, R.J. Moffett and G.J. Bailey (1988) Simulations of the seasonal variations of the thermosphere and ionosphere using a coupled, three-dimensional, global model, including variations of the interplanetary magnetic field. J. Atmos. Terr. Phys. Vol. 50, No.10/11, 903-930.

Rich F.J. and N.C. Maynard (1989) Consequences of Using Simple Analytical Functions for the High-Latitude Convection Electric Field. J. Geophys. Res. 94, 3687-3701.

Rishbeth H. (1972) Thermospheric winds and the F-region. J. Atmos. Terr. Phys. 34, 1-47.

Roble R.G., R.E. Dickinson and E.C. Ridley (1982) The global circulation and temperature structure of the thermosphere with high-latitude plasma convection. J. Geophys. Res. 87, 1599-1614.

Watkins B.J. (1978) A numerical computer investigation of the polar F-region. Planet. Space Sci. 26, 559-569.

ANNEX: NOTATION

p   pressure,

g   gravitational acceleration,

h   height of pressure surface,

ρ neutral mass density,

ω vertical velocity in pressure co-ordinate system
$\omega = \dfrac{dp}{dt}$,

$\underline{\nabla}_p.$ horizontal divergence operator in pressure coordinate system

$$\text{e.g. } \underline{\nabla}_p.\underline{V} \equiv \frac{1}{r\sin\theta}\frac{\partial}{\partial\theta}(V_\theta\sin\theta) + \frac{1}{r\sin\theta}\frac{\partial}{\partial\phi}V_\phi,$$

V_z true vertical velocity,

$\underline{V}$ horizontal vector neutral wind,

V_θ, V_ϕ meridional and zonal components of neutral wind velocity,

Φ geopotential, $d\Phi = gdh$

r   radius of earth

θ co-latitude

ϕ longitude

$\underline{k}$ unit vector in vertical direction

H   scale height $H = \dfrac{RT}{mg}$

m   mean molecular mass

T   neutral temperature

R   universal gas constant

μ_m molecular co-efficient of viscosity

μ_T turbulent co-efficient of viscosity

K_m molecular co-efficient of heat conduction

K_T turbulent co-efficient of heat conduction

ε specific enphalpy (C_pT) plus kinetic energy density $\dfrac{(\underline{V}^2)}{2}$ per unit mass of gas

$\underline{J}$ horizontal current density

$\underline{E}$ external electric field or polarization field

$\underline{\sigma}$ layer conductivity tensor

Q_{EUV} solar radiation heating

Q_{IR} infrared cooling from [NO], [CO_2] and [O]

Q_{VIS} viscous dissipation

$\underline{V}.\underline{\nabla}_p$ horizontal advection operator
$$= \frac{\underline{V}_\theta}{r}\frac{\partial}{\partial r} + \frac{\underline{V}_\phi}{r\sin\theta}\frac{\partial}{\partial\phi}$$

K   eddy diffusion coefficient

∇^2 Laplacian operator, NOTE: different for scalers and vectors.

m1   mass of species one, e.g., atomic oxygen - 16amu

m2   mass of species two, e.g., molecular nitrogen - 28amu

REVERSALS OF THE SOLAR SOURCE SURFACE MAGNETIC FIELD AND OF THE PLANETS

S.-I. AKASOFU[1], T. SAITO[2]
1- *Geophysical Institute, University of Alaska Fairbanks, Fairbanks, Alaska 99775-0800*
2- *Geophysical Institute, Tohoku University, Sendai, Japan*

ABSTRACT. The solar magnetic field on a surface of 2.5 solar radii, known as the source surface, can be approximated by a dipole field. This dipolar field rotates by 180° meridionally throughout a sunspot cycle, although the polar unipolar fields on the photosphere do not show such a shift across the equator. It is of interest to assume that the solar source surface corresponds to the surface of the magnetized planets and that the photosphere corresponds to the core surface. An advantage of the solar situation is that one can directly observe the photospheric magnetic fields which correspond to the magnetic fields of the core surface. We examine why the dipolar field on the source surface is inclined with respect to the rotation axis and why the inclination angle changes from 0° to 180° (or 180° to 0°) during a sunspot cycle. We assume that the main dipole is axially aligned with the rotation axis, because the unipolar fields in the polar regions do not shift across the equator. It can be shown that the inclination and its change arise from the growth and decay of a few dipolar sources oriented in an east-west direction near the equator. The combined field of the axially aligned dipole and the equatorial dipoles provides an inclined dipole on the source surface. The equatorial dipoles are identified as large-scale weak dipolar fields which contain active regions. The rotation of the dipole on the source surface arises from a relative change of strength of the equatorial dipoles and the axial dipole. On the basis of the above study of the solar situation, we suggest that the inclination and eccentricity of the dipole axis of the magnetized planets (including the earth) arise from the growth and decay of equatorial dipoles near the core surface. The reversal of the earth's dipole field may be explained in a way similar to the reversal of the dipole field on the source surface.

1. Rotation of the Solar Magnetic Dipolar Field on the Source Surface

The magnetic field in the solar corona, inferred on the basis of the spherical harmonic analysis of magnetic fields on the photosphere, has been found to predict reasonably well the structure of the corona.

It has also been shown by a number of researchers that the magnetic field on the surface of a sphere of radius of 2.5 solar radii from the center of the Sun can be approximated by a dipole field and that the polarity (towards/away from the sun) of the source surface field is fairly well correlated with that of the interplanetary magnetic field (IMF) observed near the earth (cf. Shatten et al., 1969; Levine, 1977; Hoeksema et al., 1982, 1983). Indeed, it is for this reason that this particular surface is called the source surface.

Hoeksema and Scherrer (1984) and Saito and Akasofu (1987) examined sunspot cycle variations of the magnetic equator (which is usually referred to as the neutral line) on the source surface and demonstrated that the neutral line varied fairly systematically during sunspot cycle 21. The neutral line lay near the ecliptic plane at the beginning of the cycle and tilted gradually as the cycle advanced, standing vertically (with respect to the equatorial plane)

45

D. B. Stone and S. K. Runcorn (eds.),
Flow and Creep in the Solar System: Observations, Modeling and Theory, 45–57.
© 1993 *Kluwer Academic Publishers. Printed in the Netherlands.*

46

during the maximum epoch of the cycle. The neutral line tilted further during the declining epoch of the cycle and lay near the ecliptic plane at the end of the cycle. Approximating the magnetic field on the source surface by a dipole, this change can be represented by a gradual rotation of the dipole by 180°, from pointing northward to pointing southward. Figure 1 shows the dipole and the neutral line at different epoch during the sunspot cycle 21. However, it is important to realize that there is no indication that the main dipole of the photospheric magnetic field shifts in a way to produce a rotation from one polarity to the other of the dipole and the neutral line on the source surface. It is well known that the so-called "unipolar region" is located near each pole throughout the sunspot cycle. The reversal of the field occurs as a result of the "migration" of a large-scale unipolar field (say, positive) from low latitudes, cancelling the pre-existing polar unipolar (negative) field in one hemisphere. Meanwhile, the cancellation of the opposite polarity occurs in the opposite hemisphere at about the same time. Thus, the reversal of the polarity does not involve a gradual shift of the polar unipolar regions from one hemisphere to the other across the equator. These observations show that one must look for causes of the rotation of the dipole field on the source surface which do not rely on the rotation of the main dipolar field on the photosphere.

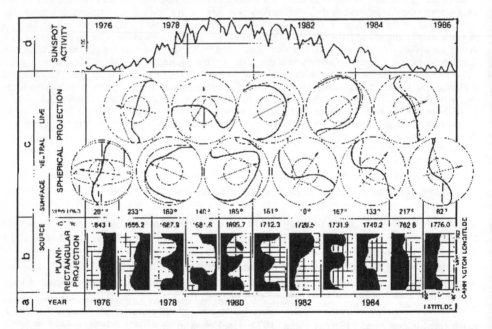

Figure 1. (a) Year. (b) The redefined Carrington rotation used for the presentation in Figure 2c; positive polarity is black and negative polarity area is hatched; the neutral line lies between the positive and negative polarity areas. (c) The neutral line on the source surface viewed from the view longitude, together with the equivalent dipole determined by Hoeksema and Scherrer (1984). (d) The sunspot numbers between 1976 and 1986.

As a solution to this puzzle, Saito et al. (1989) suggested that the major changes of the neutral line on the source surface during a sunspot cycle can be fairly well represented as a combined effect of changes of an axial dipole located at the center of the Sun and of two nearly antipodal (equivalent) dipoles near the equatorial plane on the photosphere; three photospheric dipoles are needed only in some exceptionally complicated cases.

More specifically, the observed variations of the neutral line during sunspot cycle 21 can be expressed by assuming that (1) the magnetic moment of the central dipole (parallel to the rotation axis, say, directed northward) decreases as a new sunspot cycle advances and becomes null at about the sunspot maximum, (2) subsequently, a small central dipole of the opposite polarity (directed southward) appears and its moment reaches the maximum intensity near the sunspot minimum, and (3) a pair of photospheric dipoles, located at low latitudes, increase their magnetic moments from the beginning of a sunspot cycle until about the sunspot maximum and then decrease during the declining epoch. These variations are illustrated in Figure 2. The top row of the figure shows the sunspot number during sunspot cycle 21. The second row shows the neutral line determined by the Wilcox Observatory. The next two rows show the axial and photospheric equivalent dipoles, respectively, for eight epochs. The combined magnetic fields of these dipoles result in the neutral lines shown in the bottom row, which are calculated as described in this paper. The second row shows the observed neutral line and the last row, the neutral lines computed on the basis of our three-dipole model. Their agreement is quite reasonable.

Saito et al. (1991) examined the nature of the apparent photospheric dipoles in the above modelling effort and showed that they are, indeed, large-scale dipolar fields. Figure 3 shows such an example. They modelled the source surface field in the upper left diagram by a combination of an axially parallel dipole and two photospheric dipoles (the upper right diagram and lower right diagram). The two dipoles thus determined are then transferred to the magnetic field map in the lower left diagram. One can see clearly that each of the two dipoles is fairly well located by a large-scale dipolar field. Therefore, the two dipoles, inferred from our modelling method, do exist and are not fictitious. These dipolar fields are not individual sunspot pairs, and are fluxes of larger scale dipolar fields in active regions.

It is possible, therefore, to conclude that the main dipole is axially parallel or anti-parallel, that the inclination (with respect to the rotation axis) of the dipole on the source surface is produced by a combined effect of the axially parallel (or anti-parallel) field and the photospheric dipoles. The apparent rotation of the dipole throughout the sunspot cycle is produced by a relative change of the strength of the axially parallel dipole and the photospheric dipoles, together with the reversal of the axial dipole as a result of the migration of a low latitude unipolar field to the polar region in each hemisphere.

2. Large Inclination and Eccentricity of the Dipole-like Field of Uranus and Neptune

Ness et al. (1986, 1989) suggested that the magnetic fields of Uranus and Neptune indicate that the main field can be represented, as a first approximation, by an eccentric dipole and that the dipole is greatly inclined with respect to the rotation axis; see Table 1. Their model is often referred to as the offset tilted dipole (OTD) model. Their results are based on the spherical harmonic analysis of the magnetic field observation along the fly-by trajectory of the Voyager spacecraft.

48

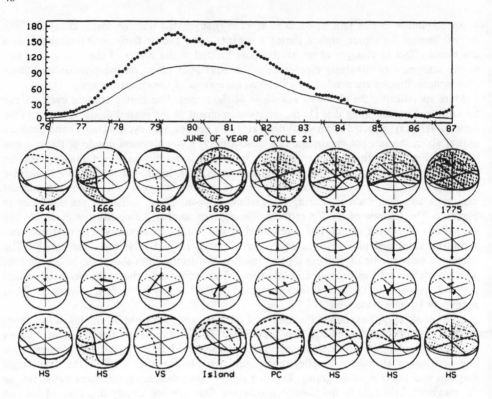

Figure 2. Variations of the neutral line on the source surface and its representations during solar cycle 21; from the top, the sunspot number, the neutral line on the source surface determined by the Wilcox Observatory (the shading representing the negative polarity), the axial (equivalent) dipole, the photospheric (equivalent) dipoles, and the neutral line reproduced from the axial dipole and the photospheric dipoles.

TABLE 1

	Inclination Angle	Location from the Center	Reference
Earth	11°30'	0.08 R_E	Chapman and Bartels, 1940
Uranus	~60°	0.3R_U	Ness et al., 1986
Neptune	~47°	0.55 R_N	Ness et al., 1989

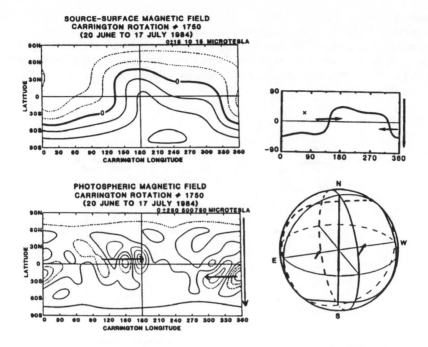

Figure 3. (Upper left) The magnetic field distribution on the source surface for Carrington rotation 1750. The line with zeros is the neutral line. Solid lines (+2.5, +5.0) indicate the positive (away from the Sun) field, and dotted lines (-2.5, -5.0) the negative (toward the Sun) field (courtesy of the Wilcox Solar Observatory). (Lower left) The smoothed photospheric magnetic field data for Carrington rotation 1750 (courtesy of the Wilcox Solar Observatory). (Upper right) The neutral line reproduced by the combined effect of a central dipole (its orientation and magnitude are indicated by an arrow just to the right-hand side) and two fictitious dipoles. (Lower right) A spherical presentation of the upper right diagram. The view longitude of the sphere is indicated by a cross in the upper right diagram.

In this section, we demonstrate that the large inclination and eccentricity of the dipole-like field of Uranus and Neptune can be described, as a first approximation, by the combined field of an axial dipole and an auxiliary dipole.

The upper left diagram of Figure 4 shows the offset tilted dipole (OTD) which is located near the surface of the core of Uranus, as proposed by Ness et al. (1986). The lower left diagram shows some magnetic field lines in the plane which contains the OTD. We determine the magnitude and the orientation of both an axial dipole and an auxiliary dipole in the same way as we examined the solar source surface field. The results are presented in the upper right diagram of Figure 4. For simplicity, we assume only one auxiliary dipole which is located at the position calculated for the single offset dipole. The parameters for the two dipoles are given in Table 2. One can see that the magnetic moment of both the axial dipole and the auxiliary dipole are similar in magnitude. Present dynamo theories suggest that the toroidal field in the core is much stronger than the poloidal field, as demonstrated by the magnetic field of a sunspot pair.

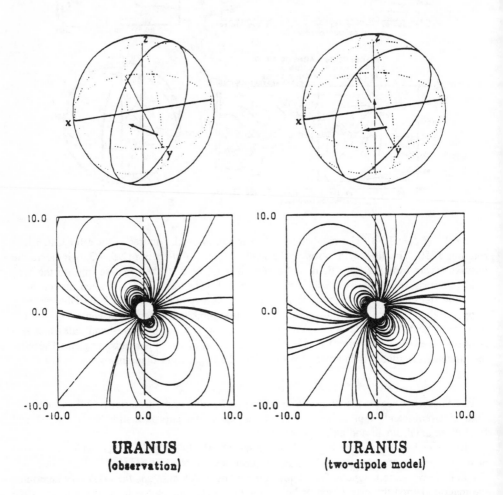

URANUS
(observation)

URANUS
(two-dipole model)

Figure 4. Uranus: Upper left: The offset tilted dipole (OTD) model. Lower left: Some of the magnetic field lines of the OTD model. Upper right: The two-dipole model. Lower right: Some of the magnetic field lines of the two-dipole model.

TABLE 2. Two-dipole Model of Uranus

	M (Gauss R_U3)	Location	Orientation
Axial dipole	0.143	x = 0 y = 0 z = 0	$\theta = 0°$
Auxiliary dipole	0.157	x = 0 y = 0 z = -0.3 R_U	$\theta = 90°$ $\theta = 0°$

Some magnetic field lines of the one- and two-dipole models are shown in Figure 4. Comparing the lower left and right diagrams, one can see that the simple two-dipole model can reproduce reasonably well the observed field which is represented by a single dipole. There is no doubt that one or two more additional dipoles can better reproduce the observed field. However, the main point is to illustrate our basic idea that even a simple two-dipole model can reproduce the observed field fairly well and thus may be able to remove the great puzzle of the large inclination angle and the large eccentricity of the main field of Uranus.

The upper left diagram of Figure 5 shows the offset tilted dipole (OTD) of Neptune, as proposed by Ness et al. (1989). The lower left diagram shows some magnetic field lines in the plane which contains the OTD. The magnitude and the orientation of an axial dipole and an auxiliary dipole giving a similar magnetic field are shown in the upper right diagram. The parameters for the two dipoles are given in Table 3. Some magnetic field lines of the one- and two-dipole models are shown in the lower part of the Figure 5.

TABLE 3. Two-dipole Model of Neptune

	M (Gauss R_N3)	Location	Orientation
Main dipole	0.0641	x = 0 y = 0 z = 0	$\theta = 0°$ $\phi = 0°$
Auxiliary dipole	0.0769	x = 0.14 y = 0.42 z = 0.24	$\theta = 60°$ $\phi = 0°$

The discovery of the large inclination angle and the eccentricity of the main field when represented by a single dipole for Uranus and Neptune has been a great puzzle. However, it is important to realize that the finding is based on the spherical harmonic analysis of the planetary fields observed by a flyby spacecraft. There is no doubt that the dipole representation based on the spherical harmonic analysis provides us with the unique mathematical description of the planetary magnetic field. However, the result thus obtained does not indicate that the field in the vicinity of the core is physically given by such a dipole. Indeed, since the dynamo process is thought to rely so strongly on the rotation of the magnetized planets, it is possible that the observed dipole field consists of the combined field of an axial dipole (parallel or anti-parallel to the rotation axis) and a few auxiliary dipoles. Our suggestion is thus physically more plausible than the mathematical representation.

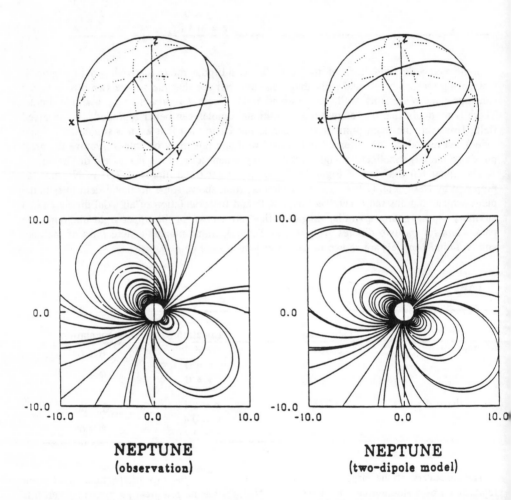

NEPTUNE
(observation)

NEPTUNE
(two-dipole model)

Figure 5. Neptune: Upper left: The offset tilted dipole (OTD) model. Lower left: Some of the magnetic field lines of the OTD model. Upper right: The two-dipole model. Lower right: Some of the magnetic field lines of the two-dipole model.

3. Is the Earth's Dipole Actually Inclined with Respect to the Rotation Axis?

A spherical harmonic analysis of the earth's magnetic field indicates that the main field can be represented, as a first approximation, by a centered dipole, and that the dipole axis is inclined with respect to the rotation axis by about 11.5° (Chapman and Bartels, 1940). Since the present dynamo theory on the generation of the earth's magnetic field relies heavily on the rotation of the earth, it may be worthwhile to examine whether or not the earth's magnetic field could consist of an axially anti-parallel dipole and a few dipoles on the surface of the core. The axially symmetric dynamo has been considered by Elsasser (1946), Bullard and Gellman (1954), Busse (1978), Moffatt (1978), Gubbins (1984), and many others.

It is our finding that three dipoles near the core surface, together with the axially anti-parallel dipole, can reproduce fairly well the magnetic equator (Figure 6). The three dipoles are found to be located at longitudes ~105°, ~210° and ~330°, respectively; thus, they are located southeast of Hawaii, at the Atlantic Ocean between Africa and South America, and at the southern part of Thailand, respectively. Thus, it is our suggestion that the main dipole is aligned with respect to the rotation axis and that the combined effect of the three dipoles tends to incline the main dipole and shifts it from the center of the earth, when the combined field of the axial dipole and the three dipoles is observed from the earth's surface or above it.

The actual location and the magnitude of the three dipoles could be a little different from what we have determined (in terms of the location, at most ±10° in longitude and latitude). Thus, unlike the spherical harmonic analysis of the earth's field, our analysis cannot provide a unique result in a strict sense. However, one should keep in mind that the spherical harmonic function provides only a mathematical representation of the earth's magnetic field, not a unique physical representation. In terms of physics, the spherical harmonic representation is simply one possibility. Thus, it is suggested that in formulating a dynamo theory of the magnetized planets, an axial main dipole with a few small dipoles near the core surface, should be considered as a possible case.

4. Does the Main Dipole of the Geomagnetic Field Rotate During the Reversals?

After a long debate, the reality of reversals of polarity of the geomagnetic field through geologic time have been established (cf. Cox, 1970; Merrill and McElhinny, 1983; Roberts and Piper, 1989). The phenomenon of reversals is considered by many researchers to be the rotation of the dipole axis, either from the normal to reversed (N $\rightarrow$ R) or from the reversed to normal (R $\rightarrow$ N). It is generally accepted:

(1) The dipole pole shifts often along a restricted sector of longitude (cf. Fuller et al., 1979).
(2) There occurs a significant reduction of the field intensity (cf. Dodson et al., 1978; Roberts and Shaw, 1984).
(3) The transition stage is relatively short, ~4500 ± 100 years (cf. Prevot et al., 1985).
(4) The frequency of reversals has been increasing from 0.5 Ma to 0.15 Ma during the last 70 Ma (McFadden and Merrill, 1984).
(5) The field becomes highly non-dipole during a transition (Williams and Fuller, 1981; Merrill and McElhinny, 1983), although the importance of the higher order terms (g2/g1, g3/g1) during the reversals is not well established, (2) is considered by some to be an indication of the growth of the higher order terms.

54

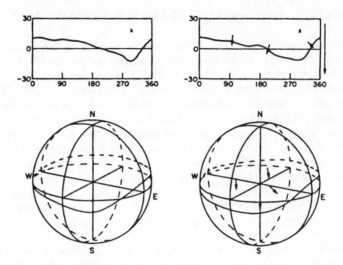

Figure 6. The upper- and lower-left presentations show the observed magnetic equator of the earth in the standard and spherical projections, respectively. The upper- and lower-right presentations show the magnetic equator reproduced by the axial and three dipoles on the core surface, which are indicated in both presentations.

In section 1, we showed that the solar magnetic field extrapolated to the source surface, a spherical surface of radius of 2.5 solar radii, can be approximated by a dipole field and that the dipole axis rotates by 180° during a sunspot cycle. We attempted to reproduce the source surface field by demanding that the main dipole axis is parallel or anti-parallel to the rotation axis (based on the fact that a polar unipolar region remains in the polar region) and that auxiliary dipoles on the photosphere can cause the apparent inclination and rotation. In the subsequent sections, we attempted to reproduce observed planetary magnetic fields by similar combinations.

Inspecting Figure 2 again, we see that at the beginning of the cycle, the magnetic equator lies nearly parallel to the heliographic equator. One can see that the axial dipole was large and was pointing approximately northward. During the ascending phase of the sunspot cycle, the axial component decreased rapidly. The magnitude of the reversed dipole grew steadily during the descending phase of the cycle.

On the other hand, the equatorial dipoles grew rapidly during the ascending phase. When the axial dipole was weakest, one of the equatorial dipoles was very large. Obviously, this particular dipole had the largest influence on the source surface. In fact, the main dipole component on the source surface was almost perpendicular to the rotation axis during Carrington Rotation 1681-1685 (see Figure 1). At the earth's distance, the IMF polarity (towards/away from the earth) was well represented by the rotation of such a dipole (which was nearly perpendicular to the rotation axis). After the reversal of the axial dipole, the magnitude of the equatorial dipoles gradually decreased.

Thus, there appears to be some similarity of the polarity reversals which are observed on the earth's surface and on the solar source surface. Both reversals occur during a relatively short period compared with the period of one polarity. A significant decrease of the main dipole field occurs. There is some indication of the growth of higher order fields. The major difference is that the solar reversals are quite regular, compared with those of the geomagnetic field. In fact, there is no physical reason why one cannot assume, as a first approximation, that the source surface corresponds to the surface of a magnetized planet and that the photosphere corresponds to the core surface. The main part of the planetary magnetic field is assumed to be produced by a dynamo process in the core, and the mantle is assumed to be a region of $\nabla \times \underline{B} = 0$.

5. Concluding Remarks

Although we must be cautious to extend the results of our solar study to planetary magnetism directly, one advantage of the solar study is that one can observe variations of magnetic fields on the photosphere; on the other hand, it is not possible to directly observe magnetic fields on the core surface. We suggest that the inclination of the dipole axis modeled as the source of the field of the magnetized planets arises from the combined results of a few equatorial dipoles with a main axially aligned dipole. We suggest that the rotation of the geomagnetic main dipole during the reversals may simply be apparent. The reduction of the axial dipole component and the simultaneous growth of the equatorial dipoles near the core surface may cause an apparent rotation.

The requirement of frequent polarity reversals has put a serious constraint to the dynamo theories (Merrill and McFadden, 1990). Although several models of the reversals have been proposed in the past (Rikitake, 1958; 1966; Allan, 1958; Levy, 1972), it is not clear what physics is really involved in some of these models; see also Merrill and McElhinny (1983). It is suggested that the physics of the reversals of the geomagnetic field could be found in the reversals of the solar photospheric field. The present understanding of the reversals of the photospheric dipole field is that weak, but large-scale, unipolar fields of a particular polarity migrate toward the polar region in each hemisphere and "cancel" and reverse the polarity of the polar unipolar field. A number of computer simulation studies of this process have been conducted (cf. Wang, Nash, and Sheeley, 1989). We should be aware at least that there is one celestial body, the Sun, which clearly demonstrates before our eyes the process involved in the reversal. We should learn as much as possible from it, in addition to considering elaborate mechanical schemes which might cause the reversal of the dynamo. This is particularly the case because it is generally believed that the dynamo process is basically the same for the Sun and the magnetized planets.

Acknowledgments. The authors would like to thank Professor Keith Runcorn and Professor Raymond Hide for their interest in our controversial work. The work reported here was supported in part by a grant from the National Science Foundation ATM 90-22819.

6. References

Allan, D.W. (1958) Reversals of the earth's magnetic field. Nature 181, 469.

Bullard, E.C. and H. Gellman (1954) Homogeneous dynamos and terrestrial magnetism. Phil. Trans. Roy. Soc. A. 247, 213.

Busse, F.H. (1978) Magnetohydrodynamics of the earth's dynamo. Ann. Rev. Fluid Mech. 10, 435.

Chapman, S. and J. Bartels (1940) Geomagnetism, II, Oxford Univ. Press, Oxford.

Cox, A. (1970) Geomagnetic reversals. Science 163, 237.

Dodson, R.F., I.R. Dunn, M.D. Fuller, I. Williams, H. Ito, V.A. Schmidt and Y.-M. Wu (1978) Paleomagnetic record of a late Tertiary field reversal. Geophys. J.R. Astr. Soc. 53, 373.

Elsasser, W.M. (1946) Induction effects in terrestrial magnetism. Phys. Rev. 69, 106 and 70, 202.

Fuller, M.D., I. Williams and K.A. Hoffman (1979) Paleomagnetic records of geomagnetic field reversals and the morphology of the transitional fields. Rev. Geophys. Space Phys. 17, 179.

Gubbins, D. (1984) The earth's magnetic field. Contemp. Phys. 25, 269.

Hoeksema, J.T. and P.H. Scherrer (1984) Harmonic analysis of the solar magnetic field, Proceedings of the 4th European Meeting on Solar Physics, The Hydromagnetics of the Sun, Eur. Space Agency Spec. Publ., ESA-SP-220, 269.

Hoeksema, J.T., J.M. Wilcox and P.H. Scherrer (1982) Structure of the heliospheric current sheet in the early portion of sunspot cycle 21. J. Geophys. Res. 87, 10331.

Hoeksema, J.T., J.M. Wilcox and P.H. Scherrer (1983) The structure of the heliospheric current sheet. J. Geophys. Res. 88, 9910.

Levine, R.H. (1977) Large scale solar magnetic fields and coronal notes. In Coronal Holes and High Speed Wind Stream, edited by J.B. Zirker, p. 103, Colorado Associated University Press, Boulder, Colo.

Levy, E.H. (1972) Kinematic reversal schemes for the geomagnetic dipole. Astrophys. J. 171, 635.

McFadden, P.C. and R.T. Merrill (1984) Lower mantle convection and geomagnetism. J. Geophys. Res. 89, 3354.

Merrill, R.T. and M.W. McElhinny (1983) The Earth's Magnetic Field, Academic Press.

Moffatt, H.K. (1978) Magnetic Field Generation in Electrically Conducting Fluids, Cambridge Univ. Press, Cambridge.

Ness, N.F., M.H. Acuña, K.W. Behannon, L.F. Burlaga, J.E.P. Connerney, R.P. Lepping and F.M. Neubauer (1986) Magnetic fields at Uranus. Science 233, 85.

Ness, N.F., M.H. Acuña, L.F. Burlaga, J.E.P. Connerney, R.P. Lepping and F.M. Neubauer (1989) Magnetic fields at Neptune. Science 246, 1473.

Prevot, M., E.A. Mankinen, R.S. Coe and C.S. Grommé (1985) The Steens Mountain (Oregon) polarity transition 2, Field intensity variations and discussion of reversal models. J. Geophys. Res. 90, 10417.

Rikitake, T. (1958) Oscillations of a system of disc dynamos. Proc. Camb. Phil. Soc. 54, 89.

Rikitake, T. (1966) Electromagnetism and the Earth's Interior, Elsevier, Amsterdam.

Roberts, N. and J. Shaw (1984) The relationship between the magnitude and direction of the geomagnetic field during the Late Tertiary in Eastern Iceland. Geophys. J. R. Astr. Soc. 76, 637.

Roberts, N. and J.D.A. Piper (1989) A description of the behavior of the earth's magnetic field. In Geomagnetism, 3, ed. by J.A. Jacobs, p. 163, Academic Press Ltd., London.

Saito, T. and S.-I. Akasofu (1987) On the reversal of the dipolar field of the Sun and its possible implication for the reversal of the Earth's field. J. Geophys. Res. 92, 1255.

Saito, T., T. Oki, C. Olmsted and S.-I. Akasofu (1989) A representation of the magnetic neutral line on the solar source surface in terms of the sun's axial dipole at the center and two equatorial dipoles in the photosphere. J. Geophys. Res. 94, 14993.

Saito, T., Y. Kozuka, T. Oki and S.-I. Akasofu (1991) The source surface and photospheric magnetic field models. J. Geophys. Res. 90, 3807.

Shatten, K.H., J.M. Wilcox and N.F. Ness (1969) A model of coronal and interplanetary magnetic fields. Sol. Phys. 9, 422.

Wang, Y.-M., A.G. Nash and N.R. Sheeley, Jr. (1989) Magneticflux transport on the sun. Science 245, 712.

Williams, I.S. and M. Fuller (1981) Zonal harmonic models of reversal transition field. J. Geophys. Res. 86, 11657.

PLANETARY MAGNETISM RE-VISITED *

RAYMOND HIDE
The Robert Hooke Institute
The Observatory
Clarendon Laboratory
Parks Road
Oxford OXI 3PU, England, UK.

ABSTRACT. This invited review paper outlines developments and outstanding questions in the study of self-exciting MHD dynamos, the Earth's core-mantle interface, sub-seismic oscillations of the Earth's core and the nature of the geomagnetic secular variation, and the origin of the magnetic fields of the outer planets.

The organisers of this meeting in memory of the late Professor T.G. Cowling have invited me to make some general introductory remarks about the nature of planetary magnetic fields, to set the stage for more detailed presentations (e.g. Soward, 1991) on mathematical theory relating to Larmor's self-exciting dynamo mechanism for generating magnetic fields by flow in a connected body of fluid. Having worked in other areas of geophysics during the past few years, I am unable to present a useful up-to-date summary of new developments in the study of planetary magnetism. So I must apologize in advance for choosing to "re-visit" a few basic topics with which I can claim some familiarity, but for the benefit of newcomers to the subject, in the written version of this talk I will include a fairly extensive list of references, some to very recent work.

1. Dynamos

Amongst Cowling's many distinguished contributions to cosmical electrodynamics is his celebrated "anti-dynamo" theorem, which has now been generalized to show that in no circumstances can a magnetic field with an axis of symmetry be maintained by fluid motions against Ohmic dissipation (see Cowling 1934, also Hide and Palmer 1982, Jacobs 1987-91, Weiss 1989). As Larmor appreciated, motional induction is the only quantitatively-viable

---

* This article was first published in a special issue of *Geophysical and Astrophysical Fluid Dynamics* (Vol **62**, pages 183-189, 1991) edited by Professor E. R. Priest, containing contributions to the Cowling Symposium on Solar System Plasmas held in April 1991 under the auspices of the European Geophysical Society. It is reproduced here with the kind permission of Gordon and Breach Science Publishers. The article summarizes some of the material discussed by Professor R. Hide in his lectures on planetary magnetic fields and the magnetohydrodynamics of rotating fluids presented at the NATO ASI in Fairbanks, Alaska.

D. B. Stone and S. K. Runcorn (eds.),
Flow and Creep in the Solar System: Observations, Modeling and Theory, 59–66.

mechanism for generating electric currents of sufficient magnitude to account for magnetic fields of the strength observed. Cowling's result effectively challenged theoreticians to (a) seek existence theorems for the case of nonaxisymmetric fields, and (b) produce realistic dynamos capable of operating under the conditions thought to prevail in the electrically conducting interiors of astronomical bodies such as the Earth and other planets. The first of these challenges was eventually met by Backus (1958), Herzenberg (1958) and others (see Moffatt 1978, Parker 1979, Jacobs 1987-91). The second challenge, an even tougher proposition, is now the subject of a considerable amount of mathematical work (see e.g. Roberts 1989, Zhang and Busse 1989, Soward 1990, 1991), in which super-computers will play an increasingly important rôle in the formidable task of dealing with the highly non-linear equations of magnetohydrodynamics (MHD), to be solved under realistic boundary conditions. As the mathematical contributions to this symposium will show, such work has a long way to go before useful detailed comparisons can be made with observational data on planetary magnetism. It is unfortunate moreover that for technical reasons these mathematical developments cannot benefit from guidance afforded by controlled laboratory experiments, for the electrical conductivities of fluids available to the experimenter are much too low.

The task of formulating, testing and improving dynamo models and demonstrating their validity will require the systematic application of diagnostic schemes based on the laws of physics. For example, certain properties of the magnetic field at the outer surface of a spherical dynamo model are implied by the laws of electrodynamics (see Hide 1979, 1981). When the magnetic permeability μ is uniform and σ, the electrical conductivity, is independent of co-latitude θ and longitude ϕ, then the integral over the whole surface of the sphere of the quantity σ^{-1} (sgn B_r)$\partial^2(r^2 B_r)/\partial r^2$ (where B_r is the radial component of the magnetic field $\mathbf{B}$) must be positive and not less than the integral of

$$-\sigma^{-1} (\mathrm{sgn}B_r)\mathrm{cosec}\theta[\partial(\sin\theta\partial B_r/\partial\theta)/\partial\theta + \mathrm{cosec}\theta\partial^2 B_r/\partial\phi^2], \tag{1}$$

a result which Cupal (1980) has used in a diagnostic study of a particular dynamo model. Topological implications of equation (1) concerning null flux curves, touch points, dip poles, etc. on the surface have not yet been systematically studied (but see Hide 1981, 1985, 1986, Proctor and Gubbins 1990, Hutcheson 1991). The equations of motion give additional constraints involving certain pseudo-scalars (e.g. relative potential vorticity $\nabla\times\mathbf{u}.\nabla\ln\rho$ (where $\mathbf{u}$ is the Eulerian flow velocity and ρ the mass density), helicity density $\mathbf{u}.\nabla\times\mathbf{u}$, and superhelicity density $\nabla\times\mathbf{u}.\nabla\times\nabla\times\mathbf{u}$ (see Hide 1989a). Also of interest in connection with diagnostic schemes are dynamical constraints expressible as integrals over cylindrical surfaces co-axial with the rotation axis (Hide 1982), including the celebrated Taylor constraint (see Taylor 1963, Skinner and Soward 1988, Roberts 1989, Jones and Roberts 1990, Soward 1991) based on angular momentum considerations.

2. Earth's Core-Mantle Interface

The best data available in the study of planetary magnetism are geomagnetic observations (see Jacobs 1987-1991). Indirect observations based on the magnetism of terrestrial rocks indicate that the main geomagnetic field has existed since the very earliest periods of geological history. The discovery of geomagnetic polarity reversals is the most striking finding of such investigations. Direct observations of the declination and inclination of the geomagnetic field at various places on the Earth's surface go back no more than a few centuries, and intensity measurements cover an even shorter period. The detailed characteristics of the geomagnetic field revealed by these direct observations provide important information about the structure and dynamics of the liquid metallic core and also of the lower mantle. Core motions are subject to thermal, mechanical and electrical boundary conditions at the core-mantle interface that are determined by motions in the lower mantle and other processes. So the temporal and spatial characteristics of the geomagnetic field should and indeed do include features that correlate strongly with geological processes (see Hide 1967, Jacobs 1987-91, Gubbins 1991, Laj *et al.* 1991, Loper 1991).

On the typical time scales of the geomagnetic secular variation, decades to centuries, the electrical conductivity σ of the Earth's liquid outer core can for certain theoretical purposes be treated as being infinite and the viscosity zero. This means that the secular variation can be regarded to a good approximation as a process of continual rearrangement of the lines of force of the geomagnetic field, but with the total number of lines of magnetic force intersecting the surface of the core remaining constant with time. The application of this principle to the inverse problem of finding the core radius from global geomagnetic secular variation data gives a value within 2% of the radius determined by seismologists. The "frozen flux" hypothesis also enables something to be said about the field of fluid flow in the upper layers of the core, though a complete description of the flow field needs further plausible assumptions to be made (see Bloxham and Jackson 1991, Gubbins 1991). Concomitant horizontal pressure gradients can be inferred from the flow field by using the geostrophic approximation, which should be valid unless the length scale associated with the vertical variation of the toroidal magnetic field in the core is very small indeed compared with the radius of the core. These pressure gradients, when combined with the hypothetical pattern of topography of the core-mantle boundary inferred from seismic tomography and other geophysical data (gravity, etc), can be used to determine the hypothetical net topographic torque exerted by core motions on the overlying mantle. This torque can then be compared with that implied by geodetic determinations of fluctuations in the rotation of the solid Earth on decadal time scales (see Hide, 1989b). Investigations along these lines are contributing to current research on the structure and dynamics of the core and lower mantle. They strongly imply planetary-scale irregular topographic features of the core-mantle boundary with a vertical amplitude of about 0.5km, but other interpretations are of course possible (see Hide 1969, 1989b, Hide and Dickey 1991, Jault and Le Mouël 1991, Aldridge *et al.* 1990, Voorhies 1991).

3. Sub-Seismic Oscillations

It is very likely that the magnetic field **B** within the core is largely toroidal in geometry, with the azimuthal (ϕ) component of **B** typically very much stronger than the other (θ, r) components. The average strength B_i of this toroidal field is an important parameter in studies of the Earth's deep interior, particularly in considerations of the energetics of the dynamo, for Ohmic dissipation in the core is proportional to B_i^2. Attempts to detect the poloidal electric field near the Earth's surface associated with the toroidal magnetic field in the core have not yet proved successful. But other approaches provide evidence that $B_i = 10^{-2}$T, which is some 10 times stronger than the poloidal field. One of these approaches involves the use of the argument that dynamo amplification ceases when Lorentz forces become comparable in average magnitude with Coriolis forces (see equation (6) below) (even though Coriolis forces are likely to be dominant in the outer reaches of the core, where the toroidal field should be comparatively weak). Another approach involves the interpretation of the time scales and other properties exhibited by the geomagnetic secular variation in terms of MHD planetary waves in the core. These have been discussed in the context of theoretical investigations of the transverse oscillations at "sub-seismic" frequencies that the core, as a rotating electrically-conducting body of fluid pervaded by a magnetic field, can support in principle, with periods ranging from less than a day to several centuries.

Such investigations are concerned with spatial and temporal characteristics of the oscillations, possible internal or external mechanisms by which the oscillations might be excited, and the extent to which their effects might be present in geomagnetic and other geophysical data (see Hide 1966, Braginsky 1967, 1980, Malkus 1967, Hide and Stewartson 1972, Acheson and Hide 1973, Acheson 1983, Melchior 1986, Jacobs 1987-91, Fearn 1989, Friedlander and Vishik 1990). Their properties depend inter alia on their wavelength λ (suitably defined) relative to the critical length scale

$$\Lambda \equiv \pi(RV / \Omega)^{1/2}, \tag{2}$$

where R is the outer radius of the core, Ω the angular speed with which the Earth rotates relative to an inertial frame, and $V \equiv B_i(\mu\rho)^{-1/2}$ is the Alfvén speed (Hide 1966). With $B_i=10^{-2}T$ we have $V=10^{-1}ms^{-1}$ and $\Lambda=200km$ (=0.05R). Associated with each spatial eigenmode are "inertial" and "magnetic" eigenperiods τ_I and τ_M respectively. When $\lambda<<\Lambda$, pure MHD oscillations arise with periods $\tau_M=\tau_I=\lambda/V$ and characterized by having magnetic and kinetic energies in equipartition. But when $\lambda>>\Lambda$, the inertial mode corresponds to a "planetary wave" with period

$$\tau_I = (\pi/\Omega)F_I^{-1} \tag{3}$$

and the magnetic mode to a "MHD planetary wave" with period

$$\tau_M = (\pi/\Omega)(\lambda^3 R/8\pi\Lambda^4)F_M^{-1} \tag{4}$$

(where F_I and F_M are dimensionless functions of order unity or less of the various length scales of the eigenmode), giving $\tau_I=0.5F_I^{-1}$ days and $\tau_M=0.15F_M^{-1}$ centuries when $\lambda=R$. Kinetic energy is now confined to the inertial mode and magnetic energy to the magnetic mode. Buoyancy driven flows in the core have the right time scales for exciting magnetic modes if $B_i=10^{-2}T$, and this is the quantitative basis of the proposal that MHD planetary waves might manifest themselves in geomagnetic secular variation data, with its implications for the energetics of the core.

4. Planetary Magnetic Fields

Slow relative motions of a rapidly-rotating fluid of low viscosity are strongly influenced by the action of Coriolis forces, which render the flow patterns highly anisotropic. It has long been generally supposed that the observed near-alignment of the Earth's magnetic and rotation axes must be a manifestation of such anisotropy, for the pattern of electric currents generated by dynamo action in the core will be related – albeit in a complicated way – to the pattern of fluid motions there. The discovery that the magnetic and rotation axes of the planets Jupiter and Saturn are in near alignment appeared to strengthen the notion, but difficulties arose when, thanks to further magnetic data from the brilliantly-successful NASA Voyager mission to the outer planets, Uranus and Neptune were found (in 1986 and 1989 respectively) to have highly eccentric magnetic fields, with the dipole inclined to the rotation axis and displaced from the centre of the planet by 60° and 0.3 times the planetary radius in the case of Uranus and 47° and 0.55 times the planetary radius in the case of Neptune (see Ness et al. 1986, 1989, Connerney et al. 1987). Theoreticians concerned with detailed models of dynamos will undoubtedly rise in due course to the challenge presented by these new results. A strategy for meeting this challenge has been proposed on the basis of a reexamination of the essential physics of the rôle of rotation in the production of magnetic fields by dynamo action. Indications are that it should be possible to account for the eccentric magnetic fields of Uranus and Neptune within the current "paradigm" (Hide 1988).

Finally, a word about the strength of planetary magnetic fields. As Stevenson (1983) has emphasized in an excellent review article, attempts to "predict" these strengths have not been particularly successful. But general physical considerations (see Hide 1965, 1974) lead to useful rough upper limits and shed light on some of the mechanisms involved. A magnetic field $\mathbf{B}$ cannot be maintained or amplified by fluid motion with Eulerian flow velocity $\mathbf{v}$ (in a system that rotates with angular velocity Ω, (see Hide, 1965) relative to an inertial frame) against the effects of Ohmic decay unless (a) the patterns of $\mathbf{B}$ and $\mathbf{v}$ are sufficiently complicated, and (b) the magnetic Reynolds number $S \equiv UL\mu\sigma$ is sufficiently large (but not too large), where U is a characteristic flow speed, L a characteristic length, and μ and σ are typical values of the magnetic permeability and electrical conductivity respectively. Dynamo action is stimulated by Coriolis forces, which promote departures from axial symmetry in the pattern of $\mathbf{v}$ when $\mathbf{B}$ is weak, and is opposed by Lorentz forces, which increase in influence as $\mathbf{B}$ grows in strength. Denote by

$$B_S \equiv [\rho\Omega + UL^{-1})/\sigma]^{1/2} \qquad (5)$$

$(=(\rho\Omega/\sigma)^{1/2}$ when $U/L\Omega \ll 1)$ the "scale magnetic field" (Hide 1974) (where ρ is the mean density of the core), and by B_i and B_e the respective average strengths of magnetic fields within the dynamo region and just outside it. If Lorentz forces build up to a strength typically no greater than the so-called "magnetostrophic" value, at which they would be comparable with the Coriolis term in the equation of motion, then

$$B_i < B_s S^{1/2} \text{ and } B_e < B_s S^{-1/2}. \qquad (6)$$

Estimates of S, B_e and B_i (see above discussion of MHD planetary waves) indicate that the Earth's magnetic field is close in magnitude to the upper limits implied by these expressions, which in turn implies that approximate magnetostrophic balance obtains throughout most parts of the core. These expressions are also consistent with what is known about the strengths of the magnetic fields of other planets.

5. References

Acheson, D.J. (1983) Local analysis of thermal and magnetic instabilities in a rapidly rotating fluid. Geophys. Astrophys. Fluid Dyn. 27, 123-136.

Acheson, D.J. and Hide, R. (1973) Hydromagnetics of rotating fluids. Rep. Prog. Phys. 36, 159-221.

Aldridge, K.D., et al. (1990) Core-mantle interactions. Surveys in Geophys. 11, 329-353.

Backus, G.E. (1958) A class of self-sustaining dissipative spherical dynamos. Annals of Physics 4, 372-447.

Bloxham, J. and Jackson, A. (1991) Fluid flow near the surface of the Earth's core. Rev. Geophys. 29, 97-120.

Braginsky, S.I. (1967) Magnetic waves in the Earth's core. Geomag. Aeron. 7, 851-859.

Braginsky, S.I. (1980) Magnetic waves in the core of the Earth II. Geophys. Astrophys. Fluid Dyn. 4, 189-208.

Connerney, J.E.P., Acuña, M.H. and Ness, N.F. (1987) The magnetic field of Uranus. Geophys. Res. 92, 15,329-15,336.

Cowling, T.G. (1934) The magnetic fields of sunspots. Mon. Not. Roy. Astron. Soc. 94, 39-48.

Cupal, I. (1980) Oscillating dynamo and the magnetic flux. Proceedings of Internal Workshop on Dynamo Theory, CCSR Academy of Sciences.

Fearn, D.R. (1989) Differential rotation and thermal convection in a rapidly rotating hydromagnetic system. Geophys. Astrophys. Fluid Dyn. 49, 173-194.

Friedlander, S. and Vishik, M.M. (1990) Non-linear stability for stratified magneto-hydrodynamics. Geophys Astrophys. Fluid Dyn. 55, 19-46.

Gubbins, D. (1991) Convection in the Earth's core and mantle. Quart. Journ. Roy. Astron. Soc. 32, 69-84.

Herzenberg, A. (1958) Geomagnetic dynamos. Phil. Trans. Roy. Soc. A 250, 543-585.

Hide, R. (1965) On the dynamics of Jupiter's interior and the origin of his magnetic field. In Magnetism and the Cosmos (ed. Hindmarsh, Lowes, Roberts and Runcorn), Oliver and Boyd, Edinburgh, pp. 378-395.

Hide, R. (1966) Free hydromagnetic oscillations of the Earth's core and the theory of the geomagnetic secular variation. Phil. Trans. Roy. Soc. A 259, 615-647.

Hide, R. (1967) Motions of the Earth's core and mantle and variations of the main geomagnetic field. Science 157, 55-56.

Hide, R. (1969) Interaction between the Earth's liquid core and solid mantle. Nature 222, 1055-1056.

Hide, R. (1974) Jupiter and Saturn. Proc. Roy. Soc. A 336, 63-84.

Hide, R. (1979) Dynamo theorems. Geophys. Astrophys. Fluid Dyn. 12, 171-176.

Hide, R. (1981) Magnetic flux linkage of a moving medium; a theorem and geophysical applications. J. Geophys. Res. 86, 11681-11687.

Hide, R. (1982) On the rôle of rotation in the generation of magnetic fields by fluid motions. Phil. Trans. Roy. Soc A 306, 223-234.

Hide, R. (1985) A note on short-term core-mantle coupling, geomagnetic secular variation impulses, and potential magnetic fields as Lagrangian tracers of core motions. Phys. Earth Planet. Interiors. 39, 297-300.

Hide, R. (1986) Frozen vector fields and the inverse problem of inferring motions in the electrically-conducting fluid core of a planet from observations of secular changes in its magnetic field. In The Physics of Planets (ed. S.K. Runcorn), Royal Astronomical Society, London. pp. 185-192.

Hide, R. (1988) Towards an interpretation of Uranus's eccentric magnetic field. Geophys. Astrophys. Fluid Dyn. 44, 207-209.

Hide, R. (1989a) Superhelicity, helicity and potential vorticity. Geophys. Astrophys. Fluid Dyn. 48, 69-79.

Hide, R. (1989b) Fluctuations in the Earth's rotation and the topography of the core-mantle interface. Phil. Trans. Roy. Soc. A 328, 351-363.

Hide, R. and Dickey, J.O. (1991) Earth's variable rotation. Science 253, 629-637.

Hide, R. and Palmer, T.N. (1982) Generalization of Cowling's theorem. Geophys. Astrophys. Fluid Dyn. 19, 301-309.

Hide, R. and Stewartson, K. (1973) Hydromagnetic oscillations of the Earth's core. Rev. Geophys. Space Phys. 10, 579-598.

Hutcheson, K.A. (1991) Geomagnetic field modelling. Ph.D. Dissertation, Cambridge University.

Jacobs, J.A. (1987-91) Geomagnetism (4 volumes), New York, Academic Press Ltd.

Jault, D. and Le Mouël, J-L. (1989) The topographic torque associated with tangentially geostrophic motion at the core surface and inferences on the flow inside the core. Geophys. Astrophys. Fluid Dyn. 48, 273-296.

Jault, D. and Le Mouël, J-L. (1991) Exchange of angular momentum between core and mantle. J. Geomag. Geoelect. 43, 111-129.

Jones, C.A. and Roberts, P.H. (1990) Magnetoconvection in rapidly-rotating Boussinesq and compressible fluids. Geophys. Astrophys. Fluid Dyn. 55, 263-308.

Laj, C., Mazaud, A., Weeks, R., Fuller, M. and Herrero-Bervera, E. (1991) Geomagnetic reversal paths. Nature 351, 447.

Loper, D.E. (1991) The nature and consequences of thermal interactions twixt core and mantle. J. Geomag. Geoelect. 43, 79-91.

Malkus, W.V.R. (1967) Hydromagnetic planetary waves. J. Fluid Mech. 28, 793-802.

Melchior, P. (1986) The Physics of the Earth's Core. Pergamon Press, Oxford.

Moffatt, H.K. (1978) Magnetic Fluid Generation by Fluid Motion. Cambridge University Press.

Ness, N.F., Acuña, M.H., Behannon, K.W., Burlaga, L.F., Connerney, J.E.P., Lepping, R.P. and Neubauer, F.M. (1986) Magnetic field at Uranus. Science 233, 85-89.

Ness, N.F., Acuña, M.H., Burlaga, L.F., Connerney, J.E.P., Lepping, R.P. and Neubauer, F.M. (1989) Magnetic fields at Neptune. Science 246, 1473-1478.

Parker, E.N. (1979) Cosmical Magnetic Fields. Oxford, Clarendon Press.

Proctor, M.R.E. and Gubbins, D. (1990) Analysis of geomagnetic directional data. Geophys. J. Int. 100, 69-77.

Roberts, P.H. (1989) From Taylor state to Model-Z. Geophys. Astrophys. Fluid Dyn. 49, 143-160.

Skinner, P.H. and Soward, A.M. (1988) Convection in a rotating magnetic system and Taylor's constraint. Geophys. Astrophys. Fluid Dyn. 44, 91-116.

Soward, A.M. (1990) A unified approach to a class of slow dynamos. Geophys. Astrophys Fluid Dyn. 53, 81-107.

Soward, A.M. (1991) The Earth's dynamo. Geophys. Astrophys. Fluid Dyn. 62, 191-209.

Stevenson, D.J. (1983) Planetary magnetic fields. Rep. Prog. Phys. 46, 555-620.

Taylor, J.B. (1963) The magnetohydrodynamics of a rotating fluid and the Earth's dynamo problem. Proc. Roy. Soc. A 274, 274-283.

Voorhies, C.V. (1991) Coupling an inviscid core to an electrically-insulating mantle. J. Geomag. Geoelect. 43, 131-156.

Weiss, N.O. (1989) Dynamo processes in stars. In Accretion Disks and Magnetic Fields in Astrophysics (ed. G. Belvedere), pp. 11-29, Kluwer Academic Publishers.

Zhang, K.-K. and Busse, F.H. (1989) Convection-driven magnetohydrodynamic dynamos in rotating spherical shells. Geophys. Astrophys. Fluid Dyn. 49, 97-116.

SOME REFLECTIONS ON SOLID STATE CONVECTION IN THE MANTLES OF THE EARTH, MOON AND TERRESTRIAL PLANETS

S. K. RUNCORN
University of Alaska Fairbanks
Physics Department, Imperial College, London

ABSTRACT. Convection in the Earth's mantle was long a controversial question in geoscience because the importance of solid state creep in the mantles of the Earth and terrestrial planets was not appreciated. Solid state convection produces low degree, non-hydrostatic gravitational anomalies: in the Moon, Mars and Mercury without causing plate motions, while in the Earth and probably Venus surface movements due to their relatively thin lithospheres result. The reconciliation of the different pictures of the convection pattern in the Earth's mantle seen from the geoid, seismic tomography and the observed plate motions remains a fundamental problem in geoscience.

1. Introduction

Convection is Nature's way of moving heat on the large scale, where diffusion of heat by molecular vibrations is ineffective and where radiative transfer is small because of absorption. The recognition of the importance of the convective transfer of heat was made by Count Rumford and in the solar system first by meteorologists (Brunt, 1941). Early this century experimental studies of convection in the laboratory established that steady convection in a fluid layer tends to form into cells, the aspect ratio being about 1. The pattern changes generally on time scales which are long when compared to the overturn time; the result that the speed of movement of the singular points of the convection pattern are orders of magnitude less than the current velocities is important in the discussion of mantle convection and is relevant to the controversy about plumes. The cellular patterns characteristic of convection have enabled its occurrence in the solar system to be discovered, for example the granulations in the solar chromosphere and rolls in the atmosphere, but the fundamental role of convection in the Earth's mantle—and in those of other terrestrial bodies—has been recognized very slowly by geologists and geophysicists. There were early speculations about its role in mountain building and a simple laboratory model was made by D. T. Griggs. There were several reasons which account for its tardy acceptance. Firstly, compelling evidence for its occurrence was not believed to exist by most geoscientists; through the half century since Wegener's first paper on continental drift only a small minority accepted his argument. Secondly, among those inclined to entertain its possibility, the difficulty of imagining its three-dimensional pattern—most representations were two dimensional—was a real stumbling block. Thirdly, almost all geophysicists were rooted in classical physics—in seismology and gravity—and they had no or very little need for recourse to a more sophisticated behavior of solids than that of classical elasticity. Furthermore most physicists were dubious until the

D. B. Stone and S. K. Runcorn (eds.),
Flow and Creep in the Solar System: Observations, Modeling and Theory, 67–82.
© 1993 *Kluwer Academic Publishers. Printed in the Netherlands.*

1950s of the reality of dislocations in solids and so the physical basis of creep in crystalline solids was not easily accessible to the geophysicist. Fourthly, there were difficulties in applying hydrodynamics—as Jeffreys (1952) remarks "at least as difficult as nuclear physics"—to natural phenomena. It was for instance widely supposed that the marginal theory of convection had no relevance to convection in the Earth's mantle because the Rayleigh number appears to be orders of magnitude greater than the critical value at which convection occurs. So it was supposed that the horizontal cell dimensions might be very different from those expected from the simple marginal convection theory. Both experimental and more recently computational models (Baumgardner, this volume) have shown that the cell dimensions are not very different from that of the simple marginal theory even at high Rayleigh numbers.

The discovery of the low degree non-hydrostatic terms in the Earth's gravitational field from satellite tracking provided the first physical evidence for mantle convection, but it was not recognized as such by geodesists. In fact they followed Jeffreys (1952) in attributing the long wavelength anomalies to density variations on horizontal surfaces in the mantle that were maintained by its finite strength and they tacitly supposed that they resulted from the early processes of accretion of the Earth or from its differentiation. In the literature there are many papers computing the resultant stress distribution in the Earth's mantle on various assumptions or showing that the gravitational anomalies could be modelled by "random" distributions of point mass sources. Jeffreys (1952) had, very much earlier than the satellite observations, maintained that the geoid had non-hydrostatic terms and he had argued that this proved that the mantle has finite strength. He pointed out that the ellipticity of the lunar figure had been shown by Laplace, from the interpretation of Cassini's laws of the Moon's rotation, to be nearly 20 times its hydrostatic value. Laplace had thought that the anomalous ellipticity had arisen during its cooling through slight inhomogeneities: Jeffreys attributed it to a fossilized tidal figure corresponding to an earlier Earth-Moon distance. Thus Laplace and Jeffreys had supposed that the non-hydrostatic bulge on the Moon was maintained by the finite strength of its cold interior. In an early example of what is now termed comparative planetology, Jeffreys (1952) concluded that the Earth's mantle also might have finite strength and if so could depart from hydrostatic equilibrium in its low degree harmonics. By analyzing 2000 values of g over the Earth's surface, he concluded that it did and produced a gravity anomaly map through the 3rd degree harmonics.

2. The Moon and Its Convecting Interior

Because such explanations in terms of finite strength require the creep rate in the lunar interior since the distortion was produced 3.0-4.5 G yr ago to have been less than 10^{-20} sec^{-1} (to be compared to the value of about 10^{-8} sec^{-1} just measurable in laboratory creep experiments), I proposed (Runcorn 1962a, 1967) that the present non-hydrostatic bulge on the Moon is produced by a convection current described by a second degree harmonic. I also argued that for this two-cell pattern to develop it was necessary to postulate the existence of an iron core in the Moon—a suggestion that up to that time had not been put forward and in fact was incompatible with the accepted theory of the cold accretion of the Moon. The return of the Apollo rocks showed that the Moon had differentiated but the concept of the "magma ocean" was developed to explain the differentiation of the highlands, without invoking the formation of a core: the Moon below depths of a few 100 km had, it was supposed on this scenario, remained unmelted.

Solid state convection in the Moon is not usually thought of by geoscientists as the fundamental example of this phenomenon in the terrestrial planets. In fact the idea of mantle convection began to make headway among this community only after the great lithospheric displacements on the Earth became accepted as a result of the palaeomagnetic evidence for continental drift (Collinson et al., 1957; Collinson and Runcorn, 1960) and after the Holmes-Hess theory of sea floor spreading (Hess, 1962) was proved by Vine and Mathews (1963) in their interpretation of the oceanic magnetic anomalies. But I would argue that the primary result of mantle convection is the departure of the figure from the hydrostatic one and this, in the case of the Moon, was an early astronomical discovery from its rotational dynamics. Thus planetary gravitational fields, apart from the high degree short wavelength anomalies arising in their lithospheres, are the primary evidence for the existence of solid state convection in their interiors and the data from which the pattern of convection ought to be determinable. Whether or not there are surface manifestations of the solid state convection occurring in planetary mantles depends not only on the parameters of convection but on the complex mechanical properties of their lithospheres. The case of the Moon is especially interesting because the impact basins and craters, being of external origin, clearly show the absence of lithospheric displacements greater than 1 km at least since the differentiation of the Moon, that formed the highlands, was completed about 4.4 G yr ago.

Another fundamental consequence of convection in spherical shells is well exemplified by the Moon: the distortion of the boundaries caused by the pressure field set up. The non-hydrostatic figure of the Moon is predominantly that of a 2nd degree harmonic, corresponding to a rising current along the Earth-Moon line and a falling current beneath the limb in the plane of the sky. If C is the lunar moment of inertia about its axis of rotation, A about the mean axis towards the Earth and B about the 3rd axis in the plane of the sky, this convection pattern would give a gravitational field potential depending on $[C - 1/2(A + B)]$ /M a^2 where M is the mass of the Moon and a its radius. For simplicity suppose A = B, then the Moon is a prolate spheroid with its axis towards the Earth. The value of (C - A) /C long known from Laplace's inference from Cassini's laws gives a "bulge" towards the Earth and a corresponding one on the lunar farside of about 1 km if the Moon were of uniform density but it has been shown from astronomical studies to have a "bulge" of about 3 km (Runcorn, 1967a): thus it has a second degree harmonic variation of density within its mantle, that is explained by two-cell convection. Thus this larger "geometrical ellipticity" contributes to the gravitational field and outweighs the effect of the gravitational low due to the negative density anomaly in the uprising convection column along the direction to the Earth. Thus for the Moon the ratio of the ellipticity of the gravitational equipotential surface to that of the outer lunar surface bulge is 1:3 roughly. This is an important parameter as will be seen later.

However it was long ago known that A and B were not equal: the Moon has a triaxial figure (Jeffreys, 1952) and thus the convection pattern is more complicated than the one described above, which is represented by a zonal harmonic about the Earth-Moon axis. The marginal theory of convection requires the Moon to have a core, so that the convection is occurring in a spherical shell where the ratio of the inner to the outer boundary is approximately 0.3, but the theory does not predict whether the convection pattern is zonal or not. In the full three-dimensional convection calculations of Baumgardner (this volume), a 2nd degree pattern, described by a tesseral harmonic, develops. Thus a prediction of the full convection theory is verified by the lunar figure. A tesseral harmonic is equivalent to a convection pattern where the horizontal cell dimension is less than one corresponding to a zonal or sectorial harmonic. In these computer calculations the initial conditions are taken to be random velocities—in the absence of other information—and therefore the preferred pattern takes time to settle down. Baumgardner further finds that a 2nd degree harmonic pattern develops if the heat driving convection is not only derived from radioactivity within the convecting mantle, but in large part arises from the lower boundary i.e. from the core. That heat is coming out of the core over the period when the convection pattern is forming fits perfectly with the inference drawn from lunar palaeomagnetism that there was, at least from over 4.0 G yr ago to 3.2 G yr ago, a lunar magnetic field generated by a dynamo process in the liquid iron core (Runcorn et al., 1977; Runcorn, 1983). Therefore vigorous convection in the core required heat to be removed at the core boundary.

Although the lunar lithosphere has not been broken into plates by mantle convection a stress pattern would be expected to have been set up in it. In fact it was held that a "lunar grid system," a name given to a global alignment—NW-SE and SW-NE—of lines of small craters, faults, etc. was visible on the lunar surface (Strom, 1964). These lineaments implied some internal tectonic activity within the Moon but were discounted when it was thought by the early workers on the impact theory of the lunar craters that the Moon was the result of cold accretion and had since been entirely dead internally. The relationship of the grid system, which has now been recognized also on the lunar far side, to the mantle convection pattern, corresponding to the low harmonics of the lunar gravitational field, needs study but it appears compatible with a second harmonic flow pattern (Runcorn, 1967a).

Laboratory experiments on the creep of rocks are of limited value in discussing the type of creep important in the interior of the terrestrial planets over times comparable with the life of the solar system (Ranalli, 1987). However a simple argument suffices that creep becomes the dominant mechanical property on the geological time scale when a certain threshold temperature is reached, equal to E/k, where E is the excitation energy of the particular creep process and k is Boltzmann's constant. The depth at which this occurs is insensitive to the exact composition of the silicates and to the small pressures at such depths. This can be applied to determine the relative thicknesses of the lithospheres of terrestrial planets, because for all of them the pressure change in E is negligible. A simple model of convecting terrestrial bodies in thermal equilibrium, in which the heat energy released per unit volume is constant, gives a temperature gradient in the lithosphere proportional to the radius. The greater thicknesses of the lithospheres of Mercury, Mars and the Moon compared to the Earth appears to explain why the smaller bodies do not show evidence for plate movements. Venus on the other hand has surface temperatures approximately a half of the threshold temperatures for creep and so its lithosphere must be only about 10 km thick. There is evidence that creep properties are sensitive to the water content so that the simple argument developed above may not be realistic, especially in the case of Venus.

The temperature gradient in the lunar surface was measured during the Apollo project and is roughly what is expected according to the simple theory given above. Thus the lunar lithosphere is about three times the Earth's, about 200 km. This is very much less than that often quoted, i.e., 700 km, which was suggested because the deep focus moonquakes occur down to this depth. However one surprising property of these moonquakes is that they occur at apogee and perogee. They are very minute releases of strain energy. The short time scale of these events suggests that they have little relevance to the creep processes associated with mantle convection. Over times long when compared to those of the quake events, in which the mantle behaves like an elastic solid, but much shorter than those associated with solid state convection, in which the rheology is approximately that of a Newtonian fluid of high viscosity, materials have transient creep properties. This is the realm of "Lomnitz law" to which Jeffreys (1959) appealed to explain the synchronous rotation of the satellites of the planets. Although he made the assumption that the parameters of the transient creep process were the same for all the varied phenomena he tried to explain by a single law, including the damping of S waves in the Earth and of the Chandler wobble, we may suppose that dissipation in solid planetary bodies must result from such transient creep processes. It was his use of this law in longer time scale phenomena that was mistaken: his argument against continental drift and mantle convection and his explanation of the geoid in terms of a primaeval distortion.

3. Solid State Convection in Terrestrial Planets

It seems probable that solid state convection, at least in part a 2nd degree or two-cell pattern, is occurring in Mercury and Mars, as both have a second degree non-hydrostatic gravitational field. Mercury rotates about its axis in exactly two-thirds of its orbital period. It was shown that to be locked into this resonance Mercury must have in its figure a tesseral or sectorial harmonic component with respect to its axis of rotation of about 10^{-5} (Goldreich and Peale, 1966). The Mariner flyby also measured a similar non-hydrostatic term (Esposito et al., 1978). As the iron core of Mercury must have a radius about 0.7 of that of the planet, the simple marginal theory of convection would indicate a more complex pattern than the 2nd harmonic that would be effective in trapping it in this resonant rotation. The Martian gravitational field is complex and better known than its topography, so that how much of the former can be explained by sources in its lithosphere is uncertain. However the second harmonic in its gravitational field was known long ago from the precession of the orbits of Demos and Phobos and its value is larger than any hydrostatic model would give (Runcorn, 1967c). Support of the Tharsis plateau is now attributed to a plume, so that some convection pattern seems necessary in any discussion of the Martian interior. However further progress depends on better determinations of its topography and gravitational field.

4. Solid State Convection in the Earth's Mantle

Convection in the Earth's mantle is more complex than in the other terrestrial planets partly because the outer boundary of moving plates cannot be simply described as free or rigid and partly because the solid state creep parameters, e.g. viscosity, vary with radius: partly as a result of the well understood phase changes at depths of 400 km and 670 km, but also because the effect of pressure on creep processes, through change in crystallographic cell dimensions, must be of great importance. Consequently while convection calculations for the Moon can reasonably assume constant viscosity, this is unlikely to be a good approximation for the Earth's mantle.

However in addition to the evidence for convection from the very accurately known geoid, Dziewonski (1984) has determined, from the seismic travel time anomalies, large scale variations of P and S wave velocities over horizontal surfaces in the Earth's mantle. Seismic tomography therefore also provides evidence of lateral inhomogeneities in both the upper and lower mantle which are attributable to mantle convection. However the two sets of data are not easily reconciled.

Using the same argument for the Earth as I had done for the Moon, I suggested (Runcorn, 1964) that the departure of the Earth's gravitational field over long wavelengths must result from solid state convection in the mantle. The order of magnitude arguments that I then used are still valid, and are less likely to obscure basic ideas than any detailed computer calculations, necessary though these are once the physics is understood. In the Earth's mantle the Navier-Stokes equation simplifies to

$$-\mu \nabla^2 v = -\nabla p + g \Delta \rho$$

where v is the velocity, p the pressure, μ the viscosity, g the gravitational acceleration and $\Delta\rho$ the varying density anomalies over horizontal surfaces giving buoyancy forces. Taking $\mu = 10^{21}$ poise and $v = 10^{-6}$ cm/sec, an order of magnitude greater than the plate velocities, and mantle wide convection cells, the viscous term is about 10^{-2} dyne/c.c. Therefore $\Delta\rho = 10^{-5}$, which would produce to a first approximation (treating the mantle as a plate in which the density $\Delta\rho$ has wavelengths greater than the mantle depth) gravity anomalies of the order of those observed i.e. 10-20 m gals. The second argument is that such convection conveys heat through the mantle (Q) consistent with the observed heat flow at the Earth's surface.

$$Q = \rho\sigma v \Delta T \text{ as } \Delta p = \rho \alpha \Delta T$$

where ΔT is the temperature anomaly and α is the volume coefficient of expansion which is 3.10^{-5} for silicates in the laboratory and decreases with increase in pressure to about 1.10^{-5} in the lower mantle. As $\rho \sigma$ for most solids equals 1, $Q = 10^{-6}$ cal/sq cm/sec., approximately what is observed.

The acceptance of the argument that the geoid was produced by convection was delayed because the early geoids showed no clearly convincing correlations (which were then expected) between the positive and negative anomalies and the trenches and the ridges, compressional and extensional features associated with plate motions. Further, gravity sliding of the plates—push at the ridges and pull at the plate edges at the trenches—began to be accepted as a likely mechanism. It was shown that such forces were of the right order of magnitude to account for plate motions, assuming a passive but flowing mantle (e.g. Runcorn, 1974). Then the theory of plate tectonics clarified the phenomena at the ridges: the topography, and the differentiation resulting in magma release, are local phenomena caused by the pulling apart of the plates, the resulting upward flow of mantle material below the lithosphere and differentiation releasing basaltic lava (McKenzie and Parker, 1967; le Pichon, 1968; Morgan, 1968). Some models of mantle convection even now assume the main currents rise under the ridges, although it was long ago realized that if this were the case the uprising flow under the mid-Atlantic ridge and the Carlsburg ridge would require compressional features in Africa rather than the extensional features of the Great Rift Valley and a pattern in which the rising currents move away from Africa.

Despite differing views concerning the exact way in which convection moves the plates, it is the basic source of motion in the Earth. Because the plates are still moving, it is likely that they have not come to an equilibrium with the present convection pattern. This is a reasonable explanation why there is no one-to-one correspondence between the phenomena associated with plate tectonics and the highs and lows of the geoid. I once argued (Runcorn, 1962b) that it is because convection is a result of instability, its pattern can change "suddenly" in spite of the fact that its parameters are changing gradually, that the recent date of the Wegener break up of the continents 100 M yr ago can be understood. I did at that time argue that the operative parameter might be the radius of the core, but it now seems certain from geochemical arguments that the core formed quickly in the very early history of the Earth. But other parameters of the convection system, e.g. the heat available to drive the convection, have been slowly changing through the Earth's life and might have reached a critical value 100-150 M yr ago at which the convection pattern changed and the continents broke up and the plates began their present motions and evolution. I also pointed out (Runcorn, 1963a,b) that the clustering of the rock ages in all shields of the world at 1000 M yr ago, 1800 M yr ago and 2600 M yr ago requires a global mechanism. I suggested that there had been earlier radical changes in convection pattern. Whether this is correct or not, no other explanation has been offered of perhaps this most striking feature of global Pre-Cambrian geology; the occurrence in all the shields of periods of enhanced metamorphic and igneous activity lasting perhaps 100-200 M yr and spaced apart at intervals of roughly 800 M yr (Runcorn, 1963a,b, 1965).

5. Mantle Plumes

One type of flow which is very widely believed to occur in the mantle are plumes: uprising columns of continuous flow from the lowest mantle to the lithosphere or detached columns of rising material, characterized by narrow horizontal dimensions of about 100 km and velocities perhaps two orders of magnitude greater than plate velocities. Whether mantle plumes exist—and no seismological evidence of their existence is available—it has to be stated that the term plumes is misleading; suggesting a false analogy between the hydro-dynamics of the mantle and the atmosphere (in the former the viscous term dominates and in the latter it is negligible). If plumes exist their material must have a viscosity a large number of orders of magnitude less than the mantle generally, otherwise the motion would quickly spread out: they must be nearly molten. Further the original argument for their existence is erroneous. To explain the age distance relationship along the Emperor-Hawaiian chain of volcanic islands and the two similar, though less well dated, linear chains in the Pacific, Morgan (1971) supposed that the Pacific plate in its motion to the northwest provided a simple explanation of the relation. He accepted Hess's two-dimensional picture of mantle convection flow rising under the ridges and carrying the plates away from the ridge, by analogy with a body floating in a stream (Hess, 1962). In fact only the simplest flow in a sphere has this property of being parallel over a considerable area (Runcorn, 1964). Morgan however with Hess's model in mind supposed that the localized sources of magma feeding the islands, the "hot spots," would move with the Pacific plate, so that he was forced to postulate that they rose from the lower mantle maintaining the small horizontal dimensions characteristic of the source of island magma. So the idea of plumes was born from a misconception. As it was then widely believed that the viscosity of the lower mantle was too large (10^{26} poise) to permit convection, plumes originating in the lower mantle would not move relative to each other, thus giving rise to the idea of an absolute frame of reference in the Earth with respect to which the plates could be assigned "an absolute" velocity. But it was only possible to argue that convection was not occurring in the lower mantle by ignoring the essential requirement of dynamo theory that heat is removed from the core so that it convects. Because the adiabatic gradients in the core and lowest mantle cannot be very different, even the heat conducted from the metallic core is always greater than can be conducted through the lower mantle.

The most significant observation about hot spots is that their relative motion is an order of magnitude less than plate velocities which does appear to be evidence of some "absolute frame of reference" in which their sources are fixed. There is also evidence of geochemical differences between the hot spots or "plumes" and mid-ocean ridge basalts (MORB). The former have more primitive composition in terms of rare earths and isotopic ratios. Again this evidence that hot spots and MORB were magma from two different reservoirs appeared to be simply explained by plumes tapping the deep mantle, which was once thought undifferentiated as it was not convecting, while the mid-ocean rifts communicate with the uppermost mantle.

An entirely different way of looking at the plume question is as follows. Differentiation and separation of magma is clearly facilitated by convection but it is a process not well understood as it is much less easy to model experimentally or by computation than is thermal convection. However the feature of convection already alluded to, that the singular points of a convection pattern move relative to each other at rates much less than the flow velocity, appears to explain why the relative hot spot motion is less than the plate velocities: the rising magma collects at the singular points of the convection pattern and these magma sources will lie just below the plates and have limited horizontal dimensions. Further these hot spots will be magma differentiated from the whole mantle, while MORB draws much more locally on upper mantle material. The geochemical evidence appears not to require plumes.

However heat carried in a convecting system must enter at the lower boundary and escape through the outer one by conduction under temperature gradients which therefore are orders of magnitude greater than the adiabatic gradients. In the earth the lithosphere is one boundary layer and the D" layer, a layer in the lowermost mantle about 200 km thick, is the other. It was recognized by the seismologists by its lower seismic velocity gradients as a zone of distinctive physico-chemical properties (Anderson, 1989). Some advocates of plumes suggest that the lower boundary layer becomes detached and rises as a plume (Loper, 1991). In addition plumes have been demonstrated in various elegant laboratory experiments. However striking such experiments are, it is far from clear that the non-dimensional parameters in the laboratory models can be made sufficiently near those in the Earth's mantle for safe conclusions to be drawn whether such flow patterns can occur in the mantle.

For the plume hypothesis to account for the geological evidence, plumes must exist for 100 M yr or more providing a more or less continuous supply of magma. If there is also conventional cellular convection in the mantle with velocities of the order of plate motions it is hard to see how plumes can retain their positions in the "absolute frame of reference" or why they are not sheared off. It seems necessary to suppose on the plume hypothesis that they are consistent only with a slow downward flow in the rest of the mantle with a velocity 10^{-2} of the upward one within the plume. However Hager and O'Connell (1978) showed that even in the absence of convection produced by buoyancy forces both toroidal and poloidal currents would be set up by the drag of the plate motions on the viscous mantle. The velocities of the currents produced will be of the order of the plate motion i.e. some cm/year. If the large number of plumes presumed to exist in the mantle are to remain stationary with respect to one another, or to move at velocities orders of magnitude less than plate motions and to exist for some 100 M yr it seems hard to understand how this stationary feature of the plume can be reconciled with the mantle flow induced by the plates during which the convection flow will move many thousands of kilometers. The fundamental reason for introducing the plume hypothesis then disappears.

6. Unresolved Questions Concerning Mantle Convection

The issue of whether mantle convection is whole mantle convection with flow unimpeded by the phase changes at 400 km and 670 km depth or whether it consists of two layer convection, i.e. one pattern of cells in the upper mantle and another in the lower, remains unresolved. There is not enough known about the thermodynamics of the phase changes to conclude whether convection currents can penetrate these barriers or would break up into two layers of cells. But Peltier has pointed out that in two-layer convection a thermal boundary layer must exist separating the two convecting regions and it follows that a considerable temperature gradient would be expected at this depth. As viscosity is very sensitive to temperature, the viscosity of the lower mantle would be expected to be less than that of the upper mantle. Attempts to determine the variation of viscosity with depth in the mantle by comparing the rate of uplift of the various regions, N.W. Canada, Fennoscandia and Lake Bonneville, and determining the viscosity down to depths comparable with their dimensions give some information (Cathles, 1975). These models do not show an effect consistent with a two-layer convection: the viscosity may be greater in the lower mantle than in the upper but only by one or two orders of magnitude.

Seismic tomography, when the anomalies are interpreted as indicative of the convection cells, does show distinctly different patterns in the upper and lower mantle, as if the convection is two layered. However the interpretation of the seismic tomography method is not entirely unambiguous. The seismic velocity anomalies determined are of the order of 10^{-2} and wavelengths the order 10,000 km. This is usually interpreted by supposing that $\Delta v/v = K \Delta \rho/\rho$ where K is a constant of about 1. Thus Woodhouse and Dziewonski (1989) assume density variations over horizontal surfaces in the mantle of roughly the same order of magnitude as the velocity anomalies, or, as the volume coefficient of expansion of silicates under laboratory conditions is 3.10^{-5} per °C and smaller (Poirier, 1991) in the deep mantle, lateral temperature variations on horizontal surfaces of the order of 100-1000°C would be expected and this seems excessive. It would imply that the mantle has a much greater viscosity than assumed in the order of magnitude calculation above. But the geoid very tightly constrains the possible variations of density over horizontal surfaces in the mantle. The large scale anomalies being 10-20 mgal, the difference in total mass in columns of unit area between the surface and the core cannot exceed 10^{-5}, neglecting geometric effects. In a convecting system the distortion of the surface produced by the flow can reduce and often change the sign of the observed gravitational anomaly which would otherwise arise from the bodily density differences creating the buoyancy forces, as was shown by Runcorn (1967a,b). The effect of the surface distortions is critical but the three orders of magnitude to be bridged seems very large.

But Richards and Hager (1984), Hager and Richards (1989) and Vigny et al. (1991) have made detailed calculations and appear to be satisfied that there is no difficulty in reconciling these two methods of studying inhomogeneities in the Earth's mantle. For the Moon it is less than one order of magnitude. But Hager and Richards (1989) do admit (on their assumptions) that "the geoid is a small difference between relatively large numbers." However solid state convection occurs by the continuous regrowth of crystals and in a stress field the axes of the crystals will align with the flow. In the lower mantle the crystals are cubic but the metallic ions in the lattice will almost certainly produce anisotropy. It is characteristic of anisotropy in crystals that sound velocities parallel and perpendicular to the axis of symmetry differ by a few per cent. Therefore the large scale anisotropy patterns expected to be present in the mantle if solid state convection is occurring would appear, on order of magnitude calculations, to be a more likely explanation of the seismic travel time anomalies than the density and temperature variations alone. However the inverse problem of determining the distribution of the axes of anisotropy in the mantle from the travel time anomalies is a very much more difficult task than the already remarkable achievement of determining the velocity variations with latitude, longitude and depth. However Woodhouse and Dziewonski (1989) showed that the general pattern obtained from P wave and S wave travel time anomalies and the shear wave velocities determined from the free oscillations in the lower mantle are similar and this would argue against any dominant effect from anisotropy. It is also held that the lower mantle anomalies correlate reasonably well with the geoid (Hager et al., 1985) while the upper mantle seismic anomalies show some expected correlations (if they are due to density and temperature variations) with processes under the mid-ocean ridges and ocean trenches. However all these anomalies have to be represented by low degree spherical harmonics, so that considerable smoothing of the data results, which makes assessment of the significance of these various correlations difficult, as Anderson (1989) also concludes. If this interpretation is accepted it certainly implies that convective models, in which the viscosity is treated as a function of radius only, are not realistic. Temperature differences of 100-1000°C on horizontal surfaces would result in considerable variation of viscosity in the uprising or falling convection currents. While this has been incorporated in computer models of mantle convection, this complication would make it more unlikely that mantle convection pattern could be uniquely determined: for more parameters are being included. Furthermore such solid state phenomena as creep depend on parameters that cannot be determined either theoretically or by laboratory experiment.

I return to the question whether the effect on the Earth's gravitational field of the contributions of the "bodily" variations of density within the mantle and those arising from the masses in the distorted boundary (of opposite sign) cancel out to within one part in 10^3. I have shown that for the Moon this does not happen but the ratio depends on the variation of g and viscosity with depth—very different for the Earth. But the ratio depends on the degrees of the harmonics in the convection pattern as I showed (Runcorn, 1967a,b). If the cancellation happened to be almost complete for one harmonic, it would not be for a harmonic say two degrees different. I think that the density differences inferred from seismic tomography lead to difficulties that have not been fully faced.

There is a further serious problem with the interpretation of seismic topography solely in terms of density variation. If density anomalies of the order of 10^{-2} exist throughout the mantle, stress differences are produced and if the wavelength of these anomalies are of the order of 10^8-10^9 km as indicated by the seismic tomography maps, the stress differences will be of about the same order as the breaking stress of rocks under laboratory conditions 10^3-10^4 bars. This conclusion is not appreciably altered by taking account of the boundary distortion produced by convection. These stress differences seem very high on other grounds. Solid state creep at high stress differences is dependent on a power law (Poirier, 1991) and the effective viscosity would be lower than for small stress differences. As the stress differences associated with the removal of the ice loads 10^4 years ago are much smaller, of the order of 10 bars, the mantle viscosity found from the rate of uplift is not likely to be the same as the effective viscosity for these large stress differences. In fact the latter must be smaller. Both the greater density anomalies and the lower effective viscosity result, through the order of magnitude calculations given above, in convection velocities many orders of magnitude greater than the plate velocities.

A further difficulty arises if the density variations on horizontal surfaces in the mantle are of the order of 10^{-2} or even 10^{-3} if K is 8 as Vigny et al. (1991) assume, for it follows that the distortions of the boundaries must be of the order of kilometers, if the gravity anomalies are to be almost completely compensated by the masses on the distorted boundaries of the convecting spherical shell. Whether such distortion is present at the lower boundary must await the refinement of seismic determinations of the shape of the core surface. Morelli and Dziewonski (1987) find undulations of a few kilometers but evidence of density variations in the D" layer is found from P and S wave velocities within it and are inferred in a theory of why the reversing geomagnetic dipole follows paths restricted to certain mantle longitudes (Runcorn, 1992a). If the D" layer is inhomogeneous this complicates the question, as in a similar way the continent-ocean distribution obscures the question whether the outer boundary of the convecting mantle is distorted. In the classical theory of isostacy, the depth of compensation lies below the Moho under the continents, and it is a surface of constant pressure, density and geopotential and is spheroidal. Solid state convection distorts this surface. Consequently if the relative positions of continents and oceans changes with respect to the convection pattern it follows that the actual heights of the continents with respect to the Earth's center and also above sea level could change by some kilometers. In the last 100 M yr this has not happened: Pitman (1978) and Wilgus (1988) suggest changes of a few hundred meters as upper limits. To be more concrete, suppose a continent 100 M yr ago is above the descending convection current, which pulls the surface of compensation downwards, and is now above the ascending convection current which distorts the surface of compensation upwards. The geoid is fixed relative to the convection pattern, both by the bodily density distribution in the mantle and the positive and negative masses of the distorted surface of compensation resulting from the pressure field arising from convection. The continent-ocean distribution, through being a potential double layer by the theory of isostacy, contributes only a few centimeters to the geoid heights. Thus the geoid has not in the last 100 M years changed with respect to the convection pattern and therefore the ocean surface would have changed by only + 100 m with respect to the freeboard of the continents during drift in the absence of the distortion of the surface of compensation. But if this effect is taken into account the continents could have changed in elevation relative to the ocean surface by several kilometers in the last 100 M yr. The question of what change is allowable is a matter for palaeogeographical experts, and is complicated by changes in the volume of the oceans by changes in the ice caps. The oceanic ridges also affect the level of the oceans. It would however seem unlikely that more than a fraction of a kilometer distortion in the surface of compensation is allowed by the geological evidence.

The way in which this problem has been removed is to choose a radical viscosity variation so that the gravity signature of the bodily density variations is reduced to that observed. Hager and Richards (1989), Forte and Peltier (1991) and King and Masters (1992) find that lower viscosities in the upper mantle will reconcile in a general way the two sets of observations. In essence what is being done is to bury the major distorted boundary at the depth where the viscosity changes, i.e., at 670 km. Suitable choice of the viscosity of the upper mantle then reduces the undulations of the outer boundary, i.e., the surface of compensation, to what may be allowable from geological data. I conclude that there is serious objection to attributing the seismic wave anomalies discovered by seismic tomography simply to density differences on horizontal surfaces in the mantle. To interpret the seismic travel time anomalies in terms of the hypothesis of large scale anisotropy is as yet an unsolved problem.

7. Mantle-Core Interactions

I have expressed some scepticism about the large temperature variations on horizontal surfaces in the Earth's mantle currently inferred from seismic tomography: the issue is whether they are in the range 1° to a few 10°C or a few 100° to 1000°C. An interesting consequence of the widespread acceptance of the higher values is that the question has been raised whether such temperature variations, associated with mantle convection, diffuse into the core and generate motions that control the core dynamo to some degree.

Two discoveries in palaeomagnetism have raised the possibility that the mantle influences the core dynamo despite the disparity of about 10^7 in the velocities of core and mantle convection. As argued above, the role of the mantle in carrying heat away from the core is crucial for magnetic field generation. One discovery of palaeomagnetism, which has gradually become rather certain, is the rather sudden change about 100 M yr ago in the frequency pattern of geomagnetic polarity reversals: before that time there were long periods of predominant polarity, in the Permian, Jurassic and Cretaceous; after it the sequence of reversals can be well described by a Poisson series—the average interval between reversals being about 1 M yr. The rough coincidence in time between this striking change and the "Wegener" break up was regarded as significant. I have argued that the latter event was brought about by a change in the mantle convection pattern, for example, in the degree of the harmonic describing it: this would affect the heat being carried away from the boundary of the core and possibly change the behaviour of the core dynamo. Though the magneto-dynamics of the core is not sufficiently well understood to develop a theoretical under-standing of this idea, it seems easy to accept as a possible scenario.

It is less easy to accept an idea more recently put forward: that the core produces fields of different characteristics in different longitudes for times of the order of mantle convection. This idea arose from further historical evidence that the geomagnetic secular variation has been smaller in the Pacific hemisphere than elsewhere (Bloxham et al., 1989), but especially from the discovery by Laj et al. (1991) that during polarity reversals over the last 11 M yr the V.G.P. moves along one of two longitude sectors, the Americas and Australia–E. Asia. They drew attention to the correlation between these two preferred paths and two roughly similar high P wave velocity anomalies in the lower mantle. I have argued (Runcorn, 1991) that the fundamental starting point in the discussion of core-mantle interactions is the explanation of the irregular changes in the length of the day by interchange of angular momentum between the core and the mantle. Whether the variable torque between them is electromagnetic, or due to topographical coupling, the rotation of the core relative to the mantle must change, for only odd degree toroidal velocities carry angular momentum. The relative rotation of the core relative to the mantle, seen clearly in the westward drift of non-axial parts of the field in historical times, seems incompatible with the idea that the dynamo produces fields with characteristics, which remain stationary with respect to the mantle for times long when compared to the lifetime of secular variation foci.

I have put forward (Runcorn, 1992a) an alternative which does not require a core dynamo generating fields with properties dependant on longitude. I assert that if the core was alone in space and the field reversed by rotation of the dipole through 180°, as the new evidence strongly suggests (Laj et al., 1991), the V.G.P.s would not, in different reversals, trace the same path in longitude in a frame of reference defined by the mean rotation of the core. Similarly, because a fluid has no memory, the secular variation and non-dipole field would have no bias towards any particular longitude zone. I have suggested that the key lies in the

D" layer because of its inhomogeneity. I suggest that it has a near metallic conductivity below the Pacific hemisphere and thus screens the secular variation generated in the core. I also suggest that this conducting hemispherical shell interacts with the reversing rotating dipole of the core to produce an electromagnetic torque that rotates the core so that the reversing dipole path lies along one of its boundaries, i.e. the Americas or Australia–E. Asia.

However the idea that the core dynamo possesses a bias towards these preferred mantle longitudes has been given apparent support by an analysis (Constable, 1992) of the stable geomagnetic field, aside from the reversal process, in the last 5 M yr. Constable has taken over 2000 spot readings of the geomagnetic field directions obtained from lava flows, which, as they cool and become magnetized in a year or so, do not smooth out the secular variation. She determined the V.G.P.s of these geomagnetic field directions and shows that they peak, when plotted against longitude, at the two preferred longitudes of the Americas and Australia–E.Asia. However the sites from which the data were obtained cluster around 0° and 180° longitudes and it can be shown (Runcorn, 1992b) that the dipole formula will produce a clustering 90° away from the site, if the palaeomagnetic directions are scattered randomly about the mean field (which is that of an axial dipole), as is assumed whenever palaeomagnetists use Fisher's "dispersion on a sphere" frequency distribution in their analysis. The geomagnetic secular variation field, unlike the main field, does not have a strong single harmonic and consequently at a particular site the scatter of directions over 10^3 yr is well described by Fisherian statistics. Consider the special case of observations at a site on the equator; the mean field is horizontal and directed either to the geographical north or south, while the field component perpendicular to it is uniformly scattered through 360°. These latter components determine the longitudes of the V.G.P.s calculated from the directions of the total field. It is easily shown that the poles calculated from the field directions are four times as densely grouped at longitudes 90° away from the site as at 0° or 180°. For if the field component perpendicular to the north makes an angle I to the E horizontal, the longitude ϕ of the V.G.P. is given by

$$\cot \phi = \frac{1}{2} \tan I$$

If these component directions are scattered between I and I + dI, the corresponding V.G.P. will be distributed between longitudes ϕ and $\phi + d\phi$ where

$$d\phi = \frac{1}{2} (1 + 3 \cos^2\phi) \, dI$$

Thus $d\phi = \frac{1}{2} dI$ at $\phi = 90°$ and 270° and $d\phi = 2 \, dI$ at $\phi = 0°$ and 180°. Constable finds a ratio of about 1.6, which is a consequence of the sites being spread in latitude and also to their much smaller spread in longitude. Thus I dissent from Jackson's (1992) emphatic statement: "Constable upturns all conventional thinking by showing, over the past 5 million years, there has been a discernible tendency for the stable field (normal or reversed) V.G.P.s to be biased towards two longitudinal bands rather than being uniformly distributed in longitude around the geographical North Pole. . . . Intriguingly, this bias points to control of the magnetic field (generated in the Earth's core) by the mantle." I rather suggest that Constable has performed a work of supererogation in carrying through this exhaustive test of the basic equation of geomagnetism.

I return to the hypothesis that the D" layer is hemispherically asymmetrical in electrical conductivity, the result of a higher FeO or FeS content under the Pacific. Is this asymmetry the result of the accidental convergence of many factors and on the Earth's time scale a transient phenomenon—the palaeomagnetic record only covers the last 11 M yr? Or has it a more fundamental significance? For other striking differences exist between the Pacific and non-Pacific hemispheres. The Pacific plate is the largest; almost all of the world's subduction zones and deep focus earthquakes ring the Pacific and result from the motion of the Pacific plate and that of two smaller plates towards the Pacific rim; this hemisphere has only one major oceanic rise while the other has a complex network of ridges. The fact that continents are largely in the non-Pacific hemisphere has, since C. H. Darwin's speculation, been thought to be of fundamental, yet enigmatic, significance in understanding the evolution of the Earth. All these plate tectonic features reflect in a complex way the mantle convection pattern. The D" layer is the thermal boundary layer of mantle convection and so it is probable that the hemispherical asymmetry of the D" layer is not merely an ad hoc assumption but is connected with the global asymmetry of the Earth's outer crust.

8. Conclusion

Convection in the mantles of the Earth, Moon and terrestrial planets requires for its study a remarkable range of scientific disciplines—solid state physics, tectonics, geodesy, seismology, astronomy, chemistry, palaeogeography—as well as hydrodynamics. The many unresolved questions, some of which I have touched upon in this paper, are among the most fundamental in geoscience.

9. References

Anderson, D. L. (1989) Theory of the Earth. Blackwell Science Pub., Oxford, 366 pp.

Bloxham, J., Gubbins, D. and Jackson, A. (1989) Geomagnetic secular variation. Phil. Trans. R. Soc. A329, 415-502.

Brunt, D. (1941) Physical and Dynamical Meteorology. Cambridge University Press, London, 428 pp.

Cathles, L. M. (1975) The Viscosity of the Earth's Mantle. Princeton University Press, Princeton.

Collinson, D. W. and Runcorn, S. K. (1960) Polar wandering and continental drift: evidence from palaeomagnetic observations in the United States. Bull. Geol. Soc. Amer. 71, 915-958.

Collinson, D. W., Creer, K. M., Irving, E. and Runcorn, S. K. (1957) Palaeomagnetic investigations in Great Britain. I. Phil. Trans. A250, 73-156.

Constable, C. (1992) Link between geomagnetic reversal paths and secular variation of the field over the past 5 M yr. Nature 358, 230-233.

Dziewonski, A. M. (1984) Mapping the lower mantle: determination of lateral heterogeneity in P velocity up to degree and order 6. J. Geophys. Res. 89, 5929-5952.

Esposito, P., Anderson, J. B. and Ng, A. T. Y. (1978) Experimental determination of Mercury's mass and oblateness. COSPAR Space Research 17, 639-649.

Goldreich, P. and Peale, S. (1966) Spin-orbit coupling in the solar system. Astron. J. 71, 425-438.

Hager, B. H. and O'Connell, R. S. (1978) Kinematic models of large-scale flow in the Earth's mantle. J. Geophys. Res. 84, 1031-1048.

Hager, B. H. and Richards, M. A. (1989) Long wavelength variations in the geoid: physical models and dynamical implications. Phil. Trans. A328, 309-327.

Hager, B. H., Clayton, R. W., Richards, M. A., Comer, R. P. and Dziewonski, A. M. (1985) Lower mantle heterogeneity, dynamic topography and the geoid. Nature 313, 541-545.

Hess, H. (1962) History of ocean basins. In Petrologic Studies, E. A. J. Engle, Ed., Geological Society of America, pp. 599-620.

Jackson, A. (1992) Still poles apart on the reversals? Nature 358, 194-195.

Jeffreys, H. (1952) The Earth. 3rd Edition. Cambridge University Press, 392 pp.

Jeffreys, H. (1959) The Earth. 4th Edition. Cambridge University Press, 438 pp.

Laj, C., Mazaud, A., Weeks, R., Fuller, M. and Herrero-Bervera, E. (1991) Geomagnetic reversal paths. Nature 351, 447.

le Pichon, X. (1968) Sea floor spreading and continental drift. J. Geophys. Res. 73, 3611-3697.

Loper, D. E. (1991) Structure and processes within the Earth's interior. Paper presented at the NATO Advanced Study Institute on Dynamic Modeling and Flow in the Earth and Planets, June 17-28, University of Alaska Fairbanks.

McKenzie, D. P. and Parker, R. (1967) N. Pacific: An example of tectonics on a sphere. Nature 216, 1276-1280.

Morelli, A. and Dziewonski, A. M. (1987) Topography of the core-mantle boundary and lateral homogeneity of the liquid core. Nature 325, 678-683

Morgan, W. J. (1968) Rises, trenches, great faults and crustal blocks. J. Geophys. Res. 73, 1959-1982.

Morgan, W. J. (1971) Convection plumes in the lower mantle. Nature 230, 42-43.

Pitman, W. C. (1978) Relationship between eustacy and stratigraphic sequences at passive margins. Geol. Soc. Amer. Bull. 89, 1389-1403.

Poirier, J.-P. (1991) Introduction to the Physics of the Earth's Interior. Cambridge University Press, 264 pp.

Ranalli, G. (1987) Rheology of the Earth. Allen and Unwin, Boston, 366 pp.

Richards, M. A. and Hager, B. H. (1984) Geoid anomalies in a dynamic Earth. J. Geophys. Res. 89, 5987-6002.

Runcorn, S. K. (1962a) Convection in the Moon. Nature 195, 1150-1151.

Runcorn, S. K. (1962b) Towards a theory of continental drift. Nature 193, 311-314.

Runcorn, S. K. (1962c) Convection currents in the Earth's mantle. Nature 195, 1248-1249.

Runcorn, S. K. (1963a) A new mechanism for convection in the mantle and continental accretion. Nature 197, 582-583.

Runcorn, S. K. (1963b) Growth of the Earth's core. Nature 197, 992.

Runcorn, S. K. (1964) Satellite gravity measurements and a laminar viscous flow model of the Earth's mantle. J. Geophys. Res. 69, 4389-4394.

Runcorn, S. K. (1965) Changes in the convection pattern in the Earth's mantle and continental drift: evidence for a cold origin of the Earth. Symposium on Continental Drift, Phil. Trans. Roy. Soc. A258, 228-251.

Runcorn, S. K. (1967a) Convection in the Moon and the existence of a lunar core. Proc. Roy. Soc. A296, 270-284.

Runcorn, S. K. (1967b) Flow in the mantle inferred from the low degree harmonics of the geopotential. Geophys. J. 14, 375-384.

Runcorn, S. K. (1967c) The problem of the figure of Mars. In Mantles of the Earth and Terrestrial Planets, Interscience Publishers, pp. 425-430.

Runcorn, S. K. (1974) On the forces not moving lithospheric plates. Tectonophysics 21, 197-202.

Runcorn, S. K. (1984) The primeval axis of rotation of the Moon. Phil. Trans. Roy. Soc. A313, 77-83.

Runcorn, S. K. (1991) In New Approaches in Geomagnetism and the Earth's Rotation, S. Flodmark, Ed., World Scientific Press, Singapore, p. 281.

Runcorn, S. K. (1992a) Polar paths in geomagnetic reversals. Nature 356, 654-658.

Runcorn, S. K. (1992b) Submitted to Nature Aug. 13; Rejected 4 Nov.

Runcorn, S. K., Collinson, D. W. and Stephenson, A. (1977) Intensity and origin of the ancient lunar magnetic field. Phil. Trans. R. Soc. London A285, 241-247.

Strom, R. G. (1964) Comm. of the Lunar and Planetary Laboratory, The University of Arizona 2, 205-216.

Vigny, C., Richard, Y. and Froidevaux, C. (1991) The driving mechanism of plate tectonics. Tectonophysics 187, 345-360.

Vine, F. J. and Mathews, D. H. (1963) Magnetic anomalies over ocean ridges. Nature 199, 947-949.

Wilgus, C., et al. (1988) Soc. Econ. Paleont. and Mineral., Special Pub. No. 42.

Woodhouse, J. H. and Dziewonski, A. M. (1989) Seismic modelling of the Earth's large-scale three-dimensional structure. Phil. Trans. R. Soc. London A328, 291-308.

ROTATING SPHERICAL CONVECTION WITH APPLICATIONS TO PLANETARY SYSTEMS

KEKE ZHANG
Department of Mathematics
University of Exeter
Exeter, EX4 4QJ, UK

ABSTRACT. Convection in the form of three dimensional thermal Rossby waves in rapidly rotating spherical fluid shells is investigated. New types of instability and the associated nonlinear phenomena are found. It is shown that the Prandtl number, defined as the ratio of the viscous to thermal diffusivity of a fluid, $P_r = \nu/\kappa$, plays a key role in determining the fundamental features of both the instabilities and the corresponding nonlinear flow. The results shed new light on the understanding of convection occurring in the Earth's core and the atmospheres of the major planets.

1. Introduction

A problem of vital importance in understanding the dynamics of planetary fluid systems like the Earth's outer core is convection in a rapidly rotating spherical fluid shell. The fluid parts of the Earth are the atmosphere, oceans and the outer core. Observations of the geomagnetic secular variation provide clear evidence of fluid motion in the Earth's liquid core (Jacobs, 1975). In contrast to the atmosphere and oceans, there is no direct access to the Earth's core, and the form of convection and the related dynamo action remain poorly understood. Our knowledge of fluid motions and dynamo processes in the remote liquid core of the Earth is based mainly on theoretical studies. Numerical investigations play an increasingly important role in understanding the essential fluid dynamic and magnetohydrodynamic processes taking place in the Earth's fluid core (e.g. Zhang et al., 1989).

The traditional model of the Earth's liquid core, which contains all the essential physics, is that of a Boussinesq fluid, uniformly rotating and with constant material properties containing a uniformly distributed heat source (Chandrasekhar, 1961). The buoyancy forces are associated with the self-gravitation of the whole mass of fluid which is confined to a sphere or a spherical shell. Theoretical investigations relevant to this problem have been carried out by Roberts (1968), Busse (1970) and Soward (1977). Roberts (1968) derived an asymptotic numerical system with a second order differential equation to describe the onset of convection. An analytic linear perturbation approach was employed by Busse (1970) to obtain the analytic expressions for the critical parameters such as the azimuthal wavenumber as a function of the Prandtl number P_r. With the assumption that the azimuthal scale of flows is much shorter than that of the radial scale, the radial dependence in the asymptotic analysis of Roberts and Busse is neglected. In both pioneering works, consequently, the structure of flows was not determined. Soward (1977) extended the linear stability analysis to a weakly

83

D. B. Stone and S. K. Runcorn (eds.),
Flow and Creep in the Solar System: Observations, Modeling and Theory, 83–96.
© 1993 *Kluwer Academic Publishers. Printed in the Netherlands.*

nonlinear regime which reveals some information on the structure of convection columns. Detailed discussions on the subject can be found in comprehensive reviews by Fearn et al. (1988) and Gubbins and Roberts (1987).

A number of laboratory experiments on rapidly rotating fluid spheres have also been undertaken to investigate the onset and structure of convection (Carrigan and Busse, 1983; Chamberlain and Carrigan, 1986). These exploit the fact that the driving force of convection in a rapidly rotating system is dominated by the component of the driving force perpendicular to the axis of rotation. An experimental investigation with radial gravity was carried out in the space shuttle Challenger, but with weaker effects of the Coriolis forces (Hart et al., 1986).

Numerical analysis in connection with the onset and structure of convection in rotating spheres or spherical shells has been carried out by Gilman (1975) and Zhang and Busse (1987). In both the studies, the Taylor number T, representing the square of the ratio between the Coriolis forces and frictional forces (see definition in section 2), is too low to extrapolate the properties and structures of the flow to higher values such as those pertaining to the Earth's core. The difficulties involved in the studies of rapidly rotating spherical convection arise mainly from the fact that a high order of spherical harmonics is required to resolve small-scale flows owing to the controlling influences of rotation. One of the most severe difficulties encountered in solving the problem of the geodynamo is also the extreme value of the Taylor number T (Zhang and Busse, 1988, 1989, 1990). The simplest and most widely used approach to avoid this type of difficulty is to employ the concept of eddy viscosity which has been successfully used in the studies of the atmosphere, oceanography and solar convection zone for large scale flows. Without any information about the Earth's core turbulence, it seems reasonable to assume that the ratio of the eddy viscosity, v, to the magnetic diffusivity, λ, is of order unity (Roberts, 1988). Accordingly, the Taylor number appears to be in the range $10^{14} < T < 10^{16}$ corresponding to $O(1)\ m^2/s < v < O(10)\ m^2/s$. It is therefore vital to study a system which is predominantly constrained by the Coriolis force and, particularly, show an asymptotic behaviour with respect to the parameter associated with rotation. In this paper, we will present a brief review of such studies recently carried out by Zhang (1991a, 1991b, 1992a, 1992b).

This paper is made up as follows. After introducing model and mathematical formulation of the problem in section 2, we discuss the onset of convection in section 3. Section 4 presents the form of convection rolls, and the properties of nonlinear drifting rolls are discussed in section 5. Symmetry-breaking and secondary convection are covered in section 6, and applications to planetary systems are presented in section 7. The paper closes in section 8 with concluding remarks.

2. Model and Formulation

We have adopted the traditional model for the Earth's core and planetary atmospheres: a homogeneous fluid spherical shell of constant thermal diffusivity κ, fluid thermal expansion coefficient α and viscosity v that is rotating uniformly with a constant angular velocity Ω in the presence of its own gravitational field

$$g = -\gamma r,$$

and in the absence of external forces. An unstable temperature gradient,

$$\nabla T_s = -\beta \mathbf{r}$$

is produced by a uniform distribution of heat sources. Using the thickness of the fluid shell, d = $r_o - r_i$, as length scale, the viscous diffusion time, d^2/ν, as scale of time, and βd^2 as scale of temperature fluctuation of the system, respectively, the Navier-Stokes equation of motion for the velocity, $\mathbf{u}$, the heat equation for the temperature deviation, Θ, from the purely conductive state, T_s, can be written as

$$\left(\frac{\partial}{\partial t} + \mathbf{u} \cdot \nabla\right)\mathbf{u} + T^{1/2}\mathbf{k} \times \mathbf{u} = -\nabla p + Rr\Theta + \nabla^2\mathbf{u}, \tag{1}$$

$$\nabla \cdot \mathbf{u} = 0, \tag{2}$$

$$\left(\nabla^2 - P_r\frac{\partial}{\partial t}\right)\Theta + \mathbf{r} \cdot \mathbf{u} = P_r\mathbf{u} \cdot \nabla\Theta, \tag{3}$$

where $\mathbf{k}$ is a unit vector parallel to the axis of rotation and the term ∇p represents the force due to the pressure gradient. By assuming a weak magnetic field, the magnetic forces are neglected in equation (1). The non-dimensional physical parameters of the problem are the Rayleigh number R, the Prandtl number P_r and the Taylor number T, defined as

$$R = \frac{\alpha\beta\gamma d^6}{\nu\kappa}, P_r = \frac{\nu}{\kappa}, T = \left(\frac{2\Omega r_o^2}{\nu}\right)^2 (1 - \eta)^4$$

where η is the radius ratio of a shell, $\eta = r_i/r_o$. While the Rayleigh number is effectively the ratio of destabilizing forces to dissipative forces, the Prandtl number provides a measure of relative importance of viscous to thermal diffusion, the properties of fluids. Equations (1-3) are solved subject to the following conditions on the spherical bounding surfaces: impenetrable, perfectly thermally conducting and stress free. The frame of reference in which the rigid body rotation of fluid motions vanishes has been adopted, but the transformation to other frames of reference is possible after a nonlinear solution is obtained. Moreover, the same thermal boundary conditions are used in the nonlinear computation because of the small volume of the inner core (see discussions in Zhang and Busse, 1989).

3. Onset of Convection

As the Rayleigh number R increases to a critical value, the conducting state becomes unstable and convection occurs. Convective instability is characterized by the following set of parameters: azimuthal wavenumber m, oscillation frequency ω, the Rayleigh number R, the Taylor number T and the Prandtl number P_r. The critical mode, denoted by subscript c, of convective instability corresponds to the smallest possible value of the Rayleigh number, R_c, with respect to all possible values of the azimuthal wavenumber and to all possible modes likely to be excited. It is shown by Roberts (1968) that there exist asymptotic dependences between the critical parameters, R_c, m_c and ω_c, on the Taylor number T,

$$R_c = C_R T^{2/3}, \; m_c = C_m T^{1/6}, \; \omega_c = C_\omega T^{1/3}$$

where the coefficients are a function of the Prandtl number. The physically realizable asymptotic relations with the correct symmetry were obtained by Busse (1970). This power-law dependence at the onset of convection appears in our numerical analysis at large T; the parameter region with asymptotically large T behaviour is apparently reached. Here is an example of the case for infinitely large Prandtl number (note that the thermal diffusion time-scale is used for this case):

C_R (T = 5.5 x 10^7) = 1.73; C_R (7.5 x 10^8) = 1.67; C_R (4.5 x 10^9) = 1.64
C_ω (T = 5.5 x 10^7) = -0.410; C_ω (7.5 x 10^8) = -0.430; C_ω (4.5 x 10^9) = -0.436
C_m (T = 5.5 x 10^7) = 0.718; C_m (7.5 x 10^8) = 0.730; C_m (4.5 x 10^9) = 0.738

Very slow variations of the value of $R_c/T^{2/3}$, $m_c/T^{1/6}$ and $\omega_c/T^{1/3}$ are clearly indicative of reaching the asymptotic region of large T. The value of the coefficient of the power-law dependence for R_c changes less than 2%, from $R_c/T^{2/3} = 1.67$ at $T = 7.5 \cdot 10^8$ to $R_c/T^{2/3} = 1.64$ at $T = 4.5 \cdot 10^9$. It is important to recognize that the geophysical significance of the solution relies on its asymptotic behaviour. The asymptotic relations at the infinite Prandtl number estimated from the numerical solution are

$$R_c = 1.63 T^{2/3}, \; \omega_c = -0.44 T^{1/3}, \; m_c = 0.74 T^{1/6}.$$

The critical coefficients, C_R, C_ω and C_m, estimated from our numerical solutions up to the Taylor number $T = O(10^{12})$ and obtained from Busse's formula (1970) (taking into account the different scale), are displayed in Figure 1. The critical Rayleigh number required for the instability of convection is much underestimated by the theory, while the frequency of oscillation is much overestimated, particularly for small Prandtl numbers. It is also important to note that the ratio of the numerical estimate to the theoretical value varies quite substantially as a function of P_r. While the critical wavenumber of the theories and the numerical results is nearly the same for the Prandtl number $P_r \leq 1$, the difference increases sharply with increasing Prandtl number. It is also interesting to note that both the theories and the numerical results show a tendency of asymptotically large P_r dependence for $P_r \geq 10$.

4. The Form of Convection Rolls

In rapidly rotating fluid systems, the Proudman-Taylor theorem in conjunction with the influences exerted by boundary geometry provide the key for understanding the structure of convection. The Proudman-Taylor theorem states that a steady and slow motion in the absence of viscous and magnetic forces satisfies the following vorticity equation

$$(\mathbf{k} \cdot \nabla)\mathbf{u} = 0. \tag{4}$$

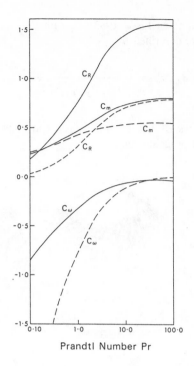

Prandtl Number Pr

Figure 1. The critical coefficients plotted against the Prandtl number. The dashed curves correspond to Busse's result (1970) and the solid curves are obtained from numerical analysis.

It follows that the flow satisfying the above equation is independent of the coordinate parallel to the axis of rotation, and the pressure acts as a stream function for the two-dimensional flow. However, the realization of such flow depends upon the geometry of the bounding surfaces of a fluid container on which

$$u_z + \chi(s)u_s = 0,$$

must be satisfied, where χ is associated with the geometry of the fluid container, assumed to be axisymmetric, and cylindrical coordinates, (s, ϕ, z) are used. In a rotating annulus with parallel top and bottom boundaries, $\partial\chi/\partial s = 0$, two-dimensional convection can be realized except in the thin Ekman layers. In a rotating annulus with the constant inclined top and bottom boundaries, $\partial\chi/\partial s = 0$, convection without violating the PT theorem cannot occur. The resulting convection is in the form of nearly two-dimensional rolls aligned parallel to the axis of rotation, and the rolls drift in the azimuthal direction with phase speed proportional to the slope of the boundaries, χ. With an addition of weak curvature on the boundaries, $\partial\chi/\partial s = \varepsilon$, where ε is a small parameter, the phase of the convection rolls shifts slightly in the azimuthal direction depending on the sign of ε (Busse, 1983). In a rotating spherical system, the effects of spherical boundary curvature described by

$$\partial \chi / \partial s = \frac{1}{r_o\left(1 - \left(s/r_o\right)^2\right)^{3/2}},$$

together with strong influences of the Coriolis force can cause a drastic distortion of convection rolls in the form of prograde spiral.

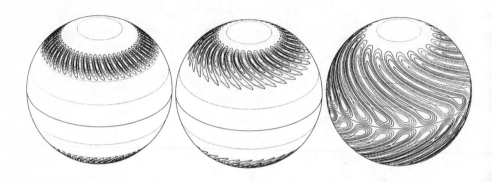

Figure 2. Stream lines of the toroidal flow on the outer surface of the fluid shell are displayed for (a) $P_r = 6.0$, (b) $P_r = 0.6$ and (c) $P_r = 0.05$ at $T = 10^{10}$.

A typical structure of the columnar convection is illustrated in Figure 2 for $T = 10^{10}$ at three different Prandtl numbers. Displayed are streamlines for the toroidal component of velocity, which is dominant over the poloidal component, on the outer surface of the fluid shell. The pattern as a whole drifts eastward with its phase speed $c = -\omega/m$. Other components, such as radial velocity, display a similar spiral structure and are therefore not shown here. For fluids of moderately small Prandtl numbers, fluid rising near a latitude of 60° is convected down at substantially lower latitudes in the neighbourhood of the equator. At the same time, in attempting to satisfy the two-dimensional constraint, the variation of flows is minimized in the direction of z. The elongated and prograde spiralling roll spans an azimuthal distance of about five wavelengths, while the inclination angle between a stretched convection roll and the radial direction, η_i, is about 50°. There are no clear centers of the vortices, as flows move all the way from the region near the equator up to a latitude corresponding to the cylinder attached to the inner boundary equator. One consequence of the spiralling is that the difference between the scales of the azimuthal and radial direction becomes negligibly small. In contrast to an assumption of the previous theories, it appears that there are only two different scales for the convection: the short length scale of convection roll in both azimuthal and radial directions, and the long length scale in the direction of rotation axis,

$$\frac{\partial}{\partial \phi} \approx \frac{\partial}{\partial s} \gg \frac{\partial}{\partial z}.$$

Another consequence, of even more significance, is that the spiralling causes a close correlation between different components of velocity, as we will discuss in the next section. As the Prandtl number increases, however, the fundamental features of the flow structure change markedly. The magnitude of the spiralling angle decreases sharply with increasing Prandtl number. As P_r is increased from $P_r = O(0.1)$ to asymptotically large values, the spiralling angle η_i decreases accordingly from about $50°$ to $15°$. At the same time, the position of convection concentration, characterized by the critical latitude at which a cylindrical convection annulus coaxial with the axis of rotation meets the outer surface, shifts from about $35°$ for $P_r = 1.0$ to approximately $45°$ for asymptotically large Prandtl numbers. When P_r is increased to $P_r > 1.0$, three different scales begin to emerge:

$$\frac{\partial}{\partial \phi} > \frac{\partial}{\partial s} \gg \frac{\partial}{\partial z}.$$

The combined effects of the strong Coriolis forces, the spherical boundary and weak viscous dissipations are responsible for the occurrence of the spiralling columnar convection. While the strong influences of rotation cause the preferred nearly two-dimensional columnar motions, the z-dimension of the column varies continuously as a function of the radial distance. The column of fluid is forced to change its z-dimension as it approaches or moves away from the axis of rotation. That the radial flow must vanish on the outer spherical boundary therefore eventually controls the form of convection; the induced vorticity arising from the z-dimension change of columnar fluid gives rise to the Rossby wave character of convection. It is important to note that the phase speed of the Rossby-type wave is approximately proportional to the inclination of the boundary, and thus increases rapidly, $\partial \chi / \partial s \gg 1$, in the case of spherical geometry as the column of fluid moves toward the equator. Consequently, the columnar convection roll in a spherical container is forced to spiral progradely. If the viscous dissipations, measured by the size of the Prandtl number, are not large enough to break it up, it can extend spirally from middle latitudes to the equatorial region and elongate enormously in the azimuthal direction.

5. Nonlinear Drifting Rolls

For capturing the fundamental physical processes associated with the different forms of convection roll, further calculation into the nonlinear regime is necessary. By expecting that the velocity deviates slightly from the state of the Proudman-Taylor two-dimensional condition, we assume, for the purpose of illustration, that the velocity of small amplitude convection in cylindrical coordinates, (s, ϕ, z), has components

$$\mathbf{u} = \left(-\frac{1}{sT^{1/2}} \frac{\partial p}{\partial \phi}, \frac{1}{T^{1/2}} \frac{\partial p}{\partial s}, u_z \right),$$

where p is the deviation of pressure from hydrostatic state in a drifting frame

$$p = A \sin(\pi\zeta) \sin m(\phi - f(\zeta)),$$

and $\zeta = 2(s - s_0)/d$ is the radial distance from the center of a convection roll scaled by the thickness of the cylindrical convective layer. The spiralling properties of the columnar rolls are described by the phase function, $f(\zeta)$, which is primarily associated with the curvature of the boundary and with the value of the Prandtl number. Taking the average of the azimuthal component of equation 2 over a cylindrical surface, and utilizing the drifting properties of finite amplitude convection and the equation of continuity, an equation for differential rotation generated by nonlinear interactions can be obtained,

$$\left[2s(u_\phi u_s) + s^2 \frac{\partial(u_\phi u_s)}{\partial s} \right] = \left(s \frac{\partial}{\partial s} s \frac{\partial}{\partial s} - 1 \right) U_\phi,$$

where [ ] represents the integral over a cylindrical surface and the mean flow is denoted by $U_\phi = [u_\phi]$. Taking into account that the s-derivative is much larger than the z-derivative, $\partial/s\partial s \gg 1$, it yields

$$\frac{\partial U_\phi}{\partial s} \sim [u_\phi u_s] \sim - \int \frac{1}{sT} \frac{\partial p}{\partial \phi} \frac{\partial p}{\partial s} d\phi.$$

The relationship between the distortion of the columnar rolls and the differential rotation is thus given by

$$\frac{\partial U_\phi}{\partial \zeta} \sim \frac{A^2 m^2 \pi}{T} \frac{\partial f}{\partial \zeta} \sin^2(\pi\zeta), \tag{5}$$

where $\partial f(\zeta)/\partial \zeta$ is approximately proportional to the inclination angle, η_i, between a stretched convection roll and the radial direction. Several important nonlinear aspects of the problem are captured by this simple illustrative case. It is evident that the Coriolis force cannot sustain the non-vanishing axisymmetric flow $U_\phi(s)$. An equilibrium state is achieved by balancing the Reynolds stress with the viscous forces of the differential rotation. The strength of the differential rotation is approximately proportional to the inclination angle, $\eta_i \approx \partial f(\zeta)/\partial \zeta$, of a columnar roll. If the phase function $f(\zeta)$ is independent of radial coordinate ζ, that is, $\partial f(\zeta)/\partial \zeta = 0$, the differential rotation cannot be maintained by the Reynolds stress. With a substantial tilt of convection rolls, $\eta_i > 45°$, a large amplitude differential rotation of the order of $U_\phi = O(u_s u_\phi)$ can be generated through nonlinear interactions between the spiralling rolls. The direction of the differential rotation is determined by the sign of $\partial f(\zeta)/\partial \zeta$. Provided that the convection roll spirals progradely, $\partial f(\zeta)/\partial \zeta > 0$, U_ϕ will be predominantly eastward in lower latitudes.

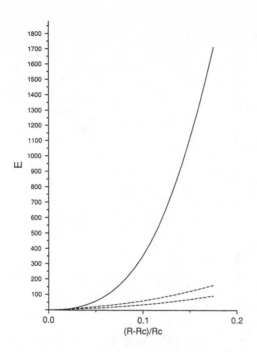

Figure 3. Kinetic energies of various components [(*i*) the differential rotation; (*ii*) non-axisymmetric toroidal flows; (*iii*) non-axisymmetric poloidal flows] are shown as a function of the supercritical Rayleigh number $(R - R_C)/R_C$ for $P_r = 0.1$.

Because of the drifting property of finite amplitude solutions, not only the axisymmetric components but also the average non-axisymmetric quantities are independent of time. The energy spectrum for $P_r = 0.1$ is displayed in Figure 3 as a function of $(R - R_C)/R_C$. The most remarkable feature is that the finite amplitude convection bifurcating from the spiralling columnar mode is primarily dominated by the component of axisymmetric azimuthal flows at the supercriticality of the order of $(R - R_C)/R_C = O(0.1)$, although non-axisymmetric motions are preferred at the onset of convection. For moderately large Prandtl numbers, by contrast, nonlinear interactions are much less important, owing largely to small values of the inclination angle η_i, and the resulting finite amplitude convection is primarily dominated by the non-axisymmetric components of flows. The differences between finite amplitude solutions of different Prandtl numbers can be most easily explained in terms of the Reynolds stress that is closely associated with the form of convection rolls.

Clear indications of the asymptotic behaviour of linear solutions strongly suggests that a similar asymptotic relationship between the typical amplitude of the convection U and the Rayleigh number R at large T may exist for finite amplitude solutions. For the nonlinear solution of the steadily drifting rolls, the typical amplitude of flow is time-independent, and can be given approximately by

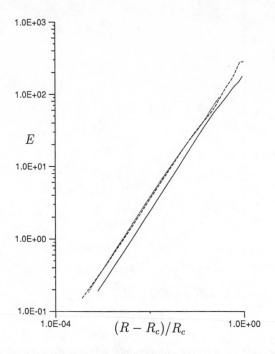

Figure 4. The total kinetic energy, E, of the convective flow is shown as a function of the supercritical Rayleigh number $(R - R_c)/R_c$ for $T = 5.5 \cdot 10^7$ (the solid line), $T = 7.5 \cdot 10^8$ (the dashed line) and $T = 4.5 \cdot 10^9$ (the dotted line) at $P_r = \infty$.

$$U = \frac{1}{V} < \mathbf{u} \cdot \mathbf{u} >^{1/2} = (2E)^{1/2}$$

where E is the density of the total kinetic energy. In Figure 4, E is displayed as a function of $(R - R_c)/R_c$ for these three different Taylor numbers. It is remarkable that the asymptotic relationship for finite amplitudes appears to exist, and, furthermore, that the nonlinear solutions at $T = 7.5 \cdot 10^8$ have reached the asymptotic region. The relationship between U, R and T is given by

$$U = 20.1 \, (R/T^{2/3} - 1.63)^{1/2} = 25.6((R - R_c)/R_c)^{1/2}$$

which has nearly the same accuracy as the asymptotic relation for the onset of convection. It is particularly significant that the asymptotic behaviour of the nonlinear convection gives strong support to the idea that solutions of the realistic geophysical system can be extrapolated.

6. Symmetry-Breaking and Secondary Convection

In contrast to the well-known Rayleigh-Bernard problem, symmetry properties of rotating spherical convection are simple, particularly with respect to numerical treatments. In principle, steadily drifting rolls bifurcating from a spherically symmetric static state possess three basic symmetries. The first symmetry is associated with the nature of the Coriolis forces,

$$(u_s, u_z, u_\phi) \, (s, z, \phi) = (u_s, -u_z, u_\phi) \, (s, -z, \phi),$$

where (s, z, ϕ) are cylindrical coordinates in a frame of reference with the axis of rotation about $s = 0$ and the equatorial plane at $z = 0$, referred to as equatorial symmetry. In view of the linear convection theories (Busse, 1970), it seems reasonable to assume that the modes with equatorial anti-symmetry

$$(u_s, u_z, u_\phi) \, (s, z, \phi) = (-u_s, u_z, -u_\phi) \, (s, -z, \phi),$$

are unimportant in the parameter range of $(R - R_C)/R_C \leq 2$, to be considered in this paper. It is thus quite safe to disregard the equatorially anti-symmetric perturbation and to focus on a subclass of equatorially symmetric solutions in the investigation of stability of the columnar rolls. The second is time-symmetry, and is related to the drifting properties of the rolls: solutions are stationary in a drifting frame of reference. The final symmetry is rotational symmetry involving an azimuthal periodicity of the rolls. Loss or change of the symmetries can be investigated through linear stability analysis (e.g. Zhang, 1991). Generally speaking, the characteristic frequencies of different wavenumbers are different, and an introduction of a secondary frequency into the system by symmetry-breaking is likely to lead to a time-dependent convection through Hopf-type bifurcation.

Two types of the secondary convection (instability) have been identified so far: vacillatory and modulational (Zhang, 1992b). While vacillatory convection changes amplitude periodically with little variation in the pattern of flow, the most interesting feature of modulational convection is a dramatic spatial modulation of the small-scale columnar roll structure. Figure 5 illustrates typical patterns of the secondary modulational convection for T = 7.5 x 10^8 and $P_r = \infty$ at the time when heat transfer by convection reaches a maximum. Surprisingly, the azimuthally regularly spaced rolls are modulated so drastically that the rolls positioned in two different longitudes, in association with the wavenumber $m = 2$, are nearly destroyed. At the same time, a pronounced large-scale feature with a phase shift of 90° from the position of the destroyed rolls is developed, as clearly shown in Figure 5(a). Undoubtedly, this modulation is a result of the strong nonlinear interactions of the two competing modes (m = 20 and 22) which have the close critical Rayleigh numbers. In comparison with the drifting rolls, the resulting large-scale mode with $m = 2$ can easily reach the outer boundary, while the small-scale motions in the form of columnar rolls hardly touch the boundary. As a result, the heat transfer by the secondary convection is much more efficient.

7. Applications to Planetary Systems

After proper scaling, the basic quantities of a nonlinear solution, such as the amplitude of flow, obtained at different Taylor numbers become independent of the Taylor number. This fact allows us to extrapolate from our nonlinear solutions to the solutions relevant to realistic geophysical and planetary systems.

For the atmospheres of the major planets, the size of the Prandtl number is likely to be small, due largely to the radiative contribution to the thermal diffusivity, though the precise values of the Prandtl number and the eddy viscosity are unknown [Busse, 1983]. If we choose $P_r = 0.1$ for the atmosphere of Jupiter, our nonlinear solutions suggest the following relation between the typical speed of the equatorial mean zonal flow, $\overline{U}_\phi$, (with respect to the frame of reference with vanishing rigid body rotation) and the relative supercritical Rayleigh number

$$\overline{U}_\phi \approx \nu 10^{-2} \ (km/s)[(R - R_c)/R_c]^{1/2}.$$

To obtain a zonal velocity of $U_\phi = 0.1 \ km/s$ exhibited on the atmosphere of Jupiter, for example, at $(R - R_c)/R_c = 2$, a value of eddy viscosity of about $5 \ km^2/s$ is required. This apparently huge value is not unreasonably large because the typical azimuthal length scale of convective flows is provided by

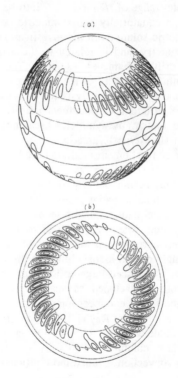

Figure 5. (a) Contours of the heat flux $-\partial\theta/\partial r$ on the outer surface of the shell, (b) streamlines of the toroidal flow on the outer surface of the fluid shell (viewed from the north pole) for $T = 7.5 \times 10^8$, $R = 1.45 \times 10^6$ and $P_r = \infty$ at the instant of maximum heat transfer.

$$L_\phi = [\nu(r_o - r_i)/(2\Omega)]^{1/3}, \qquad\qquad (6)$$

where Ω is the rate of rotation. Taking $\nu = 5\ km^2/s$, $r_o - r_i = 2 \cdot 10^4\ km$ and $\Omega = 1.7 \cdot 10^{-4}/s$ for Jupiter gives the typical length scale of about $L_\phi = 600\ km$, which is quite small compared to the scale of the Red Spot ($\approx 20000\ km$).

8. Concluding Remarks

The belief that the thermally driven convection in rotating spherical systems exhibits a similar behaviour for fluids with different values of the Prandtl number seems to be invalid. The essential features of convection, such as the form of motion, the location of waves and the group wave speed, are strongly Prandtl number–dependent. The analysis reported in this paper also comprises a part of the author's attempt to tackle the nonlinear problem of the generation of magnetic field in the Earth's fluid outer core by extending the previous model of Zhang and Busse (1988, 1989, 1990) to a geophysical parameter space. A useful qualitative guide on the choice of the parameters for the model of the geodynamo is provided by studying the problem of non-magnetic convection. Particularly, our results show that the fundamental features of a convection solution change very little as the Taylor number T increases from $T = 10^9$ to $T = 10^{12}$, indicating that the Taylor number, $T = O(10^9)$, can be considered asymptotically large, at the present stage of research, for the numerical convection and dynamo model in the Earth's fluid core.

Acknowledgment. This work is mainly supported by the Leverhulme Trust, and partially supported by NERC grant GR3/8238. I would like to thank Professor F. H. Busse for numerous discussions about the problem. I would also like to thank an anonymous reviewer for the quite constructive comments.

9. References

Busse, F. H. (1970) Thermal instabilities in rapidly rotating systems. J. Fluid Mech. 44, 441-460.

Busse, F. H. (1983) A model of mean flows in the major planets. Geophys. Astrophys. Fluid Dyn. 23, 152-174.

Carrigan, C. R. and Busse, F. H. (1983) An experimental and theoretical investigation of the onset of convection in rotating spherical shells. J. Fluid Mech. 126, 287-305.

Chamberlain, J. A. and Carrigan, C. R. (1986) An experimental investigation of convection in a rotating sphere subject to time varying thermal boundary conditions. Geophys. Astrophys. Fluid Dyn. 41, 17-41.

Chandrasekhar, S. (1961) Hydrodynamic and hydromagnetic stability. Clarendon Press, Oxford.

Fearn, D. R., Roberts, P. H. and Soward, A. M. (1988) Convection, stability and the dynamo, Energy, Stability and Convection (B. Straughan and P. Galdi, eds.), Longman, 60-324.

Gilman, P. A. (1975) Linear simulations of Boussinesq convection in a deep rotating spherical shell. J. Atmos. Sci. 32, 1331-1352.

Gubbins, D. and Roberts, P. H. (1987) Magnetohydrodynamics of the Earth's core. In Geomagnetism, Vol. 2 (J. A. Jacobs ed.).

Hart, J. E., Glatzmaier, G. A. and Toomre, J. (1986) Space-laboratory and numerical simulations of thermal convection in a rotating hemispherical shell with radial gravity. J. Fluid Mech. 173, 519-544.

Jacobs, J. A. (1975) The Earth's Core. Academic Press, New York.

Roberts, P. H. (1968) On the thermal instability of a self-gravitating fluid sphere containing heat sources. Phil. Trans. R. Soc. Lond. A 263, 93-117.

Soward, A. M. (1977) On the finite amplitude thermal instability of a rapidly rotating fluid sphere. Geophys. and Astrophys. Fluid Dyn. 9, 19-74.

Zhang, K. (1991) Vacillatory convection in a rotating spherical fluid shell at infinite Prandtl number. J. Fluid Mech. 228, 607-628.

Zhang, K. (1991a) Convection in a rapidly rotating spherical fluid shell at infinite Prandtl number: steadily drifting rolls. Phys. Earth Planet. Inter. 68, 156-169.

Zhang, K. (1991b) Parameterized rotating convection for core and planetary atmosphere dynamics. Geophys. Res. Lett. 18, 685-689.

Zhang, K. (1992a) Spiralling columnar convection in rapidly rotating spherical fluid shells. J. Fluid Mech. 236, 535-556.

Zhang, K. (1992b) Convection in a rapidly rotating spherical fluid shell at infinite Prandtl number: transition to vacillating flows. Phys. Earth Planet. Inter. (in press).

Zhang, K. and Busse, F. (1987) On the onset of convection in rotating spherical shells. Geophys. and Astrophys. Fluid Dyn. 39, 119-147.

Zhang, K. and Busse, F. (1988) Finite amplitude convection and magnetic field generation in a rotating spherical shell. Geophys. and Astrophys. Fluid Dyn. 44, 33-53.

Zhang, K. and Busse, F. (1989) Convection driven magnetohydrodynamic dynamos in rotating spherical shells. Geophys. and Astrophys. Fluid Dyn. 49, 97-116.

Zhang, K. and Busse, F. (1990) Generation of magnetic fields by convection in a rotating spherical fluid shell of infinite Prandtl number. Phys. Earth Planet. Inter. 59, 208-222.

Zhang, K., Busse, F. and Hircshing, W. (1989) Numerical models in the theory of geomagnetism. F.J. Lowes et al. (eds), Geomagnetism and Paleomagnetism, 347-358, Kluwer Academic Publishers.

GEOMAGNETISM AND INFERENCES FOR CORE MOTIONS

DAVID GUBBINS
University of Leeds
Department of Earth Sciences
Leeds LS2 9JT, UK

ABSTRACT. Measurements of the geomagnetic field made over the last few centuries provide maps of changes of magnetic flux on the core surface. This reveals a stationary, rather symmetrical, pattern of four main concentrations of flux with superimposed secular variation, much of it in the form of westward drift in low latitudes across the Atlantic hemisphere. The secular change is interpreted in terms of fluid flow at the top of the core, and models of recent secular change have been inverted to give estimates of the flow pattern. The stationary field pattern could be generated by dynamo action deep within the core, and simple kinematic models produce the observed four-fold symmetry with flux concentration beneath downwelling fluid. Much remains to be understood about the dynamo action of even simple fluid motions. The overlying mantle may exert a significant influence on both surface flow and deep convection; simple model calculations suggest temperature or heat flux variations around the core-mantle boundary will drive thermal winds which penetrate deep into the core and strongly influence the main convection.

1. Introduction

Direct observations have been made of the Earth's magnetic field for several centuries. Satellites have provided detailed coverage of the entire globe at separate intervals throughout the last 30 years, with MAGSAT providing measurements of all three components of the magnetic field in 1980. Permanent magnetic observatories have operated for over 100 years; their geographical coverage is rather limited, particularly in the oceans, but their long occupation of the same site allows us to remove the effect of crustal magnetisation and to monitor slow changes of the internal field (secular variation). There are at present about 200 magnetic observatories.

There has been intensive magnetic surveying throughout the 20th century, partly in the search for minerals and partly for general magnetic charting. These survey measurements are strongly influenced by the magnetisation of crustal rocks and it is difficult to obtain secular change from them. There has been intensive recording of the magnetic field by ships at sea, even from early times, because of its importance in navigation. These marine measurements provide an important source of data for the 17th, 18th and early 19th century. Initially, measurements of inclination were sparse and most of the measurements were of declination, but in later scientific voyages, such as those of Captain James Cook, inclination measurements were made routinely. In 1833 Gauss devised a method for measuring absolute magnetic intensity and within 10 years intensity measurements were made routinely on many voyages. There are over a million magnetic measurements in the records of the last few hundred years.

D. B. Stone and S. K. Runcorn (eds.),
Flow and Creep in the Solar System: Observations, Modeling and Theory, 97–111.
© 1993 *Kluwer Academic Publishers. Printed in the Netherlands.*

We measure the magnetic field in an insulator and therefore it can be represented as a potential field, usually as a spherical harmonic series. Spherical harmonic coefficients define the magnetic field at the Earth's surface and can be extrapolated into any adjacent region provided no electric currents flow. In order to interpret the geomagnetic field morphology and the secular variation, we must map the magnetic field at the core surface. The mathematical representation of the potential field can be continued down to the core-mantle boundary to provide a map of any component of the magnetic vector on the core surface. It is not valid to continue the solution into the core.

The International Geomagnetic Reference Field (IGRF) is produced by a least squares fit of spherical harmonic coefficients to surface observations (see, for example, Peddie, 1982). These mathematical models are used as a general guide for magnetic surveying but are not derived with a view to examining core fields. Recently, methods have been developed which impose a regularising condition on the core field (Shure *et al.*, 1982; Gubbins, 1983; Bloxham *et al.*, 1989). Bloxham and Jackson (1989), Gubbins (1983), and Gubbins and Bloxham (1984) developed a Bayesian method using *a priori* information about the core field based on the associated electrical heating which provides error estimates for the field models. It is possible, in principle, to use field models in conjunction with their error estimates to test hypotheses about core dynamics and secular change (Gubbins and Bloxham, 1984), although such statistical approaches are always open to criticism of the underlying assumptions about the prior information (Backus, 1988). Alternatively, one can disregard error estimates and examine the maps of the core field to characterise the behaviour of secular variation; this approach has led to a number of speculative ideas about core dynamics (Bloxham and Gubbins, 1985), which is outlined below.

This work has culminated in a set of models of the magnetic field at the core-mantle boundary for epochs 1650, 1715, 1777, 1845, 1882, 1905, 1915, 1925, 1935, 1945, 1955, 1960, 1966, 1969, 1975, 1980 and 1985 (Gubbins and Bloxham, 1984; Bloxham, 1986; Bloxham *et al.*, 1989; Hutcheson and Gubbins, 1990) and two time-dependent models spanning the dates 1600–1820 and 1820–1990 (Bloxham and Jackson, 1989; Hutcheson, 1990).

The maps of the vertical component of the magnetic vector on the core-mantle boundary show both stationary and drifting features, as well as standing oscillations. The principal contribution to the Earth's dipole moment resides in four main concentrations of flux, near 60°N, 120°E; 60°N, 120°W; 60°S, 120°E; and 60°S, 120°W (Figure 1). When viewed from the poles we see that these main flux lobes appear to touch the circle inscribing the inner core (Figures 2 and 3) with regions of low flux near both poles. The cylinder parallel to the rotation axis enclosing the inner sphere is important because of the Proudman-Taylor theorem, and this feature in the geomagnetic field may reflect the dynamics. Figure 1 also shows a region of low flux in the northeast Pacific and a concentration of high flux south-west of Hawaii; both these features have been stationary, as far as we can tell, for the entire historical record.

Perhaps the best known feature of secular variation is the westward drift, the tendency for contours of any component of the magnetic field to drift towards the west at a rate of about 10 kilometres per year at the Earth's surface. Figure 4 shows the vertical component of magnetic field for epoch 1845, which may be compared with the map in Figure 1 for 1980. Westward drift is evident in the Atlantic region, beneath Africa, and beneath the Indian Ocean, but is notably absent from the Pacific region. Westward drift has been confined to the one hemisphere throughout the historical period, the secular change elsewhere being eastward drift over northern Canada, having the form of standing oscillations, without drift, near Indonesia, and northward drift from Antarctica near longitude 120°E.

Secular variation is caused by fluid motions near the surface of the core. It is usually interpreted by ignoring diffusion and making the so-called frozen flux approximation (Roberts and Scott, 1965). Bloxham and Gubbins (1986) argued that diffusion was evident beneath South Africa; this can be seen by comparing the maps in Figure 1 and Figure 4.

In 1980 (Figure 1) there was a region of reversed flux (negative radial component of magnetic field in the southern hemisphere) beneath Africa and the Indian Ocean which is missing from the map in 1845 (Figure 4). This patch of reversed flux grew up quite suddenly at the beginning of the 20th century and has been increasing in size ever since. Flux of opposite sign can only be produced by diffusion and this patch violates the frozen flux hypothesis. Determinations of fluid flow from secular change are discussed in section 2.

If diffusion becomes important, then it is possible to balance the advection with diffusion and produce steady magnetic field and fluid flow. We expect downwelling to concentrate magnetic field and upwelling to disperse it; it may therefore be possible to infer slow fluid motion, in balance with diffusion, from the stationary part of the magnetic field, but this has not yet been attempted.

The existence of both stationary and drifting features suggests an influence of the solid mantle, and the stationary regions of flux concentration may be the sites of downwelling. Hide (1967) suggested that core-mantle topography would affect the flow, Bloxham and Gubbins (1987) suggested that lateral variations in heat flux would drive fluid motions and therefore affect the magnetic field, and Li and Jeanloz (1987) suggested that strong lateral variations in electrical conductivity may be present in the lower mantle because of dissolved iron, which could also affect the magnetic field. These mechanisms are now being investigated by a number of workers.

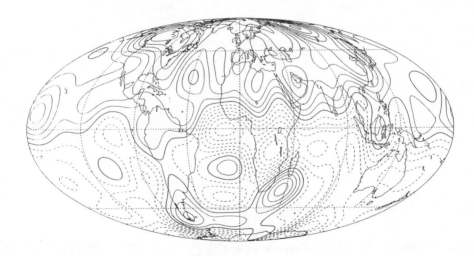

Figure 1: Vertical component of the magnetic field on the core-mantle boundary at 1980 derived from MAGSAT data. Mollweide Projection.

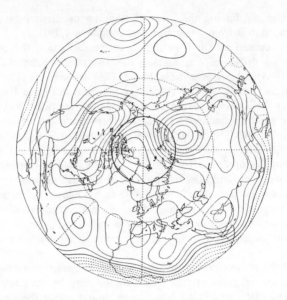

Figure 2: Radial component of magnetic field at 1980 in Lambert equal area projection viewed from the North Pole. The circle indicates the size of the inner core.

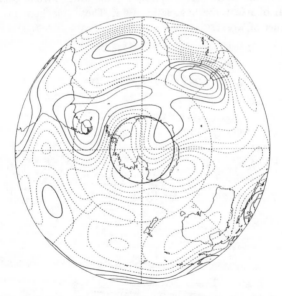

Figure 3: Radial component of magnetic field at 1980 in Lambert equal area projection viewed from the South Pole. The region of low flux near the South Pole has persisted throughout the historical record.

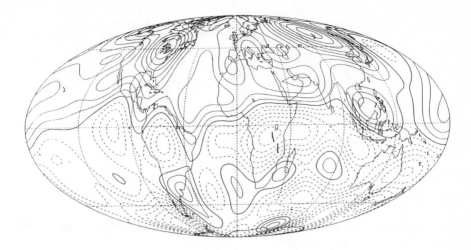

Figure 4: Vertical component of magnetic field on the core-mantle boundary at 1842, for comparison with Figure 1. Mollweide Projection.

2. Determination of Core Motions

A great deal of work has been done on determining core fluid motion, from mathematical models of the main field and secular variation, by neglecting diffusion. Roberts and Scott (1965) first proposed that one could ignore diffusion because the decay time is about 15,000 years whereas the time scale of secular change is much shorter, only a few centuries. In a perfect fluid, the magnetic field is frozen to the fluid by Alfvén's theorem (see, for example, Roberts, 1967) and therefore each field line can be used as a tracer for the flow. There is a problem, however, because we cannot label individual magnetic field lines, making the determination of the flow non-unique.

Backus (1968) explored the non-uniqueness in some detail but more recent work has resolved the ambiguity by making additional plausible assumptions about the core flow and by using all three components of the magnetic induction equation, not just the radial component.

It has been shown that steady motions can be uniquely determined (Voorhies and Backus, 1985); that toroidal motions and their radial derivatives can be determined uniquely from all three components of the induction equation (Lloyd and Gubbins, 1990); that velocities satisfying the vertical component of the vorticity equation without magnetic forces, the so-called "tangentially geostrophic" flows, can be found uniquely over about half the core surface (LeMouël, Gire and Madden, 1985; Backus and LeMouël, 1986), with one component determined everywhere; and that, by using all three components of the induction equation, tangentially geostrophic flows and their radial derivatives become over-determined (Jackson and Bloxham, 1991; Gubbins, 1991).

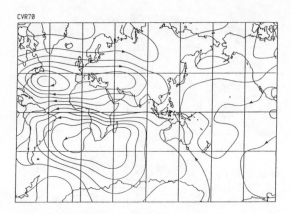

Figure 5: Streamlines of toroidal velocity derived from a model of secular change centred on 1970. From Lloyd and Gubbins (1990).

Figure 5 shows a calculation of toroidal flow for epoch 1970 (Lloyd and Gubbins, 1990). It consists of a double gyre restricted to the Atlantic hemisphere, a feature common to most determinations. Other assumptions yield flows that typically exhibit downwelling off Peru and upwelling in the Indian Ocean, as required to feed the westerly flow across the Atlantic hemisphere in equatorial regions. These flows generally explain more than 90% of the secular variation of the last two or three decades; the subject is reviewed most recently by Bloxham and Jackson (1991).

The problem of formal uniqueness, first highlighted by Backus (1968), has now been circumvented to some extent by placing additional constraints on the flow but there remains the practical problem of discrimination between different candidate flows and their underlying dynamics. This will require very careful assessment of the misfit between the model predictions and measurements, which has not yet been done. Furthermore, inversion methods used so far demand fluid flows with the largest possible length scale consistent with the observations, with the result that westward drift of rather small patches of flux is accounted for by large-scale westerly drift. This may well be correct, but we could equally well argue for a type of wave motion in which fluid flow and magnetic field had similar length scales. In fact, the fluid motions already determined are too small scale to allow discrimination between stratified (toroidal) flow and tangentially geostrophic flow (Gubbins, 1991). The best way to assess the accuracy of these core motions would be to extend their application in time and use older data, but then magnetic diffusion would become evident.

3. Kinematic Dynamos

The appearance of the main flux lobes near the inner core circle, and their symmetry across the equator, suggest they result from a global dynamo operating deep within the liquid core. The field morphology may therefore be a characteristic signature of the fluid flow responsible for the main dynamo. This idea is being investigated by a kinematic study; no dynamics are required to explore the relationship between magnetic field morphology and fluid flow. By "dynamo" we imply a balance between advection and diffusion, whereas by "frozen flux" we ignore diffusion altogether and calculate only the balance between time dependence and advection; the velocities inferred from the dynamo calculations should therefore be much smaller than those inferred from secular variation using the frozen flux hypothesis. Ascertaining fluid motion for a dynamo from a stationary surface pattern is clearly an ill-posed inverse problem and no unique solutions are possible, but some insight may be gained from simple forward calculations.

A great deal of work has been carried out with so-called "mean field" (or $\alpha\omega$) dynamos but these are of no use for the present purpose because they generate axisymmetric, large scale, magnetic fields which bear no relation to the morphology of the underlying flow. We need to study the dynamo action of large-scale flow, which unfortunately requires solution of the full induction equation without the mean field approximation.

We ignore the underlying dynamics and choose solenoidal fluid motions as candidates for dynamo action. Our only recognition of the dynamics is to select flows which might plausibly arise from magneto-convection in a rotating spherical shell. The induction equation is written in dimensionless form as

$$\frac{\partial B}{\partial t} = R_m \nabla \times (v \times B) + \nabla^2 B \tag{1}$$

where

$$\nabla \cdot B = 0 \tag{2}$$

and the magnetic Reynolds number is defined by

$$R_m = UL\mu_0\sigma \tag{3}$$

where U, L are characteristic velocity and length scales, μ_0 is the permeability of free space, and σ is the electrical conductivity.

Equation 1 is linear in B once v is chosen, and solutions vary exponentially with time: $B = B_0 \exp pt$. The equations are solved numerically and R_m is chosen so that B is either steady or oscillatory: all solutions reported here are steady. The numerical solution is effected by expansion in vector spherical harmonics, following Bullard and Gellman (1954), and the radial derivatives are represented by finite differences to give an algebraic eigenvalue problem for the growth rate p (where $B_0 \propto \exp pt$) as eigenvalue and magnetic field B_0 as eigenvector. The algebraic problem is solved by inverse iteration (see also Gubbins, 1972).

An early attempt to solve this problem by Bullard and Gellman (1954) resulted in failure, since the solution failed to converge when more harmonics were included in the representation (Gibson and Roberts, 1967). Bullard and Gellman's simple flow contained a plane of symmetry which was thought to inhibit dynamo action because of results obtained by Braginsky (1964) in an asymptotic limit of large R_m. Consequently, Lilley (1970) investigated a more complex fluid flow, which seemed to obtain satisfactory convergence but was also found unsatisfactory (Gubbins, 1972). Finally, Kumar and Roberts (1975) introduced an additional component of meridional circulation into the flow and obtained a numerical solution that has withstood further scrutiny by others.

The putative flows are expressed in terms of vector spherical harmonics. We use the form introduced by Bullard and Gellman (1954):

$$T_l^{mc} = \nabla \times \left[t_l^{mc}(r) P_l^m \cos m\phi \hat{r} \right]$$

$$S_l^{mc} = \nabla \times \nabla \times \left[s_l^{mc}(r) P_l^m \cos m\phi \hat{r} \right]$$

(4)

where $\hat{r}$ is the unit vector in the radial direction and P_l^m is an associated Legendre function; two similar equations for T_l^{ms} and S_l^{ms} define the harmonics dependent on $\sin m\phi$. Kumar and Roberts (1975) investigated the flow defined by

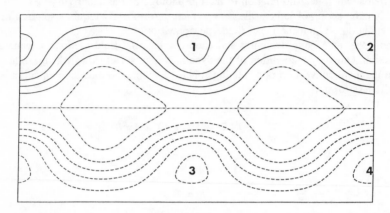

Figure 6: Radial component of the surface magnetic field generated by the Kumar and Roberts dynamo with velocity given by (5) with parameters $n = 3$, $\varepsilon_1 = 0.03$, $\varepsilon_1 = \varepsilon_3 = 0.04$. After Hutcheson (1990).

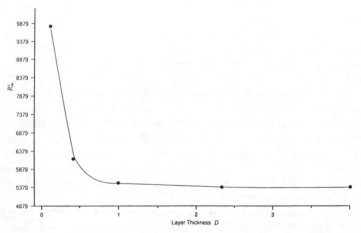

Figure 7: Variation with thickness of the outer stagnant layer, D, of the critical magnetic Reynolds number for the Kumar and Roberts dynamo with parameters as in Figure 6. After Hutcheson (1990).

$$v = T_1^0 + \varepsilon_1 S_2^0 + \varepsilon_2 S_2^{2s} + \varepsilon_3 S_2^{2c} \tag{5}$$

where the defining radial functions are

$$t_1^0(r) = r^2\left(1 - r^2\right)^2$$

$$s_2^0(r) = r^6\left(1 - r^2\right)^3$$

$$s_2^{2s}(r) = r^4\left(1 - r^2\right)^2 \cos(n\pi r)$$

$$s_2^{2c}(r) = r^4\left(1 - r^2\right)^2 \sin(n\pi r)$$

$$\tag{6}$$

where n is an integer and the $\{\varepsilon_i\}$ are variable parameters. In this notation the Lilley dynamo is defined by $\varepsilon_1 = 0$ and the Bullard-Gellman dynamo by $\varepsilon_1 = \varepsilon_2 = 0$. Kumar and Roberts (1975) found dynamo action for a restricted set of values for the $\{\varepsilon_i\}$ and $n = 3$. The integer n gives the number of cells of the motion in radius, and both Gubbins (1973) and Pekeris et al. (1973) had found that several radial cells were necessary for dynamo action by simpler fluid flows. Furthermore, Bullard and Gubbins (1977) argued that the insulating boundary caused concentration of current and enhanced diffusion for some flows, thus inhibiting dynamo action, and showed that a layer of stagnant, conducting fluid surrounding the flow led to dynamo action for single-cell, axisymmetric flows.

Hutcheson (1990) has investigated the Kumar-Roberts dynamo further and shown that downwelling in the outer cell of the flow leads to concentration of flux at the surface (Figure 6). There are two downwellings of flow in the Kumar-Roberts dynamo, leading to four flux concentrations at the surface; upwelling tends to disperse flux and is associated with regions of low flux, rather like the core-surface field beneath the Pacific (Figure 1).

Hutcheson (1990) further investigated the effect of a stagnant layer on a one-cell ($n = 1$) Kumar-Roberts dynamo. The dependence of the critical magnetic Reynolds number, R_m^c, for steady solutions, on thickness of the stagnant layer, D, is shown in Figure 7. The value of R_m^c decreases to an asymptotic value as D increases, suggesting the dynamo operates more efficiently with the stagnant fluid. The surface field becomes smoother as D increases (the "surface" being at $r = 1+D$), the upward continuation for a steady field in a conductor being the same as in an insulator. The surface field becomes arbitrarily close to that of a dipole for sufficiently large D, in contradiction of Hide's (1981) conjecture (Figure 8). Hutcheson (1990) also studied Lilley dynamos (with $\varepsilon_1 = 0$). Dynamo action was obtained for $\varepsilon_2 = \varepsilon_3$ in the range 0.04 to 0.1 and $D > 0.05$, showing that even a small change in boundary condition at the interface can produce a dramatic effect (Figure 9). The surface field is again concentrated by downwelling, as in the Kumar-Roberts dynamo. The radial component of magnetic field is shown for one set of parameters in Figure 10. The maximum flux concentrations are labelled 1-4. Figure 11 shows contours of radial flow in the equatorial plane for this model. The longitudes of the flux lobes are marked with arrows; they lie above the principal areas of downwelling (negative radial velocity).

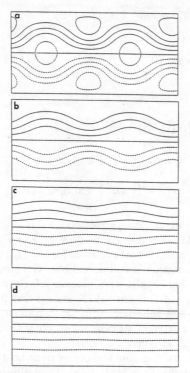

Figure 8: Changes in the surface magnetic field with D for the dynamo of Figure 6. After Hutcheson (1990).

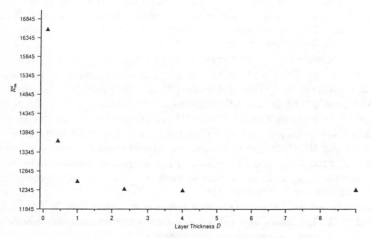

Figure 9: Variation of the critical magnetic Reynolds number for the Lilley dynamo with parameters $n = 1$, $\varepsilon_1 = 0$, $\varepsilon_2 = \varepsilon_3 = 0.04$, $D = 0.429$. After Hutcheson (1990).

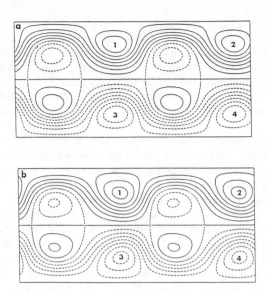

Figure 10: Radial component of field for a Lilley dynamo with $D = 0.429$, (a) $\varepsilon_2 = \varepsilon_3 = 0.4$ and (b) $\varepsilon_2 = \varepsilon_3 = 0.09$. The flux concentrations are marked 1-4.

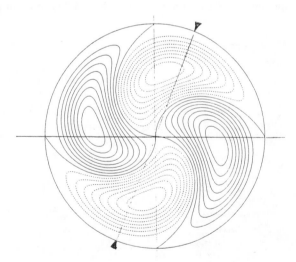

Figure 11: Contours of radial flow for the Lilley dynamo in Figure 10. The arrows mark the longitudes of maximum flux concentrations 1-4 in Figure 10; they are associated with the regions of downwelling.

4. Thermal Core-Mantle Boundary Interactions

It is possible to solve for the temperature by assuming the frozen flux core motions satisfy the entire vorticity equation: thermal winds are driven by buoyancy forces associated with lateral variations in temperature. The temperature models (Bloxham and Jackson, 1990) can satisfy a large amount of recent secular change and can be derived from tangentially geostrophic flows (Bloxham and Jackson, 1990) or, to a good approximation, from toroidal flows (Gubbins, 1991).

Seismic anomalies in the lower mantle are sensitive to temperature in the thermal boundary layer. The core provides a very accurate isothermal bottom boundary for the mantle. Therefore, cold temperatures in the boundary layer, which is a few hundred kilometres thick could imply a high heat flux out of the core, and hot temperatures could imply a low heat flux. In principle we could correlate seismic anomalies with heat flux out of the core if we knew the relationship between core surface temperatures and heat flow. Kohler and Stephenson (1990) have attempted to do the reverse process of calculating fluid motions from seismic anomalies using the thermal wind equation. Bloxham and Gubbins (1987) showed that flows driven in this way would be strong enough to generate secular change but the connection between heat flux and temperature at the core-mantle boundary depends on the governing dynamics. Zhang and Gubbins (1992a) have set up a model problem in order to examine this connection. They study a rotating shell of Boussinesq fluid which is initially either at uniform temperature or density-stratified; they prescribe a laterally varying temperature on the outer spherical surface and uniform temperature on the inner surface. There is no critical state: convection sets in for any value of the horizontal Rayleigh number, however small, and the mathematical problem is inherently non linear, unlike Bénard convection in which the fluid is heated from below and the stability problem is linear.

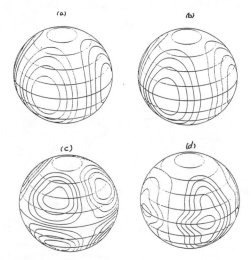

Figure 12: Fluid flow induced in a rotating sphere for applied surface temperature varying as the spherical harmonic Y_2^2. (a) gives the temperature on the boundary, (b) the heat flux, (c) toroidal flow, and (d) the upwelling. From Zhang and Gubbins (1992a).

The fluid flow for an applied temperature with the form of the spherical harmonic Y_2^2 is shown in Figure 12. The induced flow is a thermal wind which penetrates deeply into the spherical shell; fluid flows across cold regions rather than welling down beneath them, as would be expected in a non-rotating system. In this parameter regime the fluid flow increases linearly with horizontal Rayleigh number, in contrast to Bénard convection in which it varies with the square root of the supercritical Rayleigh number. For small values of the horizontal Rayleigh number the response is almost linearly dependent on the applied temperature; it is therefore possible to superimpose responses to different applied temperatures. There is a good correlation between the applied temperature and the heat flux for small horizontal Rayleigh numbers ($R_a \approx 1$) and all the flows are steady. At larger values of the Rayleigh number higher modes are excited, through the advection of heat, and, although the surface flow still satisfies the thermal wind equation, the simple relationship between heat flux and temperature is lost. If the length scale of the applied temperature is very small, the penetration of the induced convection is much less. Density stratification also reduces the penetration but, most noticeably, it suppresses the radial flow. The flow is confined close to the boundary, although not as much as might be expected from a simple comparison of vertical stratification and lateral temperature gradients because of the effect of Coriolis forces. The limit of large stratification seems to entail complete suppression of the radial motion but finite horizontal flow.

The simple correlation between temperature and heat flux and linear dependence on applied temperature means that, if the horizontal Rayleigh number in the core is small, we can apply a direct correlation between seismic anomalies in the lower mantle and core temperature. However, it is much more likely that the horizontal Rayleigh number will be large and that no such simple correlation is possible.

Temperature anomalies on the surface will also strongly influence Bénard convection driven by either heating or compositional variations. Gubbins and Bloxham (1987) and Gubbins and Richards (1986) suggested that core-mantle boundary features may have locked convection rolls responsible for the main magnetic flux lobes. The problem of Bénard convection with imperfect boundaries (i.e. variation in temperature on the outer surface) is much more difficult than simply calculating the response to an applied laterally varying surface temperature, as was done by Zhang and Gubbins (1992a). Convection with perfect boundaries is small in scale, depending on the inverse Taylor number (Roberts, 1968), which presents severe numerical difficulties. On the other hand, the response to a large scale temperature variation on the surface is itself large scale. There are therefore two conflicting driving mechanisms for convection with imperfect boundaries; these produce a very rich dynamics. Preliminary calculations (Zhang and Gubbins, 1992b) suggest that it is possible to obtain a number of interesting phenomena, including stationary convection locked to the boundaries, resonance between boundary forcing and the Bénard convection, time-dependent flow in which drifting solutions slow down as they pass favourable parts of the boundary and speed up when they pass unfavourable parts, and two layer convection with an upper layer driven from the boundary and a lower layer driven by Bénard convection from beneath. All these possibilities may have relevance for the Earth's core and are being explored by numerical calculation.

Acknowledgments. This work was partially supported by NERC grant GR3/7338.

110

5. References

Backus, G. E. (1968) Kinematics of geomagnetic secular variation in a perfectly conducting core. Phil. Trans. R. Soc. Lond., A 263, 239-263.

Backus, G. E. (1988) Bayesian inference in geomagnetism. Geophys. J. 92, 125-142.

Backus, G. E. and LeMouël, J-L. (1986) The region on the core-mantle boundary where a geostrophic velocity field can be determined from frozen-flux magnetic data. Geophys. J. R. Astr. Soc. 85, 617-628.

Bloxham, J. (1986) Models of the magnetic field at the core-mantle boundary. J. Geophys. Res. 91, 13,954-13,966.

Bloxham, J. and Gubbins, D. (1985) The secular variation of the earth's magnetic field. Nature 317, 777-781.

Bloxham, J. and Gubbins, D. (1986) Geomagnetic Field Analysis-IV Testing the frozen-flux hypothesis. J. Geophys. Res. 84, 139-152.

Bloxham, J. and Gubbins, D. (1987) Thermal core-mantle interactions. Nature 325, 511-513.

Bloxham, J. and Jackson, A. (1989) Simultaneous stochastic inversion for geomagnetic main field and secular variation-2 (1820-1980). J. Geophys. Res. 94, 15,753-15,769.

Bloxham, J. and Jackson, A. (1990) Lateral temperature variations at the core-mantle boundary deduced from the magnetic field. Geophys. Res. Lett. 17, 1997-2000.

Bloxham, J. and Jackson, A. (1991) Fluid flow near the surface of earth's outer core. Rev. Geophys. 29, 97-120.

Bloxham, J., Gubbins, D. and Jackson, A. (1989) Geomagnetic Secular Variation. Phil. Trans. R. Soc. Lond. 329, 415-502.

Braginsky, S. I. (1964) Kinematic models of the earth's hydromagnetic dynamo. Geomagn. Aeron. 4, 572-583.

Bullard, E. C. and Gellman, H. (1954) Homogeneous dynamos and terrestrial magnetism. Phil. Trans. Roy. Soc. Lond. 247, 213-278.

Bullard, E. C. and Gubbins, D. (1977) Generation of magnetic fields by fluid motions of global scale. Geophys. Astrophys. Fluid Dyn. 8, 43-56.

Gibson, R. D. and Roberts, P. H. (1967) Some comments on the theory of homogeneous dynamos, in Magnetism and the cosmos 108-120, edited by W. R. Hindmarsh, F. J. Lowes, P. H. Roberts, and S. K. Runcorn, Oliver and Boyd, Edinburgh.

Gubbins, D. (1972) Kinematic Dynamos. Nature Phys. Sci. 238, 119-122.

Gubbins, D. (1973) Numerical solutions of the dynamo problem. Phil. Trans. R. Soc. Lond., A 274, 493-521.

Gubbins, D. (1983) Geomagnetic field analysis I – Stochastic Inversion. Geophys. J. R. Astr. Soc. 73, 641-652.

Gubbins, D. (1991) Dynamics of the Secular Variation. Phys. Earth Planet. Int. 68, 170-182.

Gubbins, D. and Bloxham, J. (1984) Geomagnetic field analysis III – Magnetic fields on the core-mantle boundary. Geophys. J. R. Astr. Soc. 80, 696-713.

Gubbins, D. and Bloxham, J. (1987) Morphology of the geomagnetic field and implications for the geodynamo. Nature 325, 509-511.

Gubbins, D. and Richards, M. (1986) Coupling of the core dynamo and mantle: thermal or topographic? Geophys. Res. Lett. 13, 1521-1524.

Hide, R. (1967) Motions of the earth's core and mantle, and variations of the main geomagnetic field. Science 157, 55-56.

Hide, R. (1981) The magnetic flux linkage of a moving medium: A theorem and geophysical applications. J. Geophys. Res. 86, 11,681-11,687.

Hutcheson, K. (1990) Geomagnetic field modelling. Ph.D. Thesis, University of Cambridge.

Hutcheson, K. and Gubbins, D. (1989) A model of the geomagnetic field for the 17th century. J. Geophys. Res. 95, 10,769-10,781.

Jackson, A. and Bloxham, J. (1991) Mapping the fluid flow and shear near the core surface using the radial and horizontal components of the magnetic field. Geophys. J. 105, 199-212.

Kohler, M. D. and Stevenson, D. J. (1990) Modeling core fluid motions and the drift of magnetic field patterns at the CMB by use of topography obtained by seismic inversion. Geophys. Res. Lett. 17, 1473-1476.

Kumar, S. and Roberts, P. H. (1975) A three dimensional kinematic dynamo. Proc. R. Soc. 344, 235-258.

LeMouël, J-L., Gire, C. and Madden, T. (1985) Motions of the core surface in the geostrophic approximation. Phys. Earth Planet. Int. 39, 270-287.

Li, X. and Jeanloz, R. (1987) Measurement of the electrical conductivity of (Mg, Fe) SiO_3 perovskite and a perovskite-dominated assemblage at lower mantle conditions. Geophys. Res. Lett. 95, 5067-5078.

Lilley, F. E. M. (1970) On kinematic dynamos. Proc. R. Soc. 316, 153-167.

Lloyd, D. and Gubbins, D. (1990) Toroidal motion at the top of the earth's core. Geophys. J. Int. 100, 455-467.

Peddie, N. W. (1982) International Geomagnetic Reference Field: the third generation. J. Geomag. Geoelectr. 34, 309-326.

Pekeris, C. L., Accad, Y. and Shkoller, B., (1973) Kinematic dynamos and the earth's magnetic field. Phil. Trans. Roy. Soc. Lond. 275, 425-461.

Roberts, P. H. (1967) An Introduction to Magnetohydrodynamics. Elsevier, New York, 264 pp.

Roberts, P. H. (1968) On the thermal instability of a rotating fluid sphere containing heat sources. Phil. Trans. Roy. Soc. Lond. 263, 93-117.

Roberts, P. H. and Scott, S. (1965) On the analysis of the secular variation, I, A hydromagnetic constraint: Theory. J. Geomagn. Geoelectr. 17, 137-151.

Shure, L., Parker, R. L. and Backus, G. E. (1982) Harmonic splines for geomagnetic field modelling. Phys. Earth Planet. Interiors 28, 215-229.

Voorhies, C. and Backus, G. E. (1985) Steady flows at the top of the core from geomagnetic field models: the steady motions theorem. Geophys. Astrophys. Fluid Dyn. 34, 451-487.

Zhang, K. and Gubbins, D. (1992a) On convection in the earth's core forced by lateral temperature variations in the lower mantle. Geophys. J. Int. 108, 247-255.

Zhang, K. and Gubbins, D. (1992b) On imperfect convection in a rotating fluid shell at infinite Prandtl number. Submitted, J. Fluid Mech.

ENERGETIC ASPECTS OF THERMAL CONVECTIVE MAGNETOHYDRODYNAMIC DYNAMOS

YU-QING LOU[1]
Geophysical Institute
University of Alaska Fairbanks
Fairbanks, Alaska 99775-0800 USA

ABSTRACT. We examine qualitatively the global energetics of magnetohydrodynamic (MHD) dynamos maintained by thermal convection. A flow field for sustaining a magnetic field needs to be pressured globally to compensate Joule and viscous dissipations in the absence of net Poynting and mechanical energy influxes across boundaries; this pressure field is associated with thermal convective heat flux passing through the system; thus thermal MHD dynamos are prohibited for incompressible flows. A clear distinction is drawn between a fully consistent thermal MHD dynamo and a Boussinesq dynamo. Another possibility for sustaining both compressible and incompressible MHD dynamos is to extract mechanical energy from the core rotation via viscous stress.

1. Introduction

A magnetohydrodynamic (MHD) dynamo maintained by thermal (or radiation-driven) convection is defined as a thermal MHD dynamo in the present context. Since no fully consistent MHD dynamo model has been constructed to mimic the major characteristics of the solar and the Earth's magnetic fields, it is important to be aware of the basic energetic and dynamic properties of MHD dynamos in addition to the other known kinematic, topological, and dynamic constraints (Taylor, 1963; Gubbins, 1977; Moffatt, 1978).

It was recognized from the beginning that the kinematic formulation of the dynamo problem is only a necessary but not sufficient step toward the basic understanding of the origin and maintenance of cosmic magnetic field. The success in demonstrating numerous kinetic dynamos (Herzenberg, 1958; Backus, 1958; Childress, 1970; Roberts, 1970, 1972) and the rapid development of the mean-field or turbulent MHD dynamo formalism (Braginskii, 1964a,b; Steenbeck et al., 1966; Steenbeck and Krause, 1966; Vainshtein and Zel'dovich, 1972) have led to tacit optimism; yet almost no theoretical study has been conducted on the formidable problem of a fully consistent MHD dynamo until recent numerical studies (Busse, 1983; Glatzmaier, 1984, 1985; Jacobs, 1987; DeLuca and Gilman, 1988; Nordlund and Stein, 1989). In this note, we examine mainly thermal MHD dynamos from the perspective of global energetics and dynamics. Valuable insights can be derived from such an analysis (Braginskii, 1964c; Hewitt et al., 1975; Backus, 1975; Gubbins, 1977; Loper, 1978). For example, the simplifying assumption of an incompressible flow in the

---

[1]Now at the Department of Astronomy and Astrophysics, The University of Chicago

D. B. Stone and S. K. Runcorn (eds.),
Flow and Creep in the Solar System: Observations, Modeling and Theory, 113–120.
© *1993 Kluwer Academic Publishers. Printed in the Netherlands.*

kinematic formulation is restrictive for thermal MHD dynamos. Thus for incompressible flows, the possibility of maintaining kinematic dynamo in the framework of mean-field MHD does not imply the feasibility of fully consistent thermal MHD dynamos. For convection-driven MHD dynamos (Childress and Soward, 1972; Busse, 1973, 1975; Soward, 1974, 1980, 1983), the Boussinesq approximation differs from a fully consistent thermal MHD dynamo in terms of the global energetics. These observations raise the further question of whether non-stationary kinematic dynamos, which are possible in the mean-field MHD theory, have any direct bearing on the feasibility of nonlinear, time-dependent thermal MHD dynamos (Gilman and Miller, 1981; Gilman, 1983). Since the related topics in the study of dynamo theory are diverse, no attempt is made here to review all the previous work. For a more complete review of dynamo studies, the reader is referred to Roberts (1971), Gubbins (1974), Moffatt (1978), Parker (1979), Krause and Rädler (1980), Soward (1983), Zel'dovich et al. (1983), Melchior (1986), Jacobs (1987), and Vainshtein and Rosner (1991).

We provide an analysis in Section 2 of the full MHD equations for a thermal MHD dynamo. We indicate the essential difference between a Boussinesq dynamo and a consistent thermal MHD dynamo in Section 3. We summarize and discuss the major results of our analysis in Section 4.

2. Energetics of a Thermal MHD Dynamo

The basis of our analysis is the set of standard MHD equations, including the energy equation in the presence of thermal diffusion, Joule and viscous dissipations. The equation of state is assumed in a general form. The forms of diffusion coefficients are left unspecified for generality. Explicit in our consideration is Poisson's equation which describes the effect of self-gravity due to the system under study. For the Sun, if we limit dynamo actions to the radiative and convective zones, self-gravity is negligible because most of the solar mass is concentrated in the core. In contrast, for the Earth's core, the density of the outer melting core does not differ significantly from that of the inner solid core, and the thickness of the outer core is nearly three halves the radius of the inner core; thus, self-gravity can be significant in terms of the internal dynamics for a geodynamo.

We present the MHD equations in a form for energetic considerations of an MHD dynamo.

$$\frac{\partial \rho}{\partial t} = -\nabla \cdot (\rho \vec{v}), \tag{2.1}$$

$$\frac{\partial}{\partial t}\left(\frac{\vec{B}^2}{8\pi}\right) = -\nabla \cdot [(\eta \nabla \times \vec{B} - \vec{v} \times \vec{B}) \times \vec{B}]/4\pi - \vec{v} \cdot [(\nabla \times \vec{B}) \times \vec{B}]/4\pi - \vec{j}^2/\sigma_e, \tag{2.2}$$

$$\frac{\partial(\rho c_v T)}{\partial t} = -\nabla \cdot (\rho \vec{v} c_v T) - p(\nabla \cdot \vec{v}) + \nabla \cdot (\kappa \nabla T) + \Phi + \vec{j}^2/\sigma_e, \tag{2.3}$$

$$\frac{\partial}{\partial t}\left(\frac{\rho \vec{v}^2}{2}\right) + V_G \frac{\partial \rho}{\partial t} = -\nabla \cdot \left[\rho \vec{v}\left(\frac{\vec{v}^2}{2} + V_G\right)\right] - \vec{v} \cdot \nabla p + \nabla \cdot (\vec{v} \cdot \overleftrightarrow{\sigma})$$
$$- \Phi + \vec{v} \cdot [(\nabla \times \vec{B}) \times \vec{B}]/4\pi, \tag{2.4}$$

$$p = f(\rho, T), \qquad \nabla \cdot \vec{B} = 0, \qquad \nabla \times \vec{B} = 4\pi \vec{j}/c, \tag{2.5a,b,c}$$

$$\overleftrightarrow{\sigma} \equiv \sigma_{ik} = v \left(\frac{\partial v_i}{\partial x_k} + \frac{\partial v_k}{\partial x_i} - \delta_{ik} \frac{\partial v_l}{\partial x_l} \right) + \zeta \delta_{ik} \frac{\partial v_l}{\partial x_l}, \tag{2.6}$$

$$\nabla^2 V_G = 4\pi G \rho, \qquad \Phi = \frac{\partial v_k}{\partial x_i} \sigma_{ik}, \qquad \eta \equiv \frac{c^2}{4\pi\sigma_e}, \tag{2.7a,b,c}$$

where ρ, $\vec{v}$, $\vec{B}$, $\vec{j}$, p, V_G and T are mass density, velocity, magnetic field, electric current density, gas pressure, gravitational potential, and temperature, respectively; η, σ_e, c_v, κ, G, c, v, and ζ are magnetic diffusivity, electrical conductivity, specific heat at constant volume, thermal conductivity, gravitational constant, light speed, first dynamic viscosity, and second dynamic viscosity, respectively; $\overleftrightarrow{\sigma} \equiv \sigma_{ik}$ and δ_{ik} are the viscous stress tensor and the Kronecker δ–function, respectively; Φ and $\vec{j}^2/\sigma_e$ are the heating rates per unit volume due to viscous and Joule dissipations, respectively.

Our analysis will be carried out within a shell volume confined between two spherical surfaces—one at the interface separating the central mass sphere and the conducting fluid, and the other at a larger radius enveloping the entire fluid system. In fact, the specific shapes of the two confining surfaces are not essential as long as the possible fluid magnetism is not directly pumped by external macroscopic electromagnetic sources (radiative transfer excluded), and the mass exchange with the outside is insignificant. We require that the normal component of the Poynting flux vector associated with dynamo effects vanishes at both boundaries, and that velocity relative to the central sphere vanishes. In the absence of viscosity, only the normal component of velocity is required to vanish; the tangential components of velocity are unconstrained. The system undergoes thermal conductive (or radiative) exchange with the surroundings.

It is instructive to first look into the possibility of a steady MHD dynamo for an *inviscid* fluid. The left-hand sides of equations (2.1)–(2.4) are set to zero. By adding equations (2.2)–(2.4) and integrating the resulting equation over the volume, we see that the same amount of thermal energy entering one part of the boundary leaves the volume from the rest of the boundary, i.e.,

$$\int \nabla \cdot (\kappa \nabla T) d^3x = 0. \tag{2.8}$$

The volume integrals of the second and the third terms must balance in equation (2.2), viz., the Lorentz force must extract power from the flow in order to compensate Joule dissipation globally. The volume integral of equation (2.3) implies that the pressure force delivers positive power globally on the flow field to provide the energy source for Joule dissipation. The volume integral of equation (2.4) is a restatement of the above results. To summarize,

$$\int \frac{j^2}{\sigma_e} d^3x = \int p(\nabla \cdot \vec{v}) d^3x = -\int \vec{v} \cdot (\nabla \times \vec{B}) \times \vec{B} \, d^3x. \tag{2.9}$$

We note that Zel'dovich and Ruzmaikin (1983) made a sign error in their equation (2.24) and their following statement was incorrect (see equation [2.2]). We interpret equation (2.9) as follows. For a dynamo flow field, correlation of high pressure with diverging flow dominates correlation of low pressure with converging flow on average. Unless Lorentz force is force-free in high pressure regions, relation (2.9) is consistent with the scenario that magnetic field tends to concentrate in regions of low pressure and converging flow on average to oppose pressure force. As expected, the possibility of a stationary MHD dynamo for compressible flows is permitted from the global energetics point of view. For an incompressible and inviscid flow, a steady MHD dynamo cannot be maintained even in the presence of rotation. It is commonly argued that when magnetic field strength is weak, the back reaction of the Lorentz force can be ignored in the momentum equation. But as is clear in the energetic study, it is just this Lorentz force against which an appropriate flow field with sufficient strength must be maintained. Since exactly the same term appears in both the induction and momentum equations (2.2) and (2.4), there seems no justification to drop it in the momentum equation at any stage of magnetic field amplification or maintenance (Gilman and Miller, 1981; Gilman, 1986).

Similarly, a steady MHD dynamo is impossible for a viscous, incompressible fluid in a nonrotating system, because the middle term of equation (2.9) is now replaced by $\int p(\nabla \cdot \vec{v})d^3x - \int \Phi d^3x$ which would be negative. However for a rotating system with the no-slip boundary condition, a net mechanical power can be delivered into an incompressible flow by viscous shear stress term $\int \nabla \cdot (\vec{v} \cdot \overleftrightarrow{\sigma})d^3x$ The volume integral $\int \nabla \cdot (\kappa \nabla T)d^3x$ of equation (2.3) must now be negative, i.e., the system adjusts itself to lose the same amount of mechanical power via thermal conduction. In this context, the possibility of sustaining a steady MHD dynamo by incompressible flows cannot be excluded provided that the net mechanical power transferred by viscous shear stress due to rotation is sufficient to compensate viscous and Joule dissipations. For example, the precessional geodynamo models rely upon the core-mantle viscous coupling. For pros and cons of the precessional geodynamos, the reader is referred to Malkus (1971) and Loper (1975), respectively. In order to sustain a steady MHD dynamo for viscous flows in a nonrotating system, it is necessary that the global power delivered to a *compressible* flow by pressure force compensates both viscous and Joule dissipations in a thermal convection. For a rotating system, the pressure force and the viscous shear stress can work together to power a steady compressible thermal MHD dynamo.

From equations (2.1), (2.2), (2.4), and equation (2.7a), we obtain

$$\frac{\partial}{\partial t}\int (\frac{\vec{B}^2}{8\pi} + \frac{\rho\vec{v}^2}{2} + \frac{\rho V_S}{2} + \rho V_E)d^3x = -\int \frac{\vec{j}^2}{\sigma_e}d^3x - \int \Phi d^3x + \int p(\nabla \cdot \vec{v})d^3x$$

$$-\int (\eta\nabla \times \vec{B} - \vec{v} \times \vec{B}) \times \vec{B}/4\pi \cdot d\vec{S} - \int \rho\vec{v}\,(\frac{\vec{v}^2}{2} + \frac{p}{\rho} + V_S + V_E) \cdot d\vec{S} + \int \vec{v} \cdot \overleftrightarrow{\sigma}\, d\vec{S},$$

(2.10)

where V_S and V_E are the gravitational potentials produced by the fluid mass within the volume and by external masses, respectively; $V_G = V_S + V_E$. According to the specified boundary conditions, the fourth and the fifth terms on the right-hand side of equation (2.10) vanish. Let us suppose further that $\int \vec{v} \cdot \overleftrightarrow{\sigma} \cdot d\vec{S} = 0$, which can correspond to inviscid flows or no core rotation or the stress-free boundary condition. It is then clear that the time rate of change of the sum of the total magnetic, kinetic, and gravitational energies is equal to the total power delivered by a pressure force to a flow against Joule and viscous dissipations. For incompressible flows, this sum will decay even if internal radioactive sources are present; independently, the total internal energy will grow or decay depending on the presence of net thermal influx or sufficiently rapid heat efflux, respectively (see equation [2.3]). Thus, *it is the condition of incompressibility which cuts off the thermal connection of equation (2.10) with the energy equation (2.3).* The above conclusion is a dynamic generalization of Cowling's theorem (1934) for three-dimensional incompressible flows and is reached by the global energetic consideration.

3. Limitations of a Boussinesq Dynamo

The basic assumptions for Boussinesq convection are as follows: (i) the mass density variation is small and the equation of state is assumed in the form of $\rho = \rho_0[1 - \alpha(T - T_0)]$, where α is the coefficient of volume expansion and T_0 is the temperature at which $\rho = \rho_0$; (ii) $\nabla \cdot \vec{v} = 0$; (iii) the density variation is ignored except when coupled with gravity; self-gravity is usually ignored. When the assumption (ii) is made, the continuity equation (2.1) reduces to

$$\frac{\partial \rho}{\partial t} = \vec{v} \cdot \nabla \rho = 0. \tag{3.1}$$

Although it is possible to solve for the perturbations in velocity, density, and temperature from the standard equations in the Boussinesq framework, a physical solution should not violate mass conservation (3.1). We may write $\rho = \rho_0 + \rho_1$ where ρ_0 is independent of time and may be dependent on spatial coordinate $\vec{r}$, and $\rho_1(\vec{r},t)$ is a small density deviation and we consider an initial static equilibrium described by $-\nabla p_0 - \rho_0 \nabla V_G = 0$. The work done by gravity per unit volume per unit time on a flow is

$$-\rho_1 \vec{v} \cdot \nabla V_G = -\nabla \cdot (\rho V_G \vec{v} + p_0 \vec{v}) - V_G \frac{\partial \rho_1}{\partial t}. \tag{3.2a}$$

where the background equilibrium and the condition (3.1) are used. The volume integral of equation (3.2a) and the boundary condition that normal flow vanishes on the boundaries indicate that the total power delivered by gravity to a flow is equal to the decreasing time rate of the total gravitational potential energy which vanishes for a steady state. However,

$$-\rho_1 \vec{v} \cdot \nabla V_G = \rho_0 \alpha T_1 \vec{v} \cdot \nabla V_G \tag{3.2b}$$

in the Boussinesq approximation, where $T_1(\vec{r},t)$ is the temperature deviation. The right-hand side of equation (3.2b) does not vanish for a steady state. It is apparent that a Boussinesq dynamo contradicts the mass conservation. In other words, by ignoring the mass conservation, the total work done on a flow by gravity per unit time appears as an explicit power source which nevertheless should only be regarded as the decreasing time rate of the total gravitational potential energy in a general formalism of thermal convection (see equation [2.10]). Therefore, Boussinesq convection and thus a Boussinesq dynamo are peculiar in terms of energetics, whereas for a fully consistent thermal MHD dynamo, the fluid compressibility must be retained and the total power is derived from thermal exchange with the surroundings.

In contrast to a thermal MHD dynamo, the release of gravitational potential energy plays an active role in the context of a compositional convection for driving a geodynamo (Braginskii, 1964c; Gubbins, 1977; Loper, 1978; Loper and Roberts, 1983). For incompressible flows and the simplified boundary conditions, the sum of kinetic, magnetic, and gravitational potential energies decreases with time. However, it may be possible that the decreasing time rate of the gravitational potential energy exceeds the total dissipation rates to generate compositional convection and MHD dynamo actions. Therefore, the basic premise for incompressible compositional dynamo is that the initial storage of gravitational potential energy is sufficiently large at the formation of the Earth's outer core and a significant fraction of that storage is available for convection in the process of forming the inner core via crystallization over long geological time scales.

4. Summary and Discussion

Our analysis is summarized as follows. Firstly, in the absence of a Poynting influx across the boundaries, a flow field with sufficient strength must be maintained by some agent against the Lorentz force globally to compensate the Joule dissipation for a self-sustained MHD dynamo. Secondly, without an input of mechanical power, the only possible means to sustain a thermal MHD dynamo is that the pressure force delivers positive power to a flow globally. Consequently, for incompressible flows, a self-sustained thermal MHD dynamo is impossible. Thirdly, the core rotation coupled with shear viscosity might be a viable mechanism to drive a steady MHD dynamo. For a stationary dynamo, the total mechanical energy gained must be dissipated and thermally conducted out of the fluid system simultaneously. For incompressible flows in a rotating system, the mechanical power input might be able to sustain an MHD dynamo, solely from the energetic perspective, by compensating Joule and viscous dissipations. For compressible flows in a rotating system, the possibility of both stationary and nonsteady MHD dynamos remains. Finally, we made these points by invoking specific sets of boundary conditions which are more pertinent to a thermal MHD geodynamo; variations of these boundary conditions are allowed so long as the influence of external macroscopic electromagnetic sources is excluded. For the Sun, the formulation of the relevant boundary conditions is not as straightforward. However, energy in various forms is likely lost across "boundaries." Thus, our conclusions, derived from the energetic perspective, are necessary but far from sufficient for a thermal MHD dynamo, i.e., the specific dynamic feasibility remains to be examined.

The condition of incompressibility excludes the possibility of a pressure field for supplying positive power to a flow field globally in order to maintain a thermal MHD dynamo against Joule and viscous dissipations. In the Boussinesq approximations, the decreasing time rate of the gravitational potential energy is regarded as an explicit power source by ignoring the mass conservation and by assuming a special form of thermal coupling. Most kinematic dynamos assume incompressible flows (Parker, 1955; Braginskii, 1964b; Childress, 1970; Roberts, 1971; Roberts, 1970, 1972); in particular, the formal theory of the mean-field MHD developed by Steenbeck et al. (1966; see translations by Roberts and Stix, 1971, and Krause and Rädler, 1980) is frequently applied to incompressible turbulent dynamos (Kraichnan and Nagarajan, 1967; Roberts, 1971; Rädler, 1983; Molchanov et al., 1984). We have to resort to other processes (e.g., random forcing, Meneguzzi et al., 1981; Cattaneo and Vainshtein, 1991) than thermal convection to maintain these kinds of dynamos. As to the convection-driven dynamo in the Boussinesq framework (Childress and Soward, 1972; Busse, 1973, 1975; Soward, 1974, 1980; Gilman and Miller, 1981; Gilman, 1983), the important difference (in contrast to a consistent thermal MHD dynamo) in terms of energetics must be kept in mind. Since time-dependent and axisymmetric dynamos for compressible flows have been shown to be impossible (Hide and Palmer, 1982; Lortz and Meyer-Spasche, 1982; Busse, 1983), it seems the final verdict for the feasibility of a consistent dynamo hinges upon the explicit construction of a three-dimensional, compressible, thermal MHD dynamo model with sensible boundary conditions.

Acknowledgments. I thank R. Rosner for stimulating discussions and critical comments on the manuscript and thank the anonymous reviewer for suggestions on its presentation. This research was supported by the State Funding of Alaska, NSF grant ATM-9014888 at the University of Alaska Fairbanks, and NSF grant ASTR/ATM-9196237 at the University of Chicago.

5. References

Backus, G. E. (1958) Ann. of Phys. 4, 372.
Backus, G. E. (1975) Proc. Nat. Acad. Sci. 72, 1555.
Braginskii, S. I. (1964a) Sov. Phys. JETP 20, 726.
Braginskii, S. I. (1964b) Sov. Phys. JETP 20, 1462.
Braginskii, S. I. (1964c) Geomagn. Aeron. 4, 698.
Busse, F. H. (1973) J. Fluid Mech. 57, 529.
Busse, F. H. (1975) Geophys. J. Roy. Astron. Soc. 42, 437.
Busse, F. H. (1983) Annu. Rev. Earth Planet. Sci. 11, 241.
Cattaneo, F. and Vainshtein, S. I. (1991) Ap. J. Lett. 376, L21.
Childress, S. (1970) J. Math. Phys. 11, 3063.
Childress, S. and Soward, A. M. (1972) Phys. Rev. Lett. 29, 837.
Cowling, T. G. (1934) M. N. R. A. S. 94, 39.
DeLuca, E. E. and Gilman, P. A. (1988) Geophys. Astrophys. Fluid Dyn. 43, 119.
Gilman, P. A. (1983) Ap. J. Supp. 53, 243.
Gilman, P. A. (1986) in Physics of the Sun, eds. P. A. Sturrock, T. E. Holzer, D. Mihalas and R. K. Ulrich, vol. I. Dordrecht: Reidel, p. 95.
Gilman, P. A. and Miller, J. (1981) Ap. J. Supp. 46, 211.
Glatzmaier, G. A. (1984) J. Comp. Phys. 55, 461.
Glatzmaier, G. A. (1985) Ap. J. 291, 300.
Gubbins, D. (1974) Rev. Geophys. and Space Phys. 12, 137.
Gubbins, D. (1977) J. Geophys. 43, 453.
Herzenberg, A. (1958) Phil. Trans. Roy. Soc. 250, 543.
Hewitt, J. M., McKenzie, D. P. and Weiss, N. O. (1975) J. Fluid Mech. 68, 721.
Hide, R. and Palmer, T. N. (1982) Geophys. Astrophys. Fluid Dyn. 19, 301.
Jacobs, J. A. (1987) The Earth's Core. London: Academic Press.
Kraichnan, R. H. and Nagarajan, S. (1967) Phys. Fluids 10, 859.
Krause, F. and Rädler, K.-H. (1980) Mean-Field Magnetohydrodynamics and Dynamo Theory. Oxford: Pergamon Press.
Loper, D. E. (1975) Phys. Earth Planet. Inter. 11, 43.
Loper, D. E. (1978) Geophys. J. R. Astr. Soc. 54, 389.
Loper, D. E. and Roberts, P. H. (1983) in Stellar and Planetary Magnetism, ed. A. M. Soward. New York: Gordon and Breach, p. 297.
Lortz, D. and Meyer-Spasche, R. (1982) Math. Meth. Appl. Sci. 4, 91.

Malkus, W. V. R. (1971) in Mathematical Problems in the Geophysical Sciences, ed. W. H. Reid. American Mathematical Society, Providence, RI, p. 207.

Melchior, P. J. (1986) The Physics of the Earth's Core. Oxford: Pergamon Press.

Meneguzzi, M., Frisch, U. and Pouquet, A. (1981) Phys. Rev. Lett. 47, 1060.

Mochanov, S. A., Ruzmaikin, A. A. and Sokoloff, D. D. (1984) Geophys. Astrophys. Fluid Dyn. 30, 242.

Moffatt, H. K. (1978) Magnetic Field Generation in Electrically Conducting Fluids. Cambridge: Cambridge University Press.

Nordlund, Å. and Stein, R. F. (1989) in Solar and Stellar Granulation, eds. R. J. Rutten and G. Severino. Dordrecht: Klewer, p. 453.

Parker, E. N. (1955) Ap. J. 122, 293.

Parker, E. N. (1979) Cosmical Magnetic Fields. Oxford: Clarendon Press.

Rädler, K.-H. (1983) in Stellar and Planetary Magnetism, ed. A. M. Soward. New York: Gordon and Breach.

Roberts, G. O. (1970) Phil. Trans. R. Soc. Lond. A266, 535.

Roberts, G. O. (1972) Phil. Trans. R. Soc. Lond. A271, 411.

Roberts, P. H. (1971) in Mathematical Problems in the Geophysical Sciences, ed. W. H. Reid. American Mathematical Society, Providence, RI, p. 129.

Roberts, P. H. and Stix, M. (1971) The Turbulent Dynamo: a translation of a series of papers by F. Krause, K. H. Rädler and M. Steenbeck. NCAR Tech. Note 1A 60, Boulder, CO.

Ruzmaikin, A. A. (1985) Solar Phys. 100, 125.

Soward, A. M. (1974) Phil. Trans. R. Soc. Lond. A275, 611.

Soward, A. M. (1980) Phys. Earth Planet. Interior 20, 134.

Soward, A. M. (1983) Stellar and Planetary Magnetism. New York: Gordon and Breach.

Steenbeck, M. and Krause, F. (1966) Z. Naturforsch. 21a, 1285.

Steenbeck, M., Krause, F. and Rädler, K.-H. (1966) Z. Naturforsch 21a, 369.

Taylor, J. B. (1963) Proc. R. Soc. Lond. A274, 274.

Vainshtein, S. I. and Rosner, R. (1991) Ap. J., 376, 199.

Vainshtein, S. I. and Zel'dovich, Y. B. (1972) Soviet Phys. 15, 159.

Zel'dovich, Y. B. and Ruzmaikin, A. A. (1983) in Astrophysics and Space Physics Reviews, Vol. 2, ed. R. A. Syunyaev. Amsterdam: OPA, p. 333.

Zel'dovich, Y. B., Ruzmaikin, A. A. and Sokoloff, D. D. (1983) Magnetic Fields in Astrophysics. New York: Gordon and Breach.

PREFERRED BANDS OF LONGITUDE FOR GEOMAGNETIC REVERSAL VGP PATHS: IMPLICATIONS FOR REVERSAL MECHANISMS

C. LAJ[1], A. MAZAUD[1], M. FULLER[2], E. HERRERO-BERVERA[3]

1- Centre des Faibles Radioactivités, Laboratoire Mixte CEA-CNRS, 91198
   Gif-sur-Yvette Cedex, France
2- Department of Geological Sciences, University of California Santa Barbara,
   Santa Barbara CA 93106, USA
3- Hawaii Institute of Geophysics, University of Hawaii, USA

ABSTRACT. Different analysis of new records of recent polarity reversals and of many records published in the last years show a remarkable confinement of the transitional VGP's to the longitudinal sector over the Americas and, to a lesser extent, over its antipode. This structure has persisted in time for at least 11 million years. The two longitudinal sectors defined by the VGP's are of particular significance in reconstructions of the present day geomagnetic field and of fluid motions in the surface layers of the core. They are also found to coincide with anomalously fast regions in the mantle and at the core-mantle boundary documented by seismic tomography. Therefore the observed VGP paths link features of the core fluid motion to the temperature in the mantle, fluid motions in the core being driven by temperature patterns at the core mantle boundary during reversals.

1. Introduction

New paleomagnetic records of the transitional fields during geomagnetic reversals are bringing about a radical revision of ideas on the possible morphologies of these fields. More than a decade ago, comparison of the paleomagnetic records of the Brunhes-Matuyama reversal from Japan and the Western USA appeared to indicate that the field could not be dominantly dipolar (Hillhouse and Cox, 1976). Later, it appeared that the Brunhes-Matuyama records could best be explained by axisymmetric fields of low order zonal harmonics (Hoffman, 1977). Zonal models were developed (Hoffman and Fuller, 1978) and records were found consistent with the model. Subsequently, however, more and more non-axisymmetric records emerged, showing in some cases the possibility of a memory of the geodynamo from one reversal to the next. (Valet et al., 1988; Laj et al., 1988).

A new compilation of published geomagnetic reversal studies demonstrate a surprising aspect of the transitional fields, namely that there is a remarkable preponderance of Virtual Geomagnetic Poles (VGP) paths over the longitude of the Americas and its antipode for the last 10 million years. In this paper we show that these new data may reflect key processes in the generation of the earth's magnetic field.

D. B. Stone and S. K. Runcorn (eds.),
Flow and Creep in the Solar System: Observations, Modeling and Theory, 121–129.
© 1993 Kluwer Academic Publishers. Printed in the Netherlands.

122

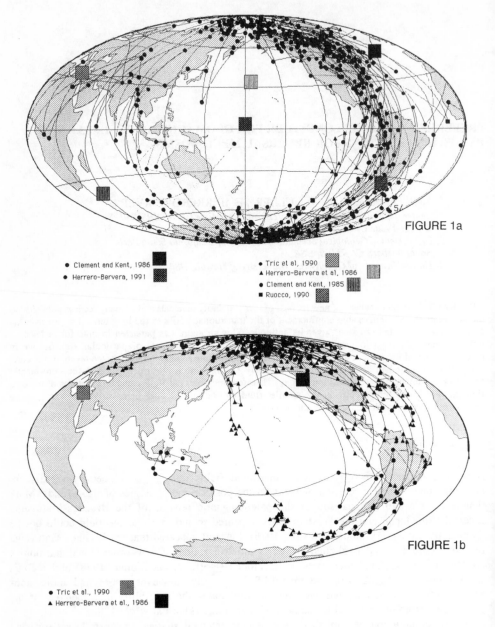

FIGURE 1a

● Clement and Kent, 1986
● Herrero-Bervera, 1991

● Tric et al, 1990
▲ Herrero-Bervera et al, 1986
● Clement and Kent, 1985
■ Ruocco, 1990

FIGURE 1b

● Tric et al., 1990
▲ Herrero-Bervera et al., 1986

Figure 1. a) Reversal paths obtained by our group and several authors for the Upper Olduvai polarity reversal, b) reversal paths obtained by our group and by Herrero-Bervera et al. for the Blake event.

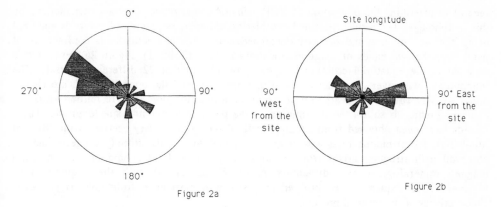

Figure 2a

Figure 2b

Figure 2. a) Longitudinal distribution of the mean equatorial crossing point of 53 VGP paths of 22 different reversals or events of age < 12 My reported by different authors from different sites (see Table 1). b) longitudinal distribution of the paths relative to the sites.

2. Existence of Preferred Longitudinal Bands for Transitonal VGP

2.1. NEW DATA

The new paleomagnetic data which gave us the idea that it was necessary to reconsider the geometry of the transitional fields during reversals and events are: a recent record of the same reversal from high sedimentation rate marine sediments from northern Italy (Tric *et al.*, 1991a), and a recent detailed record of the Blake event from a high sedimentation marine core in the eastern Mediterranean (Tric *et al.*, 1991b). The VGP paths of these records are shown in Figure la for the Upper Olduvai transition and lb for the Blake event, together with other records obtained by several authors. These are a new record of the Upper Olduvai polarity transition obtained by E. Herrero-Bervera from equatorial Pacific sediments, four records of the same reversal from the northern Atlantic (Clement and Kent, 1986), the southern Indian Ocean (Clement and Kent, 1985), and the mid northern Pacific (Herrero-Bervera and Theyer, 1986), and from continental deposits in Argentina (Ruocco, 1990), and finally a record from a lacustrine section from Oregon which, although not precisely dated, is probably the Blake event (Herrero-Bervera *et al.*, 1989). It appears clearly that the VGPs are preferentially confined on a longitudinal band over the Americas and to a lesser extent its antipode.

2.2. COMPILATION OF PUBLISHED DATA

Since this confinement is also observed in others' studies of reversals, in particular of Middle and Upper Miocene (12 and 6.5 my) successive reversals from Zackinthos and Crete respectively (Valet *et al.*, 1988; Laj *et al.*, 1988), we have examined the different published records to determine the proportion of VGP paths of geomagnetic reversals characterized by this morphology. The results, obtained considering sedimentary records of reversals analyzed using discrete sampling and stepwise demagnetizations, reveal a remarkable confinement of paths over the two bands of longitude mentioned above (Table 1). Figure 2a shows a rose diagram of the equatorial crossing longitudes of 53 paths of 22 different reversals. The longitudinal band over the Americas, and to a lesser extend the opposite band, are clearly preferred. This situation appears to have persisted in time for more than 10 million years. By contrast, Figure 2b shows the distribution of the paths in a common site longitude. These records have been obtained from geographically distant sedimentary sections with different lithologies, sedimentation rates and magnetic mineralogies. This strongly suggests that the observed path structure arises from genuine geomagnetic field behaviour and not from magneto-mineralogical or sedimentary processes associated with the acquisition of remanence. Consequently the observed path structure must arise from non-axisymmetric characteristics of the reversal process.

3. Transitional Field Morphology

When only our records for the Upper Olduvai or the Blake event are considered, it is tempting to return to the suggestion of dipole dominated transition field (Creer and Ispir, 1970). Transition fields are, however, likely to be more complicated than that. For example, the multiple records of the Upper Olduvai reveal not only similarities with our results, but also differences (Clement and Kent, 1985; Clement and Kent, 1986). Also, for the Brunhes-Matuyama transition a non-zonal term $h3^1$ appears to account for several features of the transitional field (Clement, 1991), although the same preferred longitudinal bands are involved. Clearly, it is still somewhat premature to describe precise and detailed field morphology; many more records are needed. The main fact is the emergence of the two preferred longitudinal bands for the VGP path of recent (<12 my) geomagnetic reversals. This is a new global observation which has some bearing on the possible mechanisms of the reversal process.

First, these results are clearly inconsistent with axially symmetric transition fields. Models predicting axially symmetric fields and site dependence of the transitional VGP paths thus appear inadequate for a correct description of the reversal process. Curiously, the two key records which demonstrated that the transitional field during the Brunhes-Matuyama reversal could not be dipolar (Hillhouse and Cox, 1976) are consistent with the zonal model and also confined to the two preferred longitudinal bands. These results are also inconsistent with pure statistical models for geomagnetic reversals. Recently two of these models have been proposed. In one case (Mazaud and Laj, 1989) a system of interacting point dipoles, each characterized by a finite lifetime, is used to produce Monte Carlo simulation of the morphology and temporal distribution of reversals. In the other model developed by Constable (1990), reversals are obtained in a very elegant way by allowing the dipole to decay and grow back in an opposite direction while letting the non-dipole part, described using a Gaussian statistics consistent with the characteristics of the present field (Constable and Parker, 1998), be as usual. In these models there is no requirement for the VGP's to be confined to particular meridional regions, the paths are determined only by the field configuration at the onset of the reversal and vary for different reversals and for any given reversal from site to site, contrary to the observations discussed here.

Table 1.

Authors	Reversal	Long. path	References
Herrero-Bervera et al.	B Ev	300/180	Phys. Earth. Planet. Int. 56, 112, 1989
Tric et al.	B Ev	120/280	Earth Planet. Sci. Lett., 102, 1-13, 1991
Clement and Kent	B-M	280	DSDP Init Reports 94, 831-852, 1986
Clement et al.	B-M	250/150	Ph. Trans R. Soc. Lond. A306, 113, 1982
Clement et al.	B-M	170	Ph. Trans R. Soc. Lond. A306, 113, 1982
Valet et al.	B-M	210	Earth Planet. Sci. Lett. 87, 436, 1989
Valet et al.	B-M	75	Earth Planet. Sci. Lett. 94, 371, 1989
Theyer et al.	B-M	140	J. Geophys. Res. 90, 1963-1980, 1989
Clement and Kent	B-M	290	Geophys. Res. Lett. 18, 81-84, 1991
Koci and Sibrav	B-M	320	Quat Glac. in N. Hemisp., Pragua, 1976
Niitsuma	B-M	120	N. Tohuku Univ. Sci. Rept 2, 43, 1, 1971
Herrero-Bervera and Thayer	Up Jar	285	Nature 322, 161, 1986
Herrero-Bervera and Thayer	Lo Jar	270	Nature 322, 161, 1986
Clement and Kent	Lo Jar	160	DSDP Init Reports 94, 831-852, 1986
Clement and Kent	Lo Jar	210	J. Geophys. Res. 89, 1049, 1984
Gurarii	Lo Jar	300	J. Geomag. Geoelec. 28, 295-307, 1976
Ruocco	Lo Jar	290	Ruocco Thesis Univ. Stockholm, 1990
Clement and Kent	Up Od	30/330	Phys. Earth Plan Int. 39, 301-313, 1985
Herrero-Bervera and Theyer	Up Od	285	Nature 322, 161, 1986
Tric et al.	Up Od	280	Phys. Earth Plan Int. 65, 319-336, 1991
Clement and Kent	Up Od	80/275	DSDP Init Reports 94, 831-852, 1986
Ruocco	Up Od	290	Ruocco Thesis Univ. Stockholm, 1990
Herrero-Bervera and Thayer	Lo Od	230	Nature 322, 161, 1986
Herrero-Bervera et al.	Lo Od	270	Phys. Earth Plan. Int. 49, 325, 1987
Liddicoat	G-M	320	Ph. Tr R. Soc. Lond. A306, 121-128, 1982
Gurarii	G-M	120	J. Geomag Geoelect 28, 295-307, 1976
Ruocco	G-M	290	Ruocco Thesis Univ. Stocholm, 1990
Linssen	Up Nu	300	Linssen Thesis Univ. Utrecht, 1991
Linssen	Lo Si	80/300	Linssen Thesis Univ. Utrecht, 1991
Linssen	Up Th	90/280	Linssen Thesis Univ. Utrecht, 1991
Linssen	Up Ma	210	Linssen Thesis Univ. Utrecht, 1991
Gurarii	G-K	260	J. Geomag. Geoelect 28, 295-307, 1976
Gurarii	Mio	30	J. Geomag. Geoelect 28, 295-307, 1976
Clement and Kent	Up Co	200	Earth Planet. Sci. Lett. 81, 253, 1986
Clement and Kent	Lo Co	330	Earth Planet. Sci. Lett. 81, 253, 1986
Clement and Kent	Gilsa	0/310	Earth Planet. Sci. Lett. 81, 253, 1986
Clement	Cobb	0/180	EOS
Valet and Laj	KP102	180	J. Geophys. Res. 93, 1131, 1988
Valet and Laj	KS 06	290	J. Geophys. Res. 93, 1131, 1988
Valet and Laj	KS 05	110	J. Geophys. Res. 93, 1131, 1988
Valet and Laj	KS 02	300	J. Geophys. Res. 93, 1131, 1988
Laj et al.	ZM 01	290	J. Geophys. Res. 93, 11655, 1988
Laj et al.	ZM 02	120	J. Geophys. Res. 93, 11655, 1988
Laj et al.	ZM 03	140	J. Geophys. Res. 93, 11655, 1988

*B Ev: Blake Event; B-M: Brunhes-Matuyama; Up Jar: upper Jaramillo; Lo Jar: lower Jaramillo; Up Od: upper Olduvai; Lo Od: lower Olduvai; G-M: Gauss Matuyama; Up Nu: upper Nunivak; Lo Si: lower Sidufjall; Up th: upper Thvera; Up Ma: upper Mammoth; G-K Gauss Kaena; Mio: Miocene; Up Co: upper Cobb; Lo Co: lower Cobb

This gives a new insight to the behaviour of the non-dipole field during reversals. It is now well established that the intensity of the transitional field is reduced to about 10-15% of the non transitional value. Consequently, if the non-dipole field retained its present day non-transitional characteristics, as assumed in Constable's model, one would expect the direction of the geomagnetic field to vary in an erratic way during the transitions. Longitudinal confinement of paths frequently observed in sedimentary studies has often been ascribed to smoothing by magnetization acquisition processes. However, even assuming strong smoothing, it is not clear how clustering of different paths about a preferred longitude could be generated, if the nondipole field remains undiminished during reversals. Thus the hypothesis of a non-dipole field which would persist unchanged and identical to the present day field during reversals is not sustained by these data. However, the data do suggest that there is a fundamental link between transitional and nontransitional features of the geomagnetic field.

4. VGP Paths, Outer Core Fluid Flow, and Thermic Structure of the Lower Mantle

These two bands of longitude have particular significance in models of the present magnetic field on the Core Mantle Boundary (CMB). For instance, the main patches of the radial field which appear to be standing features of the historical field calculated at the CMB by Bloxham and Gubbins (Gubbins and Bloxham, 1987) lie on longitudinal bands defined by the transitional VGP's. In the hypothesis of a significant dipolar component these bands correspond to sectors where the radial field is large during reversals. These two bands of longitude also appear to be those along which north-south flow is predominantly seen in the models (Bloxham and Jackson, 1991) (Figure 3a).

The observed confinement of transitional VGP paths has implications for interactions between the core and the mantle. The MHD equations governing the geodynamo give rise to a near axisymmetric field, which averaged over time scales of thousands of years closely approximates the rotation axis of the earth. The very asymmetric pattern of reversal paths observed for the last 12 million years must therefore arise from boundary conditions which have persisted over this interval of time. In our opinion, a most plausible cause for these non uniform boundary conditions is the presence of thermal anomalies at the base of the mantle. Indeed, the VGP paths appear to coincide with anomalously rapid (cold) regions in the lower mantle documented by seismic tomography (Figure 3b) (Dziewonski and Woodhouse, 1987; Giardini et al., 1987; Morelli and Dziewonski, 1987), so that the geometry of the transitional field, a matter of discussion among paleomagnetists over the last 15 years, appears to be to a large extent determined by the thermal structure of the mantle. Because the time constants of mantle convection are orders of magnitude longer than those in the core, this persistence of two preferred bands may constitute experimental evidence that the fluid motion in the core, and consequently the geodynamo, is driven by temperature patterns at the core mantle boundary established by mantle convection, as recently suggested by Bloxham and Gubbins from present day and historical magnetic observations (Bloxham and Gubbins, 1987). This is the reason for the persistence of the two preferred VGP paths for over 10 million years, which is clearly too long to be determined by core phenomena alone. The thermal structure at the base of the mantle appears to control the geodynamo during both transitional and non-transitional periods.

It also appears (Olson *et al.,* 1990) that the distribution of anomalously rapid regions in the lower mantle correlates with the downward continuation of subducted slabs in the upper mantle below convergent plate margins. Thus the heat engine of the earth manifests itself in the related internal motions of the planet from the core to the lithosphere, and geomagnetic phenomena in the core appear to be related to plate tectonics, although the nature of the link is still a question of debate.

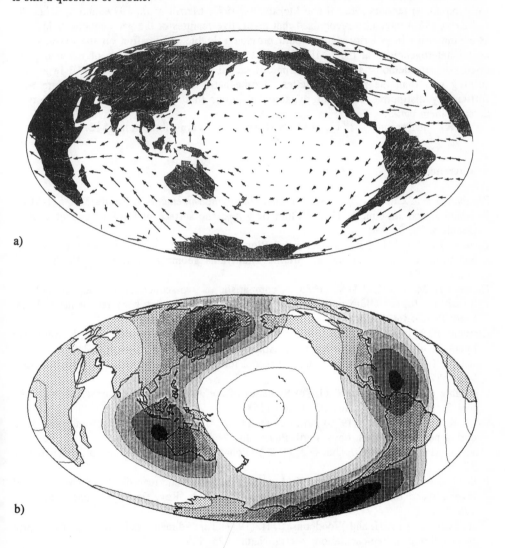

Figure 3. a) Fluid motion at the top of the core (adapted from Bloxham and Jackson, 1991), and b) high velocity zones calculated by Giardini et al. (1987) at a depth of 2300 km in the mantle (adapted from Dziewonski and Woodhouse (1987)).

5. Conclusion

We do not yet know what the actual connection between the pattern of the VGP paths and the field generating mechanism is. Nevertheless, the paths are drawing attention to lateral changes at the core-mantle boundary which play a particular role in the reversal process. Such changes are similar to those proposed by Gubbins (1988) to explain the historical secular variation. Other authors (Merrill and McElhinny, 1977; Merrill et al., 1979a and b; Schneider and Kent, 1988) have also recognized that distinctive features of the geomagnetic field, such as asymmetries in polarity states, could provide valuable information on the existence of lateral heterogeneities at the CMB. We think, however, that the pattern of reversal paths discussed here demonstrates the existence of lateral heterogeneities at the CMB in a more striking way than the previous studies. Geomagnetic field reversals may therefore be a sufficiently distinctive aspect of dynamo behaviour that they warrant a major effort to describe them better.

CFR contribution no. 1317; DBT INSU Instabilités contribution no. 512.

6. References

Bloxham J. and Gubbins D. (1987) Thermal core-mantle interactions. Nature 325, 511-513.

Bloxham J. and Jackson A. (1991) Fluid flow near the surface of Earth's outer core. Rev. Geophys. 29, 97-120.

Clement B. M. (1991) Geographical distribution of transitional VGPs: evidence for non-zonal equatorial symmetry during the Brunhes-Matuyama geomagnetic reversal. Earth Planet. Sci. Lett. 104, 48-58.

Clement B. M. and Kent D.V. (1985) A comparison of two sequential geomagnetic polarity transitions (upper Olduvai and lower Jaramillo) from the Southern Hemisphere. Phys. Earth Planet. Int. 39, 301-313.

Clement B. M. and Kent D.V. (1986) Geomagnetic polarity transition records from five hydraulic piston core sites in the north Atlantic. DSDP Init Reports 94, 831-852.

Constable C. A. (1990) Simple statistical model for geomagnetic reversals. J. Geophys. Res. 95, 4587-4596.

Constable C. and Parker L. R. (1988) Statistics of the geomagnetic secular variation for the past 5 m.y. J. Geophys. Res. 93, 11569-11581.

Creer K. M. and Ispir Y. (1970) An interpretation of the behaviour of the geomagnetic field during polarity reversal. Phys. Earth Planet. Int. 2, 283-293.

Dziewonski A. M. and Woodhouse J. (1987) Global images of Earth's interior. Science 236, 37-48.

Fuller M., Williams I. and Hoffman K. A. (1979) Paleomagnetic records of geomagnetic field reversals and the morphology of the transitional fields. Rev. Geophys. Space Phys. 17, 179-203.

Giardini X. D., Li J. H. and Woodhouse J. H. (1987) Three-dimensional structure of the Earth from splitting in free-oscillation spectra. Nature 325, 405.

Gubbins D. (1988) Thermal core-mantle interactions and time averaged paleomagnetic field. J. Geophys. Res. 93, 3416-3420.

Gubbins D. and Bloxham J. (1987) Morphology of the geomagnetic field and implications for the geodynamo. Nature 325, 510-511.

Herrero-Bervera E. and Theyer, F. (1986) Non-axisymmetric behaviour of Olduvai and Jaramillo polarity transitions recorded in Northcentral Pacific deep sea sediments. Nature 322, 159-162.

Herrero-Bervera E., Helshey C. E., Hammond R. and Chitwood L. A. (1989) A possible lacustrine paleomagnetic record of the Blake episode from Pringle Falls, Oregon, USA. Phys. Earth Planet. Int. 56,112-123.

Hillhouse J. W. and Cox A. (1976) Brunhes-Matuyama polarity transition. Earth and Planet. Sci. Lett. 29, 51-64.

Hoffman K. A. (1977) Polarity transition records and the geomagnetic dynamo. Science 196, 1329-1332.

Hoffman K. A. and Fuller M. (1978) Transitional field configuration and geomagnetic reversal. Nature 273, 715-718.

Laj C., Guitton S., Kissel C. and Mazaud A. (1988) Complex behaviour of the geomagnetic field during three successive polarity reversals, 11-12 m.y. B.P. J. Geophys. Res. 93, 11655-11666.

Mazaud A. and Laj C. (1989) Simulation of geomagnetic polarity reversals by a model of interactive dipole sources. Earth Planet. Sci. Lett. 92, 299-306.

Merrill R. T. and McElhinny N. W. (1977) Anomalies in the time-averaged paleomagnetic field and their implication for the lower mantle. Rev. Geophys. Space Phys. 15, 309-323.

Merrill R. T., McElhinny N. W. and Stevenson D. J. (1979a) Evidence for long term asymmetries in the Earth's magnetic field and possible implication for the dynamo theory. Phys. Earth Planet. Int. 20, 75-82.

Merrill R. T., McFadden P. L. and McElhinny M. W. (1979b) Paleomagnetism and the nature of the geodynamo. Phys. Earth Planet. Int. 64, 87-101.

Morelli A. and Dziewonski A. M. (1987) Topography of the core-mantle boundary and lateral homogeneity of the liquid core. Nature 325, 678-683.

Olson P., Silver P. G. and Carlson R. W. (1990) The large scale structure of convection in the Earth's mantle. Nature 344, 209-215.

Ruocco M. (1990) Paleomagnetic analyses of continental deposits of the last 3 Ma from Argentina: Magnetostratigraphy and fines structures of reversal. Doctoral Thesis, University of Stockholm.

Schneider D. and Kent D. (1988) The paleomagnetic field from equatorial deep-sea sediments: axial symmetry and polarity asymmetry. Science 242, 252-256.

Tric E., Laj C., Jehanno C., Valet J. P., Kissel C., Mazaud A. and Iaccarino S. (1991a) High resolution record of the Upper Olduvai Transition from Po Valley (Italy) sediments: support for dipolar transition geometry? Phys. Earth Planet. Int. 65, 319-336.

Tric E., Laj C., Valet J. P., Tucholka P., Paterne M. and Guichard F. (1991b) The Blake geomagnetic event: transition geometry, dynamical characteristics and geomagnetic signifiance. Earth Planet. Sci. Lett. 102, 1-13.

Valet J. P., Laj C. and Langereis C. (1988) Sequential geomagnetic reversals recorded in Upper Tortonian marine clays in western Crete (Greece). J. Geophys. Res. 93, 1131-1151.

PARAMETERIZATION OF TEMPERATURE- AND STRESS-DEPENDENT VISCOSITY CONVECTION AND THE THERMAL EVOLUTION OF VENUS

V. S. SOLOMATOV
Seismological Laboratory 252-21
California Institute of Technology
Pasadena, CA 91125 U.S.A.

ABSTRACT. While many laboratory and numerical experiments have been done on variable viscosity convection the theoretical interpretation of the results is almost absent. This work suggests a theoretical interpretation of the results of the numerical modeling of temperature- and stress-dependent viscosity convection in a wide parameter range. The relationships found between the parameters of convection are summarized into a simple parameterization scheme and can be applied to various geophysical problems. The thermal evolution of Venus is considered as an example.

1. Introduction

Rheology of many geophysical systems can not be described with the help of a constant viscosity model. Subsolidus planetary interiors, partially or entirely molten mantle reservoirs such as magma chambers, volcanic lavas or magma oceans have more complicated rheology. The simplest rheological model for these systems at large deformations and time scales is the Arrhenius temperature-dependent and power-law stress-dependent viscosity:

$$\eta = \frac{b}{\tau^{n-1}} exp\left(\frac{A}{T}\right) = \frac{b^{1/n}}{\dot{e}^{(n-1)/n}} exp\left(\frac{A}{nT}\right) , \qquad (1)$$

where b, A and $n \geq 1$ are considered as constants, $\tau = (1/2\tau_{ij}^2)^{1/2} = \eta\dot{e}$ is the second invariant of the deviatoric stress tensor, $\dot{e} = (1/2\dot{e}_{ij}^2)^{1/2}$ is the second invariant of the strain rate tensor, T is the temperature. Convection in such systems can not be described just with the help of the results obtained for the well-studied constant viscosity convection. This produces uncontrolled errors as follows from the first systematic investigation of steady-state variable viscosity convection by Christensen (1984a,b, 1985). A review of early laboratory and numerical experiments and some theoretical results can be found there. The absence of any satisfactory theory which would explain the experimental results is the main motivation for the present study. Some results have already been published (Solomatov and Zharkov, 1990). They were obtained for the small viscosity contrasts non-Newtonian convection. In the present paper the parameterization of temperature- and stress-dependent viscosity convection is developed for higher viscosity contrasts.

D. B. Stone and S. K. Runcorn (eds.),
Flow and Creep in the Solar System: Observations, Modeling and Theory, 131–145.
© 1993 *Kluwer Academic Publishers. Printed in the Netherlands.*

2. Nondimensional Parameters and Integral Relationships

We consider convection in an infinite horizontal shallow layer of the thickness d. The temperature difference $\Delta T = T_1 - T_0$ between the lower and upper boundaries is fixed, the boundaries are free slip and the Prandtl number is equal to infinity. Equations and discussion of nondimensional parameters can be found in Christensen's (1984a,b, 1985) works. The problem has only two nondimensional parameters if the viscosity law (1) is approximated as follows:

$$\eta = \eta_0 \left(\frac{\tau_0}{\tau}\right)^{n-1} exp(-p\theta) , \qquad (2)$$

where τ_0 is determined by the formulae:

$$\eta_0 = \eta(T = T_0; \ \tau = \tau_0) = \frac{b}{\tau_0^{n-1}} exp\left(\frac{A}{T_0}\right) , \qquad (3)$$

$$\tau_0 = \eta_0 \dot{e}_0 , \qquad (4)$$

$$\dot{e}_0 = \frac{d^2}{\kappa} . \qquad (5)$$

Parameter p is equal to $A\Delta T/T_0^2$, κ is the thermal diffusivity, $\theta = (T-T_0)/(T_1-T_0)$ is the nondimensional temperature.

The nondimensional parameters which are chosen to represent the results are the "surface" Rayleigh number

$$Ra_0 = \frac{\alpha g \rho \Delta T d^3}{\kappa \eta_0} \qquad (6)$$

and the "averaged" Rayleigh number

$$Ra_T = \frac{\alpha g \rho \Delta T d^3}{\kappa \eta_T} , \qquad (7)$$

where

$$\eta_T = \eta(T = \overline{T}; \ \tau = \tau_0) = \eta_0 \, exp(-p\overline{\theta}) , \qquad (8)$$

η_0 and τ_0 are defined by Eqs.(3)-(5), α is the coefficient of thermal expansion, g is the gravity acceleration, ρ is the density, $\overline{T}$ and $\overline{\theta}$ mean the volumetrically averaged values.

The main equation which will be used is the integral balance between the viscous dissipation and the capacity of buoyancy forces which can be derived from the general equations of thermal convection:

$$\int_V \tau_{ij} \frac{\partial u_i}{\partial x_j} dV = -\int_V \alpha g \rho (T - T_0) u_z \, dV , \qquad (9)$$

where

$$\tau_{ij} = \eta \dot{e}_{ij} = \eta \left(\frac{\partial u_i}{\partial x_j} + \frac{\partial u_j}{\partial x_i} \right) ,$$

u_z is the vertical component of the velocity vector.

The integral capacity of buoyancy forces can be exactly found in terms of the heat flux F (Golitsyn, 1980) and Eq. (9) is reduced to

$$\int_V \tau_{ij} \frac{\partial u_i}{\partial x_j} \, dV = \frac{\alpha g d}{c_p} F S , \tag{10}$$

where c_p is the thermal heat capacity at constant pressure and S is the surface area. Each boundary contributes one half to the capacity of buoyancy forces and in a general case the contribution from the boundary is equal to $(\alpha g d / 2 c_p) F_i$ where F_i is the heat flux through this boundary.

Thus we don't need to estimate the integral capacity of buoyancy forces but we must estimate the dissipation integral.

3. Convective Regimes and Estimate of the Dissipation Integral

We suppose that the convective structure can be approximated with the help of a periodic structure of the convective rolls and so only one two-dimensional convective cell is considered. The aspect ratio is chosen to be equal to one as in the numerical experiments. The estimate of the dissipation integral is essentially different for small viscosity contrasts and large viscosity contrasts. These two asymptotic cases are connected with the two different regimes of convection which can be distinguished in Christensen's experiments. In the first regime the convective flow structure is uniform and the convective pattern is very close to that of the constant viscosity case. In the second regime the cold boundary layer is almost stagnant, the convective flow is asymmetric, the velocities near the surface are much lower than the velocities in the central region of the convective cell.

In the case of small viscosity contrasts the velocity distribution can be approximated with any trial stream function which gives the velocity gradients of the order of u_0/d where u_0 is the velocity amplitude of the flow. This stream function is chosen to be as follows:

$$\psi = \psi_m \sin \frac{\pi x}{d} \sin \frac{\pi z}{d} , \tag{11}$$

where

$$\psi_m = \frac{u_0 d}{\pi} \tag{12}$$

is the maximum of the stream function.

The temperature distribution can be approximated by a constant value because in the case of highly developed convection most of the convective cell is well-mixed and the variations of temperature are small with the exception of the thin thermal boundary layers near the boundaries. Supposing that the thermal boundary layers contribute a negligible amount to the dissipation integral and taking into account the symmetry of the problem we can write:

$$T(x, z) \approx \overline{T} \approx \frac{T_1 + T_0}{2} . \tag{13}$$

Now the dissipation integral can be calculated as follows:

$$\int_V \tau_{ij} \frac{\partial u_i}{\partial x_j} \, dV = \int_V \eta_0 \left(\frac{\dot{e}_0}{\dot{e}}\right)^{(n-1)/n} exp\left(-\frac{p\theta}{n}\right) \frac{1}{2} \dot{e}_{ij}^2 \, dV$$

$$\approx \eta_0 \dot{e}_0^{(n-1)/n} exp\left(-\frac{p\overline{\theta}}{n}\right) \int_V \dot{e}^{(n+1)/n} \, dV$$

$$= 2^{(n+1)/n} \pi^{1/n} \left[\frac{\Gamma(1 + 1/2n)}{\Gamma(3/2 + 1/2n)}\right]^2 \eta_0 \dot{e}_0^{(n-1)/n} exp\left(-\frac{p\overline{\theta}}{n}\right) \left(\frac{u_0}{d}\right)^{(n+1)/n} d^2 l , \tag{14}$$

where V is the volume of the convective roll, l is the length of the convective roll, $\Gamma(x)$ is the gamma function.

The estimate of the dissipation integral in the case of large viscosity contrasts is based on the following physical idea. At small viscosity variations the dissipation in the thermal boundary layers is small in comparison with the dissipation in the major part of the convective cell. When the viscosity contrasts increase, the dissipation in the cold thermal boundary layer eventually becomes comparable with the dissipation in the central part of the convective cell. At this point the structure of the small viscosity contrast convective regime is changed to the structure of the large viscosity contrast convective regime. After this transition the dynamics of the cold boundary layer is determined mainly by the balance between the dissipation in the cold boundary layer and the capacity of buoyancy forces. The part of the capacity of buoyancy forces which is spent on the dissipation in the cold boundary layer must be of the same order of magnitude as the total one. If we suppose that the main buoyancy force which drives the cold boundary layer is its own buoyancy force then we get that about 1/2 (each boundary contributes exactly one half of the total capacity of buoyancy forces) of the total capacity of buoyancy forces is spent on the dissipation in the cold boundary layer. The rest of the capacity of buoyancy forces is spent on the dissipation in the internal region of the convective cell. If we suppose that the main buoyancy force which drives the flow in the internal part of the convective cell is the buoyancy force of the hot boundary layer then we get that about 1/2 of the total capacity of buoyancy forces is spent on the dissipation in the internal part of the convective cell. In any case the three integrals—the dissipation in the cold boundary layer, the dissipation in the internal part of the convective cell and the capacity of buoyancy forces—must be of the same order of magnitude.

The flow structure in the internal region can be taken as for the small viscosity contrasts regime, i.e.

$$\frac{\partial u_i}{\partial x_j} \approx \frac{u_1}{d} , \tag{15}$$

where u_1 is the velocity amplitude in the internal region.

The flow structure in the cold boundary layer can be described as

$$\frac{\partial u_i}{\partial x_j} \approx \frac{u_0}{\delta_0} ,$$

(16)

where u_0 and δ_0 are the characteristic velocity and the characteristic thickness of the cold boundary layer.

The dissipation in the cold boundary layer region of the volume V_0 can be estimated by analogy with the steepest descents method of integration: the main part of the integral is calculated as the product of the subintegral function in the point of its maximum (at the surface of the convective layer) and the square of the space scale of the characteristic changes of this function. This space scale with logarithmical accuracy can be taken as δ_0. The capacity of buoyancy forces in the cold boundary layer region is of the order of the total one with the same accuracy. The dissipation integral is now estimated as:

$$\int_{V_0} \tau_{ij} \frac{\partial u_i}{\partial x_j} dV \approx \eta_0 \left(\frac{d^2 u_0}{\kappa \delta_0}\right)^{-(n-1)/n} \left(\frac{u_0}{\delta_0}\right)^2 \delta_0^2 l .$$

(17)

The dissipation in the internal region of the volume $V_1 \approx V$ is:

$$\int_{V_1} \tau_{ij} \frac{\partial u_i}{\partial x_j} dV \approx \eta_0 exp(-p\bar{\theta})(\frac{du_1}{\kappa})^{-(n-1)/n} (\frac{u_1}{d})^2 d^2 l ,$$

(18)

where the temperature of the internal region was substituted by $\bar{\theta}$. It is correct if the boundary layer thicknesses are small in comparison with the thickness of the layer.

4. The Thermal Boundary Layer Theory

The thermal boundary layer theory allows us to find the relationships between the thermal boundary layer thickness, the characteristic velocity of the boundary layer and the heat flux (Turcotte and Oxburgh, 1967. The only difference from the constant viscosity case is that the boundary layer thicknesses and the velocities are different at the upper and the lower boundaries.

The main results of the thermal boundary layer theory can be represented in the following form:

$$F = k\frac{\Delta T_0}{\delta_0} = k\frac{\Delta T_1}{\delta_1} ,$$

(19)

$$\Delta T_0 = T_C - T_0 \approx \overline{T} - T_0 ,$$

(20)

$$\Delta T_1 = T_1 - T_C \approx T_1 - \overline{T} ,$$

(21)

$$\delta_0 = \frac{\pi^{1/2}}{2}(\frac{\kappa d}{u_0})^{1/2} ,$$

(22)

$$\delta_1 = \frac{\pi^{1/2}}{2}\left(\frac{\kappa d}{u_1}\right)^{1/2}, \tag{23}$$

where u_0 and u_1 are characteristic velocities near the boundaries, T_C is the temperature of the isothermal central part of the convective cell, which is approximately equal to the averaged temperature of the convective cell $\bar{T}$, δ_0 and δ_1 are the thicknesses of the upper and lower thermal boundary layers, k is the thermal conductivity. The characteristic velocities u_0 and u_1 can be approximately calculated as the amplitude velocities near the boundaries when the velocity varies along a boundary in accordance with Eqs. (11), (12).

In the case of small viscosity contrasts

$$u_0 = u_1, \quad \delta_0 = \delta_1. \tag{24}$$

In the case of large viscosity contrasts

$$u_0 \ll u_1, \quad \delta_0 \gg \delta_1. \tag{25}$$

5. Parameterization

Now we have a complete system of equations from which all the relationships are found. In the case of small viscosity contrasts the dependence of boundary layer thicknesses on the Rayleigh numbers is obtained from Eqs. (10), (14), (19)-(24):

$$\delta_0 = \delta_1 = \delta = a\,Ra_0^{-\beta_0}\,Ra_T^{-\beta_T}, \tag{26}$$

$$a = 2^{-1/(n+2)}\pi\left[\frac{\Gamma(1+1/2n)}{\Gamma(3/2+1/2n)}\right]^{2n/(n+2)}, \tag{27}$$

$$\beta_0 = \frac{n-1}{n+2}, \tag{28}$$

$$\beta_T = \frac{1}{n+2}, \tag{29}$$

The function $a(n)$ is weak:

$$a \approx 2.3,\ 2.8,\ 4.0 \quad when \quad n = 1,\ 3,\ \infty \tag{30}$$

and slightly varies in dependence on the chosen flow structure.

In the case of high viscosity contrasts after using Eqs. (10), (17)-(23), substituting

$$\Delta T_0 \approx \Delta T \tag{31}$$

and

$$\Delta T_1 = \Delta T_0\frac{\delta_1}{\delta_0} \approx \Delta T\frac{\delta_1}{\delta_0} \tag{32}$$

and neglecting small parameters of the order of δ_1/δ_0 we arrive at the following expressions:

$$\delta_0 = a_0 \, Ra_0^{-\beta_0'} , \tag{33}$$

$$\delta_1 = a_1 \, Ra_0^{-\beta_{01}} \, Ra_T^{-\beta_{T1}} , \tag{34}$$

$$a_0 \approx a_1 = O(1) , \tag{35}$$

$$\beta_0' = \frac{n}{3} , \tag{36}$$

$$\beta_{01} = \frac{n^2 + 3(n-1)}{6(n+1)} , \tag{37}$$

$$\beta_{T1} = \frac{1}{2(n+1)} . \tag{38}$$

For both cases the other parameters are expressed in terms of the found functions $\delta_0(Ra_0, Ra_T)$ and $\delta_1(Ra_0, Ra_T)$:

$$Nu = \frac{Fd}{\kappa \Delta T} = \frac{1}{\delta_0 + \delta_1} , \tag{39}$$

$$\overline{T} = \frac{1}{1 + \delta_1/\delta_0} , \tag{40}$$

$$< u_{sf} > = a_{sf} \frac{1}{\delta_0^2} , \tag{41}$$

$$\psi_m = a_m \frac{1}{\delta_1^2} , \tag{42}$$

$$a_{sf} = \frac{1}{2}, \quad a_m = \frac{1}{4} . \tag{43}$$

The transition between the convective regimes is determined with the help of the requirement that the dissipation integral for the cold boundary layer (17) becomes of the same order of magnitude as the dissipation integral for the entire convective cell (14). In Ra_T–Ra_0 space the transition curve $Ra_{T,tr}(Ra_0)$ is:

$$Ra_{T,tr} \sim Ra_0^{1 + \frac{n(n-1)}{3}} . \tag{44}$$

6. Comparison with Numerical Experiments

The formulae found are used as the fitting functions to the experimental data published by Christensen (1985) in the form of numerical tables. The tables contain the results of the numerical modeling of Newtonian viscosity convection ($n=1$) obtained by Christensen (1984b) and of non-Newtonian viscosity convection ($n=3$) obtained by Christensen (1985). The fitting functions for Nu, $\overline{T}$, $< u_{sf} >$ and ψ_m are expressed in terms of the boundary layer thicknesses as in formulae (39)-(42) but not in terms of the Rayleigh numbers and so only a minimal number of fitting coefficients were used. For the first convective regime four coefficients $(\beta_T, a, a_{sf}, a_u)$ are used in the case of $n=1$ and five coefficients $(\beta_0, \beta_T, a, a_{sf}, a_u)$ are used in the case of $n=3$. The coefficient β_0 in the case of $n=1$ is essentially equal to zero. For the second convective regime seven coefficients $(\beta_0', \beta_{0T}, \beta_{T1}, a_0, a_1, a_{sf}, a_u)$ are used in both cases.

The results are shown in Figs. 1,2 and in Table 1. The dashed curves with solid boxes are the results of the numerical experiments of Christensen (1984b, 1985). The solid curves are the fitting functions. The curves are shown at constant values of the "surface" Rayleigh numbers $Ra_0=2000$, 4000, 8000, 16000, 32000, 64000, 128000 for $n=1$ and $Ra_0=100$, 200, 500, 1000, 2000, 4000 for $n=3$. The coincidence between the theoretical and the experimental results is quite good. For the second regime in the case of $n=3$ the experimental data are almost absent and the fit can be improved with new data. These curves are shown very approximately. A deviation between the results at low values of the Rayleigh numbers is possibly due to decreasing of the accuracy of the thermal boundary layer theory in approaching the lowest values of the Nusselt number. A deviation at the highest Rayleigh numbers possibly reflects the transition to the third convective regime. This regime is due to an influence of the logarithmical terms in the parameterization which were ignored in the derivation. Physically it means that the space scale of the viscosity drop becomes much smaller than the cold boundary layer thickness and so this space scale must control the convective regime near the cold surface.

The transition between the first and the second convective regimes can be determined as the curve at which two asymptotic fitting functions intersect each other. The transition curve is slightly different for the different parameters because the fitting coefficients are not exactly equal to the theoretical ones: $\beta_{tr}=1.1-1.3$ if $n=1$ and $\beta_{tr}=2.1-2.6$ if $n=3$. Eq. (44) gives the values 1 and 3 respectively.

7. Extrapolation of the Parameterization

Can the parameterization be extrapolated to other cases? The small viscosity contrast convection is qualitatively the same as constant viscosity convection. In the case of constant viscosity convection the results obtained for the problem considered above can be used as approximate solutions to some other problems: the different kinds of heating—constant temperature difference, constant bottom heat flux, constant internal heating, combined internal and bottom heating and so on; the different geometrical variations—two- or three-dimensional layers, horizontal or spherical layers; the different boundary conditions—fixed or free boundaries; the time-dependent problems with time-dependent heating when the oscillations are small. In these cases experimental and theoretical results justify this conclusion and only slight changes of the parameters are required (see reviews, e.g., in Busse, 1979; Schubert, 1979; Turcotte et al., 1979). We suppose that the same extrapolation is valid in the case of the small viscosity contrast convective regime. For example the results obtained by Parmentier et al. (1976) for the fixed boundaries, $n=3$, without temperature dependence of the viscosity, can be represented in terms of the above parameterization if $\alpha \approx 3$, $\beta_0+\beta_T \approx 0.5$. The correspondent values for the case of free boundaries are 2–2.8 and 0.55–0.60 (Table 1).

The same extrapolation of the large viscosity contrasts parameterization possibly will give qualitatively reasonable results but additional investigations are needed.

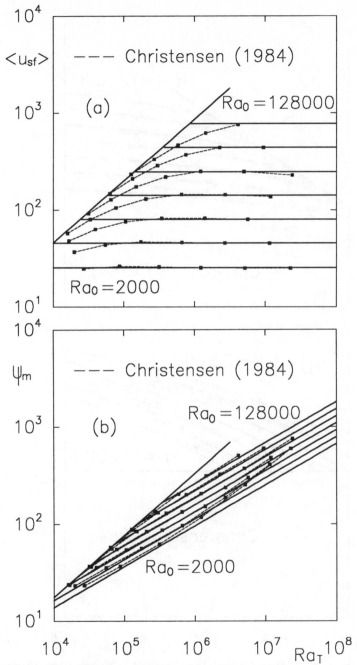

Fig. 1: The averaged surface velocity (a) and the maximum of stream function (b) as functions of Ra_T and Ra_0; n=1.

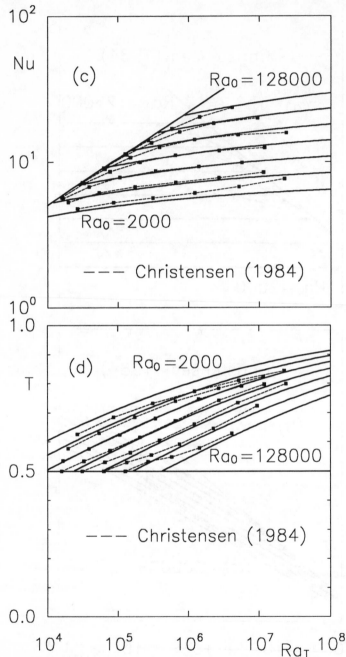

Fig. 1: The Nusselt number (c) and the averaged temperature (d) as functions of Ra_T and Ra_0; n=1.

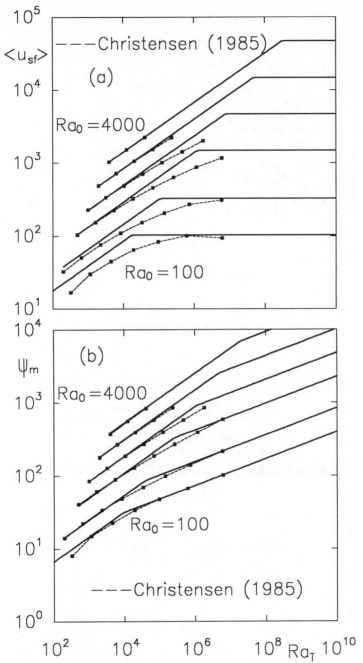

Fig. 2: The averaged surface velocity (a) and the maximum of stream function (b) as functions of Ra_T and Ra_0; n=3.

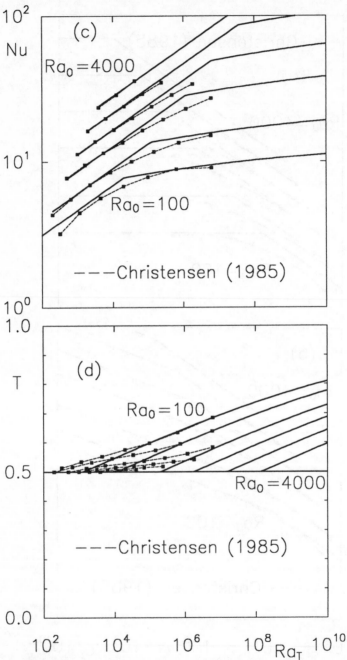

Fig. 2: The Nusselt number (c) and the averaged temperature (d) as functions of Ra_T and Ra_0; n=3.

Coefficient	Regime 1			
	$n = 1$		$n = 3$	
	Theory	Fit	Theory	Fit
β_0	0	0	0.40	0.38
β_T	0.33	0.32	0.20	0.17
a	2.3	1.85	2.8	2.0
a_{sf}	0.5	0.45	0.5	0.45
a_m	0.25	0.17	0.25	0.17
Coefficient	Regime 2			
	$n = 1$		$n = 3$	
	Theory	Fit	Theory	Fit
β'_0	0.33	0.41	1.0	0.83
β_{01}	0.083	0.12	0.63	0.54
β_{T1}	0.25	0.21	0.13	0.092
a_0	O(1)	3.2	O(1)	3.3
a_1	O(1)	1.6	O(1)	1.69
a_{sf}	0.5	0.52	0.5	0.54
a_m	0.25	0.12	0.25	0.11

Table 1: Comparison of theoretical and fitting coefficients.

The pressure dependence of the parameter A in (1) is important for the dynamics of planetary interiors (Zharkov and Trubitsyn, 1978). In the first approximation this dependence can be taken into account by calculating A at a mean depth of the convective layer. This leads to small errors (about 10–30%) for the Nusselt number (Christensen, 1985).

The suggested parameterization doesn't depend on a specific form of the temperature dependence of the viscosity and so the parameterization is valid for both Eq. (1) and (2).

8. The Thermal Evolution of Venus

The thermal evolution of Venus was discussed recently by Solomatov and Zharkov (1990). A development of these results follows below.

In contrast with the Earth, Venus is supposed to have no Earth-like plate-tectonics. Thus the convection in Venus' mantle takes place beneath the crust. This means that the rheology of the convective part of the Venusian interiors is more or less uniform and a rheology law (1) can be used. The small contrast viscosity approximation turns out to be a sufficient one for this problem.

The temperature of the upper boundary of the convective mantle is supposed to be close to the solidus temperature and was varied in the range (1300–1700) K. The number of the convective layers in the mantle was taken to be equal to one or two. The present-day radiogenic heat production in the undifferentiated mantle was varied in the range $(2.4–9.6) \cdot 10^{-12}$ W kg^{-1}. The parameter A in Eq. (1) was varied in the range $(6–9) \cdot 10^4$ K for the upper mantle and in the range $(1.2–1.6) \cdot 10^5$ for the lower mantle in the case of two-layered convection and in the range $(1.2–1.6) \cdot 10^5 \ K$ in the case of one convective layer (the models and parameters were discussed by Zharkov, 1983). The influence of the core is unessential.

The thermal evolution is calculated with the help of the parameterized convection scheme which supposes that the parameterization obtained for the steady-state convection can be applied to a time-dependent convection if the cooling of the planet is considered as an additional heat production. Substituting the new effective heat production into the parameterization results in an ordinary differential equation for the averaged temperature of the convective layer (McKenzie and Weiss, 1975; Sharp and Peltier, 1978, 1979; Turcotte et al., 1979; McKenzie and Richter, 1981).

The important result is that Venus cools down in an asymptotic regime. This means that the present-day parameters don't depend on the initial temperature distribution. The characteristic time for the transition to the asymptotic regime is about $(3–4) \cdot 10^9$ years. This regime can be described analytically (Solomatov et al., 1987; Solomatov and Zharkov, 1990).

One of the properties of the asymptotic regime is the following. If we decompose the surface heat flux F_S into the part F_Q due to the radiogenic heat production and the part F_T due to the cooling of the planet then the last one at all considered variations of the model is about

$$F_T \approx 10 - 15 \ erg \, cm^{-2} s^{-1} \ . \tag{45}$$

For a preferable value of the total radiogenic heat production in the undifferentiated mantle of Venus $4.8 \cdot 10^{-12} \ W \ kg^{-1}$ corresponding to $F_Q = 36 \ erg \ cm^{-2} \ s^{-1}$ the surface heat flux is

$$F_S = F_Q + F_T \approx 46 - 51 \ erg \, cm^{-2} s^{-1} \ . \tag{46}$$

Other parameters can be less constrained. The mantle temperature beneath the upper thermal boundary· layer is about $(1500–1800) \ K$, the temperature at the core-mantle boundary $(3400–4000) \ K$, the velocities in the upper mantle $(1–5) \ cm \ y^{-1}$, the velocities in the lower mantle $(0.5–2) \ cm \ y^{-1}$, the shear stresses in the upper mantle $(5–20) \ bar$, the shear stresses in the lower mantle $(50–200) \ bar$.

The models predict partial melting in the uppermost part of the mantle. The connected effects of chemically induced buoyancy lead to a chemically driven convection near the surface or in the mantle depending on the density differences between the crust, partially molten mantle, crystalline residue and the unmolten mantle.

9. Conclusion

Parameterization of temperature- and stress-dependent viscosity convection is derived using the integral balance between the viscous dissipation and the capacity of buoyancy forces, thermal boundary layer theory and some physical assumptions about the flow structure. The parameterization is obtained for the two asymptotic cases—the small viscosity contrast limit and the large viscosity contrast limit. The theoretical description is in good agreement with the experimental data. The new parameterization can be used to calculate the variable viscosity convection in various geophysical applications. As an example the thermal evolution of Venus is studied. The suggested qualitative approach can be used to find the parameterization for other rheological laws.

Acknowledgments. The author wishes to thank Vladimir Zharkov for fruitful collaboration on Venus' models and David Stevenson for stimulating discussions of convection in the Earth's mantle and for helpful comments on an earlier version of this work. This work was supported by the National Science Foundation grant EAR-89-16611.

10. References

Busse, F. H. (1979) High Prandtl number convection. Phys. Earth Planet. Inter. 19, 149-157.

Christensen, U. R. (1984a) Convection with pressure- and temperature-dependent non-Newtonian rheology. Geophys. J. R. Astron. Soc. 77, 343-384.

Christensen, U. R. (1984b) Heat transport by variable viscosity convection and implications for the Earth's thermal evolution. Phys. Earth Planet. Inter. 35, 264-282.

Christensen, U. R. (1985) Heat transport by variable viscosity convection II: Pressure influence, non-Newtonian rheology and decaying heat sources. Phys. Earth Planet. Inter. 37, 183-205.

Golitsyn, G. S. (1980) A study of convection with geophysical applications and analogies, Gidrometeoizdat, Leningrad (in Russian).

McKenzie, D. P. and Weiss, N. O. (1975) Speculations on the thermal and tectonic history of the Earth. Geophys. J. R. Astron. Soc. 42, 131-174.

McKenzie, D. P. and Richter, F. M. (1981) Parameterized thermal convection in a layered region and the thermal history of the Earth. J. Geophys. Res. 86, 11667-11680.

Parmentier, E. M., Turcotte, D. L. and Torrance, K. E. (1976) Studies of finite amplitude non-Newtonian thermal convection with application to convection in the Earth's mantle. J. Geophys. Res. 81, 1839-1846.

Schubert, G. (1979) Subsolidus convection in the mantles of terrestrial planets. Ann. Rev. Earth Planet. Sci. 7, 289-342.

Sharp, H. N. and Peltier, W. R. (1978) Parameterized mantle convection and the Earth's thermal history. Geophys. Res. Lett. 5, 737-740.

Sharp, H. N. and Peltier, W. R. (1979) A thermal history model for the Earth with parameterized convection. Geophys. J. R. Astron. Soc. 59, 171-205.

Solomatov, V. S., Leontjev, V. V. and Zharkov, V. N. (1987) Models of thermal evolution of Venus in the approximation of parameterized convection. Gerlands Beitr. Geophys. 96, 73-96.

Solomatov, V. S. and Zharkov, V. N. (1990) The thermal regime of Venus. Icarus 84, 280-295.

Turcotte, D. L. and Oxburgh, E. R. (1967) Finite amplitude convection cells and continental drift. J. Fluid Mech. 28, 29-42.

Turcotte, D. L., Cook, F. A. and Willeman, P. J. (1979) Parameterized convection within the Moon and the terrestrial planets. Lunar Planet. Sci. Conf. 10, 2375-2392.

Zharkov, V. N. (1983) Models of the internal structure of Venus. Moon and Planets 29, 139-175.

Zharkov, V. N. and V. P. Trubitsyn (1978) Physics of Planetary Interiors. Edited by W. B. Hubbard. Pachart Pub. House, Tucson, Arizona.

COMPLEX FLOW STRUCTURES IN STRONGLY CHAOTIC TIME-DEPENDENT MANTLE CONVECTION

DAVID A. YUEN, WULING ZHAO and ANDREI V. MALEVSKY
Minnesota Supercomputer Institute,
Army High Performance Computing and Research Center, and
Dept. of Geology and Geophysics
University of Minnesota
Minneapolis, MN 55415
U.S.A.

ABSTRACT. Large-scale numerical simulations of two-dimensional thermal convection have been conducted in the strong time-dependent regime for infinite Prandtl number fluids, as applied to the Earth's mantle. Both Newtonian and non-Newtonian (strain-rate proportional to the third power of the deviatoric stress) rheologies have been studied. We have also investigated for Newtonian fluids the effects of multiple phase transitions in mantle convection. The transition from soft to hard turbulence is studied for linear and nonlinear rheologies. In non-Newtonian convection with internal heating, high temperature can be produced in stagnant regions. A non-Newtonian mantle can tolerate less internal-heating than a Newtonian mantle. Non-Newtonian plumes behave quite differently from Newtonian ones in that noticeable curvatures are developed in their ascent. The transition to the disconnected-plume regime takes place at much lower Nusselt numbers for non-Newtonian rheology. For the Earth's mantle this would have strong implications, as the upper-mantle rheology is probably non-Newtonian. At high Rayleigh number, greater than 10^7, convection with a single olivine to spinel phase transition becomes intermittently layered. The effects of depth-dependent thermal expansivity and internal-heating are to increase the propensity toward layering. With increasing Ra the system becomes more layered, irrespective of the sign of the Clapeyron slope and various types of phase transitions. In an early Earth with a hotter temperature and greater amounts of radiogenic heating, mantle convection would have a greater tendency for layering.

1. Introduction

In the last several years with the increasing availability of supercomputers and powerful graphics workstations there has been considerable progress made in understanding time-dependent mantle convection (e.g. Machetel and Yuen, 1986, 1987, 1989; Christensen, 1987; Vincent and Yuen, 1988; Weinstein et al., 1989; Bercovici et al., 1990; Travis et al., 1990; Hansen et al., 1990) mainly in two-dimensional configurations. However, many of these new ideas have not yet been spread to the rest of the geoscience community in which concepts drawn from steady-state convection, such as continuous plumes, cellular structures, are still in common use. Some of these steady-state notions, such as steady continuous plumes with mushroom head, are being employed to explain events such as the continental flood basalts (e.g. Richards et al., 1989; Watson and McKenzie, 1991). Yet three-dimensional seismic tomographic data, notably the recent high-resolution tomographic work by Zhang and Tanimoto (1992) and van der Hilst et al. (1991), have revealed the richness in the complexity

147

D. B. Stone and S. K. Runcorn (eds.),
Flow and Creep in the Solar System: Observations, Modeling and Theory, 147–173.

of mantle heterogeneities which defies explanation by simple steady-state pictures of convection (e.g. Turcotte and Oxburgh, 1967; Peltier and Jarvis, 1982). There is, however, a growing recognition that many geological phenomena, such as back-arc spreading (Karig, 1974), fluctuations in sea-level (Gurnis, 1990), polar-wandering (Goldreich and Toomre, 1968), and geochemical mixing (Zindler and Hart, 1986) must be explained within the framework of time-dependent convection.

In this paper we will discuss three recently studied topics which are important for understanding complex structures in time-dependent mantle convection. They will be concerned with the strongly chaotic regimes for both Newtonian and non-Newtonian convection and the influences of multiple phase transitions in highly time-dependent Newtonian convection. All of these situations will reveal the presence of complex thermal and flow field structures, which are not commonly found in numerical and laboratory experiments conducted in the quasi-steady or weakly chaotic regimes.

We will begin with a discussion of strongly time-dependent Newtonian and non-Newtonian convection and the transition to the hard-turbulent regime from the chaotic domain. Then we will discuss the dynamics of strongly time-dependent mantle convection with multiple phase transitions and the effects on the style of layered mantle convection from increasing the amount of the driving forces in convection due to greater internal-heating and lower viscosity from hotter interior temperatures.

2. High Rayleigh Number Convection: Newtonian and Non-Newtonian Rheologies

We will now discuss the phenomenon of the transition from soft to hard turbulence in thermal convection at the infinite Prandtl number limit. In the last few years the topic of hard turbulence in thermal convection has been a growing subject of investigation by physicists and other fluid dynamicists (Heslot et al., 1987; Castaing et al., 1989; Sano et al., 1989; Hansen et al., 1990). This transition is characterized by the appearance of disconnected plume structures in contrast to continuous plumes with mushroom-shaped tops found at lower Rayleigh numbers, below 10^6 for Newtonian rheology. The mode of heat transfer also changes in this turbulent convective regime, as the large-scale flow becomes the dominant agent to the global heat transport with the small-scale disconnected plumes playing a minor role (Krishnamurti and Howard, 1981; Solomon and Gollub, 1991; Hansen et al., 1992). This type of heat transport may be prevailing in the Earth's mantle today (Davies, 1988). Therefore, the phenomenon of hard turbulent convection is very relevant for earth scientists in order to better appreciate the complexities present in the dynamics of planetary convection.

2.1. MATHEMATICAL EQUATIONS USED IN MODELLING

In this subsection we will present the equations used in studying two-dimensional convection for both Newtonian and non-Newtonian rheologies. With the present generation of vector supercomputers, modelling three-dimensional hard turbulent convection still requires much computational time resources and disk-storage requirements. Thus only a few runs have been attempted for finite Prandtl number high Ra convection (Balachandar and Sirovich, 1991). By restricting the computational domain to two dimensions, we can reach higher Rayleigh numbers, conduct much longer runs to obtain meaningful statistics (Hansen et al., 1992) and explore the much larger parameter space inherent in mantle convection (Leitch et al., 1991).

We will make very simplifying physical assumptions in order to focus our attention on fundamental aspects of hard turbulent convection. Effects of mantle compressibility (Leitch et al., 1991; Machetel and Yuen, 1989), internal heating (Weinstein et al., 1989) and temperature-dependent viscosity (Christensen, 1984) have been neglected. The dimensionless equations governing mantle convection for an incompressible Boussinesq fluid in the infinite Prandtl number limit, appropriate for mantle substances are:

$$\nabla \cdot u = 0 \tag{1}$$

$$\frac{\partial \sigma_{ij}}{\partial x_j} - \nabla p - RaTe_z = 0 \tag{2}$$

$$\frac{\partial T}{\partial t} + \underset{\sim}{u} \cdot \nabla T = \nabla^2 T + R \tag{3}$$

Eqn. (1), (2), and (3) are the mass conservation, momentum and energy equations. The internal-heating parameter (Leitch and Yuen, 1989) is given by R. T is temperature, u is velocity, e_z is the unit vector aligned with gravity, p is dynamical pressure, and t is time non-dimensionalized by the thermal diffusion time across the layer depth. The Rayleigh number Ra is the only dimensionless parameter in this model problem. The deviatoric stress tensor σ_{ij} is related to the velocity components given by

$$\sigma_{ij} = \frac{\eta}{2}\left(\frac{\partial u_i}{\partial x_j} + \frac{\partial u_j}{\partial x_i}\right) \tag{4}$$

where η is the dynamic viscosity. For Newtonian or linear rheology η is taken as a constant and is included in the definition of Ra. On the other hand, laboratory experiments (Goetze and Kohlstedt, 1973) indicate that mantle rocks can be described by a non-linear or non-Newtonian power-law rheology (e.g. Ranalli, 1987) which is characterized by a non-linear dependence of the viscosity on the strain-rate elements. Because of constraints due to limited time-duration of the experiments, geophysicists have traditionally taken a power-law dependence in numerical modelling (e.g. Parmentier et al., 1976) of the mantle. This takes the form

$$\varepsilon_{ij} = A^{-1}\sigma^{n-1}\sigma_{ij} \tag{5}$$

Where A is the viscosity for n=1, and n is the power-law index. For mantle substances n lies between 3 and 4 (Ranalli, 1987) and σ is the second invariant of the deviatoric stress tensor. From a computational point of view σ_{ij} must be rewritten in terms of the velocity components. We have then

$$\sigma_{ij} = A^{\frac{1}{n}}\varepsilon^{\frac{1-n}{n}}\varepsilon_{ij} \tag{6}$$

where ε is the second invariant of the strain-rate tensor, defined here to be

$$\varepsilon = \left[\varepsilon_{ij} : \varepsilon_{ij}\right]^{\frac{1}{2}} \tag{7}$$

The simplicity of this type of formulation for non-linear mantle rheology is due to the type of laboratory experiments capable of being conducted today. The effective Ra is defined as

$$Ra = \frac{2^{\frac{n-1}{n}} \alpha g d^{\frac{n+2}{n}} \Delta T \rho}{K^{\frac{1}{n}} A^{\frac{1}{n}}} \tag{8}$$

where d is the depth of the layer, K is the thermal diffusivity, ρ is the average density of the mantle, ΔT is the temperature drop across the mantle, g is the gravitational acceleration and α is the thermal expansivity. We note that there are differences in the definition of Ra, depending on the power-law index n. Henceforth, Ra will be used for Newtonian convection, while Ra' will be employed for non-Newtonian flows. Thus one must use another criterion for comparing Newtonian and non-Newtonian convection. A good candidate would be the time-averaged surface Nusselt (Nu) number, which measures the amount of convective heat transport and is an output of the nonlinear convective system.

The results reported below are taken from two-dimensional numerical simulations in a rectangular domain for a configuration heated purely from below. Stress-free boundary conditions appropriate for mantle convection in the Earth (Olson and Corcos, 1980) are applied at the top and bottom boundaries, which are impermeable (vanishing vertical velocity). Reflective boundary conditions on the shear stress, horizontal velocity and horizontal temperature gradient are imposed along the vertical boundaries. The temperature is set to 0 at the top and 1 at the bottom.

The stream function approach is employed to satisfy the continuity equation (eqn. 1). By taking the curl of the momentum equation we obtain a single scalar partial differential equation for the stream function ψ. For non-Newtonian rheology this is a nonlinear elliptic partial differential equation because η depends on the derivatives of ψ. It takes the form:

$$\left(\frac{\partial^2}{\partial z^2} - \frac{\partial^2}{\partial x^2}\right)\eta\left(\frac{\partial^2 \psi}{\partial z^2} - \frac{\partial^2 \psi}{\partial x^2}\right) + \frac{4\partial^2}{\partial x \partial z}\eta\frac{\partial^2 \psi}{\partial x \partial z} = Ra\frac{\partial T}{\partial x} \tag{9}$$

In the case of constant Newtonian viscosity, we have the usual biharmonic equation

$$\nabla^4 \psi = Ra\frac{\partial T}{\partial x} \tag{10}$$

The effective viscosity η depends on the shear-deformation and is

$$\eta = \frac{1}{2}\left[\frac{1}{2}\left(\frac{\partial^2 \psi}{\partial z^2} - \frac{\partial^2 \psi}{\partial x^2}\right)^2 + 2\left(\frac{\partial^2 \psi}{\partial z \partial x}\right)^2\right]^{\frac{1-n}{2n}} \tag{11}$$

The energy equation is

$$\frac{\partial T}{\partial t} = \nabla^2 T + \frac{\partial \psi}{\partial x}\frac{\partial T}{\partial z} - \frac{\partial \psi}{\partial z}\frac{\partial T}{\partial x} \tag{12}$$

From eqn. (9) or (10), ψ can be solved in terms of T. Thus the nonlinearity of the temperature equation is made more evident by rewriting (12) as

$$\frac{\partial T}{\partial t} = \nabla^2 T + a(T)\frac{\partial T}{\partial z} + b(T)\frac{\partial T}{\partial x} \tag{13}$$

where a(T) and b(T) are derived from ψ in terms of T from the elliptic momentum equation, eqn. (9).

The numerical methods for solving eqn. (9), (10) and (11) are based on bi-cubic splines and the details can be found in Malevsky and Yuen (1991, 1992). In the case of nonlinear rheology ψ must be solved by means of iterative techniques, because the viscosity depends nonlinearly on the derivatives of ψ. Eqn. (11) is a nonlinear time-dependent advection-diffusion equation. We have employed a finite-element scheme based on the Lagrangian formulation of the total time derivative in the energy equation (Malevsky and Yuen, 1991). This scheme is fourth-order in space and second-order in time. This method is very efficient for strong time-dependent thermal convection, which is dominated by advection. Mesh sizes range from 120 x 28 grid points for the lowest Ra to 800 x 170 for the highest Ra cases. The grid is unevenly distributed along the vertical and equally spaced in the horizontal direction.

2.2. RESULTS FOR NEWTONIAN CONVECTION

We will now show the changes in the temperature fields, when Ra is increased from 10^6 to 10^8 for Newtonian (n=1) rheology. In Fig. 1 are shown the temperature fields for base-heated convection in an aspect-ratio 1.8 box for Ra=10^6 (top) and Ra=10^8. These snapshots are taken after several overturns.

We can observe regular cellular temperature fields for the lower Ra. The hot (white) and cold (black) plumes are continuous for Ra=10^6 and are along the edges. The shape of the bent cold plume at the left shows the possibility of complex structures developing in time-dependent flows. The presence of a large-scale flow can also be detected. As Ra is increased beyond 10^7, disconnected plumes begin to appear for Ra=10^8. There is still vertical symmetry at this point but will be broken eventually because of numerical roundoff. The plumes develop multiple branches in this hard-turbulent regime. These branches are much more fragile and can be torn off easily by the vigorous large-scale flow. At a higher Ra of 10^{10} (Fig. 2) we can see thinner plumes with smaller diapirs emerging. The diapiric bursts have characteristic frequencies in the hard-turbulent regime (Vincent et al., 1991, 1992). Such a high Ra may be characteristic of the Earth's early mantle, when the internal temperatures were much hotter and the viscosity considerably lower (e.g. Sharpe and Peltier, 1978; Stevenson, 1989).

The diapirs, which are isolated thermal anomalies, are characteristic of hard-turbulent convection and are not commonly found in lower Ra's, such as 0(10^6) for Newtonian rheology. Most of the heat-transfer takes place by the large-scale circulation (Hansen et al., 1992). The interior becomes much more isothermal (green color) as a consequence of the strong convective mixing.

2.3. RESULTS FOR NON-NEWTONIAN CONVECTION

In non-Newtonian convection the time-dependent flow structures are different from the ones found for Newtonian rheology because of the stress-dependent viscosity. This nonlinear attribute induces significant curvatures in the plume trajectories of non-Newtonian flows (see Fig. 3). Sinking and rising plumes are changing their positions continuously. Separate diapirs are being constantly ejected from the boundary layers. We see that the cell interior is filled with both detached blobs and continuous plumes.

152

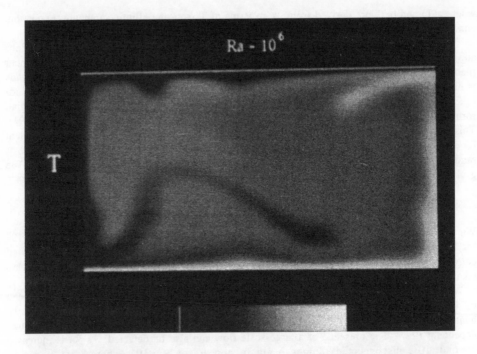

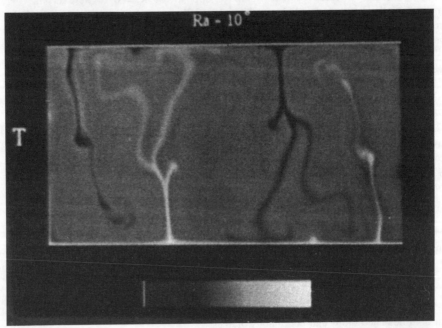

Fig. 1. Temperature T fields for Ra=10⁶ and 10⁸ in an aspect-ratio 1.8 box. Grid points used are 160 x 48 for Ra=10⁶ and 400 x 140 for Ra=10⁸.

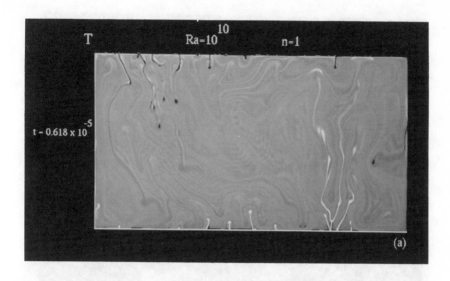

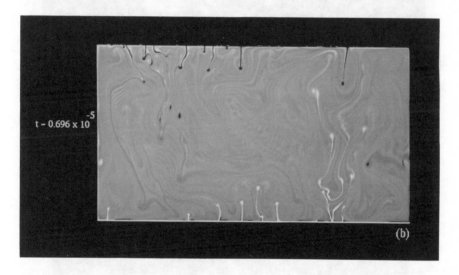

Fig. 2. Temperature field for Ra=10^{10}. Gridpoints used are 800 x 170. Temperature color scales are linearly divided into 25 intervals.

154

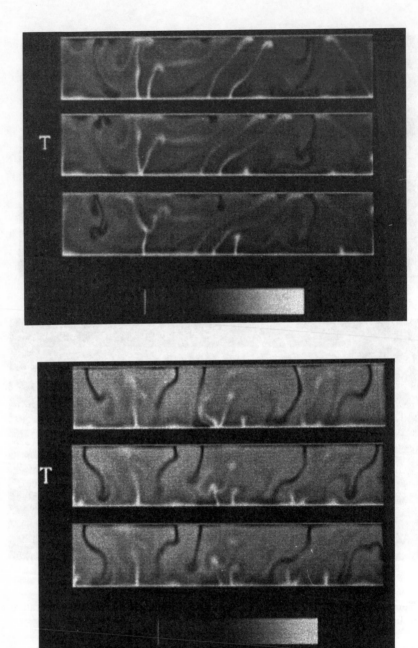

Fig. 3. Evolution of temperature fields for Ra'=4500, n=3 (top) base-heated only, R=0 (bottom) internally heated R=7. An aspect-ratio of five is used. A mesh with 300 x 48 elements is employed. Time instants are after five overturns. Averaged dimensionless surface heat-transport are 27 and 23 respectively for R=0 and 7.

The contrast between internally heated and base-heated configurations is brought out in Fig. 3. The descending plumes of the internally heated case come down quite vigorously and the hot plumes do not develop long-trailing swarms of diapirs as in the base-heated situation. There are some regions which are substantially overheated in non-Newtonian convection with internal-heating. The interior is much hotter than in Newtonian internally heated convection with the same convective vigor. Non-Newtonian convection with Ra'=2500 roughly corresponds to Newtonian with Ra=10^6 in terms of the surface heat transport. An upper limit of R=16 was found for Newtonian whole mantle convection to avoid melting (Leitch and Yuen, 1989). A value of R=10 would cause wholesale melting for non-Newtonian whole mantle convection. Hence, non-Newtonian mantle will tolerate less internal heating than a Newtonian mantle in order to avoid large-scale melting. The averaged surface heat flow is only around 17% higher for the internally heated case. There are still many open questions left regarding the relative efficiency of heat-transport between base-heated and internally heated configurations in strongly time-dependent non-Newtonian convection. At low Ra' viscosity fields of non-Newtonian convection are cellular in structure and correlate well spatially with the smooth flow structure (Malevsky and Yuen, 1992; van den Berg et al., 1992). At higher Ra' the effective viscosity begins to assume sharper spatial heterogeneities (Malevsky and Yuen, 1992). In the hard turbulent regime viscosity field of non-Newtonian convection becomes granular in appearance and has a multiplicity of spatial-scales, as shown below in Fig. 4.

The strong contrast between rapid motion in one part of the convecting medium and stagnation in some other part is a very distinct feature of time-dependent non-Newtonian convection (Christensen and Yuen, 1989). The sharp spatial heterogeneity can be found for turbulent non-Newtonian convection (Fig. 5) but the areas of vigorous motion are not long-lasting. They can appear and disappear anywhere randomly. We have displayed the $\sqrt{\text{speed}}$ in Fig. 5 in order to show how dramatically the fluid motions change from instant to instant. Speed is defined as the modulus of the velocity field vector. It is a much better quantity than streamlines for displaying the evolution of the flow field intensity in non-Newtonian convection.

Strong temporal fluctuations are found in the time series of the root-mean-squared velocity v_{rms}. The amplitude of these fluctuations is almost an order of magnitude higher than the averaged value of v_{rms}. Fluctuations of v_{rms} in Newtonian convection with the same averaged intensity of motion have much smaller magnitudes (Malevsky and Yuen, 1991). These results would imply that rapid shifts in plate speeds and the attendant shifts in moment of inertia and polar motions (Goldreich and Toomre, 1968) are much more likely for a non-Newtonian mantle than for a Newtonian one. In contrast, the spatially averaged value of the averaged value of the inverse of viscosity η_{inv} fluctuates by a factor of three and is less than v_{rms}. We have used a root-mean-squared average to calculate η_{inv}. The inverse of viscosity was used because of singularities in the viscosity in the stagnant regions. In Fig. 6 we show the temporal development of the root-mean-squared velocity and the spatially averaged value of the inverse of the effective viscosity. The smaller Ra' of 3000 with R=7 has been taken.

Lateral viscosity variations can influence the inference of mantle viscosity from postglacial rebound data. At low Ra' the low viscosity zones are associated with sinking and rising currents (Malevsky and Yuen, 1992). The amplitudes of the interior viscosity variations are less than those near the vertical flow structures. Viscosity field of turbulent non-Newtonian convection becomes granular and has very small-scale heterogeneities. These small-scale

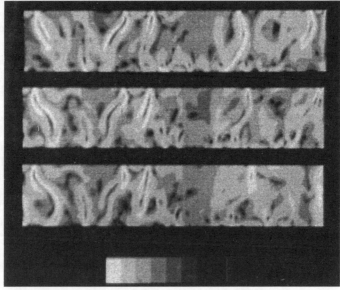

Fig. 4. Logarithmic effective viscosity of base-heated non-Newtonian convection with Ra'=4500, R=0 and n=3 (top panel) and internally heated convection with Ra'=4500, R=7, n=3 (bottom panel) at the same times as for the previous figure. The intensity of the greyscale is linearly proportional to the logarithmic viscosity. Dark areas denote high viscosity while light areas indicate low viscosity. Logarithm of viscosity ranges from -3.75 to 0.55 for R=0 and from -3.5 to 0.8 for R=7.

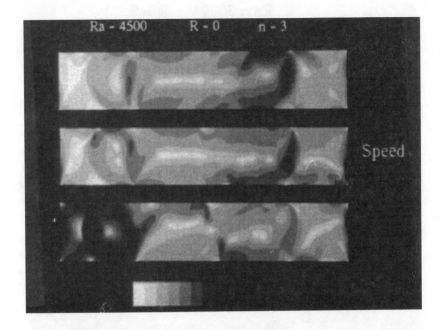

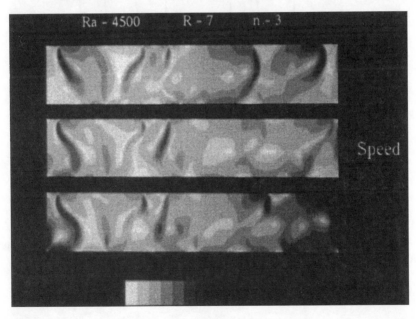

Fig. 5. Square of the speed of base-heated non-Newtonian convection with Ra'=4500, R=7, n=3 (bottom panel) at the same times as in Fig. 4. The intensity of the greyscales is linearly related to the √speed. Dark and low areas denote respectively high and low speeds. Square root ranges from 0 to 100.

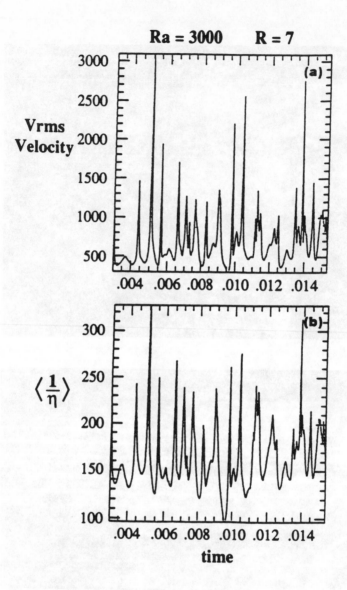

Fig. 6. Root-mean-squared velocity time series (top panel) and the evolution of the spatially averaged inverse of the effective viscosity (bottom panel) for Ra'=3000, R=7 and n=3.

fluctuations are associated with the many diapirs. In Fig. 7 the horizontal Fourier spectra of $\frac{1}{\eta}$ at the depth of z=0.1 below the top boundary layer are shown for contrasting Ra's of 500 and 4500.

For high Ra' of 4500 the spectrum of viscosity heterogeneities shows little decay over nearly two decades in wavelength (top curve in Fig. 7). For the most part the deviations of the spectral components from the average value (k=0) do not exceed an order in magnitude for both cases. Thus the viscosity field of strongly chaotic non-Newtonian thermal convection appears to be more uniform for processes averaged over long wavelengths, than at lower Ra'. This has important implications for postglacial rebound and the retrieval of mantle viscosity.

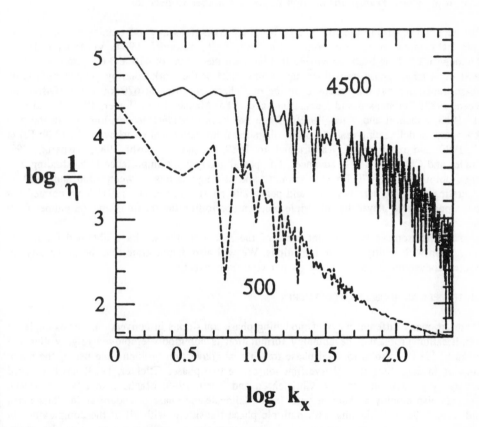

Fig. 7. Horizontal Fourier spectra of the inverse of the effective viscosity for Ra=500 (dashed line) and Ra'=4500 (dotted line), n=3 at a time instant after several overturns. Depth is z=0.1 below the top boundary layer. The horizontal wavenumber is given by k_x.

Horizontal spectra of viscosity in turbulent non-Newtonian convection exhibits a power-law decay with k far away from the horizontal boundary layers (Malevsky et al., 1992). This power-law decay indicates a self-similar structure of the viscosity isolines over many different scales. A fractal dimension of 1.6 to 1.8 for the isoviscosity lines can be derived from the power-law dependence (Malevsky et al., 1992). A fractal dimension of 1.6 to 1.8 would indicate the presence of large-scale coherent structure in these chaotic non-Newtonian flows.

In general, there is great richness in the complexity of strongly chaotic non-Newtonian convective flows. Much is still left to be discovered. But the journey there will be much more arduous than the Newtonian path because of the much greater computational demands, by factors of 3 to 5, than the Newtonian counterparts.

3. Multiple Phase Transitions in High Rayleigh Number Convection

The question regarding the role played by phase transitions in geodynamics is not new and goes back many years (e.g. Vening-Meinesz, 1956; Knopoff, 1964; Verhoogen, 1965). Although it has long been recognized that between the depths of 410 and 670 km there are at least three major phase transitions, up to now most of the work done by geodynamicists on phase transitions has only focused on the one phase transition at 670 km depth (Olson and Yuen, 1982; Christensen and Yuen, 1984 and 1985; Machetel and Weber, 1991; Sabadini et al., 1983; Sabadini and Yuen, 1989). There has been considerable development in the past few years on delineating realistic phase diagrams from theoretical (Kuskov et al., 1989; Fei et al., 1990) and experimental (Katsura and Ito, 1989; Ito and Takahashi, 1989; Gasparik, 1990; Böhler and Chopelas, 1991) advances. Of special geodynamic importance is the finding of a triple point near the 670 km discontinuity and at temperatures between 2000 and 2300 K, separating the β-spinel, γ-spinel, and perovskite phases. Liu et al. (1991) have pointed out the important role played by this triple point in controlling the traffic of mantle plumes from the deep mantle.

In this section we will present some of the recent results we have obtained for mantle convection with multiple phase transitions. We will also demonstrate that the propensity for layered convection is enhanced by greater vigor in convection.

3.1. NUMERICAL MODELLING PROCEDURES

There are many different ways of modelling phase transitions in convection. For example, in solidification problems, the enthalpy formulation is commonly employed (e.g. Voller and Prakash, 1988). In solid to solid phase transitional convection problems the use of the phase function in describing the relative function of the two phases (Richter, 1973) has been used greatly by geodynamicists (e.g. Christensen and Yuen, 1985; Machetel and Weber, 1991). One can also employ a chain of particles to delineate the phase boundary as in Christensen and Yuen (1984). In dealing with multiple phase transitions with all of the complexities in triple point and diverse segments of Clapeyron slopes we have employed the phase distribution method (Liu et al., 1991) which basically prescribes a probability function for a particular Clapeyron curve to exist on the (T,p) plane. For any segment of an equilibrium phase boundary, governed by the Clapeyron slope γ, we may write the phase distribution function f as

$$f(T,p) = \exp\left[-(\gamma T - p + p_0)^2 / \tau^2\right] \tag{14}$$

where p_0 is the pressure intercept and τ is the dimensionless pressure width of the Gaussian peak used to control the thickness of the transition zone. This Gaussian function, by virtue of its local properties, delineates the location of the phase transition zone, whose width is taken to be around 10 km. By means of this phase distribution function, we can generalize this approach to multiple phase transitions buoyancy, expressed in terms of an effective thermal expansivity α. In dimensional form it is given by

$$\alpha(T,p) = \alpha_0(p) + \sum_{i=1}^{k} \frac{\Delta \rho f i}{\rho_0 \delta T} \tag{15}$$

where ρ_0 is the background mantle density, $\Delta \rho_i$ is the density change across the phase change, and δT is the temperature change over which the phase transition takes place. It is positive for positive Clapeyron slope and negative for negative (endothermic) phase slope. The background thermal expansivity $\alpha_0(p)$ decreases with depth in accordance with recent laboratory findings (Chopelas and Böhler, 1989). The index k represents the number of phase transition segments under consideration. An effective heat capacity can likewise be written as

$$C_p(T,p) = C_p \left(1 + \sum_{i=1}^{k} \frac{|1 \Delta H_i| f_i}{C_p \delta T} \right) \tag{16}$$

where the latent heat release per unit mass is given by

$$\Delta H_i = \frac{\gamma_i T \Delta \rho_i}{\rho_0^2} \tag{17}$$

The background heat capacity is taken to be 1kJ/kg K. T is in degrees Kelvin; and γ_i is the Clapeyron slope of the individual phase-boundary.

The extended-Boussinesq approximation (Christensen and Yuen, 1985) will be assumed to study the problem of convection through multiple phase transitions. This means that the background density is assumed to be a constant of 3,500 kg/m³ model throughout the mantle. This approximation enables us to treat much easier multiple phase transitions with triple points and other complexities easier than in the anelastic compressible models (e.g. Machetel and Weber, 1991; Solheim and Peltier, 1992) and also to reduce the computational time.

The dimensionless momentum equation in the stream-function-vorticity (ψ, ω) formulation as

$$\nabla^2 \omega = -Ra_0 \bar{\alpha}(T,z) \frac{\partial T}{\partial x} \tag{18}$$

where x and z are, respectively, the dimensionless horizontal and vertical coordinates with the z-axis pointing upward, and Ra_0 is the Rayleigh number based on the depth of the layer, and the surface values of the physical properties. The dimensionless form of thermal expansivity $\bar{\alpha}(T,z)$ has been converted from the corresponding hydrostatic pressure. The dimensionless time-dependent temperature equation is

$$\frac{\partial T}{\partial t} = \frac{1}{\bar{C}_p(T,z)} \left[\nabla^2 T + w D_0 \bar{\alpha}(T,z)(T + T_0) + R + \frac{D_0}{Ra_0} \tau_{ij} \frac{\partial u_i}{\partial x_j} \right] - \underset{\sim}{u} \cdot \nabla T - \sum_{i=1}^{k} w \frac{\overline{\Delta H_i} f_i}{\delta z} \tag{19}$$

where w is the vertical velocity, u_i is the component of velocity vector, τ_{ij} is the deviatoric stress tensor, D_0 is the dissipation number (Jarvis and McKenzie, 1980) based on the surface properties, δz is a characteristic width of a phase transition zone, T_0 is the non-dimensional surface temperature and R is the strength of the internal heating parameter for a bottom-heated configuration (Weinstein et al., 1989). The dimensionless enthalpy change is given by

$$\overline{\Delta H_i} = \frac{\Delta H_i}{C_p \Delta T} \tag{20}$$

There are three free parameters in this phase-change formulation, namely $\tau, \delta T$ and δz. We have used $\tau = 1.05$ GPa, and $\delta z = 15$km in these calculations. The width τ and δz can be bounded by the grid resolution. The temperature change δT in this formulation assumes a value of 50K. This can be bounded from above by the product of the inverse Clapeyron slope and the width δz. The momentum equation includes the effect of locally enhanced thermal expansivity from density changes across phase transitions. In the energy equation latent heat effects from advection of the phase boundary enhanced heat capacity and enhanced adiabatic heating produced by phase changes have been included. The effects of the phase boundary distortion are modelled by the movement of the peaks of the probability functions f in response to variations of the thermal anomalies. Eqn. (4) is a simplified approach which does not take into account the nonlinearity of the thermal inertia term due to Cp (Christensen and Yuen, 1985). The basic physics of convection with phase transitions can still be captured with this level of approximation. This facilitates numerical solution of the governing partial differential equations with three unknowns, ψ, ω and T.

The boundary conditions at the top and bottom are constant temperature and stress-free. Periodic boundary conditions for the temperature and velocity fields are employed along the vertical boundaries. A model mantle which has been scaled to 1500 km depth and a width of 4500 km. For Ra lower than $O(10^7)$ an equidistant grid of 100 x 300 points was used, but above Ra~$O(10^7)$ 150 x 450 points were used. Other technical details can be found in Zhao et al. (1992).

3.2. RESULTS FOR CONVECTION WITH MULTIPLE PHASE TRANSITIONS

In this subsection we will focus our attention on the effects of multiple phase transitions, depth-dependent thermal expansivity $\alpha(z)$ and increasing vigor of convection in the problem of phase-transitional mantle convection.

In the last few years the role played by $\alpha(z)$ in mantle convection has received greater attention mainly because of the experimental demonstration of a strong pressure-dependence of α by Chopelas and Böhler (1989) and Chopelas (1990). Its geodynamical importance has been emphasized by Leitch et al. (1991), Yuen et al. (1991), and Hansen et al. (1991). The analytical formula for $\alpha(z)$ can be found in Zhao and Yuen (1987) and Zhao et al. (1992). Multiple phase transitions, which include the olivine to spinel, β-spinel to γ-spinel, both β-spinel and γ-spinel to perovskite, are considered. Fig. 8 shows a comparison between constant α and $\alpha(z)$ for about the same effective Ra of 3 x 10^6. In Fig. 9 the effects of increasing Ra are illustrated by studying convection through the olivine to spinel transition in a mantle with constant thermal expansivity.

In Fig. 8 we can see from examining the grayscale temperature fields that with the inclusion of $\alpha(z)$ the effects of enhanced adiabatic cooling from the phase transition prevail and a stronger barrier is produced in that locally the temperature gradient steepens. Note that the upper mantle is cooler with $\alpha(z)$, because of the sharp decrease of the mantle adiabat from $\alpha(z)$ (Leitch et al., 1991). Although no hot plumes pass through the transition zone in both cases, it is clear that $\alpha(z)$ acts as a stronger barrier to flows passing through.

From linear theory (Schubert et al., 1975) the olivine to spinel transition was predicted to promote whole-mantle convection. But finite amplitude effects from phase boundary distortion and the latent heat absorption tend to arrest the ascending plumes. Some blockage of flows does occur for Ra=3x10^6 and even more for Ra=2x10^7, as can be seen from the sharper interface formed for Ra=2x10^7. There appears an intermediate zone in the mid-mantle at higher Ra. Thus a multiplicity of vertical scales is produced by the strong chaotic flow.

In Fig. 10 and 11 are shown the effects of increasing Ra on both constant α and $\alpha(z)$ phase-transitional convection.

Fig. 10 shows the effects of increasing Ra for depth-dependent thermal expansivity in the multiple phase-transitions (olivine to spinel and (β, γ) spinel to perovskite) comparing $Ra_0=10^7$ with $Ra_0=3\times10^7$. We can observe that the sharpness of the thermal interface between upper and lower mantle is increased with greater convective strength. Many thermal instabilities are developed at the transition zone (Liu et al., 1991). At higher Ra the cold blobs, breaking through the phase-change barrier, soon become thermally absorbed by the hot lower mantle. At lower Ra the cold descending plumes can reach the bottom because of their thicker boundary layers.

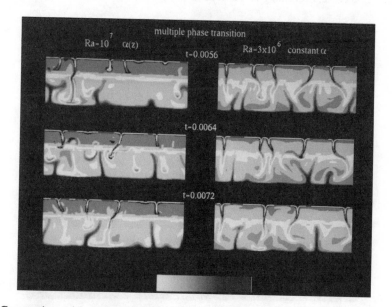

Fig. 8. Convection with multiple phase-transitions for an effective Ra of around 3 x 10^6. Both constant α and $\alpha(z)$ were employed. An internal-heating strength R=10 was used. D_0 was 0.3 and $T_0=0.2$ Time is thermal-diffusion time, 0.001 is equivalent to 72 Myr. Temperature grayscales are divided linearly into 16 intervals.

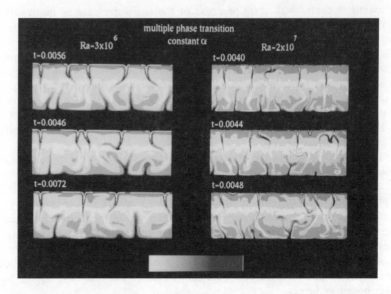

Fig. 9. Convection with only the olivine to spinel phase transition. Constant thermal expansivity. Otherwise, everything is the same as in Fig. 8.

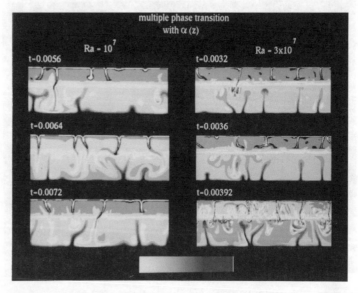

Fig. 10. Convection with multiple phase transitions and $\alpha(z)$. The surface Ra's are 10^7 (left) and 3×10^7 (right). Other parameters are the same as for Fig. 8. 150 x 450 grid points were used for 3×10^7.

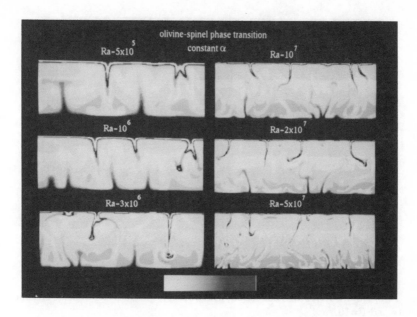

Fig. 11. Effects of increasing Ra on the convection with only the olivine to spinel phase transition. Constant thermal expansivity and R=10. 150 x 450 grid points were used for Ra=5x10^7.

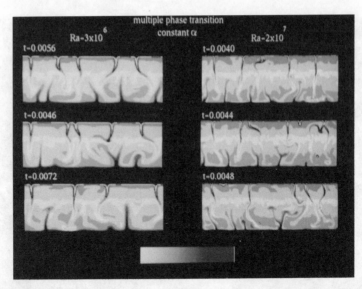

Fig. 12. Temperature fields of convection with multiple phase transitions. Constant thermal expansivity and R=10.

The phenomenon of hard turbulence discussed in section two provides an explanation for the propensity of phase-transitional convection to become more layered at higher Ra. This is due to the greater vulnerability of the disconnected plumes at high Ra to the phase-transition barrier. This result has far-reaching ramifications and implications regarding thermal evolution of the Earth. Conceivably the style of mantle convection might have changed with time from a strongly layered system in an early Earth to one which is only intermittently layered today. Our results show that with any sort of phase-transition present in the upper mantle some degree of obstruction to mantle flows would take place.

In Figs. 12 to 15 we will display the evolutions of the thermal fields, streamfunctions, vertical mass fluxes and local heat transport for Ra=3×10^6 and 2×10^7 in an internally heated system with constant α, and multiple phase-transitions.

Snapshots of the thermal fields and streamfunctions are shown in Figs. 12 and 13. Although the thermal interface appears to be sharper for the higher Ra, the patterns of the streamfunction are quite similar. They reveal many small cells with no large-scale circulation. Some cells are found on top of one another. These results show that the circulation patterns of phase-transitional convections at high Ra are quite different from regular Rayleigh Bénard convection in that small cells are favored over large-scale circulations found convection without phase transitions (e.g. Hansen et al., 1990). This may have implications on geochemical mixing, as such small cells may facilitate lateral mass transport of chemical heterogeneities by their frequent interactions. The relative efficiency in the lateral mass transport between the different styles of convection remains an important question to be answered.

Following Solheim and Peltier (1992), we have employed the dimensionless average vertical mass flux at a given depth to study the effects of phase-transitions on mass transport upwards. As the vertical velocity would scale in normal convection as Ra$^{2/3}$ we would expect an increase in the mass flux with a factor of 3.5. Inspection of Fig. 14 shows that for Ra=2x10^7 the increase of the vertical mass flux falls short of the expected value by around 30 to 40%. This shows that there is a non-negligible amount of blockage from the phase-transitions at these high Ra's.

In Fig. 15 we show the time-series of the local heat-transport Nu=$\frac{\partial T}{\partial z}$-wT at four different depth levels for the two Ra's.

We observe that in all cases the flows, as reflected by the time-dependence of Nu, are more chaotic in the phase-transition zone (400 and 600 km in depth) than at the top and bottom boundaries. This shows that phase-transitions can intrinsically promote chaotic behavior locally in the transition zone and can generate thermal instabilities there (Liu et al., 1991). There can be instants when Nu is amplified by a large factor, 0(10). This is an indication of local turbulence (Grossmann and Lose, 1991). This has important ramifications for geochemical mixing. One can correlate the thermal events between the two depths in the transition zone, but not between the top and bottom of the mantle. Although these solutions have not yet reached a statistically stationary state, the trend enforced by the multiple phase transitions should be evident. Locally in the transition zone heat-transfer is reduced. We should remark here that the time-dependence, as measured by Nu(t), exhibits quite a large vertical variation between the top and bottom boundary layers and the transition zone. Thus these results suggest that the dynamical timescales in the transition zone are much shorter than those at the top and bottom boundary layers.

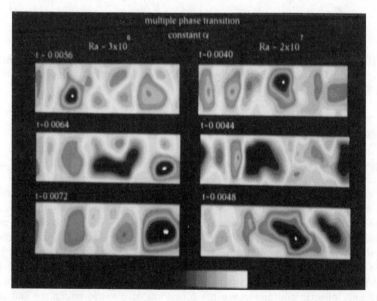

Fig. 13. Streamfunctions of convection with multiple phase changes. Dark areas indicate regions with local maxima. Fast velocities are found in regions with sharp variations in the greyscales.

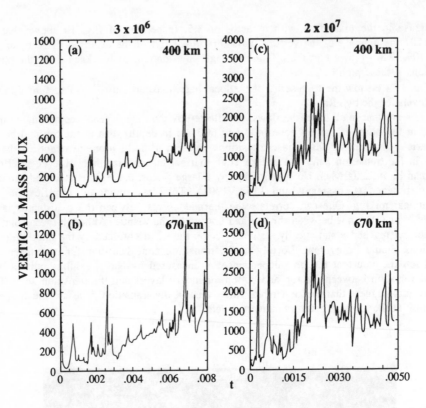

Fig. 14. Time histories of the vertical mass flux at two different depths. The vertical mass flux is calculated from integrating the absolute value of the vertical velocity across the width of the box (Zhao et al., 1992).

4. Concluding Remarks

This paper discusses recent developments in strongly time-dependent thermal convection with applications for the Earth's mantle. A key message of these results is that very complex structures with a multiplicity of spatial-temporal scales can exist in high Ra convection. These structures are not obvious at all from a steady-state perspective. Diapiric structures or disconnected plumes are rather common features associated with hard-turbulent convection. The transition to hard turbulence for non-Newtonian convection takes place at much lower Nu than for Newtonian flows. The development of these disconnected plumes at high Ra is responsible for the propensity toward layered convection with phase transitions. Both non-Newtonian rheology and phase-transitions increase the tendency of the convective system to become strongly chaotic and encourage the formation of diapiric structures and small-scale instabilities. Dynamics with shorter timescales are introduced by the increase in nonlinearity from non-Newtonian rheology and phase transitions.

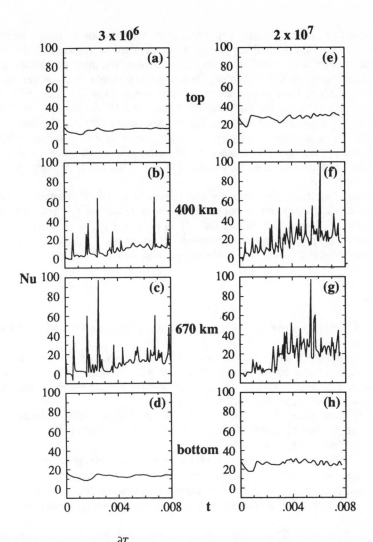

Fig. 15. Local heat-transport, $\frac{\partial T}{\partial z}$ -wT, computed at different depths for the two models. Other parameters are the same as in Fig. 12.

We have found that both non-Newtonian rheology and mantle convection with phase transitions are computationally much more intensive than Rayleigh Bénard convection. Until the next generation of massively parallel computers becomes available, it will be hard to conduct these calculations in three-dimensional configurations and to acquire robust statistics on the properties of these types of strongly time-dependent flows.

Our understanding of strongly time-dependent mantle convection is greatly dependent on our understanding of fundamental aspects of hard turbulent convection. The concepts of hard turbulent convection are still something of a novelty for most geoscientists. From a geophysical perspective, nonlinear rheology with phase transitions in the upper mantle may be sufficient to produce hard turbulence in the upper mantle in the form of intermittent layered convection with many diapiric instabilities generated off the phase boundaries (Liu et al., 1991). In fact, the presence of cold and hot patches can be detected (Morin et al., 1992) and visualized from seismic tomographic data (Zhang and Tanimoto, 1992). Such observational findings are very encouraging, as they offer new vistas into the Earth's interior, which run counter to the steady-state images that still rest in the minds of many geophysicists.

Acknowledgments. We thank our colleagues Ulli Hansen and Alain P. Vincent for stimulating conversations and their continual contributions to works related to this research. We thank also M. Lundgren, L. Weyer, J. Smedsmo for help in preparing this manuscript. This research has been supported by the Innovative Research Program of N.A.S.A., National Science Foundation and Army High Performance Computing and Research Center.

5. References

Balachandar, S. and L. Sirovich (1991) Probability distribution functions in turbulent convection. Phys. Fluids A 3, 919-930.

Bercovici, D., G. Schubert, and G.A. Glatzmaier (1990) Influence of heating mode on three-dimensional mantle convection. Geophys. Res. Lett. 16, 617-620.

Böhler, R. and A. Chopelas (1991) A new approach in laser heating for high pressure mineral physics. Geophys. Res. Lett. 18, 1147-1150.

Castaing, B., G. Gunaratne, F. Heslot, L. Kadanoff, A. Libchaber, S. Thomae, X.Z. Wu, S. Zaleski, and G. Zanetti (1989) Scaling of hard thermal turbulence in Rayleigh-Bénard convection. J. Fluid Mech. 204, 1-30.

Chopelas, A. (1990) Thermal properties of forsterite at mantle pressures derived from vibrational spectroscopy. Phys. and Chemistry of Minerals 17, 149-156.

Chopelas, A. and R. Böhler (1989) Thermal expansion measurements at very high pressure, systematics and a case for a chemically homogeneous mantle. Geophys. Res. Lett. 16, 1347-1350.

Christensen, U.R. (1984) Convection with pressure- and temperature-dependent non-Newtonian rheology. Geophys. J.R. Astr. Soc. 77, 343-384.

Christensen, U.R. and D.A. Yuen (1984) The interaction of a subducting lithospheric slab with a chemical or phase boundary. J. Geophys. Res. 89, 4389-4402.

Christensen, U.R. and D.A. Yuen (1985) Layered convection induced by phase transitions. J. Geophys. Res. 90, 10291-10300.

Christensen, U.R. and D.A. Yuen (1989) Time-dependent convection with non-Newtonian viscosity. J. Geophys. Res. 94, 814-820.

Davies, G.F. (1988) Ocean bathymetry and mantle convection 1. Large-scale flow and hotspots. J. Geophys. Res. 93, 10408-10420.

Fei, Y., S.K. Saxena, and A. Navrotsky (1990) Internally consistent thermodynamic data and equilibrium phase relations for compounds in the system MgO. SiO_2 at high pressure and high temperature. J. Geophys. Res. 95, 6915-6928.

Gasparik, T. (1990) Phase relations in the transition zone. J. Geophys. Res. 95, 15751-15769.

Goetze, C. and D.L. Kohlstedt (1973) Laboratory study of dislocation climb and diffusion in olivine. J. Geophys. Res. 78, 5961-5971.

Goldreich, P. and A. Toomre (1968) Some remarks on polar wandering. J. Geophys. Res. 74, 2555-2567.

Grossmann, S. and D. Lose (1991) Fourier-Weierstrass mode analysis for thermally driven turbulence. Phys. Rev. Lett. 67, 445-448.

Gurnis, M. (1990) Bounds on global dynamic topography from Phanerozoic flooding of continental platforms. Nature 344, 754-756.

Hansen, U., D.A. Yuen, and A.V. Malevsky (1992) A comparison of steady-state and strongly chaotic thermal convection at high Rayleigh number. Phys. Rev. A, in press.

Heslot, F., B. Castaing, and A. Libchaber (1987) Transitions to turbulence in helium gas. Phys. Rev. A 36, 5870-5873.

Ito, E. and E. Takahashi (1989) Postspinel transformations in the system Mg_2SiO_4-Fe_2SiO_4 and some geophysical implications. J. Geophys. Res. 94, 10637-10646.

Jarvis, G.T. and D.P. McKenzie (1980) Convection in a compressible fluid with infinite Prandtl number. J. Fluid Mech. 96, 515-583.

Karig, D.E. (1974) Evolution of arc systems in the western Pacific. Ann. Rev. Earth Planet. Sci. 2, 51-75.

Katsura, T. and E. Ito (1989) The system Mg_2SiO_4-Fe_2SiO_4 at high pressures and temperatures: precise determination of stabilities of olivine, modified spinel and spinel. J. Geophys. Res. 94, 15663-15670.

Knopoff, L. (1964) The convection current hypothesis. Rev. Geophys. 2, 89-123.

Krishnamurti, R. and L.N. Howard (1981) Large-scale flow generation in turbulent convection. Proc. Natl. Acad. Sci. 78(4), 1981-1985.

Leitch, A.M. and D.A. Yuen (1989) Internal heating and thermal constraints on the mantle. Geophys. Res. Lett. 16, 1407-1410.

Leitch, A.M., D.A. Yuen, and G. Sewell (1991) Mantle convection with internal-heating and pressure-dependent thermal expansivity. Earth Planet. Sci. Lett. 102, 213-232.

Liu, M., D.A. Yuen, W. Zhao, and S. Honda (1991) Development of diapiric structures in the upper mantle due to phase transitions. Science 252, 1836-1839.

Machetel, P. and P. Weber (1991) Intermittent layered convection in a normal mantle with an endothermic phase change at 670 km. Nature 350, 55-57.

Machetel, P. and D.A. Yuen (1986) The onset of time-dependent convection in spherical shells as a clue to chaotic convection in the Earth's mantle. Geophys. Res. Lett. 13, 1470-1473.

Machetel, P. and D.A. Yuen (1987) Chaotic axisymmetrical spherical convection and large-scale mantle circulation. Earth Planet. Sci. Lett. 86, 93-104.

Machetel, P. and D.A. Yuen (1989) Penetrative convective flows induced by internal heating and mantle compressibility. J. Geophys. Res. 94, 10609-10626.

Malevsky, A.V. and D.A. Yuen (1991) Characteristics-based methods applied to infinite Prandtl number thermal convection in the hard turbulent regime. Phys. Fluids A3(9), 2105-2115.

Malevsky, A.V. and D.A. Yuen (1992) Strongly chaotic non-Newtonian mantle convection. Geophys. Astro. Fluid Dyn., in press.

Malevsky, A.V., D.A. Yuen, and L.M. Weyer (1992) Viscosity and thermal fields associated with strongly chaotic non-Newtonian thermal convection. Geophys. Rev. Lett. 19, 127-130.

Morin, P.J., T. Tanimoto, D.A. Yuen, and Y.S. Zhang (1992) Hotspots in the Earth's upper mantle. Pixel 3(1), 20-26.

Olson, P.L. and G.M. Corcos (1980) A boundary layer model for mantle convection with surface plates. Geophys. J.R. Astr. Soc. 62, 195-219.

Olson, P.L. and D.A. Yuen (1982) Thermochemical plumes and mantle phase transitions. J. Geophys. Res. 87, 3993-4002.

Parmentier, E.M., D.L. Turcotte, and K.E. Torrance (1975) Numerical experiments on the structure of mantle plumes. J. Geophys. Res. 80, 4417-4425.

Peltier, W.R. and G.T. Jarvis (1982) Whole mantle convection and the thermal evolution of the Earth. Phys. Earth Planet. Inter. 29, 281-304.

Ranalli, G. (1987) Rheology of the Earth: Deformation and Flow Processes in Geophysics and Geodynamics, Chapter 10, Allen and Unwin, Boston.

Richards, M.A., R.A. Duncan, and V.E. Courtillot (1989) Flood basalts and hot-spot tracks: Plume heads and tails. Science 246, 103.

Richter, F.M. (1973) Finite amplitude convection through a phase boundary. Geophys. J. Roy. Astr. Soc. 35, 265-287.

Sabadini, R. and D.A. Yuen (1989) Mantle stratification and long-term polar wander. Nature 339, 373-375.

Sabadini, R., D.A. Yuen, and E. Boschi (1983) Dynamical effects from mantle phase transitions on true polar wandering ice ages. Nature 303, 694-696.

Schubert, G., D.A. Yuen, and D.L. Turcotte (1975) Role of phase transitions in a dynamic mantle. Geophys. J.R. Astr. Soc. 42, 705-735.

Sharpe, H.N. and W.R. Peltier (1978) Parametized mantle convection and the Earth's thermal history. Geophys. Res. Lett. 5, 737-740.

Solheim, L.P. and W.R. Peltier (1992) Mantle phase transitions and layered convection. Canadian J. Earth Sciences, in press.

Solomon, T.H. and J.P. Gollub (1991) Thermal boundary layers and heat flux in turbulent convection: the role of recirculating flows. Phys. Rev. A 43, 6683-6693.

Stevenson, D.J. (1989) Formation and early evolution of the Earth. In Mantle Convection, ed. by W.R. Peltier, pp. 817-873, Gordon and Breach Inc., New York.

Travis, B., P. Olson, and G. Schubert (1990) The transition from two-dimensional to three-dimensional planforms in infinite-Prandtl-number thermal convection. J. Fluid Mech. 216, 71-91.

Turcotte, D.L. and E.R. Oxburgh (1967) Finite amplitude convective cells and continental drift. J. Fluid Mech. 28, 29-42.

van den Berg, A.P., P.E. van Keken, and D.A. Yuen (1992) Mantle convection with a combined Newtonian and non-Newtonian rheology. submitted to J. Geophys. Res.

van der Hilst, R., R. Engdahl, W. Spakman, and G. Nolet (1991) Tomographic imaging of subducted lithosphere below northwest Pacific island arcs. Nature 353, 37-43.

Vening-Meinesz, F.A. (1956) Convective instability by phase transition in the mantle. Proc. Kon. Ned. Akad. B59, 1-12.

Verhoogen, J. (1965) Phase changes and convection in the Earth's mantle. Phil. Trans. R. Soc. London A 258, 276-283.

Vincent, A.P. and D.A. Yuen (1988) Thermal attractor in chaotic convection with high Prandtl number fluids. Phys. Rev. A., 38, 328-334.

Vincent, A.P., U. Hansen, D.A. Yuen, A.V. Malevsky, and S.E. Kroening (1991) On the origin of a characteristic frequency in hard thermal turbulence. Phys. Fluids A3(8), 2222-2226.

Vincent, A.P., A.V. Malevsky, U. Hansen, and D.A. Yuen (1992) On the existence of a spectral buoyancy subrange in thermal turbulence at infinite Prandtl number. Phys. Fluids, in press.

Voller, V.R. and C. Prakash (1990) A fixed grid numerical modelling methodology for convection-diffision mushy region phase-change problems. Int. J. Heat Mass Transfer, part B, 17, 25-41.

Watson, S. and D. McKenzie (1991) Melt generation by plumes: a study of Hawaiian volcanism. J. Petrology 32, 501-537.

Weinstein, S.A. and P.L. Olson (1990) Planforms in thermal convection with internal heat sources at large Rayleigh and Prandtl numbers. Geophys. Res. Lett. 17, 239-242.

Yuen, D.A., A.M. Leitch, and U. Hansen (1991) Dynamical influences of pressure-dependent thermal expansivity on mantle convection. In Glacial Isostasy, Sea-level and Mantle Rheology, ed. by R. Sabadini and K. Lambeck, pp. 663-701, Kluwer Acad. Publ.

Zhang, Y.S. and T. Tanimoto (1992) Ridges, hotspots and their interaction as observed in seismic velocity maps. Nature 355, 45-49.

Zhao, W. and D.A. Yuen (1987) The effects of adiabatic and viscous heatings on plumes. Geophys. Res. Lett. 14, 1223-1227.

Zhao, W., D.A. Yuen, and S. Honda (1992) Multiple phase transitions and the style of mantle convection. Phys. Earth Planet. Int. 72, 185-210.

Zindler, A. and S.R. Hart (1986) Chemical geodynamics. Ann. Rev. Earth Planet. Sci. 14, 493-571.

AN EXPLICIT INERTIAL METHOD FOR THE SIMULATION OF VISCOELASTIC FLOW: AN EVALUATION OF ELASTIC EFFECTS ON DIAPIRIC FLOW IN TWO- AND THREE- LAYERS MODELS

A.N.B. POLIAKOV[1]
HLRZ, KFA-Jülich
Postfach 1913, D-5170 Jülich, Germany

P.A. CUNDALL
Itasca Consulting Group Inc.
1313 5th Street, Minneapolis, MN 55414, USA

Y.Y. PODLADCHIKOV
Institute of Experimental Mineralogy, Chernogolovka
Moscow District, 142 432, Russia

V.A. LYAKHOVSKY
Department of Geophysics and Planetary Sciences
Tel-Aviv University, Ramat-Aviv
69978 Tel Aviv, Israel

ABSTRACT. The explicit finite-difference approach used in the FLAC (Fast Lagrangian Analysis of Continua) algorithm is combined with a marker technique for solving multi-component problems. A remeshing procedure is introduced in order to follow the viscoelastic flow when a Lagrangian mesh is too distorted. Dimension analysis for the case of Maxwell rheology is made. The adaptive density scaling for increasing time step of explicit scheme and influence of inertia are explained.

Analytical and numerical examples of Rayleigh-Taylor instability with different Deborah, and Poisson's ratios are given. A three-layer model with a high viscous upper layer representing the lithosphere has been studied. Amplification of stresses in the upper layer due to unrelaxed elastic stresses and topography elevation for different *De* number and viscosity contrasts is calculated.

1. Introduction

Modelling of viscoelastic flow in geophysics is still a very difficult problem which is quite different from classic numerical modelling of purely viscous or purely elastic media. The fundamental problem in modelling viscoelastic flow is the mixed rheological properties which result in a dependence of the stress on the history of loading. Some finite-element models of viscoelastic behavior are shown by Melosh and Raefsky (1980) for a fluid with a non-Newtonian viscosity, and by Chery et al. (1991) for coupled viscoelastic and plastic behavior.

---

[1]Present address: Hans Ramberg Tectonic Laboratory Institute of Geology, Uppsala University, Box 555, 751 22 Uppsala, Sweden

D. B. Stone and S. K. Runcorn (eds.),
Flow and Creep in the Solar System: Observations, Modeling and Theory, 175–195.
© 1993 *Kluwer Academic Publishers. Printed in the Netherlands.*

Both techniques are very powerful, but they simulate relatively small deformations and thus are limited by the distortion of the Lagrangian grid.

Therefore, it is important to introduce new non-traditional methods to model complex rheologies over long periods of time and which are convenient for remeshing. Numerical methods using the explicit form of the constitutive relation between stress and strain are most appropriate for these purposes (Cundall and Board, 1988). Explicit methods have very short time increments which are chosen to be small enough that perturbations can not physically propagate from one element to the next within one time step. However, the computational effort per time step is very small due to the fact that no system of equations needs to be formed and solved. By performing many short time steps, it is easy to model flows with non-linear rheologies provided that certain stability criteria are satisfied.

However, the simulation of viscoelastic systems requires modelling processes on both short time scales (elastic behavior) as well as on very long ones (viscous flow) simultaneously. Thus explicit methods require a large number of time steps for full simulation. It is important, therefore, to maximize time step during the calculations. As shown by Cundall (1982), this can be accomplished via a density scaling, provided that inertial forces are negligible.

Furthermore, a typical difficulty for the large strain problems is that deformable Lagrangian mesh is required. At some point in the simulation, the distortion of the mesh is so great that it is impossible to continue calculations. A combination of marker tracers which are moving with the grid is found most preferable and fast (Poliakov and Podladchikov, 1992). Markers are used during remeshing for interpolation of physical properties of the system with sharp material discontinuities. However the problem of interpolating the stress state in during the remeshing procedure remains open.

A combination of the FLAC technique and a remeshing procedure allows us to simulate the Rayleigh-Taylor (RT) instabilities in viscoelastic media, but with some limitations explained in a later section.

On the basis of dimensional analysis and numerical calculations we show that the Deborah number De (equal to the ratio of the Maxwell relaxation time of viscoelastic material to the characteristic time of viscous flow) and Poisson's ratio ν control different types of viscoelastic behavior. De number is also equal to the ratio of the stress magnitude to the shear moduli G for Maxwell type rheology. We note that the Maxwell relaxation time τ_{relax} (ratio of viscosity η to shear moduli G) affects only the time scale over which flow occurs and does not affect the qualitative behavior.

We show that the RT instability grows faster as the Deborah number increases and Poisson's ratio decreases.

The estimated limits for the Deborah number of the upper crust are 10^{-4}–10^{-3} and 10^{-3}–10^{-2} for the upper mantle. We show that the flow exhibits viscous behavior for $De = 10^{-4}$–10^{-3} and viscoelastic behavior for $De > 10^{-2}$.

We show that the elasticity has a strong influence on the time evolution of topography and stress in the lithosphere. This occurs when the viscosity contrast between the lithosphere and the underlying mantle is greater than 10^4 for $De = 10^{-2}$–10^{-3}.

2. Numerical Method

2.1. THE CONCEPTUAL BASIS OF FLAC

The method used in FLAC (Fast Lagrangian Analysis of Continua) employs an explicit, time-marching solution of the full equations of motion (Cundall and Board, 1988; Cundall, 1989).

The general procedure basically involves solving a force balance equation for each gridpoint in the body

$$\frac{\partial v_i}{\partial t} = \frac{F_i}{m} \tag{1}$$

where v_i is velocity and F_i is force applied to a node of mass m. Or in its general form,

$$\rho \frac{\partial v_i}{\partial t} = \frac{\partial \sigma_{ij}}{\partial x_j} + \rho g_i, \qquad i, j = 1, 2 \tag{2}$$

where ρ is density, g_i is acceleration due to gravity, and σ_{ij} is the stress tensor.

Solution of the equations of motion provide velocities at each of the gridpoints which are used to calculate internal element strains. These strains are used in the constitutive relation to provide element stresses and equivalent gridpoint forces. These forces are the basic input necessary for the solution of the equations of motion on the next calculation cycle.

Although the dynamic motion equation is implemented, the mechanical solution is limited to equilibrium or steady condition through the use of damping to extract oscillation energy from the system.

2.2. GENERAL NUMERICAL PROCEDURE

The computational mesh consists of quadrilateral elements, which are subdivided into pairs of constant-strain triangles, with different diagonals. This overlay scheme ensures symmetry of the solution by averaging results obtained on two meshes (Cundall and Board, 1988).

Linear triangular element shape functions L_k can be defined as follows (e.g. Zienkiewicz, 1989)

$$L_k = \alpha_k + \beta_k x_1 + \gamma_k x_2, k = 1, 3 \tag{3}$$

where α_k, β_k and γ_k are constants and (x_1, x_2) are grid coordinates. These shape functions are used to linearly interpolate the nodal velocities $v_i^{(k)}$ within each triangular element (e). This yields the following equation for velocity $v_i^{(e)}$ at any point (x,y) within an element

$$v_i^{(e)}(x, y) = \sum_{k=1}^{3} v_i^{(k^{th} node)} \cdot L_k \tag{4}$$

This formula enables the calculations of the strain increments $\Delta \varepsilon_{ij}^{(e)}$ in each triangle (e) as

$$\Delta \varepsilon_{ij}^{(e)} = \frac{1}{2} \left(\frac{\partial v_i^{(e)}}{\partial x_j} + \frac{\partial v_j^{(e)}}{\partial x_i} \right) \Delta t \tag{5}$$

where

$$\frac{\partial v_i^{(e)}}{\partial x_1} = \sum_{k=1}^{3} v_i^{(k)} \cdot \beta_k, \qquad \frac{\partial v_i^{(e)}}{\partial x_2} = \sum_{k=1}^{3} v_i^{(k)} \cdot \gamma_k. \tag{6}$$

At this stage, a mixed discretization scheme is applied in order to overcome the "mesh locking" problem associated with the satisfying incompressibility condition of viscous or plastic flow (Marti and Cundall, 1982). The isotropic part of strain is averaged over each pair of triangles, while the deviatoric components are treated separately for each triangle. This procedure decreases the number of incompressibility constraints by two times and prevents the mesh from locking.

Element stresses are computed invoking a constitutive law

$$\sigma_{ij}^{(e)} = M(\sigma_{ij}^{(e)}, \Delta e_{ij}^{(e)}, S_i) \tag{7}$$

where the operator M is the specified constitutive model, and S_i are state variables which vary with constitutive models.

When the stresses in each triangle are known, the forces at node n, $F_i^{(n)}$ are calculated by projecting the stresses from all elements surrounding that node. The projection of stresses adjacent triangles onto the n'th node is given by

$$F_i^{(n)} = -\sum \frac{1}{2} \sigma_{ij}^{(e)} (n_j^{(1)} \Delta l^{(1)} + n_j^{(2)} \Delta l^{(2)}) \tag{8}$$

where n_j is j's component of the unit vector normal to the each of two element sides adjacent to node n. The length of each side is denoted by Δl. The minus sign is a consequence of Newton's Third Law. After the stresses are projected the gravitational force acting on each node is determined and the force on each node is updated as follows

$$F_i^{(n)} = F_i^{(n)} + m^{(n)} g \tag{9}$$

where $m^{(n)}$ is an equivalent mass of node (n) obtained by distributing continuous density field to discrete nodes.

Once the forces are known, new velocities are computed by integrating over a given time step Δt

$$v_i^{(n)}(t + \Delta t) = v_i^{(n)}(t) + [F_i^{(n)} - \alpha |F_i^{(n)}| sign(v_i)] \frac{\Delta t}{m_{inert}} \tag{10}$$

where m_{inert} is inertial mass of the node which can vary during calculations (see Section 2.3), and α is a damping parameter.

If a body is at mechanical equilibrium, the net force $F_i^{(n)}$ on each node is zero; otherwise, the node is accelerated. This scheme allows the solution of quasi-static problems by damping the oscillation energy. The damping term $\alpha |F_i^{(n)}| sign(v_i)$ is proportional to the accelerating (out-of-balance) force and a sign opposite to velocity to ensure the dissipation of energy. This term vanishes for the system in steady-state.

New coordinates of the grid nodes can be computed by

$$x_i^{(n)}(t + \Delta t) = x_i^{(n)}(t) + v_i^{(n)}\Delta t \tag{11}$$

and then calculations are repeated for new configuration.

This method has an advantage over implicit methods because it is computationally inexpensive for each time step and it is memory efficient because matrices storing the system of equations are not required.

2.3. TIME STEP AND ADAPTIVE DENSITY SCALING

The choice of the proper time step for the time-dependent calculations is a crucial point for stability, precision and run time of the calculations. The time step must be chosen in such a way that information cannot physically propagate from one element to another during one calculation cycle. For elastic and viscoelastic models the critical time step dt_{crit} is the minimum of the Maxwell relaxation time and propagation of the elastic compression wave across a distance equal to local grid spacing Δx. This statement can be written as follows

$$dt_{crit} = min(\frac{\Delta x}{\sqrt{K/\rho_{inert}}}, \frac{\eta}{G}) \tag{12}$$

where K and G are bulk and shear elastic moduli, η is shear viscosity. The inertial density ρ_{inert}, can be treated as relaxation parameter, and can be adjusted during a calculation in order to obtain a desired effect.

If we assume reasonable values for the density and elastic moduli then the time step Δt will be very small and equal to only a few seconds for typical geophysical problems. Therefore, the simulation of creeping flow, which occurs over hundreds of thousands years, will require too many time steps for a full simulation.

One means of circumventing this problem is the adaptive density scaling (Cundall, 1982). For quasi-static problems, the acceleration of the system is nearly zero. Thus, it is possible to increase the value of inertial density, providing that inertial forces $m_{inert}\dot{v}_i$ are small compared to the other forces in the system (i.e. gravitational body force). From eq. 12 we can see that

$$dt_{crit} \propto \sqrt{\rho_{inert}} \tag{13}$$

and therefore in order to increase the time step it is necessary to scale inertial density properly, preserving the stability of the scheme. Note that ρ_{inert} is different from the density used for calculation of the gravitational body force. The algorithm is designed in such a way that if the accelerating (i.e. out-of-balance) forces are smaller then a certain value, then time step and inertial density are increased (Cundall, 1982).

For creeping flow simulations it is necessary to ensure that inertial forces remain small compared to viscous forces (Last, 1988). The Reynolds number is a measure of the ratio of these two forces. We choose to write the Reynolds number as follows

$$Re = \frac{\rho_{inert}VL}{\eta} \tag{14}$$

where V and L are the characteristic velocity and length and η is viscosity of creeping flow. This number is estimated in at each time step cycle and constrains the growth of inertial density. We will show below how this parameter affects the dynamics of the simulations.

2.4. METHOD OF MARKERS AND REMESHING

There are many problems in geophysics which require simulating the dynamics of several phases with different material properties and rheologies simultaneously. For example, the case of Rayleigh-Taylor instability when a lower density fluid rises up and displaces another fluid of higher density.

Lagrangian methods, where the mesh deforms with the fluid, are very fast and are easier to implement than other methods. However, this approach fails when the mesh becomes too distorted.

Fixed Eulerian meshes combined with the method of markers (Hirt and Nichols, 1981; Weinberg and Schmeling, 1992) avoid this problem. This method is robust for finite-difference algorithms on rectangular grids but requires a lot of computational time for non-regular triangular meshes.

The combination of a moving Lagrangian mesh and a method of markers was found to be optimum (Poliakov and Podladchikov, 1992). The idea of this technique can be explained as follows. At the initial stage, material properties of the different layers are assigned to each element and to the markers. Also, within each element the Cartesian coordinates of the markers are converted to local coordinates (area coordinates for triangular elements).

At each time step the grid nodes are then updated according to eq. 11. This Lagrangian movement is very fast because it is only necessary to move the mesh nodes with known nodal velocities.

When the mesh becomes too deformed, it is necessary to remesh. Since the local coordinates of the markers remain unchanged during the Lagrangian movement of the mesh, the Cartesian coordinates of the markers can be obtained by simple interpolation from the nodes of the elements. Only at this stage is it necessary to interpolate from the markers to the array containing material properties of each element. This is in contrast to the Eulerian method where this interpolation must be performed at every time step.

The advantage of this procedure becomes very important in the case of explicit methods which require many time steps and only few remeshing procedures. Another essential advantage of this method is that there are only substantial derivatives on time in constitutive laws (compared to the partial derivatives in space and time required on a Eulerian mesh).

In the case of the viscoelastic rheology we face the additional problem of interpolating the stress field during remeshing. For triangular elements, stresses are piece-wise discontinous across elements. Thus considerable interpolation error can occur after remeshing which can lead to unbalanced stresses within the system. Because of the strong elastic response of the system these unbalanced stresses result in undesirable acceleration and oscillations of nodes. Damping of these non-physical oscillations causes the loss of the history of loading. In other words, stresses and velocities will have jumps and oscillations after each remeshing. Therefore, the results of this paper are partly based on calculations where remeshing is delayed as long as possible to characterize the initial response.

3. Algorithm for the Simulation of Maxwell Behavior

It is convenient to study the response of a viscoelastic material in shear and dilatation separately. Thus the stress σ_{ij} and strain ε_{ij} tensors are decomposed to their deviatoric parts s_{ij}, ε_{ij}^d and isotropic parts σ_{ii}, ε_{ii} as follows

$$\sigma_{ii} = \sigma_{11} + \sigma_{22} + \sigma_{33}, \qquad s_{ij} = \sigma_{ij} - \frac{1}{3}\sigma_{ii}\delta_{ij} \qquad (15)$$

$$\varepsilon_{ii} = \varepsilon_{11} + \varepsilon_{22} + \varepsilon_{33}, \qquad \varepsilon_{ij}^d = \varepsilon_{ij} - \frac{1}{3}\varepsilon_{ii}\delta_{ij}, \quad i,j = 1,3. \qquad (16)$$

The rheological constitutive relations are also separated into their deviatoric and volumetric parts. Recalling that for a linear Maxwell viscoelastic material elastic and viscous strains add and stress components are identical, the constitutive relation for the deviators is

$$\frac{\partial s_{ij}}{\partial t \cdot 2G} + \frac{s_{ij}}{2\eta} = \frac{\partial \varepsilon_{ij}^d}{\partial t} \qquad (17)$$

where G is the elastic shear modulus and η is the shear viscosity. Due to the fact that bulk viscosity does not play an important role and rocks respond elastically in dilatation the constitutive law between isotropic stresses and strains is purely elastic,

$$\sigma_{ii} = 3K\varepsilon_{ii} \qquad (18)$$

where K is the elastic bulk modulus.

Equations 17-18 are solved at each time step (i.e. stresses are updated from the previous time step) as follows.

First, the isotropic and deviatoric components of the initial stress σ_{ij} and strain increments $\Delta\varepsilon_{ij}$ are calculated for the current time step using eq. 5 and 16.

The finite-difference discretization in time of eq. 17-18 gives

$$\frac{s_{ij}' - s_{ij}}{\Delta t \cdot 2G} + \frac{s_{ij}' + s_{ij}}{2 \cdot 2\eta} = \frac{\Delta e_{ij}^d}{\Delta t} \qquad (19)$$

$$\sigma_{ii}' - \sigma_{ii} = 3K\varepsilon_{ii} \qquad (20)$$

where prime accent (') denotes the variables at the end of the time step Δt. Note the semi-implicit approximation of stress in the term corresponding to the viscous strain increment.

Thus the deviatoric and volumetric stresses are updated as

$$s_{ij}' = (s_{ij} \cdot (1 - \frac{\Delta t G}{2\eta}) + 2G \cdot \Delta e_{ij}^d)/(1 + \frac{\Delta t G}{2\eta}) \qquad (21)$$

$$\sigma_{ii}' = \sigma_{ii} + 3K\varepsilon_{ii} \qquad (22)$$

and then full stress tensor will be

$$\sigma_{ij}' = \sigma_{ii}' + s_{ij}'. \qquad (23)$$

When the stresses in each element are known, the net forces acting on each node, updated velocities and coordinates are calculated as described by eq. 8-11 following to the general FLAC algorithm (Section 2.2). This algorithm is applied for the plane strain formulation, therefore $\varepsilon_{33} = 0$ but σ_{33} must be calculated during the calculations.

4. Dimension Analysis for Maxwell Rheology

In order to ensure *rheological and dynamic* similarity between numerical models and simulated natural phenomena, the correct scaling of the constitutive rheological model (eq. 17) and momentum equation (eq. 2) is required (e.g. Weijermars and Schmeling, 1986). Equations are reduced to a convenient form containing scaling parameters or non-dimensional numbers equal to the numbers estimated from nature.

To nondimensionalize the rheological law (eq. 17) we scale stress, time and physical properties as follows

$$s_{ij} = S s'_{ij}, \quad t = T t', \quad \eta = \eta_0 \eta', \quad K = K_0 K', \quad G = G_0 G' \tag{24}$$

where a prime indicates non-dimensional quantities, and S, T are some characteristic values, which will be introduced below. Substituting these expressions into eq. 17-18 gives

$$\frac{\partial s'_{ij}}{\partial t' G'}\left(\frac{\eta_0}{T G_0}\right) + \frac{s'_{ij}}{\eta'} = 2 \cdot \frac{\partial \varepsilon'^d_{ij}}{\partial t'}\left(\frac{\eta_0}{TS}\right) \tag{25}$$

$$\frac{S}{K_0}\sigma'_{ii} = 3K'\varepsilon_{ii} \tag{26}$$

Here we need to choose a characteristic time scale T and there are two possibilities, either to choose a characteristic "viscous" time where

$$\tau_{visc} = \frac{\eta_0}{S} \tag{27}$$

or Maxwell relaxation time

$$\tau_{relax} = \frac{\eta_0}{G}. \tag{28}$$

If we choose the "viscous" time scale then eq. 25-26 become

$$De\frac{\partial s'_{ij}}{\partial t' G'} + \frac{s'_{ij}}{\eta'} = 2 \cdot \frac{\partial \varepsilon'^d_{ij}}{\partial t'} \tag{29}$$

$$\frac{S}{K_0}\sigma'_{ii} = 3K'\varepsilon_{ii} \tag{30}$$

where De is the Deborah number, equal to the ratio of the viscoelastic to the viscous characteristic time

$$De = \frac{\tau_{relax}}{\tau_{visc}} = \frac{S}{G_0}. \tag{31}$$

It is interesting to note that the scaling viscosity factor η_0 is excluded from eq. 30 and affects only the time scale of the process but not its qualitative behavior. In other words, the *rheological* behavior of two Maxwell bodies with two different scaling viscosity factors is similar and differs only in the time scale.

Since variations in density often drive the flow in geophysical problems, the characteristic stress S is chosen to be the hydrostatic pressure

$$S = \Delta\rho_0 g L \qquad (32)$$

where $\Delta\rho_0$ and L are scaling density and length factors defined as follows,

$$\rho = \Delta\rho_0\rho', \quad x_i = Lx'_i. \qquad (33)$$

Using these relations, the Deborah number can be rewritten as

$$De = \frac{\Delta\rho_0 g L}{G_0}. \qquad (34)$$

This definition will be used in the present work. Weijermars and Schmeling (1986) performed the scaling analysis for the momentum eq. 2 and showed that if the characteristic time scale is chosen to be "viscous" (eq. 27) then

$$Re \cdot \frac{\partial v'_i}{\partial t'} = \frac{\partial \sigma'_{ij}}{\partial x'_j} + \rho' \qquad (35)$$

where Re is the Reynolds number (see eq. 14). For a system with low inertia $(Re \ll 1$ in nature) the left-hand side of eq. 35 can be neglected. Then the remaining part of the equation does not contain any scaling parameters or non-dimensional numbers. It means the model and its natural analog are *dynamically* similar since they can be described by the same non-dimensional equation of motion (eq. 35) and the same density field.

Finally we arrive at the conclusion that the numerical solution for a viscoelastic fluid is similar to the modelling of natural phenomena under the same boundary conditions and geometry of the modelling object if: 1) inertia forces in the model are small compared to other forces *(dynamic similarity)*, 2) *De* numbers are equal and distribution of elastic moduli, viscosity and density field are similar *(rheological similarity)*.

Note that two viscoelastic bodies with different relaxation times will behave similarly, if their ratio of stresses to elastic moduli are equal. In this case the difference in relaxation time will affect only the time scale.

5. Analytical Analysis of the Bottom Boundary Layer with Viscoelastic Rheology: Long Wavelength Case

In this section we will consider the behavior of a viscoelastic medium with buoyantly driven flow because it is a common mechanism of flow in the Earth. The analysis of a simple two-layer system can help us to understand the physical behavior of the viscoelastic media in a gravity field.

Biot (1965) performed an analytical stability analysis for layered viscous and viscoelastic media. One of the cases he considered is the stability of a low density layer overlain by an infinite viscous layer of higher density. A generalization of his analysis can easily be done for a viscoelastic rheology, but only for the case of two layers with the same viscosity and elastic moduli.

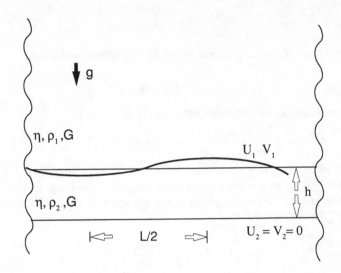

Figure 1: Viscoelastic layer with rigid base lying under an infinite viscoelastic fluid in a gravity field.

Fig. 1 shows the geometry and variables used in our analysis (following Biot, 1965). The bottom layer of density $\rho - \delta\rho$, thickness h lies under a semi-infinite fluid of density ρ. Both layers have viscosity η and elastic shear moduli G. A small sinusoidal perturbation of wavelength L is applied on the interface with displacements U_1, V_1. Displacements on the bottom boundary are zero $U_2 = V_2 = 0$. Hence the problem may be formulated entirely in terms of the two displacement components U_1, V_1 on the top of the layer, and reduced to a system of differential equations.

The characteristic solutions for stresses and displacements of differential equations are proportional to the same exponential factor, $exp(pt)$, where p is the growth rate factor. The displacement field u_i and stresses σ_{ij} are then written

$$u_i(x, t) = u_i(x)exp^{pt}, \qquad \sigma_{ij}(x, t) = \sigma_{ij}(x)exp^{pt} \qquad (36)$$

where the amplitudes $u_i(x)$, σ_{ij} are functions only coordinates x_i, while the time t appears only in the exponential factor.

These characteristic solutions are obtained by substituting eq. 36 into the rheological Eq. 17 and the momentum equation and the application of boundary conditions for the viscoelastic medium. Because the equations are homogeneous, the exponential term is factored out.

An important advantage of this approach is that the characteristic equation is obtained immediately by treating the derivatives as algebraic quantities.

Substituting the stresses and displacements into the constitutive eq. 17 for a Maxwell body, we obtain

$$\sigma_{ij}(x) = \frac{p\eta}{p\tau_{relax} + 1} \cdot 2\varepsilon_{ij}^d(x) \tag{37}$$

where $\sigma_{ij}(x)$ and ε_{ij}^d are the deviatoric stress and strain which are only functions of position.
For the case of pure viscous flow, Biot (1965) gives the expression of growth factor p as

$$p = \frac{\Delta\rho g h}{\eta\sigma} \tag{38}$$

where σ is a nondimensional parameter that depends on boundary conditions and the wavelength of applied perturbation.
Following Biot's derivation and making the following substitution

$$p\eta \Rightarrow \frac{p\eta}{p\tau_{relax} + 1} \tag{39}$$

we obtain the following expression for p

$$p = \frac{\Delta\rho g h}{\eta(\sigma - \frac{\Delta\rho g h}{G})}. \tag{40}$$

The difference in growth factors p between the viscous and the viscoelastic cases is

$$\frac{\Delta\rho g h}{G} \tag{41}$$

which means that if G goes to infinity then the influence of elasticity will be excluded and the system will have a purely viscous behavior. If this ratio increases then the instability will grow faster than for a simply viscous instability. In other words, the elasticity term accelerates the instability. Theoretically a resonance can be reached when this ratio equals σ. This case is irrelevant geophysically.
Note this analysis only applies in the small deformation limit and for the isoviscous case. Therefore the influence of elasticity can be much higher when the deformations become non-linear and for fluids with a high viscosity contrast.

6. Numerical Modelling of Viscoelastic Diapirism

6.1. INFLUENCE OF INERTIA, TIME STEP AND GRID SIZE: TWO-LAYER CASE

A viscoelastic analog of the classic Rayleigh-Taylor instability is the problem of a viscoelastic layer overlain by a viscoelastic layer of greater density. The model geometry is shown in Fig. 2. The two layers are described by their density, viscosity, shear modulus and thickness; ρ, η, G, v and h, respectively. Along all sides the free-slip condition was chosen. The aspect ratio of the box is equal to one. An initial sinusoidal perturbation of magnitude $0.05(h_1 + h_2)$ is superimposed on the boundary between layers. The physical properties of the upper layer were chosen as scaling parameters (eq. 24, 32, 33).

$$\eta_0 = \eta, \quad G_0 = G, \quad K_0 = K, \quad L = h_1 + h_2, \quad \Delta\rho_0 = \rho_1 - \rho_2, \quad T = \frac{\eta_0}{\Delta\rho_0 g L} \qquad (42)$$

Because we use the inertial method for studying non-inertial systems it is necessary to show the influence of inertia on our results. It is always desirable to increase the time step in explicit simulations because of the strong limitation imposed by the stability criteria. It was shown that using the adaptive density scaling the time step could be increased by increasing the inertial density (see eq. 13).

This relation shows that the influence of inertial forces can be controlled by limiting the Reynolds number (see eq. 14). In our calculations we define a maximum Reynolds number that limits the maximum time step and keeps the magnitude of the inertial forces relatively low compared to the viscous forces.

The time-evolution of the vertical velocity on the perturbated interface between two Maxwell layers is shown in Fig. 3 for $Re = 0.001 - 1$. The thicknesses of the two layers are equal ($h_1 = h_2 = 0.5$). The viscosities and elastic moduli are the same for both layers. The Poisson's ratio is $v = 0.25$, the density contrast is $\Delta\rho/\rho_1 = 0.1$ and $De = 0.01$.

As can be seen in Fig. 3 velocities at a given time become smaller as the Reynolds number is increased because it takes more time to overcome inertial effects.

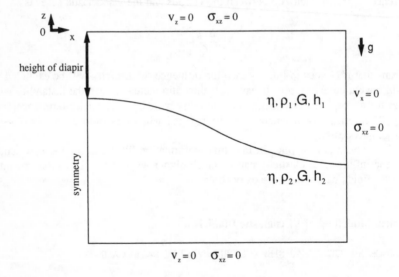

Figure 2: Description of the model for the viscoelastic RT instability. In this section, we consider a model where the rheological parameters of the two layers are equal but the densities are different.

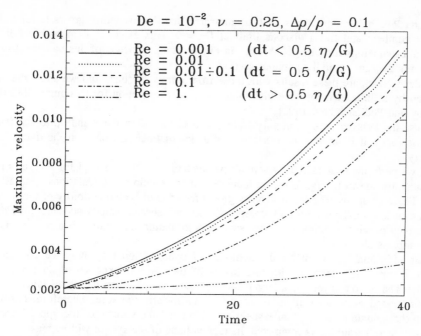

Figure 3: Growth rate of *RT* instability versus time for viscoelastic fluid at different $Re = \rho_{inert}VL/\eta$. It can be seen that the solutions with Reynolds number close to 0.01 show little inertial effect. Physical parameters of the system are shown on the top of the box.

Increasing inertial effects cause the time step to increase and to become close to the relaxation time. The time step exceeds the relaxation time for $Re > 0.1$. Thus there are two limitations on the time step (a maximum Re number and the relaxation time). Depending on the problem either of these limits is more strict and controls the time step.

A comparison of solutions at various grid spacings indicates that a 21×21 grid yields satisfactory results.

6.2. Influence of the *De* Number and Poisson's Ratio

In eq. 40, we showed analytically the influence of elastic moduli on the growth of RT instability. This equation predicts that the incompressible RT instability will grow faster in a viscoelastic medium than in a purely viscous one. However, our analytical solutions assumed that the thickness of the upper layer was infinite. We also assumed that the medium was incompressible. Therefore our numerical solutions with a two-layer system of finite thicknesses and with finite compressibility can not be directly compared with our analytical formulas. Through our analysis of the nondimensional equations we found that the behavior of a viscoelastic body depends upon the Deborah number (see eq. 31), the density contrast

$\Delta\rho/\rho_1$, and Poisson's ratio v. This analysis indicated that the instability grows faster at high Deborah numbers and has a viscous limit at $De = 0$. This effect is demonstrated in our numerical calculations for $De = 10^{-3} - 10^{-1}$ in Fig. 4. The parameters of this numerical model were chosen the same as in the previous section.

In order to compare these results with those from purely viscous fluid we show the curve computed by a finite element code that solves the Stokes equation for incompressible flows (Poliakov and Podladchikov, 1992).

Because the Poisson's ratio can vary from $v = 0.25$ for sedimentary and up to $v = 0.4$ for ultramafic rocks, it is interesting to see the influence of Poisson's ratio on the dynamics of instability.

Thus we performed calculations where all parameters were fixed except for Poisson's ratio. Our results are shown in Fig. 5. As we increase Poisson's ratio our calculations approach the viscous FE calculations. The RT instability grows faster as the compressibility of the material increases (at low v). Thus both bulk compressibility and shear elasticity accelerate the diapir because they provide additional mechanisms of deformation (compared to a purely viscous and incompressible diapir).

As an additional comment on the behavior of compressional systems we consider an unstable compressible system with two layers of the same thickness and an open upper boundary. The bottom layer has a larger uncompressed volume than the upper layer because it is compressed more than upper layer due to hydrostatic pressure. Therefore, during an overturn the volume of the bottom layer will expand and the volume of the upper layer will contract. When overturning is completed the total volume of the system will increase.

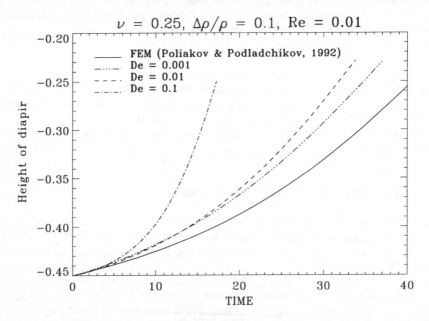

Figure 4: Height of the viscoelastic diapir for different values of Deborah number ($De = \tau_{relax}/\tau_{visc} = S/G$). For comparison with a pure viscous simulation of RT instability the FEM calculations are shown (Poliakov and Podladchikov, 1992).

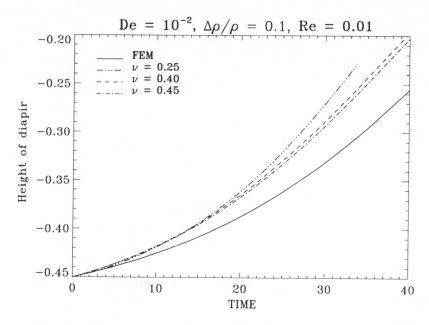

Figure 5: Height of the diapir as a function of time. Each curve corresponds to a solution with a different Poisson's ratio v. Note the convergence of the results to the incompressible FEM calculations with increasing v.

6.3. THREE-LAYER MODEL WITH HIGH RELAXATION TIME FOR UPPER LEVEL: THE INTERACTION OF THE DIAPIR WITH THE HIGH VISCOUS LITHOSPHERE

In this section a three-layer model with a highly viscous upper layer was studied. This third layer can approximate a more viscous lithosphere overlying two gravitationally unstable layers (see Fig. 6). For simplicity the viscosities of two lower layers are chosen to be equal. This choice makes the calculations much faster because the critical time step in both layers is the same. If there were a viscosity contrast between two lower layers then the characteristic velocity in the region would be controlled by the layer with the highest viscosity. However, the time step is limited by lower viscosity as shown above. Therefore the time of calculations is proportional to the viscosity contrast between two layers.

In contrast the viscosity of the upper layer does not influence the characteristic velocity of the diapir and has little effect on the speed of calculations. Thus the viscosity of the upper layer can be greatly increased compared to the viscosity of the bottom layer (up to six orders of magnitude in our calculations) with almost no change in the computational time. In this section we examine the influence of the viscosity contrast η_1/η_2 and De number on the growth of the diapir, topography and stress evolution.

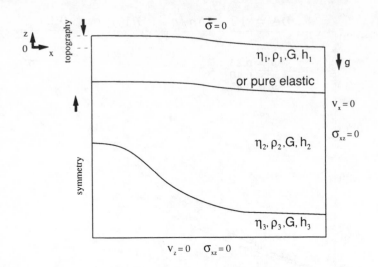

Figure 6: A model representing the interaction of a diapir with viscoelastic lithosphere with high relaxation time $\eta_1/G \gg \eta_2/G$.

The geometry of this problem is shown in Fig. 6. The three layers are described by their densities, viscosities, shear moduli and thicknesses ρ, η, G, v and h, respectively. The upper boundary is stress free and the other sides are free-slip. The aspect ratio is equal to one and an initial sinusoidal perturbation of magnitude $0.05(h_1 + h_2 + h_3)$ was superimposed on the boundary between layers 2 and 3. The physical properties of the intermediate layer were chosen as scaling parameters (eq. 24, 32, 33)

$$\eta_0 = \eta_2, \quad G_0 = G, \quad K_0 = K, \quad L = h_1 + h_2 + h_3, \quad \Delta\rho_0 = \rho_2 - \rho_3, \quad T = \frac{\eta_0}{\Delta\rho_0 g L}. \quad (43)$$

Poisson's ratio was set equal to 0.25 for all models and $\rho_1 = \rho_2$.

Fig. 7 shows the evolution of the velocity field for the following two cases: $\eta_1/\eta_2 = 1$ (left column) and $\eta_1/\eta_2 = 100$ (right column). According to our simulations, the velocity field does not significantly depend on the De number. Calculations with viscosity contrast η_1/η_2 greater than 10^2 show velocities which are very similar (compare Fig. 7a and 7b). This effect can be observed in Fig. 8 (a,b). The evolution of the diapiric growth does not strongly depend on the De number. At the same time we can see that curves representing $\eta_1/\eta_2 \geq 100$ are very similar to each other and are very distinctive from the case $\eta_1/\eta_2 = 1$. For $\eta_1/\eta_2 > 100$ the top layer is effectively rigid and has little participation in the overall flow. The presence of this layer changes the effective boundary conditions on the flow field and also the dimensions of the diapir cell.

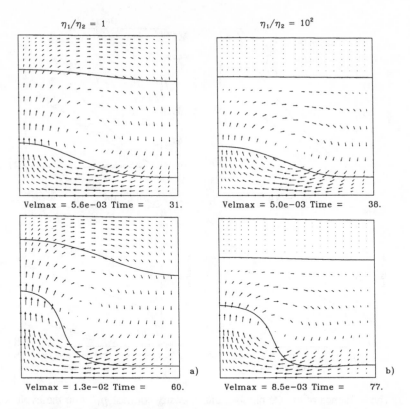

Figure 7: Velocity field evolution for a three-layer system for $De = 10^{-2}$, $\Delta\rho_1/\rho_2 = 0.1$ and $\rho_1 = \rho_2$. a) $\eta_1/\eta_2 = 1$, b) $\eta_1/\eta_2 = 100$. Note that the high viscosity upper layer is excluded from the diapiric cell in the right column.

A different dependence was found for topography, which drastically changes only at the $\eta_1/\eta_2 > 10^4$. This effect can be explained only by the high relaxation time of the upper layer. Unrelaxed elastic stresses in the upper layer resist the growth of topography in this case.

This observation may be supported by comparing the elastic and viscous terms in the rheological equation 30 for the upper layer. Using the nondimensional time interval Δt equal to 25 (taken from Fig. 8) as a characteristic viscous time, we can derive an "effective" Deborah number in layer 1

$$(De\frac{\partial s'_{ij}}{\partial t' G'})/(\frac{s'_{ij}}{\eta'}) \approx \frac{De\eta_1/\eta_2}{\Delta t} = De_1^{(eff)} \tag{44}$$

where G' is taken to be unity and $\eta' = \eta_1/\eta_2$.

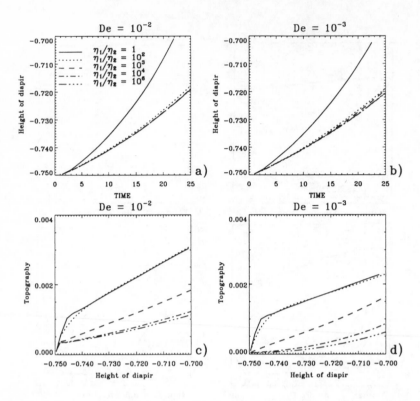

Figure 8: The influence of the *De* number and viscosity contrast η_1/η_2 on the evolution of the diapir and topography. The evolution of diapiric height with time a,b) and evolution of the topography above diapir versus height of diapir c,d). The topography and the diapiric growth rates decrease with increasing viscosity contrast. Note the drastic change at $10^2 < \eta_1/\eta_2 < 10^4$) for the topography whereas for growth rate: $1 < \eta_1/\eta_2 < 10^2$).

Substituting the *De* number (defined for the whole model) equal to the 10^{-2} gives us viscous-like behavior $(De_i^{(eff)} << 1)$ when $\eta_1/\eta_2 < 10^2$, elastic behavior $(De_i^{(eff)} >> 1)$ when $\eta_1/\eta_2 > 10^4$, and viscoelastic behavior for intermediate viscosity contrast.

The topography increases as the *De* number increases for fixed η_1/η_2 (see Fig. 8 c,d). A higher *De* number assumes that the elastic modulus is "softer" (other parameters being fixed). Thus elastic deformations are greater when the Deborah number is lower. The same dependence was outlined analytically and numerically in Section 5 for a two-layer model. A perturbation grows faster and elastic deformations are larger for higher *De* number. In the limits $\eta_1/\eta_2 \Rightarrow 10^4$ and $De \Rightarrow 0$ the topography goes to zero (rigid upper layer) which is consistent with the observed numerical dependencies.

The difference between nearly viscous and nearly elastic behavior can be demonstrated by the two-dimensional distribution of the principal stresses as well (Fig. 9). Strong differences in the magnitudes and orientations of the principal stresses are observed between models with

$\eta_1/\eta_2 = 10^2$ and $\eta_1/\eta_2 = 10^4$. Stresses in the two bottom layers for both cases are approximately the same and differences are observed only in the upper layer. There are two contributions to this difference, one due to viscous stresses and one to unrelaxed elastic stresses. The elastic component can be seen from stress distribution in the upper layer directly above the diapir on the right column. At the bottom of the upper layer the principal stresses change directions because of the effect of bending of an elastic plate. This change is up to 90 degrees at the left boundary. Again, as for topography, these two examples represent two types of the mechanical behavior of the upper layer: "viscous" for $(\eta_1/\eta_2 < 10^2)$ and "elastic" type for $(\eta_1/\eta_2 > 10^4)$.

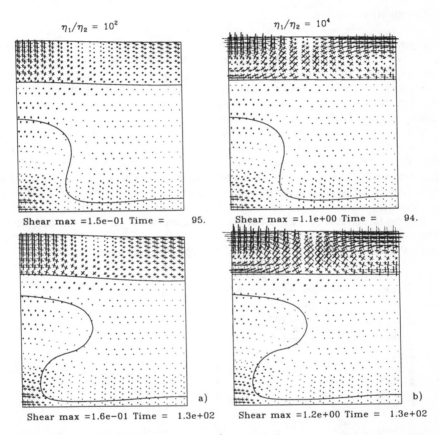

Figure 9: Principle deviatoric stresses at $De = 10^{-2}$. Maximum stress for each case is shown at the bottom of each picture. Scaling stress for all pictures is 1.1×10^{-2}. Thick lines represent compressing stresses. Note the distribution is nearly viscous for $\eta_1/\eta_2 = 10^2$ on the left and non-relaxed elastic for $\eta_1/\eta_2 = 10^4$ on the right.

For geophysical applications it is important to know the magnitude of the stresses on the surface for the determination of the different tectonic mechanisms. In Fig. 10 the evolution of the horizontal surface stress above the center of the diapir is shown. Magnitude of the surface stresses is higher for the higher viscosity ratio η_1/η_2 because the relaxation time is longer in the upper layer. It is interesting to see again the transition between two types of the behavior: elastic for $\eta_1/\eta_2 > 10^2$ and *viscous* for $\eta_1/\eta_2 < 10^2$. Stress is viscoelastic at the viscosity contrast $10^2 - 10^3$. Initially the magnitude of stress grows rapidly because of the rapid elastic response of layer 1 on the upwelling diapir and then the stress exponentially relaxes.

7. Conclusions

We show how the explicit inertial technique FLAC can be applied to geophysical problems with low inertia. This method can easily simulate phenomena with a viscoelastic rheology.

The problem which remains open is remeshing for large strains. It occurs due to problematic interpolating discontinuous stress field from one mesh to another. Combining Eulerian and Lagrangian meshes at the same time and accumulating solution on the non-moving Eulerian mesh may help to solve problem.

Numerical simulations and analytical estimations for the initial stages of the Rayleigh-Taylor instability show that for higher De numbers instability grows faster than for purely viscous (where $De = 0$). Estimates of the Deborah number for sedimentary basins and salt diapirism yields $De = 10^{-4} - 10^{-2}$. From our results this implies that influence of elasticity on the diapirism in the crust is insignificant for isoviscous cases. If we estimate De for mantle diapirism, for example diapirs from the 670 km boundary, then we arrive $De = 10^{-3} - 10^{-2}$. According to our results the elasticity plays a considerable role in the interaction of the lithosphere and underlying mantle and can decrease the surface elevation and increase extensional stresses on the surface above the rising diapir up to one order of magnitude.

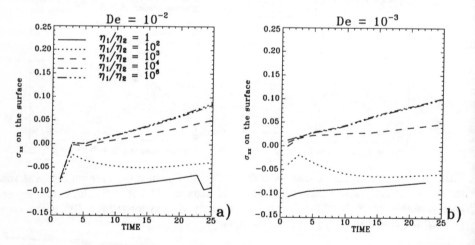

Figure 10: Evolution of the horizontal deviatoric stress at the surface above the center of the diapir for $De = 10^{-2}$ a) and for $De = 10^{-3}$ b).

The study of the morphology of diapiric flow including high viscosity lid shows the decomposition of the diapiric cell into two parts (upper rigid lithosphere and inner diapir cell) for viscosity contrast greater than 100. Decomposition in terms of stresses (unrelaxed stresses in the lithosphere and viscous-like distribution in the diapiric cell) occurs only for viscosity contrast higher than 10^4.

Acknowledgments. A. Poliakov thanks David Stone and the NATO travel fund for making it possible to take part in the meeting. Hans Herrman is greatly thanked for the discussion of the model and providing the excellent HLRZ facilities during completion of the programming, calculations, and preparing the manuscript. We thank Dave Yuen for the review. Christopher Talbot is thanked for the helpful discussion. We are grateful to the Minnesota Supercomputer Institute and Dave Yuen who supported A. Poliakov during writing of the first version of the code. Matthew Cordery and Ethan Dawson are greatly thanked for patient revising of the English of our manuscript. Without Catherine Thoraval and Valentina Podladchikova it would be impossible to complete this work.

Y. Podladchikov was greatly supported by the Swedish Academy of Science during his visit to Uppsala University.

8. References

Biot, M.A. (1965) Mechanics of Incremental Deformations, John Wiley & Sons, New York.

Chery, J., Vilotte, J.P. and Daignieres, M. (1991) Thermomechanical evolution of a thinned continental lithosphere under compression: Implications for the Pyrenees. J. Geophys. Res. 96, 4385-4412.

Cundall, P.A (1982) Adaptive density-scaling for time-explicit calculations. In 4th Int. Conf. Numerical Methods in Geomechanics, Edmonton, Canada, Vol. 1, pp. 23-26.

Cundall, P.A. (1989) Numerical experiments on localization in frictional materials. Ingenieur-Archiv 59, 148-159.

Cundall, P.A and Board, M. (1988) A microcomputer program for modelling largestrain plasticity problems. In G. Swoboda (ed.), Numerical Methods in Geomechanics. Balkema, Rotterdam, pp. 2101-2108.

Last, N.C. (1988) Deformation of a sedimentary overburden on a slowly creeping substratum. In G. Swoboda (ed.), Numerical Methods in Geomechanics. Balkema, Rotterdam.

Marti, J. and Cundall, P.A. (1982) Mixed discretization procedure for accurate modelling of plastic collapse. Int. J. for Num. and Anal. Meth. in Geomech. 6, 129-139.

Melosh, H.J. and Raefsky, A. (1980) The dynamical origin of subduction zone topography. Geophys. J.R. Astr. Soc. 60, 333-354.

Poliakov, A.N. and Podladchikov, Yu.Yu. (1992) Diapirism and topography. Geophys. J. Int. (in press).

Weijermars, R. and Schmeling, H. (1986) Scaling of Newtonian and non-Newtonian fluid dynamics without inertia for quantitative modelling of rock flow due to gravity (including concept of rheological similarity). Phys. Earth Planet. Inter. 43, 316-330.

Weinberg, R.B. and Schmeling, H. (1992) Polydiapirs: Multiwavelength gravity structures. J. Struct. Geol. (in press).

Zienckiewicz, O.C. (1989) The Finite Element Method, 4th edition. McGraw-Hill.

DYNAMICALLY SUPPORTED TOPOGRAPHY AT THE EARTH'S SURFACE AND THE CORE-MANTLE-BOUNDARY: INFLUENCES BY A DEPTH-DEPENDENT THERMAL EXPANSIVITY AND A CHEMICAL BOUNDARY LAYER

D. C. OLBERTZ and U. HANSEN
Department of Geophysics
Institute of Earth Sciences
University of Utrecht
P.O. Box 80.021
3508 TA Utrecht
The Netherlands

ABSTRACT. Two-dimensional time-dependent numerical simulations of the Earth's mantle flow show that a chemical dense layer influences the dynamical behavior in the lower mantle. Decreasing thermal expansion coefficient α (as presumed by recent high-pressure experiments) enables dense material to remain within the transition zone between mantle and core (D''-layer) without being entrained completely by the above mantle flow. Both phenomena—composition and depth-dependent α—affect the flow structure and therefore shape and amplitude of the topography at the surface and at the bottom of the convecting layer.

1. Introduction

Convective currents in the Earth's mantle are now commonly believed to be the ultimate reason behind large scale tectonic activity. Despite this, many features of mantle convection are still debated controversially. Recent progress in various fields like mineral physics, fluid dynamics, seismology and geomagnetism has contributed to a better understanding of mantle dynamics. For a review the reader is referred to the article of Lay (1989).

The dynamical processes in the Earth's mantle are manifested in surface phenomena like plate tectonics or more local features like volcanic activity. In order to relate surface observables, such as heat flow, topography or geoid anomalies, to mantle convection, one has to understand first the fundamental relationship between such surface observables and the actual flow beneath. In this study we have concentrated on the topographic signal as generated by time-dependent convection. Investigations of dynamically supported topography have been conducted previously but were restricted to steady state thermal convection (McKenzie, 1977; Christensen, 1980; Jarvis and Peltier, 1989). Today, mantle convection is assumed to be a highly time-dependent complex flow (Hansen and Ebel, 1988). A reinvestigation of topography signals with regard to such types of flows seems therefore sensible.

The main purpose of this paper is to address the more specified problems of the generation of topography by flows with depth-dependent thermal expansion coefficient, and of the deformation of a chemical boundary layer. Both topics came to the attention of researchers

197

D. B. Stone and S. K. Runcorn (eds.),
Flow and Creep in the Solar System: Observations, Modeling and Theory, 197–206.
© 1993 *Kluwer Academic Publishers. Printed in the Netherlands.*

recently, within the context of the study of the deep interior of the Earth. In order to familiarize the reader with the problem of topography generation by time-dependent convection we introduce the basic features before we move to the mentioned question. Since this study was motivated at least partially by the growing interest in phenomena at the Core-Mantle-Boundary (CMB in the following) we review a few specific features of the CMB after this introduction. More detailed information about topography signals in time-dependent convection can be found in Olbertz (1991).

2. Core-mantle-boundary

The CMB is the transition zone between the Earth's mantle and the outer core. This region is 2900 km below the Earth's surface and its thickness is about 200 to 300 km. The CMB separates two media with crass difference in their physical properties (Lay, 1989). Because of reduced velocity gradients and increased lateral variability relative to the overlying homogeneous mantle this region was identified as the D''-layer. The analysis of seismological data led to the assumption that at the CMB longer wavelength topography exists (Morelli and Dziewonski, 1987). It has been a widespread view that a thermal boundary layer is responsible for the anomalous behavior in the lower mantle. In the following we have outlined some hints which actually question the hypothesis of the D''-layer being a pure thermal boundary layer. If the estimated conductivity (Brown, 1986; Osako and Ito, 1991), internal heat sources (Bercovici et al., 1988), a decreasing thermal expansivity (Chopelas and Böhler, 1989; Anderson et al., 1990; Wang et al., 1991) and an increasing viscosity (Lliboutry, 1987; Ricard and Bai, 1991) are taken into consideration, the strength of the flow in the lower part would be diminished and undulations would be flattened. Recently there has been increased consideration of possible chemical heterogeneities with respect to various seismic and mineral physics results (Lay, 1989). The investigation in different disciplines led to the supposition that localized compositional stratification at the bottom of the mantle may be also responsible for the assumed heterogeneities.

The possibility that thermally and chemically induced density differences are responsible for heterogeneities in the lower mantle let us investigate the effect of thermal and thermo-chemical boundary layer instabilities with regard to dynamical effects on the topography. As mentioned before, the thermal expansion coefficient α has been found to decrease with depth (Chopelas and Böhler, 1989; Anderson et al., 1990; Wang et al., 1991). Decreasing α reduces the thermal buoyancy. In a thermally convecting flow this has strong influence on the shape and amplitude of the dynamically supported topography. Moreover the depth-dependence of the thermal expansivity would affect a layer of dense material in the deep mantle, so that the chemically dense material would be enabled to remain within the transition zone without being entrained completely by the mantle flow above. This compositional boundary now determines much of the bottom layer.

Four numerical experiments are presented as follows: first, time-dependent convection, second the effect of a chemical boundary layer is investigated, third a pure thermally driven flow with decreasing α is examined and fourth both effects are taken into consideration. The following paragraph describes briefly the numerical model with its shortcomings.

3. Numerical Method and Model Set-up

The equations governing the transport of momentum, energy and the concentration of a dense component have been solved by means of a finite element method (Hansen and Yuen, 1989). As a model we have chosen a rectangular box with aspect ratio $\lambda=3$ (λ: ratio of width to depth), filled with a Boussinesq fluid with infinite Prandtl number. The domain is subject to stress-free condition around all sides. Temperature was fixed at the bottom and the top while adiabatic conditions were assumed at the sidewalls. In those cases where the thermo-chemical problem is treated we have assumed no-flux conditions for the concentration on all boundaries. For the initial conditions the reader is referred to the results section. It may be noteworthy to mention that also the evolution of the concentration field was calculated by an Eulerian field approach. The compositional diffusivity was assumed to be smaller than the thermal one by a factor of 100. Thus, we allowed for an unrealistic high compositional diffusion which is however sensible as long as time scales are considered which are small compared to the diffusive time scale. This has been successfully employed in modelling double diffusive phenomena (Hansen and Yuen, 1989).

In all cases where a depth-dependent thermal expansion coefficient a was used we solved the "extended Boussinesq equation" (e.g. Steinbach et al., 1989) taking into account heat exchange by adiabatic compression or decompression and by viscous dissipation. As shown by Steinbach et al. (1989) this approach forms a sensible alternative to fully compressible models.

Instead of calculating the deformation of a free boundary we used a linear approach as proposed by McKenzie et al. (1977) to calculate the topographical signal from the flow fields. The underlying assumption is that the flow determines the deformation of the boundary, but the mantle flow is not affected by the topography. This linearization seems justified as long as the topography anomalies are small compared to the layer depth.

The topography is then determined by the condition that the normal stress should vanish on the deformed boundary. The original method as proposed by McKenzie et al. (1977) requires the calculation of third order derivatives of the streamfunction ψ. In order to avoid the numerical difficulties associated with the calculation of high order derivatives, we introduce the y- component of the vorticity ω_y as given by temperature field through:

$$\nabla^2\omega_y = Ra\frac{\partial T}{\partial x}$$

The dynamical pressure p_D can then be determined from

$$P_D(x_1) = \int\limits_{x_0}^{x_1}\frac{\partial\omega_y}{\partial z}dx$$

and finally dimensionless topographic deformation according to:

$$\Delta h(x,z) = \frac{\alpha_o\Delta T\rho_o}{Ra_o\Delta\rho}\left[2\frac{\partial^2\psi}{\partial x\partial z}(x,z) + \int\limits_{x_o}^{x}dx\left(\frac{\partial\omega_y}{\partial z}(x,z)\right)\right]$$

with

$$z = \begin{cases} 1 \text{ at the surface} \\ 0 \text{ at the bottom} \end{cases}$$

where Ra, T, ρ and α are the Rayleigh number, the temperature, the density, and the thermal expansion coefficient respectively, the subscript o refers to the reference depth at the surface. α_o is set to be 3. x 10^{-5} K^{-1}, ΔT is the temperature difference across the layer, set to be 4000 K, $\Delta\rho$ is the density difference between mantle and adjacent material for which we assumed:

$$\frac{\rho_o}{\Delta\rho} = \begin{cases} 1 \text{ at the surface} \\ -1 \text{ at the bottom} \end{cases}$$

The results have been successfully tested against the standard benchmark as given in Blankenbach et al. (1989).

To our knowledge topography has not been calculated from thermo-chemical convection flows with the exception of Hansen and Yuen (1989). In that case, where the influence of a compositionally dense layer on the topography is considered, the vorticity is calculated according to

$$\nabla^2 \omega_y = Ra \frac{\partial T}{\partial x} - Ra_c \frac{\partial C}{\partial x}$$

where C denotes the concentration of the dense component and Ra_c is the compositional Rayleigh number.

Once ω_y is calculated from the T- and C-fields, the topography can be determined as described above.

4. Results and Discussion

In all four of the experiments described in the following, a Rayleigh number of 10^6 has been chosen. This value is on the low end of the Rayleigh number range which is representative for whole mantle convection. Before we describe the specific numerical experiments we want to familiarize the reader with the general problem of dynamically supported topography in time-dependent thermal convection. In Fig. 1 a few snapshots display the evolution of the temperature field, together with the associated topography at the surface (h(s)) and at the bottom (h(b)). In general, an upflow is correlated with an upward deformation relative to the vicinity at both the top- and bottom boundary (Fig. 1a). Accordingly a negative topography anomaly is formed beneath and above a downwelling current.

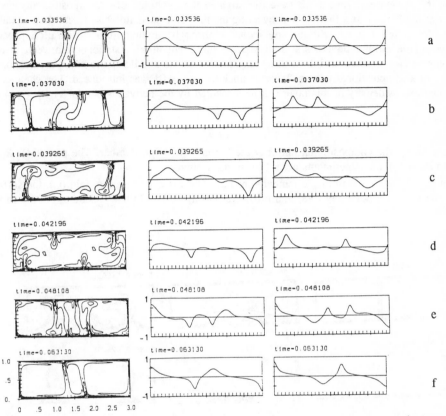

Figure 1. Experiment 1 - Series of thermal convection pattern, α constant, $Ra=10^6$; evolution of temperature (T), surface and bottom topography $h(s)$ and $h(b)$ respectively; normalization is with respect to the maximum magnitude, which is both at the surface and the bottom ± 24000 m.

Boundary layer instabilities which are drifting with the main flow but are not connected from the top to the bottom or vice versa (Fig. 1d at x= 1 shows an instability at the top, at x=2 an example for an instability at the bottom) do not contribute much to the topography signal at the opposite boundary. In other words: the topography at the bottom boundary remains nearly unaffected by drifting instabilities from the top. Fig. 1 also displays a change in the global structure of the flow when it is changing from a 5 cell to a 3 cell pattern, which is accompanied by a drastic change in the topography. This example sheds some light on the general problem of investigating a highly time-dependent phenomenon. The analysis of single snapshots does not allow reconstruction of the evolution of the system. It is even unclear if the single snapshots reflect any typical feature of the system rather than exceptional ones. Thus it seems necessary to understand the dynamics of convecting viscous fluids, also in order to be able to relate snapshots as provided by various descriptions better to the evolution of the Earth.

In the second experiment we have investigated the evolution of a compositionally dense layer at the bottom of a thermally convecting mantle. This scenario shall resemble a situation in which a relict layer without any recharge through subduction or infiltration process between core and mantle exists at the CMB. As initial condition, a stationary convective flow displaying two circulation cells with upward motion in the middle of the box was chosen. At time t=0 a compositionally dense layer of thickness $d' = 0.01$ was introduced. The chemically induced excess density of this layer can be measured by the buoyancy parameter R_ρ given by

$$R_\rho = \frac{Ra_c}{Ra_T}$$

where Ra_c is the compositional and Ra_T the thermal Rayleigh number. The parameter R_ρ is set to $R_\rho = 2$ in the experiments described in this section.

In Fig. 2 the evolution of the temperature field (T), the concentration field (C), the surface and the bottom topography (h(s) and h(b)) are portrayed. The initially existing upflow in the

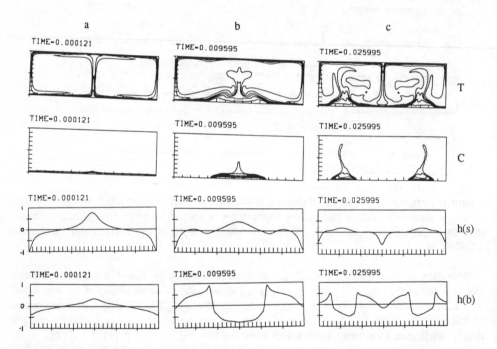

Figure 2. Experiment 1 - α constant, $Ra=10^6$, $R_\rho=2$; evolution of temperature (T), compositional (C) field, surface and bottom topography $h(s)$ and $h(b)$ respectively; normalization of the topography is with respect to the maximum magnitude, which is at the surface $\pm$ 20000 m and the bottom $\pm$ 22800 m.

middle of the box results in a clear topographical high at both the top and the bottom boundary. Conversely the downward directed pressure in the descending limbs of the flow induces a depression at the boundaries (Fig. 2a).

The following snapshots show that the dense material is accumulated beneath the upflow. While there is still a relative high of the topography above the upflow at the surface, a clear depression in topography has been formed at the bottom. Within the hill of accumulated dense material counterrotating currents are driven by viscous stresses thus leading to an internally downward directed pressure (Hansen and Yuen, 1989). After convection has endured for several hundreds of millions of years the flow pattern has drastically changed (Fig. 2c). A downstream has now succeeded in the middle of the box and leads to the topographical lows at the top and at the bottom. The dense material is dragged towards the position of the new ascending plumes resulting in relative topographical depression at these locations (Fig. 2c).

Obviously the presence of a compositionally dense layer at the CMB can lead to anticorrelated topography signals at the top- and bottom boundary of the Earth's mantle. It may be possible in the future to determine the nature of the CMB (pure thermal or thermo-chemical boundary layer) on the basis of this finding.

The recently discovered strong dependence of the thermal expansion coefficient on density (Chopelas and Böhler, 1989) can also potentially lead to a different topographic signal at the bottom and at the top. Any depth-dependence of material properties will lead to a break of symmetry and thus leads to different topographical signals at the bottom and the

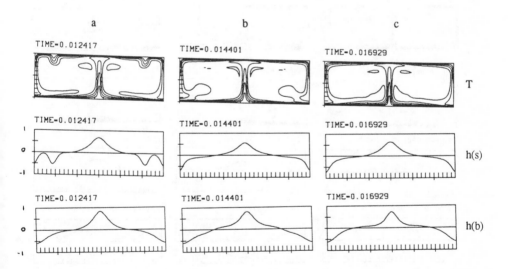

Figure 3. Experiment 3 - α depth-dependent, $Ra=10^6$; evolution of temperature (T), field, surface $h(s)$ and bottom topography $h(b)$; normalization of the topography is with respect to the maximum magnitude, which is at the surface $\pm$ 18050 m and the bottom $\pm$ 5600 m.

top. In the experiment which is displayed in Fig. 3 a decrease of the thermal expansion coefficient α by a factor of 10 has been assumed. This seems to be a sensible assumption if judged on the basis of the quoted high pressure experiments (Chopelas and Böhler, 1989). A polynomial functional dependence of α on depth as proposed by Zhao and Yuen (1987) has been employed.

The effect of the decreasing α can clearly be observed in Fig. 3. The isotherms reveal a thickening of the bottom boundary layer as a consequence of the reduced effective Rayleigh number. Besides the effect of decreasing α on the thermal field and on the spatial structure of the flow (Hansen and Yuen, 1989; Hansen et al., 1991) the topography signal also is substantially influenced. Compared to the initial conditions with constant material properties, the topography amplitudes at the bottom are reduced to 1/4 of its initial value while the surface amplitude is reduced to 2/3. At a first glance the influence of decreasing α on the surface topography seems striking because $\alpha(z)=\alpha_0$ at the surface. However the entire flow structure is changed by the depth-dependence of α and thus also the surface topography.

The actual values for the topography seem too high to be considered. A further increasing Rayleigh number, however, seems to decrease the topography anomalies (Olbertz, 1991) and a further reduction can be expected from taking into account a substantial amount of internal heating (Bercovici et al., 1988). In general the more realistic case of a thermal expansivity decreasing with density (i.e. depth) is characterized by a flattening of bottom- and top topography, in spite of the localization of the available buoyancy into a few strong upwellings as described by Hansen et al. (1991).

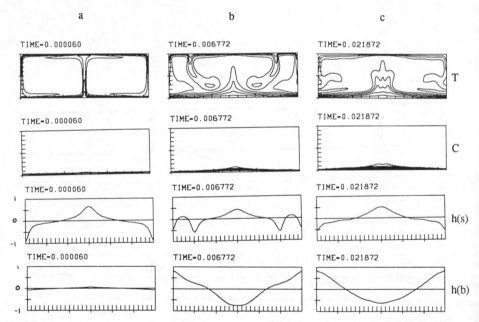

Figure 4. Experiment 4 - depth-dependent α, $Ra=10^6$, $R_\rho=2$; evolution of temperature (T), compositional (C) field, surface and bottom topography; normalization of the topography is with respect to the maximum magnitude, which is at the surface ± 14500 m and the bottom ± 16400 m.

The combined effect of a chemically dense layer and a depth-dependent α has been investigated in a further experiment. The same initial condition as described in Fig. 2 has been used. Due to the smaller α the thermally driving force is weakened at the bottom. Thus only small "compositional hills" can be piled up beneath the upwelling. Contrary to pure thermal convection with depth-dependent α the bottom topography is on average higher than the top values. This is mainly caused by the fact that only one thermal upflow and, consequently, only one but stronger compositional hill with a strong return flow within exists. In that sense the result agrees well with those observed in fluids with depth-dependent properties, namely to form only a few upflows but more comparatively small downflows.

5. Conclusions

Time-dependent mantle flow results in a time-dependent, dynamically supported topography at the top and the bottom (and at any possible interior) boundary of the mantle. In thermal convection major up- and downwelling currents lead to significant up- and downwards deflections at both boundaries, while smaller drifting boundary layer instabilities only result in a local deformation of their attached boundary. In general a thermal expansion coefficient α which is decreasing with depth acts to flatten the topography at both the top and the bottom boundary although the reduction is more pronounced at the bottom. Since only the surface value of α_o, which is kept constant, relates directly to the topography, the reduction of the topographic signal is complicated to explain, but is a result of the impact of a reduced thermal expansivity on the dynamics of the flow. The entrainment of compositionally dense material beneath upflows can lead to an anticorrelated topography, i.e. thermal upflows can be associated with a positive anomaly at the top and a negative anomaly at the bottom. A conclusion as drawn from seismological study, namely the fact that regions of high velocity are correlated with regions of topographic depressions (Morelli and Dziewonski, 1987), would mean that high velocity regions would necessarily be correlated with downwellings, and needs to be reconsidered. If the D''-layer resembles a thermo-chemical boundary layer then an accumulation of dense material beneath an upflow can lead to a correlation of regions with high velocity and topographic depression with up- instead of downwelling currents.

The combining effect of a decreasing thermal expansion coefficient and the presence of a compositionally dense layer lead to a complex behavior. Generally a reduced α tends to reduce and to smooth the topographic signal at both boundaries. Simultaneously, due to the reduced amount of available buoyancy at the bottom, a reduced α acts to stabilize the compositional heterogeneities, thus preventing them from being entrained by the thermally driven flow. Differing from the pure thermal case the topography at the bottom can be more developed than at the top, despite the depth-dependence of α.

Acknowledgments. The authors wish to thank V. Steinbach for fruitful discussions. This work was supported by the Deutsche Forschungsgemeinschaft under grant Eb 56/13-1. D. Olbertz was a recipient of a NATO stipend during her stay in Fairbanks.

6. References

Anderson, O. L., A. Chopelas and R. Böhler (1990) Thermal expansivity versus pressure at constant temperature: a re-examination. Geophys. Res. Lett. 17, 685-688.

Bercovici, D., G. Schubert and A. Zebib (1988) Geoid and topography for infinite Prandtl number convection in a spherical shell. J. Geophys. Res. 93, 6430-6436.

Blankenbach, B., F. Busse, U. Christensen, L. Cserepes, D. Gunkel, U. Hansen, H. Harder, G. Jarvis, M. Koch, G. Marquart, D. Moore, P. Olson, H. Schmeling and T. Schnaubelt (1989) A benchmark comparison for mantle convection code. Geophys. J. Int. 98, 23-38.

Brown, J. M. (1986) Interpretation of the D" zone at the base of the mantle: dependence on assumed values of thermal conductivity. Geophys. Res. Lett. 13, 1509-1512.

Chopelas, A. and R. Böhler (1989) Thermal expansion measurements at very high pressure, systematics and a case for a chemically homogeneous mantle. Geophys. Res. Lett. 16, 1347-1350.

Christensen, U. R. (1980) Numerische Modelle zur Konvektion im Erdmantel mit verschiedenen Eindringtiefen. Ph.D. Thesis, Techn. Universität Braunschweig.

Hansen, U. and A. Ebel (1988) Time-dependent thermal convection - a possible explanation for a multi-scale flow in the Earth's mantle. Geophys. Journal 94, 181-191.

Hansen, U. and D. A. Yuen (1988) Numerical simulation of thermal-chemical instabilities and lateral heterogeneities at the core-mantle boundary. Nature 334, 237-240.

Hansen, U., D. A. Yuen and S. E. Kroening (1991) Dynamical effects of depth-dependent thermal expansivity on mantle convection. Geophys. Res. Lett. 18, 1261-1264.

Jarvis, G. T. and W. R. Peltier (1989) Convection models and geophysical observations. In Mantle Convection, edited by W. R. Peltier, Vol. 4. Gordon and Breach Science Publishers, New York, London, Paris, Montreux, Tokyo, Melbourne, 479-593.

Lay, T. (1989) Structure of the core-mantle transition zone: A chemical and thermal boundary layer. EOS 70, 49-59.

Lliboutry, L. (1987) Very Slow Flows of Solids. Marinus Nijhoff Publishers, Dordrecht.

McKenzie, D. P. (1977) Surface deformation, gravity anomalies and convection. Geophys. J. R. Astr. Soc. 48, 211-238.

Morelli, A. and A. M. Dziewonski (1987) Topography of the core-mantle boundary and lateral heterogeneity of the liquid core. Nature 325, 678-683.

Olbertz, D. C. (1991) Topographie als messbare Folge zeitabhängiger Mantelkonvektion - eine numerische Studie. Diplomarbeit, Inst. für Geophysik und Meteorologie, Universität Köln.

Osako, M. and E. Ito (1991) Thermal diffusivity of $MgSiO_3$ perovskite. Geophys. Res. Lett. 18, 239-242.

Ricard, Y. and W. Bai (1991) Inferring the viscosity and the 3-D structure of the mantle from geoid, topography and plate velocities. Geophys. J. 105, 561-571.

Steinbach, V., U. Hansen and A. Ebel (1989) Compressible convection in the Earth's mantle: A comparison of different approaches. Geophys. Res. Lett. 16, 633-636.

Wang, Y., D. J. Weidner, R. C. Liebermann, X. Liu, J. Ko, M. T. Vaughn, Y. Zhao, A. Yeganeh-Haeri and R. E. G. Pacolo (1991) Phase transition and thermal expansion of $MgSiO_3$. Science 251, 410-413.

Zhao, W. and D. A. Yuen (1987) The effects of adiabatic and viscous heatings on plumes. Geophys. Res. Lett. 14, 1223-1226.

3-D NUMERICAL INVESTIGATION OF THE MANTLE DYNAMICS ASSOCIATED WITH THE BREAKUP OF PANGEA

JOHN R. BAUMGARDNER
MS B-216
Los Alamos National Laboratory
Los Alamos, New Mexico 87545, U.S.A.

ABSTRACT. Three-dimensional finite element calculations in spherical geometry are performed to study the response of the mantle with platelike blocks at its surface to an initial condition corresponding to subduction along the margins of Pangea. The mantle is treated as an infinite Prandtl number Boussinesq fluid inside a spherical shell with isothermal, undeformable, free-slip boundaries. Nonsubducting rigid blocks to model continental lithosphere are included in the topmost layer of the computational mesh. At the beginning of the numerical experiments these blocks represent the present continents mapped to their approximate Pangean positions. Asymmetrical downwelling at the margins of these nonsubducting blocks results in a pattern of stresses that acts to pull the supercontinent apart. The calculations suggest that the breakup of Pangea and the subsequent global pattern of seafloor spreading was driven largely by the subduction at the Pangean margins.

1. Introduction

A question persisting from the early days of plate tectonics has been what pattern of flow in the mantle is responsible for the breakup of Pangea and the subsequent migration of most of the fragments toward the Pacific hemisphere. Analytical and numerical studies of thermal convection in spherical shells (Chandrasekhar, 1961; Zebib et al., 1983; Baumgardner, 1983, 1985, 1988; Bercovici et al., 1989) indicate that for whole mantle convection, harmonic degrees three and higher should prevail. Current plate motions, however, display strong degree one and two components (Peltier, 1985). Seismic tomography studies (Clayton and Comer, 1983; Dziewonski, 1984; Dziewonski and Woodhouse, 1987; Inoue et al., 1990) also indicate a strong degree two component in the lower mantle's seismic velocity structure. This low degree pattern for the lower mantle is a robust feature of essentially all global seismic models (Fig. 1). Similarly, the non-hydrostatic geoid contains a large degree two component (Chase and Sprowl, 1983). A key to resolving this difference between convection models and the actual observational data seems to be that the mantle's stiff upper boundary layer, the lithosphere, plays a critical role in the mantle's overall dynamics (Christiansen, 1985; Davies, 1988a; Gurnis, 1988, 1990, 1991) and is largely responsible for enforcing the observed low degree behavior. The details of the coupling and interactions between the lithosphere and deeper mantle are as yet, however, far from being understood.

Current generation computers are providing the capabilities to perform three-dimensional dynamic simulations (Baumgardner, 1988; Glatzmaier, 1988; Bercovici et al., 1989a) of the resolution and sophistication required to address such issues. This paper describes a

D. B. Stone and S. K. Runcorn (eds.),
Flow and Creep in the Solar System: Observations, Modeling and Theory, 207–224.
© 1993 *Kluwer Academic Publishers. Printed in the Netherlands.*

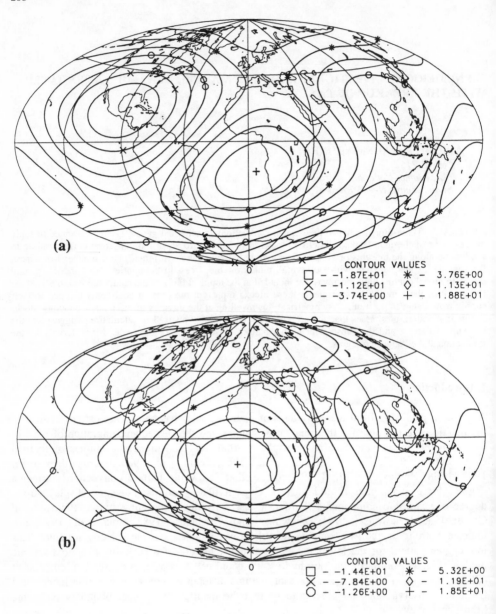

Figure 1. Vertically integrated lower mantle P-wave slowness variations for harmonic degrees up to 3 from seismic tomography models of (a) Dziewonski and Woodhouse (1987) and (b) Clayton and Comer (1983). Note similarity of two models with maximum slowness beneath the south Atlantic and slowness minima beneath North America, Antarctic, and southeast Asia.

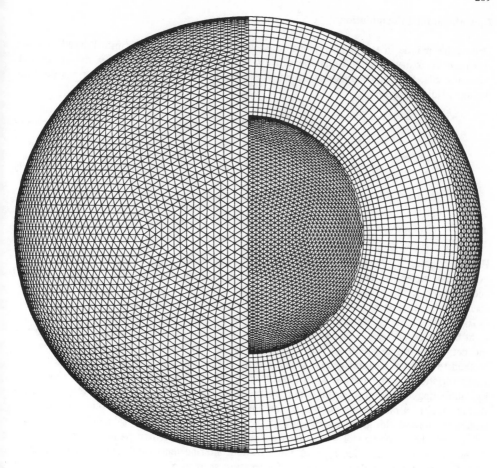

Figure 2. Finite element mesh used for the numerical model of the earth's mantle, with section removed to display the interior. Mesh consists of 17 layers of 10242 nodes each. Nonuniform thickness of layers in radial direction provides 100 km resolution at the boundaries. Lateral resolution at the top boundary is 250 km.

numerical experiment performed on a Cray Y-MP using the spherical finite element multigrid code TERRA. The experiment was aimed at investigating the effects of rigid patches, or plates, at the surface on the convecting system. Special initial conditions were chosen to model subduction at the margins of a supercontinent that is comprised of rigid subunits and has a geometry resembling that of Pangea.

2. Mathematical Formulation

The earth's mantle in this experiment is modeled as an irrotational, infinite Prandtl number, constant viscosity, linear Boussinesq fluid within a spherical shell with isothermal, undeformable, free-slip boundaries. Under these conditions the following equations describe the local fluid behavior:

$$\mu\nabla^2 u = \nabla p + \alpha(T - T_0)\rho_0 g \tag{1}$$

$$\nabla \bullet u = 0 \tag{2}$$

$$\frac{\partial T}{\partial t} = -\nabla \bullet (Tu) + \left[\nabla \bullet (k\nabla T) + H\right]/\rho_0 c_v \tag{3}$$

Here u denotes fluid velocity, p pressure, T absolute temperature, μ dynamic viscosity, α volume coefficient of thermal expansion, T_0 reference temperature, ρ_0 density, g gravitational acceleration, k thermal conductivity, H specific radiogenic heat production rate, and c_v specific heat at constant volume.

Rigid plates are included in this framework in a simple manner by adding forces to the right hand side of Eq. (1) that yield rigid rotation of each plate about a Euler pole such that the net torque on the plate is zero. These forces are obtained by first applying Newton's method to find the rotation vector ω that results in zero net torque τ on the plate. By taking small variations in ω about the x-, y-, and z-axes, one can compute $\partial\tau/\partial\omega$ and then modify ω by subtracting $\tau/(\partial\tau/\partial\omega)$ such that τ approaches zero. By creating a temporary velocity field v that is equal to u outside the volume of the plates but which has the rigid rotations within the plates, the force field, given by $f^* = \mu\nabla^2 v$, required to produce the strain field implied by v can readily be computed. These forces are added to the right hand side of Eq. (1) at the nodes enveloping and including each plate without altering the larger solution method or the boundary conditions. Other approaches for treating surface plates are described by Gable et al. (1991), King et al. (1992), and Weinstein et al. (1992).

3. Numerical Approach

The fluid equations are solved using the finite element method on a mesh constructed from the regular icosahedron (Fig. 2). The elements have the form of triangular prisms. Piecewise linear spherical finite element basis functions provide second-order spatial accuracy (Baumgardner, 1983; Baumgardner and Frederickson, 1985). The mesh has 10242 nodes in each of 17 radial layers. The 160 nodes around the equator imply a tangential spatial resolution at the outer surface of 250 km. Nonuniform spacing of nodes in the radial direction assists in resolving the boundary layers.

The calculational procedure on each time step is first to apply a two-level conjugate gradient algorithm (Ramage and Wathen, 1992) to solve for the velocity and pressure fields simultaneously from Eq. (1) and (2) and then to update the temperature field according to

Eq. (3) using a forward-in-time interpolated donor cell advection scheme. An iterative multigrid solver (Baumgardner, 1983) is employed inside the conjugate gradient procedure, and its high rate of convergence is responsible for the method's overall efficiency. The finite element operators are used for all the gradient and divergence terms in Eq. (1)-(3) except for the heat conduction, which employs a finite difference technique. A particle method is used to define the plate locations and to track their motions. Four particles per node are used to represent the plate distribution. Piecewise linear basis functions are employed to map particle data to the nodes. The particles are moved in a Lagrangian manner at each time step using the piecewise linear basis functions to interpolate the nodal velocities to the particles. The advantage of this particle method is low numerical diffusion and hence the ability to minimize the smearing of the plate edges.

4. Parameterization

The Boussinesq approximation yields a model for the mantle that involves but a few parameters, as is evident from Eq. (1)-(3). However, because computational costs preclude calculations with spatial resolution sufficient to treat the dynamics at the actual Rayleigh numbers appropriate to the earth, it was necessary to scale some of the parameters to give less vigorous convection. The choice made in this study was to scale the thermal conductivity k and the radiogenic heat production rate H to larger values to reduce the effective Rayleigh number to what could be treated with the mesh resolution described earlier. The scaling factor selected was 6.25. A benefit of this scaling is that velocities and time scales can approximate those of the physical earth. Values for the other parameters then were chosen as suitable estimates for the actual earth within the limitations of the Boussinesq formulation.

Parameters that relate to the thermal aspects of the model include the boundary temperatures, the thermal conductivity, the radiogenic heating rate, the volume coefficient of thermal expansion, and the specific heat. Temperatures chosen for the top and bottom boundaries were 300 K and 2300 K, respectively. For the earth this implies a bottom temperature of about 3200 K if one takes a typical estimate for the adiabatic temperature drop across the mantle of 900 K. Choice of this relatively low bottom boundary temperature is influenced by observations that the heat flux associated with hot spots is a small fraction of the earth's total heat flux (Davies, 1988b) and that the temperature anomalies in the hot spots themselves appear to be only on the order of 200-300 K (Richards et al., 1988). Since previous modeling has shown that most of the base heating of the mantle is transported to the surface by plumes (Baumgardner, 1985; Bercovici et al., 1989b), it is inferred that the fraction of the total heating from the core is probably small and that most of the heating is due to internal radioactivity. Consistent with this conclusion is the choice of an unscaled radiogenic heating rate of 5×10^{-12} W kg^{-1} (which translates to about 40 mW m^{-2} of surface heat flux). The unscaled thermal conductivity is assumed constant through the mantle at a value of 4.0 W m^{-1} K^{-1}. The volume coefficient of thermal expansion was chosen to be 2.5×10^{-5} K^{-1} and the specific heat was taken to be 1000 J kg^{-1} K^{-1}. The mean temperature of the computational domain was used as the reference temperature T_0. The density was chosen to be 4500 kg m^{-3}, the gravitational acceleration 10 m s^{-2}, and the dynamic viscosity 1.5×10^{22} Pa s.

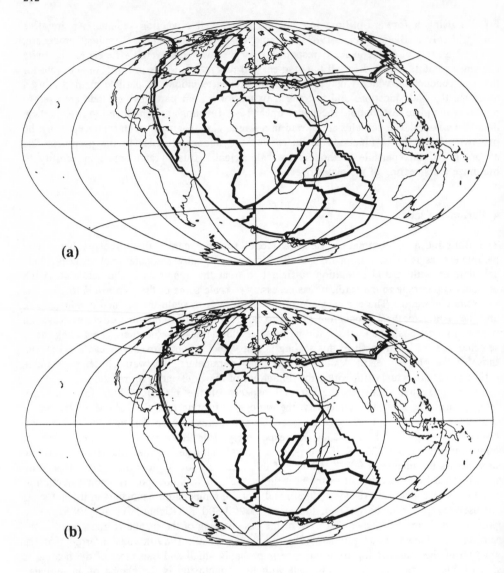

Figure 3. Initial configuration of continental blocks with fine contours representing the slablike temperature perturbations at (a) 250 km depth and (b) 1000 km depth. Magnitude of the temperature perturbations is -400 K. The fine contour lines represent 33% and 67% of the peak value. Outlines of present day shorelines are included for reference.

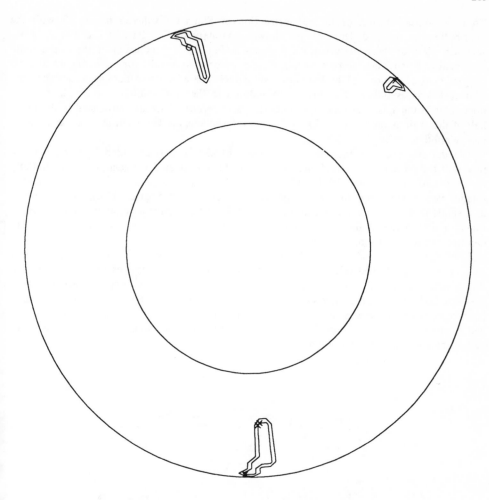

Figure 4. Cross-section through the spherical shell showing the initial slablike temperature perturbations. Contour values are -100 K and -300 K relative to the mean value at a given radius. Shape at 6 o'clock is wider because section slices obliquely through the perturbation.

5. Initial Conditions

To explore the conjecture that subduction at the margins of Pangea might be related to the fashion in which Pangea rifted and its fragments drifted apart, a set of slablike initial temperature perturbations is prescribed to represent circum-Pangea subduction. The perturbations have an amplitude of 400 K, a width that corresponds to a single finite element basis function (about 250 km), and depths that assume values of either 315 or 1450 km.

They lie at the margin of the reconstruction of Pangea displayed in Fig. 3 with the distributions at the two depth ranges shown separately in Fig. 3a and Fig. 3b. A cross-sectional view of these perturbations is given in Fig. 4. The inclusion of the deeper cold perturbation is motivated in part by the seismic tomography results for the lower mantle that suggest distinct regions of high P-wave velocity beneath North America, Antarctica, and southeast Asia as indicated in Fig. 1. In addition to the cold slablike perturbations in the upper half of the spherical shell, four circular warm perturbations are introduced in the lower half of the shell as indicated in Fig. 5. These perturbations are Gaussian in shape and have a peak amplitude of 50 K.

The definition of the individual continental blocks is also provided in Fig. 3. These represent the present continental areas mapped to their estimated Pangean locations. Initially the North American, Greenland, and Eurasian blocks are constrained to have a common rotation vector. Similarly, the South American, African, and Madagascar blocks initially rotate as a single unit. Likewise the remaining four blocks initially have a common rotation vector. Later in the course of the calculation, these composite blocks are allowed to break into constituent parts with separate rigid motions.

Since the velocity and pressure fields are computed together from the temperature field according to Eq. (1) and (2), no initial specification of velocity or pressure is needed. The initial state then consists of a relatively blank mantle as far as its temperature structure is concerned. Upon the spherically symmetric radial temperature profile of Fig. 6 with its large boundary gradients there is imposed the pattern of cold slablike perturbations in the upper half of the shell shown in Figs. 3 and 4 and the distribution of warm columns in the lower half of the shell as displayed in Fig. 5.

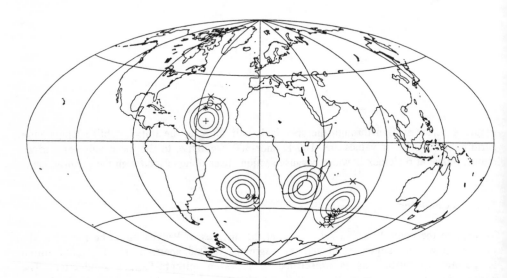

Figure 5. Initial temperature perturbations between 1445 km depth and bottom of the shell. Distributions are Gaussian in shape and have a peak value of 50 K. Contours are 10 K apart.

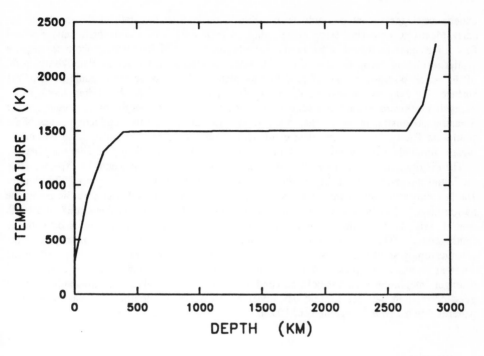

Figure 6. Spherically averaged initial radial temperature profile.

6. Results

Fig. 7 contains a sequence of snapshots at times of 20, 40, 60, and 80 Myr showing the locations of the continental blocks and the velocities at the surface of the spherical shell. A notable feature in the velocity fields of Fig. 7 is the motion of the nonsubducting blocks toward the adjacent zones of downwelling flow. This motion is primarily a consequence of the drag exerted on a nonsubducting block by the material below it as this material drifts toward the downwelling zone. Such a general pattern of flow is evident in the cross-sectional slices of Fig. 8. The translation of a nonsubducting block in this manner leads to a backward, or oceanward, migration of the location of the downwelling. This oceanward translation of the continental blocks and subduction zones therefore acts to pull the supercontinent apart. This behavior is a basic fluid mechanical result and not the consequence of any special initial conditions or unusual geometrical specifications other than the asymmetrical downwelling at the edge of a nonsubducting portion of the surface.

That the blocks move apart without colliding and overrunning one another, on the other hand, depends very strongly on the initial distribution of thermal perturbations, the shapes of the blocks, and timing of their breakup. A moderate amount of trial and error was involved in finding the special set of conditions that leads to the results shown in Fig. 7 and 8. This experience suggests that it should be possible to reconstruct the state of the mantle in the early Mesozoic or earlier from present day observational data and a sufficiently realistic forward numerical model. Clearly the calculations reported here suffer from inadequate spatial resolution, from the lack of variable viscosity effects, and from no platelike treatment of the ocean portions of the surface. Adding platelike behavior to the oceanic parts of the surface, for example, would no doubt improve the tendency for the Indian block to be dragged northward toward the subduction zone on the south margin of Asia and for the Sumatra and Australian blocks also to be rafted northward toward the subduction zones of the southwest Pacific. These numerical challenges appear to be readily addressable within the present finite element framework with the resources available on existing parallel machines.

The distribution of cold material in the lower mantle follows what would be expected from the initial temperature field. Fig. 9 shows contours of temperature at a depth of 2790 km, or 100 km above the bottom shell boundary, at a time of 60 Myr. Broad cold areas lie below the narrow zones of deep cold initial temperatures beneath the Americas, Antarctica, and south central Asia. There is also a conjunction of three belts of warm temperatures beneath the south Atlantic. This distribution of temperatures displays a qualitative similarity with the seismic tomography models of Fig. 1. A higher viscosity lower mantle, as opposed to the constant viscosity case presented here, would yield more sluggish lower mantle velocities and a longer integration time for cold descending flow. This is a plausible explanation for the good correlation between present day lower mantle seismic velocities and past subduction patterns (Richards and Engebretson, 1992).

7. Conclusions

Numerical 3-D global modeling of the earth's mantle and lithosphere is reaching enough maturity to begin testing some ideas about the earth's tectonic history. The numerical experiment reported here suggests that the essential physics associated with the breakup of Pangea was the buoyancy and strength of the continental lithosphere. The fact that continental lithosphere has sufficient buoyancy to resist subduction and sufficient strength to resist significant lateral deformation arising from drag forces at its base means that blocks of continental lithosphere tend to be dragged toward nearby zones of downwelling flow and then tend to override such zones. These basic fluid mechanics principles imply that subduction along the margins of Pangea naturally results in the tendency for Pangea to be pulled apart. The observational evidence that subduction zones today ring much of what was formerly the Pangean margin, that vast geosynclinal troughs of sediments also reside on this former Pangean margin, and that seismic tomography reveals broad zones of high velocity material reside in the lower mantle beneath major segments of the Pangean margin supports the general thesis that persisting subduction through the Mesozoic and Cenozoic, primarily along this evolving margin, is largely responsible for the observed plate motions and pattern of seafloor spreading during this recent portion of earth history. These arguments also imply that the lithosphere plays a crucial role in the overall dynamics of the earth's interior. Finally, the prospects for greatly improved 3-D numerical modeling of this strongly nonlinear system by exploiting the machine resources now available in parallel supercomputers suggest fruitful opportunities for research in many areas of earth science lie ahead.

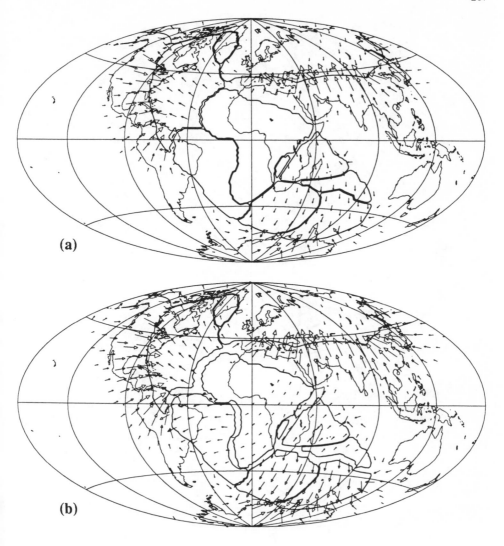

Figure 7. Surface velocities and positions of continental blocks at (a) 20 Myr, (b) 40 Myr, (c) 60 Myr, and (d) 80 Myr. Arrows are scaled with the maximum velocity equal to 5.36 cm/yr in (a), 7.05 cm/yr in (b), 7.25 cm/yr in (c), and 8.17 cm/yr in (d).

218

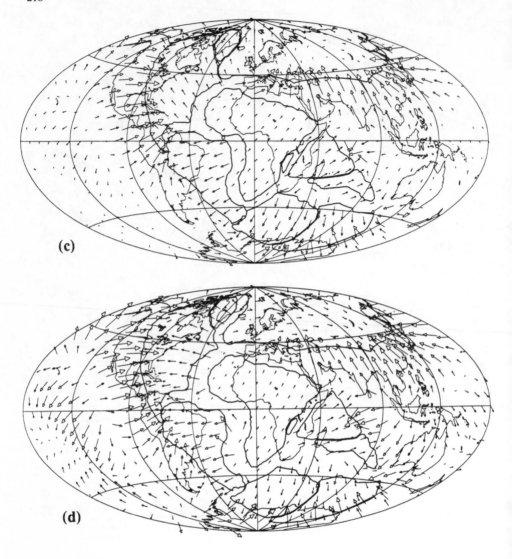

(c)

(d)

Figure 7. Continued.

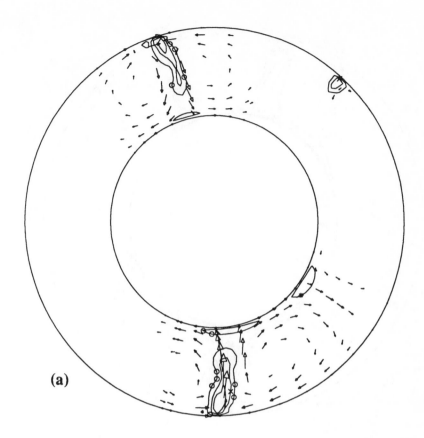

(a)

Figure 8. Cross-sectional slices showing temperature deviations from the mean radial value and components of velocity in the plane of the slice at (a) 20 Myr, (b) 40 Myr, (c) 60 Myr, and (d) 80 Myr. Temperature contours are marked with squares for the -500 K contour, 'x' for the -300 K contour, 'o' for the -100 K contour, '*' for the +100 K contour, diamonds for the +300 K contour, and '+' for the +500 contour. Arrows are scaled with the maximum velocity equal to 7.44 cm/yr in (a), 13.9 cm/yr in (b), 18.0 cm/yr in (c), and 18.2 cm/yr in (d). Downwelling flow at 11 o'clock is off the west coast of North America, at 2 o'clock is north of Africa, and at 6 o'clock is in the Antarctic. The upwelling at 4 o'clock is in the south Indian Ocean.

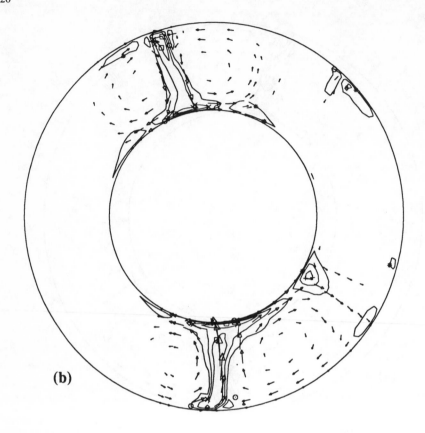

(b)

Figure 8. Continued.

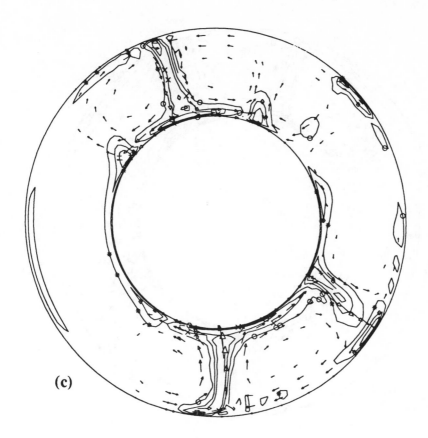

(c)

Figure 8. Continued.

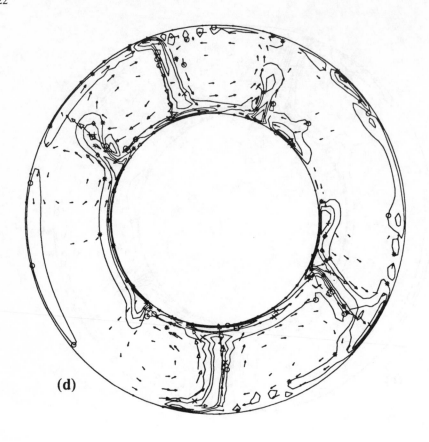

(d)

Figure 8. Continued.

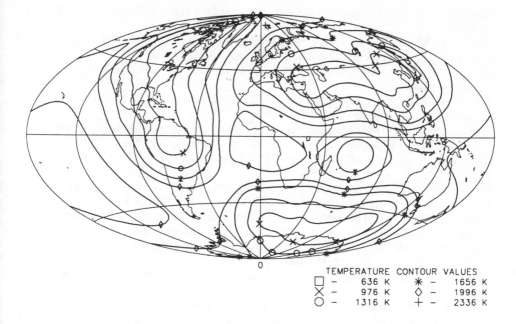

	TEMPERATURE CONTOUR VALUES		
□ –	636 K	✳ –	1656 K
✕ –	976 K	◇ –	1996 K
○ –	1316 K	+ –	2336 K

Figure 9. Temperature distribution at 2790 km depth at 60 Myr. Broad low temperature regions exist below the Americas, south central Asia, and the Antarctic region, and a conjunction of three high temperature belts occurs in the south Atlantic.

8. References

Baumgardner, J. R. (1983) A three-dimensional finite element model for mantle convection. Ph.D. thesis, UCLA.

Baumgardner, J. R. (1985) Three-dimensional treatment of convective flow in the earth's mantle. J. Stat. Phys. 39, 501-511.

Baumgardner, J. R. (1988) Application of supercomputers to 3-D mantle convection, in The Physics of the Planets, S. K. Runcorn, ed., John Wiley and Sons, 199-231.

Baumgardner, J. R. and P. O. Frederickson (1985) Icosahedral discretization of the two-sphere. SIAM J. Numer. Anal. 22, 1107-1115.

Bercovici, D., Schubert, G., and G. A. Glatzmaier (1989a) Three-dimensional spherical models of convection in the earth's mantle. Science 244, 950-955.

Bercovici, D., Schubert, G., and G. A. Glatzmaier (1989b) Influence of heating mode on three-dimensional convection. Geophys. Res. Lett. 16, 617-620.

Chandrasekhar, S. (1961) Hydrodynamic and Hydromagnetic Stability. Clarendon Press, Oxford.

Chase, G. C. and D. R. Sprowl (1983) The modern geoid and ancient plate boundaries. Earth Planet. Sci. Lett. 62, 314-320.

Christiansen, U. (1985) Thermal evolution models for the earth. J. Geophys. Res. 90, 2995-3007.

Clayton, R. W. and R. P. Comer (1983) A tomographic analysis of mantle heterogeneities from body wave travel times (abstract). EOS Trans. AGU 62, 776.

Davies, G. F. (1988a) The role of the lithosphere in mantle convection. J. Geophys. Res. 93, 10451-10466.

Davies, G. F. (1988b) Ocean bathymetry and mantle convection, 1, Large-scale flow and hotspots. J. Geophys. Res. 93, 10467-10480.

Dziewonski, A. M. (1984) Mapping the lower mantle: determination of lateral heterogeneity in P velocity up to degree and order 6. J. Geophys. Res. 89, 5929-5952.

Dziewonski, A. M. and J. H. Woodhouse (1987) Global images of the Earth's interior. Science 236, 37-48.

Gable, C. W., O'Connell, R. J., and B. J. Travis (1991) Convection in three dimensions with surface plates: generation of toroidal flow. J. Geophys. Res. 96, 8391-8405.

Glatzmaier, G. A. (1988) Numerical simulations of mantle convection: time-dependent, three-dimensional, compressible, spherical shell. Geophys. Astrophys. Fluid Dyn. 43, 223-270.

Gurnis, M. (1988) Large-scale mantle convection and the aggregation and dispersal of supercontinents. Nature 332, 695-699.

Gurnis, M. (1990) Ridge spreading, subduction, and sea level fluctuations. Science 250, 970-972.

Gurnis, M. (1991) Continental flooding and mantle-lithosphere dynamics, in Sabadini et al., eds., Glacial Isostasy, Sea Level, and Mantle Rheology. Kluwer Publ., Dordrecht, Netherlands, 445-492.

Inoue, H., Fukao, Y., Tanabe, K., and Y. Ogata (1990) Whole mantle P-wave travel time tomography. Phys. Earth Planet. Int. 59, 294-328.

King, S. D., Gable, C. W., and S. A. Weinstein (1992) Models of convection-driven tectonic plates: a comparison of methods and results. Geophys. J. Int. 109, 481-487.

Peltier, W. R. (1985) Mantle convection and viscoelasticity. Ann. Rev. Fluid Mech. 17, 561-608.

Ramage, A. and A. J. Wathen (1992) Iterative solution techniques for finite element discretisations of fluid flow problems. Copper Mountain Conference on Iterative Methods Proceedings, Vol. 1.

Richards. M. A. and D. C. Engebretson (1992) Large-scale mantle convection and the history of subduction. Nature 355, 437-440.

Richards, M. A., Hager, B. H., and N. H. Sleep (1988) Dynamically supported geoid highs over hotspots: observation and theory. J. Geophys. Res. 93, 7690-7708.

Weinstein, S. A., Olson, P. L., and D. A. Yuen (1992) Thermal convection with non-Newtonian plates. Geophys. J. Int. (in press).

Zebib, A., Schubert, G., Dein, J. L. and R. C. Paliwal (1983) Character and stability of axisymmetric thermal convection in spheres and spherical shells. Geophys. Astrophys. Fluid Dyn. 23, 1-42.

SUBDUCTION ZONES, MAGMATISM, AND THE BREAKUP OF PANGEA

LAWRENCE A. LAWVER
LISA M. GAHAGAN
Institute for Geophysics
8701 N. Mopac #300
Austin, Texas   78759-8345

ABSTRACT. Paleoreconstructions of the lithospheric plates with major subduction zones indicated are reviewed for every 40 million years from 200 Ma to Present. Relative major plate motions are based on marine magnetic anomalies, seafloor lineations including fracture zones and transform faults, and paleomagnetic data. Pangea as an entity existed for only about 100 million years, until seafloor spreading in the central North Atlantic began at ~175 Ma and it separated into Gondwana and Laurentia. Although Gondwana was ringed on its southern margin by continentward subduction zones, eventual breakup mostly occurred thousands of kilometers from the subduction zones. Major plate reorganizations are often associated in time with continental flood basalts, although it is unclear whether the flood basalts represent a cause or are an effect of major plate breakup. Around 130 Ma, there were three significant rifting events: the South Atlantic opened between South America and Africa, India rifted from Antarctica, and the Canada Basin in the Arctic opened between Arctic Canada and Arctic Alaska. At ~120 Ma, the nearly continuous subduction zone that had existed along the southern margin of Pangea since at least 200 Ma began to break apart with the subduction of the Pacific-Aluk spreading center along eastern Australia and northern New Zealand. By 40 Ma, the major subduction zones seem to have migrated from high latitude regions to lower latitudes.

Introduction

The breakup of Pangea and the subsequent breakup of its two component pieces, Gondwana and Laurentia, dominate the Mesozoic and Cenozoic tectonic evolution of the Earth. This cycle, though, is only the latest in a history of continental agglomerations and subsequent dispersal. Moores (1991), Dalziel (1991), and Hoffman (1991) have recently produced new reconstructions for a Late Precambrian to Early Paleozoic supercontinent. The Late Precambrian assemblage of the continents had a much different configuration than that of Mesozoic Pangea and may have also had a different orientation with respect to the spin axis of the Earth (Dalziel, 1992). If Archean magmatic belts indicate paleo-subduction zones then it is apparent that plate tectonics was operative prior to 2.5 Ga (timescale from Palmer, 1983). Borg and DePaolo (1991) use isotopic data (Nd, Sr, O) to define crustal basement age boundaries and subduction polarities to identify suspect terranes along the Transantarctic margin of East Antarctica. They present a case for subduction that took place at ~1.7 Ga off that margin. They propose that the same margin was again compressive between ~750 Ma to 570 Ma. At ~500 Ma they suggest a subduction zone that extended along virtually the whole present-day Transantarctic margin and into the East Australian craton region. If orogenic and magmatic belts are valid indicators of early plate convergence, then it is difficult to define a

D. B. Stone and S. K. Runcorn (eds.),
Flow and Creep in the Solar System: Observations, Modeling and Theory, 225–247.
© 1993 *Kluwer Academic Publishers. Printed in the Netherlands.*

precise start-up time for even the recent Mesozoic plate motions, and we should assume that plate tectonics, subduction, and continental agglomeration have been active through time.

Relative plate motions between the major continental masses are well constrained for only the last 160 million years and then only for the plates that have correlatable marine magnetic anomalies. Paleomagnetic data, in conjunction with hotspot tracks for certain major plates, can give a paleolatitudinal framework for the reconstructed continents back to at most 130 Ma, although reliable data prior to 84 Ma is scarce (Müller et al., in press). For paleogeographic reconstructions of Mesozoic Gondwana, conjugate geological information and the alignment of continental margins must also be used to get a close fit of the pre-breakup continents. Paleomagnetic measurements are the primary constraints on paleo-positions of the continents prior to the oldest marine magnetic anomalies [Chron M29 for Gondwana, Oxfordian, 160 Ma]. Paleomagnetic data are dependent not only on inclination and declination measurements but also on the age-date of the sample. Because of the inherent limitations involved in Paleozoic and Early Mesozoic plate motions, we restrict ourselves to a discussion of the Jurassic to recent breakup of Pangea. Unfortunately, Mesozoic Pangea [~200 Ma] can not be fixed in a whole mantle framework for the reasons cited above and for lack of information concerning true polar wander.

A fit of the major continental masses that has eliminated all post-rift extension between the continental fragments is referred to as a tight fit. We model the Mesozoic breakup of Pangea starting with a tight fit of the continents at 200 Ma and show them every 40 million years thereafter. In actual fact, our tight fit of the continents may have only existed as shown during the Late Paleozoic since there is evidence of Triassic sediments in wells that have been drilled along coastal Madagascar and eastern Africa (Coffin and Rabinowitz, 1988). The amount of stretching that occurred between the end of the Permian and 200 Ma would not substantially change the configuration of the Gondwana continents. The continental plates are reconstructed to their relative positions at subsequent times at 160 Ma and afterwards based on identified marine magnetic anomalies, seafloor lineations derived from bathymetric data and satellite altimetry information, paleomagnetic data, the subsidence of continental margins based on drilling information and seismic data, and geological information concerning the opening of seaways based on ocean drilling information. In addition, paleomagnetic data can constrain the positions of plates in both a relative and absolute framework when sufficient data are available.

The subduction zones shown in the paleoreconstructions are taken from the published literature. Evidence for them is generally the presence of dated magmatic arcs that have subduction-related geochemical and petrological affinities. Considering the sparseness of the sampling, particularly of Mesozoic magmatic arcs both temporally and spatially, it must be emphasized that these figures are indicative of the subduction zone locations, but are not definitive. The subduction zones were probably not as continuous as they are shown because collisions of continents or continental fragments, island arcs, or spreading centers with the subduction zone would temporarily or permanently disrupt the subduction process at those locations. The subduction of a spreading center would generally produce a hiatus in subduction although the previously subducted plate would have continued to descend into the mantle. In the event of a collision of either continental fragments or island arcs, the subduction zone might reverse polarity with time, or a new subduction zone oceanward of the old one might commence with the polarity of the former one. There may be an overlap in time with two subduction zones apparently active in the case of a newly reversed subduction zone, or there may be a lengthy hiatus between the time of accretion of the continental fragment and the development of a new subduction zone.

White and McKenzie (1989) discuss the role of mantle plumes with respect to continental rifting and the formation of flood basalts. Campbell and Griffiths (1990) and Hill (1991) expanded on the work of White and McKenzie (1989) to relate continental breakup to the extra gravitational potential produced by the uplift resulting from mantle plumes that arise under continental masses. Hill (1991) proposed that mantle plumes play an important role in determining both where and when continental breakup occurs. Unfortunately, the timing of the formation of continental flood basalts with respect to the initiation of plate separation and the start of seafloor spreading is not consistent. In some instances, continental "Karoo" type (Cox, 1988) mantle plume basalts clearly predate the formation of ocean crust, for example the central North Atlantic and the breakup plume associated with Gondwana (White and McKenzie, 1989) and in other cases the excess continental volcanism postdates the formation of the first ocean crust. The latter case is exemplified by the Paraná and Etendeka flood basalts associated with the inital opening of the South Atlantic (Rabinowitz and LaBrecque, 1979; Austin and Uchupi, 1982), and may also be true of the Rajmahal Traps of India and the formation of ocean crust between India and Antarctica.

Separation of Pangea and Gondwana

The 200 Ma reconstruction of Pangea is shown in Figure 1. From Permian or perhaps earlier time until Cretaceous, nearly continuous subduction occurred from eastern Australia (Cawood, 1984; Bradshaw, 1989) to South America (Forsythe, 1982; Dalziel and Forsythe, 1985; Hervé et al., 1987). Gust et al. (1985) suggested that basement rocks in southern South America indicate episodic subduction and accretion along the present day western margin of Gondwana even earlier than Permian time. Off central California, Nevada and Arizona, a cratonward subduction zone operated from Triassic to early Middle Jurassic time [~180 Ma] according to Busby-Spera (1988). Farther north, off western North America, subduction polarity was reversed, away from the Pangea margin and under the offshore continental fragment of Stikinia (Tempelman-Kluit, 1979). Tempelman-Kluit (1979) suggests that an Anvil Ocean closed between 220 Ma and 165 Ma, followed by arc-continent collision [165 to 150 Ma] and later arc, forearc and trench material was obducted onto the North American continent between 150 and 125 Ma.

North America had only collided with NW Africa and fused to Pangea during the Alleghanian orogeny that ended at about 276 Ma (Hatcher et al., 1989). Hill (1991) mentions that initiation of spreading within the central North Atlantic followed a period of considerable basaltic magmatism in eastern North America and possibly within western Africa 200 Ma ago. One example of the volcanism is the Palisades Sill which was emplaced during the Late Triassic to Early Jurassic period [202±10 to 193 Ma; dates from de Boer, 1968]. The northernmost member of the Triassic to Jurassic dikes is a single large dike in southern Nova Scotia (May, 1971). DeBoer et al. (1988) discuss the classification of the Lower Jurassic eastern North America diabase dikes into five separate groups, including one of high-Ti, quartz-tholeiite dikes (HTQ) that are characterized by $TiO_2 > 0.9\%$. They consider the chemical composition of the HTQ types to be comparable to that of the Karoo type continental basalts. The HTQ dikes are most common among the early Mesozoic dike swarmsoof the central and northern Appalachians, with most of them occurring from Virginia

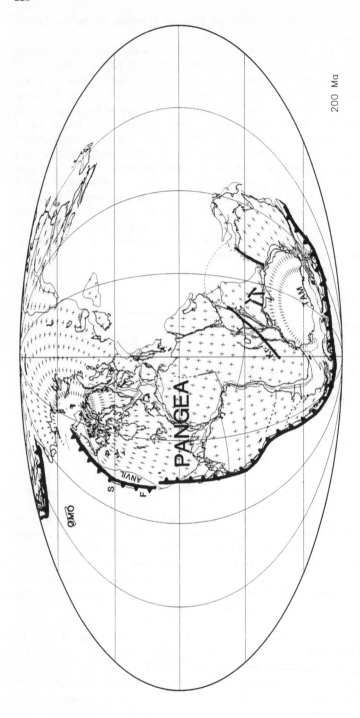

Figure 1. Mollweide projection of Pangea at 200 Ma. Continental margins, where shown, are generally taken to be 2000-meter isobath. Five-degree tick marks are based on present-day latitude and longitude of the continents. Regions such as the northern extent of the Indian continent are devoid of tick marks, indicative of its having been subducted or distorted during collision with Eurasia. The present-day equator is shown as a solid line across northern South America and central Africa. Other lines on South America and Africa are intraplate rifts. The heavy solid line with teeth is the approximate location of subduction zones that ringed the southern margin of Pangea in the Early Jurassic. Teeth indicate direction of subduction. Medium weight lines are Permian tectonic features on India, Africa, and western Australia. Gondwana is in a paleolatitudinal framework based on the work of Ziegler et al. (1983). K=Kenya, Kr=line of Permo-Triassic deposits, TAM=Transantarctic Margin of East Antarctica, S=Stikinia, F=Franciscan, OMO=Omolon block.

northward to Nova Scotia and east into Morocco, an area equivalent in size to the Gondwana breakup plume of White and McKenzie (1989). The HTQ dikes were emplaced in the "Newark" rift system that roughly parallels the later continental margin of the central North Atlantic. According to de Boer et al. (1988), the "Newark" rift zones originated as synforms and were almost half filled with clastic sediments before tholeiitic igneous activity was initiated. Sedimentation in the "Newark" rift basins was initiated in Ladinian [230-235 Ma] time in the north (Nadon and Middleton, 1984) and in middle Carnian [225-230 Ma] time in the south (Olsen et al., 1982). There was only a relatively short time [~40 Ma] between final collapse of the ocean between North America and Africa at 276 Ma, and the initiation of sedimentation in the "Newark" rift basins. While the direction of the pre-Alleghanian orogeny subduction is uncertain, it might be inferred to have been to the southeast based on the composite stratigraphic cross sections shown in Hatcher et al. (1989). If the subsequent breakup of Pangea was directly related to the initiation of a mantle plume, then it is difficult to relate the location of the plume to a nearby active subduction zone since the closest apparent one was 3000 km distant (Figure 1). In addition, if the pre-Alleghanian subduction was directed to the southeast then the Pangea breakup may have occurred where there was thinning of the crust that had resulted from the break off of the subducted slab. It can be assumed, though, that the continents were not stationary in a mantle framework for a period of more than 40 Ma if at all, before initial breakup, as indicated by the "Newark" rifts and the later Palisades Sills was seen.

Klitgord and Schouten (1986) date the start of seafloor spreading in the central North Atlantic, Gulf of Mexico, and Ligurian Tethys as 175 Ma. If Tempelman-Kluit (1979) is correct that the Anvil Ocean that separated Stikinia and western North America started to subduct under Stikinia as early as 220 Ma, then "westward" facing subduction may have produced a westward directed extensional force (trench-pull) on Pangea that contributed to the separation of Laurentia from Gondwana. On the other hand, if the Anvil Ocean was not a very large ocean, and more closely resembled a minor back-arc basin, then it is unreasonable to suggest westward subduction as the sole cause of the breakup of Pangea. Opening of the central North Atlantic may have instead resulted in westward motion of North America in a mantle framework and led to the subsequent collision of cratonic North America with Stikinia and the Franciscan blocks. It is probable, though, that multiple causes may have led to the breakup of Pangea and seafloor spreading in the central North Atlantic.

Breakup of Gondwana

Neither separation of Pangea into Laurentia and Gondwana nor the later breakup of Mesozoic Gondwana appear to have resulted from initiation of classic back-arc spreading. Creation of new oceanic crust in most cases occurred many thousands of kilometers from the nearest subduction zone, and in the case of the separation of India from Antarctica, the nearest subduction zone was at least 4000 km distant. In fact, pre-drift stretching between East and West Gondwana formed a zone nearly orthogonal to the Triassic or older subduction zone along the southern Gondwana margin (Lawver et al., 1991). Initial seafloor spreading in the Somali and Mozambique basins and in the Southwest Weddell Sea may have been subparallel to the subduction zone but the spreading centers were offset by very long transform faults that connected the near-to-the-subduction zone Southwest Weddell Sea/Rocas Verdes Basin with the quite-distant-from-the-subduction zone (~5000 km) Somali Basin.

Modern backarc basins trend roughly parallel to subduction zones, but only a few hundred kilometers at most separate the trench from the back-arc seafloor spreading center.

Some evidence suggests that the Mesozoic rifting of Gondwana was frequently preceded by continental volcanism. In at least two cases, East/West Gondwana breakup and South America/Africa breakup, the accompanying volcanism is attributed to the initiation of a mantle plume (White and McKenzie, 1989). They argued in the first case that the initiation of rifting was the result of a plume that produced the Karoo volcanics of southern Africa and similar-aged volcanic outcrops of Antarctica. Cox (1988) has argued, though, that Mesozoic flood basalts and doleritic intrusions form a belt 4500 km across Antarctica, which extends far beyond the edge of White and McKenzie's (1989) plume. Cox (1988) suggested the term "hot line" to explain the Ferrar dolerites of Antarctica and Australia.

If the Ferrar dolerites, which roughly parallel the Middle Jurassic subduction zone along Gondwana, are considered to be part of the stretching phase between East and West Gondwana, then one might argue that they were produced by a very large scale version of "back-arc spreading," in the sense that a deeply subducted slab perturbed normal mantle flow. If the "cold" slab reached a depth substantially greater than 700 km, it may have induced the formation of a mantle roll or "hot line." Once the mantle roll became established, it may have produced zones of weakness in the already stretched continental crust which in turn led to actual seafloor spreading and oceanic crust production. On the other hand, if, as Hill (1991) suggests, the plumes produce extra gravitational potential, then they may in fact "fracture" the old cratonic crust and the Ferrar dolerites may simply be the result of an exceptionally long fracture.

By the Late Permian to Early Triassic, a marine barrier existed between Madagascar and the African continent (Battail et al., 1987). Coffin and Rabinowitz (1988) discuss evidence of Carboniferous-Permian faulting that suggests even earlier extension. This implies that there was some component of continental stretching that predates the Karoo volcanics by 50 to 100 million years and most of the Ferrar dolerites by 120 to 130 million years [dates for igneous rocks from Brewer et al., 1991]. Figure 1 shows the location of the Permo-Triassic Karoo rocks (de Wit et al., 1988), the Permo-Triassic sedimentary basins of western Australia (Wopfner, 1991), and the Permian coal deposits of India (Mishra, 1991) superimposed on the 200 Ma tight fit reconstruction of Pangea. It is apparent that there were some roughly parallel zones of weakness in Gondwana prior to evidence suggestive of a plume in the Dronning Maud Land/South Africa region. Reeves et al. (1987) discuss development of the Jurassic-aged Anza Trough of Kenya which appears to be an aulocogen, and is assumed to have formed during the stretching phase of the initial breakup of Gondwana. It appears that the junction of the Anza Trough with the coast coincides with the deviation of the subsequent seafloor spreading phase of Gondwana breakup from the earlier, linear Permo-Triassic rocks of the Karoo. The Permo-Triassic sedimentary basins of western Australia were a zone of weakness in Gondwana that did not break up until Early Cretaceous time (130 Ma) while the Permian basins of India never progressed to seafloor spreading.

160 Ma

Stretching between East and West Gondwana (Lawver et al., 1991) took place between the time of the tight fit reconstruction of Gondwana which we date at 200 Ma and the initiation of seafloor spreading at about 160 Ma. The time of the stretching phase is thought to be 175±10

Ma. It is interesting to note that when seafloor spreading began in the western Somali Basin (Rabinowitz et al., 1983) and the Mozambique Basin (Simpson et al., 1979; Segoufin, 1978), the direction of opening was at quite an angle to the stretching direction that had prevailed during the first phase of the breakup of Gondwana. It is important to note that the seafloor spreading that is documented in the Somali and Mozambique basins must have been accompanied by seafloor spreading in the Southwest Weddell Sea, which may have extended into the Rocas Verdes Basin of southern South America (Dalziel, 1981). If seafloor spreading in the true back-arc Rocas Verdes Basin predates seafloor spreading between East and West Gondwana then it may have propagated into the region of stretched crust and forced the direction of opening. It is difficult, though, to relate the small but connected regions of initial oceanic crust formation in the Somali and Mozambique basins to mantle convection induced by long-term subduction of the Pacific Ocean plate. It is quite plausible that a number of different breakup mechanisms may have all interacted to produce the Jurassic split between East and West Gondwana. Back-arc basin formation functioned in the west to produce the Rocas Verdes Basin and possibly the Southwest Weddell Sea, initiation of a mantle plume rifted cratonic Africa from Antarctica in the vicinity of the Lebombo Monocline and Mozambique Plain of Southeast Africa, and seafloor spreading followed an older zone of weakness in the continental crust as evidenced by the split between Madagascar and Africa along the trend of the Permo-Triassic Karoo rocks. There may have been a second mantle plume that produced the Anza Trough of Kenya or that may have simply been a result of the change from the rift following the Permian zone of weakness and it connecting directly to the location of the mantle plume in the Mozambique Plain region.

At roughly the same time, seafloor spreading is recorded in the Northwest Australian Basin (Heirtzler et al., 1978). This Late Jurassic spreading center rifted the Southern Tibet (Lhasa) block from Australia. The spreading center is of such a direction and spreading rate that it could have been connected to the contemporaneous seafloor spreading in the western Somali Basin. Since there are no data to indicate that Eurasia moved northward at this time, the Late Jurassic seafloor spreading in the Northwest Australian Basin implies that contemporaneous subduction occurred, either northward beneath Eurasia or southward beneath the southern Tibet block as it moved northward. Since most of the Late Jurassic southern margin of Eurasia was subsequently modified when the Tethyean Seaway closed with the collision of India with Eurasia, it is difficult to ascertain the exact time of initiation of Jurassic subduction or the direction of subduction.

By 160 Ma (Figure 2), the inboard terranes of western North America (Howell et al., 1985) had begun to collide with North America. During collision and obduction of Stikinia (Monger et al., 1982) onto the North American craton [165 to 125 Ma], terrane material was also obducted onto the Brooks Range province of Alaska (Early Cretaceous; Patton and Box, 1989) and the Franciscan formation was accreted onto southwestern North America (Robertson, 1989). The subduction zone shown in Figure 1 that extended northward along the South American coast all the way to central California probably only extended through southern Central America because subduction along the northern part ceased at 180 Ma (Busby-Spera, 1988). The Omolon massif was a microplate that accreted to Siberia in the Jurassic (Fujita and Newberry, 1982), which presumes an active subduction zone off Northeast Siberia during that time. Seafloor spreading that had begun at 175 Ma in the central North Atlantic and Gulf of Mexico connected to the Pacific, but north of the eastward facing subduction zone under West Gondwana (South America/Africa).

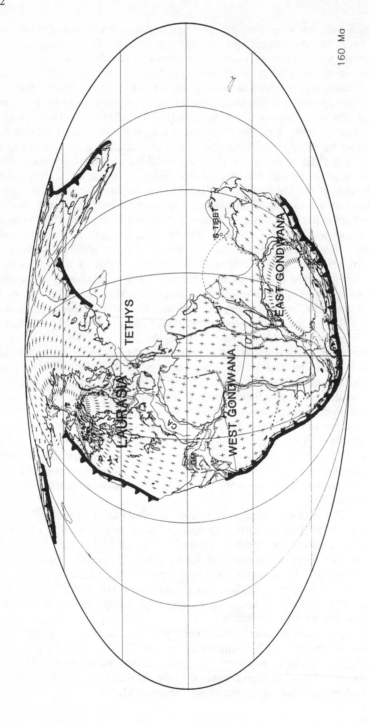

Figure 2. Reconstruction of the continents at 160 Ma in a Mollweide projection. See Figure 1 for explanation of lines and symbols. GM=Gulf of Mexico, CA=central North Atlantic.

120 Ma

Numerous changes in major plate motions occurred during the Early Cretaceous (Figure 3a). Three significant new continental breakups occurred at about 130 Ma: Africa and South America split apart to form the South Atlantic, India and Antarctica split apart, and Arctic Alaska rotated away from Arctic Canada to form the Canada Basin (Figure 3b). White and McKenzie (1989) discuss the evidence for the initiation of seafloor spreading in the South Atlantic Ocean as related to a mantle plume evidenced by the Walvis and Rio Grande rises and the present day Walvis hotspot. Unlike the Jurassic central North Atlantic break which followed related continental volcanism, the eruption of the 1.2 million km^3 Paraná continental tholeiites of eastern South America, the emplacement of the 780,000 km^3 Serra Geral formation of southwestern Africa (White and McKenzie, 1989), and the initiation of ocean crust formation in the South Atlantic (Austin and Uchupi, 1982), all occurred virtually simultaneously.

The timing of the initiation of seafloor spreading between India and Antarctica [130 Ma] is circumstantial at best, dependent mostly on inferences from marine geological data (Lawver et al., in press). The Rajmahal Traps of India are an example of a continental flood basalt province and are dated as 118-115 Ma (^{40}Ar/^{39}Ar date from Baksi, 1986). A separate hotspot that began at about the same time created the Kerguelen Plateau (Davies et al., 1989) and the Ninety East Ridge (Müller et al., in press). Since the southernmost Kerguelen Plateau is not continuous with East Antarctica then it is most probable that both the Rajmahal Traps and Kerguelen Plateau were created after seafloor spreading commenced between India and East Antarctica. Kent (1991) argues from paleodrainage (sediment dispersal) patterns that this area of India/East Antarctica was uplifted in response to a mantle plume that may have persisted for 150 m.y. prior to the commencement of igneous activity. With the initiation of seafloor spreading between India and East Antarctica, the simplest assumption is that subduction was active along the southern margin of Eurasia to compensate for the northward motion of India.

Simultaneous with the initiation of seafloor spreading in the South Atlantic, and between India and East Antarctica, was the opening of the Canada Basin (Figure 3b) between the Arctic Alaska-Chukotka block and Arctic Canada (Lawver et al., 1990). Based on paleomagnetic work, Halgedahl and Jarrard (1987) date the opening of the Canada Basin as post-Valingian [131 Ma]. Geological information (Grantz et al., 1990) indicates that the initiation of seafloor spreading occurred during the Hauterivian [131-124 Ma]. Lawver et al. (1983) suggested that the Mendeleev and Alpha ridges may have been formed by a hotspot that originated on the Mendeleev Ridge side of the opening basin, that switched to the Alpha Ridge side. There is no available information, though, to date the age of the hotspot, or to indicate that there are continental flood-type basalts that either pre-date or are simultaneous with the opening.

Eastward subduction beneath the North American continent resumed at about 125 Ma (Templeman-Kluit, 1979). According to Severinghaus and Atwater (1990), there was fairly steep, rapid subduction under North America during much of the Cretaceous. The Pacific-Aluk spreading center in the South Pacific was close to the East Australian margin and the northern margin of New Zealand but probably did not get subducted until the latest Early Cretaceous or the Late Cretaceous (Bradshaw, 1989). Subduction persisted off western South America and the Antarctic Peninsula (Hervé et al., 1987). Shortly after 120 Ma, the seafloor

234

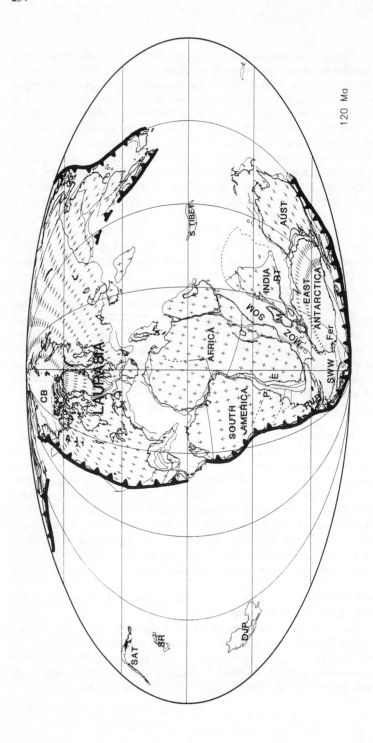

Figure 3a. Reconstruction of the continents at 120 Ma in a Mollweide projection. For this figure and the following figures, the continents are fixed in an absolute framework based on a reconciliation of African, Indian, and North and South American Plate hotspots with the relative motion of those plates (Müller et al., 1991). CB=Canada Basin, E=Etendeka volcanics, Fer=Ferrar Dolerites of East Antarctica, M=Madagascar, MOZ=Mozambique Basin, P=Paraná volcanics of Brazil, RT=Rajmahal Traps of India, RVB=Rocas Verdes Basin, SAT=Southern Alaska Superterrane, SOM=Somali Basin, SWW=Southwest Weddell Sea, SR=Shatsky Rise; OJP=Ontong Java Plateau, last two locations from hotspot tracks.

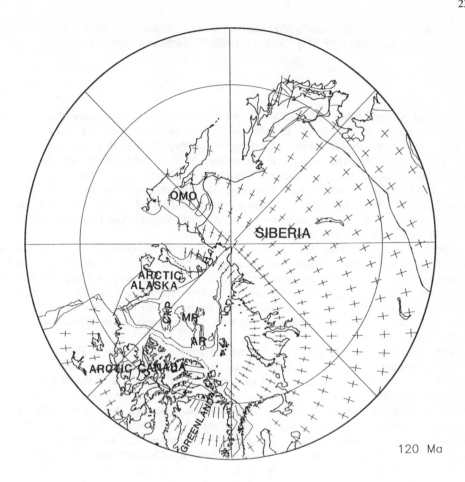

Figure 3b. Polar stereographic projection of the continents to 50°N. The Canada Basin is shown partially opened with Arctic Alaska having rotated away from Arctic Canada about a pole of rotation near 65°N 55°W for this figure. CkP=Chukchi Plateau, MR=Mendeleev Ridge, AR=Alpha Ridge, OMO=Omolon block.

spreading in the Somali Basin which had begun the breakup of Gondwana ceased, just younger than Chron M0 [118 Ma] (Cochran, 1988; Segoufin and Patriat, 1980).

80 Ma

Pindell and Barrett (1989) indicate that at about 80 Ma there was a major change in the subduction regime in the vicinity of northern South America (Figure 4a). They show the initiation of a northward- to northeastward-dipping subduction zone that formed to the west of northern South America along a Panama-Costa Rica Arc that effectively trapped the present-day Caribbean plate. The trapped "Caribbean" seafloor had been part of the Farallon plate of the Pacific Ocean. A northward direction of subduction of the Pacific plate is supported by evidence for contemporaneous northward-directed subduction under Alaska. This motion resulted in the accretation of the Alaskan Superterrane (Panuska and Stone, 1985) to the margin of North America.

Although subsidence may have begun as early as 130 Ma (Hinz et al., 1979), initial seafloor spreading in the Labrador Sea region of the North Atlantic began at about 84 Ma (Roest and Srivastava, 1989; Srivastava and Roest, 1989). Subduction along the Siberian margin was oblique but had developed into an Andean-type margin by this time (Parfenov and Natal'in, 1986). According to Sengör (1984), the northward subduction of the Tethyean Ocean beneath Eurasia also continued at this time, having begun during Albian-Aptian time [118-97 Ma].

In the mid-Cretaceous (Figure 4b), the tectonic regime in the New Zealand and eastern Australia region underwent a significant change. From the Permian to the Early Cretaceous, New Zealand was the site of convergent margin tectonics that are recorded as remnants of magmatic arcs, forearc basins, trench slope basins, and accretionary complexes (Bradshaw, 1989). At 105±5 Ma, the tectonic regime in New Zealand changed abruptly with the oblique collision of the Pacific-Aluk spreading center with the subduction zone. Soon after the subduction of the spreading center to the north of New Zealand, seafloor spreading began in the Bounty Trough region between Campbell Plateau and Chatham Rise and extended into the Great South Basin immediately east of South Island (Carter, 1988). Seafloor spreading originated between Campbell Plateau and Marie Bryd Land at about 84 Ma (Stock and Molnar, 1987) and extended into the Tasman Sea region (Weissel and Hayes, 1977). Although pre-rift stretching and rapid subsidence of the southern Australian margin began at 125 Ma (Hegarty et al., 1988), actual seafloor spreading between Australia and Antarctica did not begin until 96 Ma (Cande and Mutter, 1982; Veevers et al., 1990). Rifting began very slowly (<4 mm/yr) between 96 Ma and 45 Ma but then increased in speed to 20 mm/yr. There has been no mention of any plume related activity with regard to the rifting of Australia from East Antarctica, the opening of the Tasman Sea, or the rifting of New Zealand and the Campbell Plateau from Marie Byrd Land. While the latter two rifts might be related to the subduction of the Pacific-Aluk spreading center, no plausible cause for the initiation of the rifting between Australia and Antarctica has been proposed. The dramatic change in spreading rate between 49 and 44.5 Ma may be attributed to the collision of India with Eurasia and the switch from spreading on the ridge between the Australia-India spreading center that is now almost all subducted beneath the Sunda Arc, to spreading on the Australia-Antarctica spreading center.

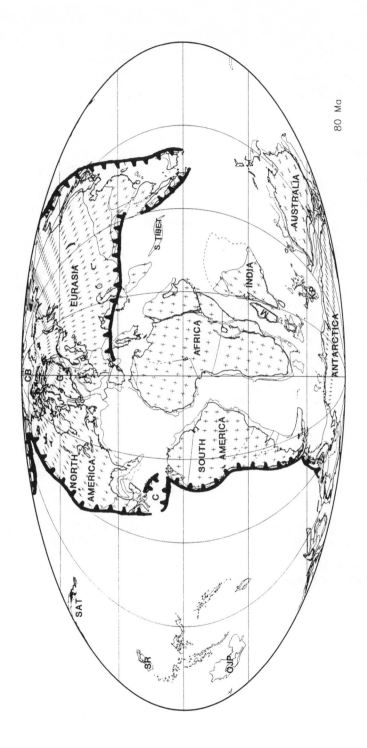

Figure 4a. Reconstruction of the continents at 80 Ma in a Mollweide projection. CB=Canada Basin, G=Greenland, KP=Kerguelen Plateau, others as in Figure 3a.

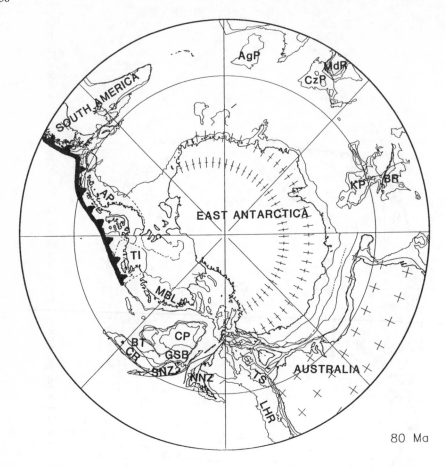

Figure 4b. Polar stereographic projection of the continents to 50°S. AgP=Agulhas Plateau, AP=Antarctic Peninsula, BR=Broken Ridge, BT=Bounty Trough, CP=Campbell Plateau, CR=Chatham Rise, CzP=Crozet Plateau, GSB=Great South Basin, KP=Kerguelen Plateau, LHR=Lord Howe Rise, MBL=Marie Byrd Land, MdR=Madagascar Rise, NNZ=North New Zealand, SNZ=South New Zealand, TI=Thurston Island, TS=Tasman Sea.

40 Ma

The final major rift to develop was the extension of seafloor spreading from the central North Atlantic (Figure 5) into the North Atlantic (Srivastava and Tapscott, 1986) between Greenland and Eurasia at about 56 Ma (Talwani and Eldholm, 1977), although seafloor spreading continued in the Labrador Sea until 30 Ma (Srivastava and Roest, 1989. White and McKenzie (1989) relate the opening of the North Atlantic to initiation of the Iceland plume. Lawver and Müller (in press) suggest that the present Iceland hotspot can be tracked in a mantle framework to have been active in the Canada Basin at 130 Ma. It was located under Cenozoic volcanics on the east coast of Greenland at 40 Ma and was located under the Early Tertiary volcanics on the west coast of Greenland at 60 Ma. This is supported by the isotopic work of Holm et al. (1991) who indicate that the West Greenland lavas show only little interaction between rising magmas and the lithosphere while the first-erupted East Greenland lavas display a strong influence of lithosphere-derived components. When the North Atlantic opened at 56 Ma, the Iceland hotspot was apparently still under the west coast of Greenland. The early magmas erupting on the East coast were travelling through the initially undepleted mantle and lithosphere beneath Greenland. Only with time did the range of erupted compositions in East Greenland narrow and become more like present-day Icelandic lavas (Holm et al., 1991).

Along the Pacific margin of Antarctica, the subduction zone that was still operative along the margin to the east of Marie Byrd Land began to shut off in stages as the Pacific-Aluk spreading center was subducted. Barker (1982) discusses the northward progression of the subduction of the Pacific-Aluk spreading center along the Antarctic Peninsula, with the spreading center between the Tharp and Heezen fracture zones having been subducted about 50 million years ago. South of the Tharp fracture zone, there are insufficient magnetic anomaly data to determine when or even if the Pacific-Aluk spreading center was subducted off the Thurston Island block of West Antarctica. From data presented in Mayes et al. (1990), 65 Ma [Anomaly 28 time] is not an unreasonable estimate for when that spreading center may have been subducted off Thurston Island. The last remnant of subduction off the Antarctic Peninsula ceased active subduction about 4 Ma (Barker, 1982) when the Pacific-Aluk spreading died. It actually left a remnant of the Aluk Plate off the South Shetland Islands between 57°W and 64°W and may be one of the few places on the Earth where the subducted slab was left hanging.

With the collision of India with Eurasia during the Eocene, there was a major world-wide reorganization of major plate motions. The dramatic bend in the Hawaii-Emperor seamount chain at 43 Ma indicates that the Pacific Plate changed from virtually northward motion to more northwestward motion. A comparison of the major subduction zones shown at 160 Ma in Figure 2 with the present-day subduction zones shown in Figure 6 shows a general change from mostly northward-directed subduction at 160 Ma to mostly westward- or eastward-directed subduction at present. There is an apparent secondary change from predominantly high latitude subduction at the start of Pangea and Gondwana breakup to mostly low latitude subduction (<45°) at present.

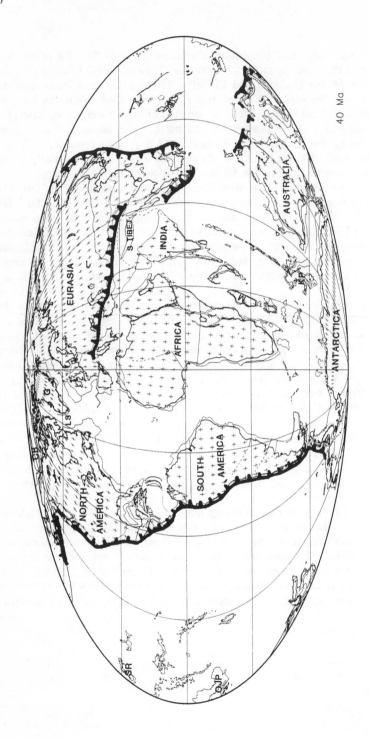

Figure 5. Reconstruction of the continents at 40 Ma in a Mollweide projection. CB=Canada Basin, G=Greenland, KP=Kerguelen Plateau, LS=Labrador Sea, M=Madagascar, SR=Shatsky Rise; OJP=Ontong Java Plateau, last two locations from hotspot tracks.

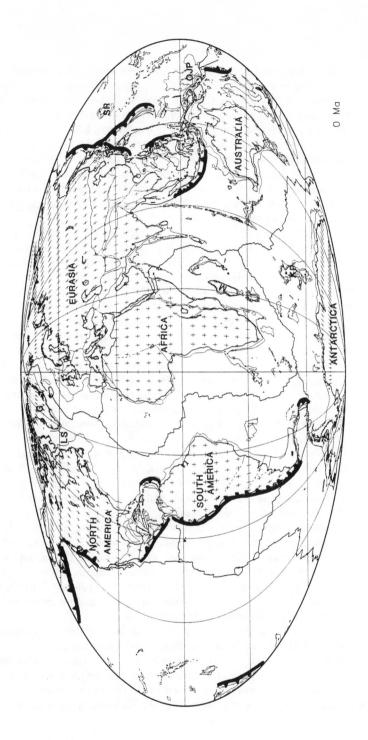

Figure 6. Present-day configuration of the continents in a Mollweide projection. See Figure 5 for abbreviations.

Conclusions

The breakup of Pangea (and Mesozoic Gondwana) only marks the latest supercontinent to be dispersed in a tectonic history that extends back to the Archean. Pangea did not finally agglomerate until the Permian [~275 Ma] and even at that time there were pieces being accreted on one side as breakup was beginning elsewhere. Within 40 million years there was evidence in the "Newark" rifts that Pangea was beginning to break apart. Late Triassic to Early Jurassic "Karoo" type volcanism paralleled the later rift that produced the central North Atlantic. If continental breakup between North America and Africa was related to the initiation of a mantle plume, then it seems merely fortuituous based on the short time between final agglomeration and initiation of breakup that the continents were located where they were with respect to a plume fixed in a mantle framework. It is difficult to relate the breakup of Pangea to forces directly related to subduction.

The Early to Middle Jurassic breakup of Gondwana may have been related to the initiation of a plume as White and McKenzie (1989) suggest. It is possible though that the pattern of the breakup was influenced by earlier zones of weakness, specifically those indicated by Permo-Triassic sedimentation. The mantle plume as suggested for the Dronning Maud Land/South Africa region produced stretching of the continental crust that connected into the older zone of weakness, possibly via a second thermal plume that produced extension in the Anza Trough of Kenya during the Early Jurassic. It is also possible that the long-term, ongoing subduction beneath the western margin of Gondwana contributed to the Late Jurassic seafloor spreading, either as a direct cause of true back-arc basin production in the Rocas Verdes basin, or as cause of the "hot-line" that may have caused the intrusion of the Ferrar dolerites. Quite likely, all three mechanisms contributed to the breakup of Gondwana.

Both the breakup between India and Antarctica and the breakup between South America and Africa seem to have occurred slightly before the associated continental volcanism. If the Early Cretaceous Rajmahal Traps or some other episode of continental volcanism initiated the breakup of India with Antarctica, the seafloor extension that they produced connected into the earlier Permo-Triassic zone of weakness along the western margin of Australia. No evidence of pre-rift uplift with respect to the Walvis hotspot has been noted although White and McKenzie (1989) discuss evidence that the margins remained abnormally elevated during rifting. It is conceiveable that in these two cases, plumes were initiated as part of the continental breakup and were not the cause of the breakup.

No plume activity has been related to extension between Australia and Antarctica. It has been suggested that the extension between Campbell Plateau and Marie Byrd Land of West Antarctica and the seafloor spreading in the Tasman Sea were a result of the subduction of the Pacific-Aluk spreading center shortly before seafloor spreading began in the two zones (Bradshaw, 1989). In both of these cases as well as in the case of the opening of the Gulf of California, there seems to be a direct link between subduction of a spreading center and the initiation of marginal basin extension within 10 to 20 million years.

The last major rift occurred in the North Atlantic. This rift has been related to the initiation of a mantle plume (White and McKenzie, 1989), but there is evidence that the Icelandic hot spot existed prior to the breakup and may in fact have produced the Alpha and Mendeleev Ridges in the Arctic Ocean. Whether or not the Icelandic hot spot originated as a mantle plume at that time [~130 Ma] is not known, and unfortunately there is no paleo-latitudinal framework that can be used to determine the position of the Icelandic hotspot earlier than 130 Ma.

With the possible exception of the Gondwana breakup plume between Africa and Antarctica, there are no good correlations between a long established (>>100 Ma) supercontinent, a particular part of the mantle, and pre-breakup continental flood basalts. It seems more likely, given the shape of the subduction zones around the southern margin of Pangea, that subduction and pre-existing crustal weaknesses are also important factors in supercontinent breakup.

Acknowledgments. This work was supported by the Division of Polar Programs, DPP 90-19247 and by the Plates project of the Institute for Geophysics. The Plates project is supported by a consortium of industry sponsors to develop rotation software and to support research in plate reconstructions. Don Anderson, Ian Dalziel, Mike Coffin and Jan Garmany provided helpful comments regarding this manuscript. This is Contribution number 903 of the Institute for Geophysics.

References

Austin, J.A. and E. Uchupi (1982) Continental-oceanic crustal transition off Southwest Africa. Am. Assoc. Petrol. Geol. Bull. 66, 1328-1347.

Baksi, A.K. (1986) ^{40}Ar-^{39}Ar incremental heating study of whole-rock samples from the Rajmahal and Bengal traps, eastern India (abstr). Terra Cognita 6, 161.

Barker, P.F. (1982) The Cenozoic subduction history of the Pacific margin of the Antarctic Peninsula: ridge crest-trench interaction. J. Geol. Soc. London 139, 787-801.

Battail, B., Beltan, L. and J-M. Dutuit (1987) Africa and Madagascar during Permo-Triassic time: the evidence of the vertebrate faunas. *In:* Gondwana Six: Stratigraphy, Sedimentology, and Paleontology, ed. G.D. McKenzie, American Geophysical Union Geophysical Monograph 41, 147-156.

Borg, S.B. and D.J. DePaolo (1991) A tectonic model of the Antarctic Gondwana margin with implications for southeastern Australia: isotopic and geochemical evidence. Tectonophysics 196, 339-358.

Bradshaw, J.D. (1989) Cretaceous geotectonic patterns in the New Zealand region. Tectonics 8, 803-820.

Brewer, T.S., Rex, D., Guise, P. and B. Storey (1991) ^{40}Ar-^{39}Ar age determinations from the Theron and Pensacola Mountains, Antarctica: implications for the age of Mesozoic magmatism in Antarctica. *In:* Programme and abstracts, Magmatism and the Causes of Continental Breakup, ed. B. Storey, Geological Society of London, 6-7.

Busby-Spera, C.J. (1988) Speculative tectonic model for the early Mesozoic arc of the southwest Cordilleran United States. Geology 16, 1121-1125.

Campbell, I.H. and R.W. Griffiths (1990) Implications of mantle plume structure for the evolution of flood basalts. Earth Planet. Sci. Lett. 99, 79-93.

Cande, S.C. and J.C. Mutter (1982) A revised identification of the oldest sea-floor spreading anomalies between Australia and Antarctica. Earth Planet. Sci. Lett. 58, 151-160.

Carter, R.M. (1988) Post-breakup stratigraphy of the Kaikoura Synthem (Cretaceous-Cenozoic), continental margin, southeastern New Zealand. New Zealand J. Geol. Geophys. 31, 405-429.

Cawood, P.A. (1984) The development of the SW Pacific margin of Gondwana: correlations between the Rangitata and New England orogens. Tectonics 3, 539-553.

Cochran, J.R. (1988) The Somali Basin, Chain Ridge and the origin of the Northern Somali Basin gravity and geoid low. J. Geophys. Res. 93, 11985-12008.

Coffin, M.F. and P.D. Rabinowitz (1988) Evolution of the conjugate East African-Madagascan margins and the western Somali Basin. Geological Society of America Special Paper 226, 78pp.

Cox, K.G. (1988) The Karoo Province. *In:* Continental Flood Basalts, ed. J.D. Macdougall. Dordrecht: Kluwer, 239-272.

Dalziel, I.W.D. (1981) Back-arc extension in the southern Andes: a review and critical reappraisal. Phil. Trans. R. Soc. Lond. A 300, 319-335.

Dalziel, I.W.D. (1991) Pacific margin of Laurentia and East Antarctica/Australia as a conjugate rift pair: Evidence and implications for an Eocambrian supercontinent. Geology 19, 598-601.

Dalziel, I.W.D. (1992) Antarctica; a tale of two supercontinents? Annu. Rev. Earth Planet. Sci. 20, 501-526.

Dalziel, I.W.D. and R.D. Forsythe (1985) Andean evolution and the terrane concept. *In:* Tectonostratigraphic terranes of the Circum-Pacific region, ed. D.G. Howell, CPCEMR ESS no. 1, 565-581.

Davies, H.L., Sun, S-S., Frey, F.A., Gautier, I. et al. (1989) Basalt basement from the Kerguelen Plateau and the trail of a Dupal plume. Contrib. Mineral. Petrol. 103, 457-469.

de Boer, J.Z. (1968) Paleomagnetic differentiation and correlation of the Late Triassic volcanic rocks in the central Appalachians (with special reference to the Connecticut Valley). Geol. Soc. Am. Bull. 79, 609-626.

de Boer, J.Z., McHone, J.G., Puffer, J.H., Ragland, P.C. and D. Whittington (1988) Mesozoic and Cenozoic magmatism. *In:* The Geology of North America, Volume I-2, The Atlantic Continental Margin, U.S., eds. R.E. Sheridan and J.A. Grow. Geological Society of America, 217-241.

de Wit, M.J., Jeffery, M., Bergh, H. and L.O. Nicolaysen (1988) Geological map of sectors of Gondwana reconstructed to their deposition—150 Ma. American Association of Petroleum Geologists (Tulsa). Map.

Forsythe, R. (1982) The late Palaeozoic to early Mesozoic evolution of southern South America: a plate tectonic interpretation. J. Geol. Soc. 139, 671-682.

Fujita, K. and J.T. Newberry (1983) Accretionary terranes and tectonic evolution of northeast Siberia. *In:* Accretion Tectonic in the Circum-Pacific Regions, eds. M. Hashimoto and S. Uyeda. Tokyo: Terra Scientific Publishing Co., 43-57.

Grantz, A., May, S.D. and P.E. Hart (1990) Geology of the Arctic Continental Margin of Alaska. *In:* The Arctic Ocean region: The Geology of North America, Volume L, eds. A. Grantz, L. Johnson and J.F. Sweeney. Geological Society of America, 257-288.

Gust, D.A., Biddle, K.T., Phelps, D.W. and M.A. Uliana (1985) Associated Middle to Late Jurassic volcanism and extension in southern South America. Tectonophysics 116, 223-253.

Halgedahl, S. and R. Jarrard (1987) Paleomagnetism of the Kuparuk River formation from oriented drill core: Evidence for rotation of the North Slope block. *In:* Alaskan North Slope Geology, eds. I.L. Tailleur and P. Weimer. Los Angeles: Soc. Econ. Paleont. and Min., Pacific Section, 581-617.

Hatcher, R.D., Jr., Thomas, W.A., Geiser, P.A., Snoke, A.W., Mosher, S. and D.V. Wiltschko (1989) Alleghanian orogen. *In:* The Appalachian-Ouachita Orogen in the United States: The Geology of North America, Volume F-2, eds. R.D. Hatcher, Jr., W.A. Thomas and G.E. Viele. Geological Society of America, Boulder, Colorado, 233-318.

Hegarty, K.A., Weissel, J.K. and J.C. Mutter (1988) Subsidence history of Australia's southern margin: constraints on basin models. AAPG Bulletin 74, 615-633.

Heirtzler, J.R., Cameron, P.J., Cook, T., Powell, H.A. et al. (1978) The Argo Abyssal Plain. Earth Planet. Sci. Lett. 41, 21-31.

Hervé, F., Godoy, E., Parada, M.A., Ramos, V. et al. (1987) A general view on the Chilean-Argentine Andes, with emphasis on their early history. *In:* Circum-Pacific Orogenic Belts and Evolution of the Pacific Ocean Basin, eds. J.W.H. Monger and J. Francheteau. Geodynamics Series Volume 18, AGU, 97-114.

Hill, R.I. (1991) Starting plumes and continental breakup. Earth Planet. Sci. Lett. 104, 398-416.

Hinz, K., Schlüter, H.-U., Grant, A.C., Srivastava, S.P., Umpleby, D. and J. Woodside (1979) Geophysical transects of the Labrador Sea: Labrador to Southwest Greenland. Tectonophysics 59, 151-183.

Hoffman, P.F. (1991) Did the breakout of Laurentia turn Gondwanaland inside out? Science 252, 1409-1412.

Holm, P.M., Hald, N. and T.F.D. Nielsen (1991) Contrasts in composition and evolution of Tertiary CFBS in West and East Greenland: tectonic effects during the establishment of the Icelandic mantle plume. *In:* Programme and Abstracts, Magmatism and the Causes of Continental Breakup, ed. B. Storey, Geological Society of London, 14-15.

Howell, D.G., Jones D.L. and E.R. Schermer (1985) Tectonostratigraphic Terranes of the Circum-Pacific region. *In:* Tectonostratigraphic Terranes of the Circum-Pacific Region, ed. D.G. Howell. Circum-Pacific Council for Energy and Mineral Resources Earth Science Series, no. 1, Houston, 3-30.

Kent, R. (1991) Lithospheric uplift in eastern Gondwana: evidence for a long-lived mantle plume system. Geology 19, 19-23.

Klitgord, K.D. and H. Schouten (1986) Plate kinematics of the central Atlantic . *In:* The Geology of North America, Vol. M, The Western North Atlantic Region, eds. R. Vogt and B.E. Tucholke. Geological Society of America, 351-404.

Lawver, L.A. and R.D. Müller (in press) A hotspot origin for the Canada Basin and the path of the Iceland Hotspot. EOS, Spring (1992) AGU meeting abstract.

Lawver, L.A., Grantz, A. and L. Meinke (1983) The tectonics of the Arctic Ocean. *In:* Arctic Technology and Policy, eds. C. Chryssostomidis and I. Dyer. Washington, D.C.: Hemisphere Publishing Company, 147-158.

Lawver, L.A., Müller, R.D., Srivastava, S.P. and W.R. Roest (1990) The opening of the Arctic. *In:* Arctic versus Antarctic Geology, eds. U. Bleil and J. Thiede. NATO ASI Series, Series C, V. 308. Amsterdam: Kluwer Academic Publishers, 29-62.

Lawver, L.A., Sandwell, D.A., Royer, J.-Y. and C.R. Scotese (1991) Evolution of the Antarctic Continental Margins. *In:* Antarctic Earth Science, eds. M.R.A. Thomson, J. Thomason, and J.A. Crame. Cambridge University Press, 533-539.

Lawver, L.A., Gahagan, L.M. and M.F. Coffin (in press) The development of paleoseaways around Antarctica. *In:* The Role of the Southern Ocean and Antarctica in Global Change: an Ocean Drilling Perspective, eds. J.P. Kennett and J. Barron. AGU Antarctic Research Series.

May, P.R. (1971) Pattern of Triassic-Jurassic diabase dikes around the North Atlantic in the context of predrift positions of continents. Geol. Soc. Am. Bull. 82, 1285-1292.

Mayes, C.L., Lawver, L.A. and D.T. Sandwell (1990) Tectonic history and new isochron chart of the South Pacific, J. Geophys. Res. 95, 8543-8567.

Mishra, H.K. (1991) A comparison of the petrology of some Permian coals of India with those of western Australia. In: Gondwana Seven Proceedings, eds. H. Ulbrich and A.C. Rocha Campos, Instituto de Geociencias, Universidade de Sao Paulo, 261-271.

Monger, J.W.H., Price, R.A. and D.J. Tempelman-Kluit (1982) Tectonic accretion and the origin of the two major metamorphic and plutonic welts in the Canadian Cordillera. Geology 10, 70-75.

Moores, E. (1991) Southwest U.S.-East Antarctic (SWEAT) connection: A hypothesis. Geology 19, 425-428.

Müller, R.D., Royer, J-Y. and L.A. Lawver (submitted) Revised plate motions relative to the hotspots from combined Atlantic and Indian Ocean hotspot tracks. Geology.

Nadon, N.C. and G.V. Middleton (1984) Tectonic control of Triassic sedimentation in southern New Brunswick: Local and regional implications. Geology 12, 619-622.

Olsen, P.E., McCune, A.R. and K.S. Thomson (1982) Correlation of the Early Mesozoic Newark supergroup by vertebrates, principally fishes. Am. J. Sci. 282, 1-44.

Palmer, A.R. (1983) The decade of North American geology 1983 geologic time scale. Geology 11, 503-504.

Panuska, B.C. and D.B. Stone (1985) Latitudinal motion of Wrangellia and Alexander terranes and the southern Alaska Superterrane. In: Tectonostratigraphic Terranes of the Circum-Pacific Region, ed. D.G. Howell. Houston: Circum-Pacific Council for Energy and Mineral Resources Earth Science Series, no. 1, 109-120.

Parfenov, L.M. and B.A. Natal'in (1986) Mesozoic tectonic evolution of northeastern Asia. Tectonophysics 127, 291-304.

Patton, W.W. and S.E. Box (1989) Tectonic setting of the Yukon-Koyukuk Basin and its borderlands, western Alaska. J. Geophys. Res. 94, 15807-15820.

Pindell, J.L. and S.F. Barrett (1989) Geological evolution of the Caribbean: a plate-tectonic perspective. In: The Caribbean Region, The Geology of North America, Volume H, eds. J.E. Case and G. Dengo. Geological Society of America, Boulder, Colorado, 405-432 plus Plate 12.

Rabinowitz, P.D. and J.L. LaBrecque (1979) The Mesozoic South Atlantic Ocean and evolution of its continental margins. J. Geophys. Res. 84, 5973-6002.

Rabinowitz, P.D., Coffin, M.F. and D.A. Falvey (1983) The separation of Madagascar and Africa. Science 220, 67-69.

Reeves, C.V., Karanja, F.M. and I.N. Macleod (1987) Geophysical evidence for a failed Jurassic rift and triple junction in Kenya. Earth Planet. Sci. Lett. 81, 299-311.

Robertson, A.H.F. (1989) Palaeoceanography and tectonic setting of the Jurassic Coast Range ophiolite, central California: evidence from the extrusive rocks and the volcaniclastic sediment cover. Mar. Petrol. Geol. 6, 194-220.

Roest, W.R. and S.P. Srivastava (1989) Seafloor spreading in the Labrador Sea: A new reconstruction. Geology 17, 1000-1004.

Segoufin, J. (1978) Anomalies magnetiques mesozoiques dans le bassin de Mozambique. Comptes Rendes Sceances, Academic Science Series 2, 287D, 109-112.

Segoufin, J. and P. Patriat (1980) Existence d'anomalies mesozoîques dans le bassin de Mozambique. Comptes Rendus de l'Academie des Sciences, Paris, 287, 109-112.

Sengör, A.M.C. (1984) The Cimmeride orogic system and the tectonics of Eurasia. Geological Society of America, Special Paper 195, 82 pp.

Severinghaus, J. and T. Atwater (1990) Cenozoic geometry and thermal state of the subducting slabs beneath western North America. Geological Society of America, Memoir 176, 1-22.

Simpson, E.S.W., Sclater, J.G., Parsons, B., Norton, I.O. and L. Meinke (1979) Mesozoic magnetic lineations in the Mozambique Basin. Earth Planet. Sci. Lett. 43, 260-264.

Srivastava, S.P. and C.R. Tapscott (1986) Plate kinematics of the North Atlantic. In: The Geology of North America, Vol. M, The Western North Atlantic Region, eds. P.R. Vogt and B.E. Tucholke. Geology Society of America, 379-405.

Srivastava, S.P. and W.R. Roest (1989) Seafloor spreading history II-VI. In: East Coast Basin Atlas Series: Labrador Sea, J.S. Bell (co-ordinator). Atlantic Geoscience Centre, Geological Survey of Canada, Map sheets L17-2 – L17-6.

Stock, J. and P. Molnar (1987) Revised history of early Tertiary plate motion in the Southwest Pacific. Nature 325, 495-499.

Talwani, M. and O. Eldholm (1977) Evolution of the Norwegian-Greenland Sea. Geol. Soc. Am. Bull. 88, 969-999.

Tempelman-Kluit, D.J. (1979) Transported cataclasite, ophiolite and granodiorite in Yukon: Evidence of arc-continent collision, Geolological Survey of Canada, Paper 79-14, 27 pp.

Veevers, J.J., Stagg, H.M.J., Willcox, J.B. and H.L. Davies (1990) Pattern of slow seafloor spreading (<4mm/year) from breakup (96 Ma) to A20 (44.5 Ma) off the southern margin of Australia. BMR Journal of Australian Geology & Geophysics 11, 499-507.

Weissel, J.K. and D.E. Hayes (1977) Evolution of the Tasman Sea reappraised. Earth Planet. Sci. Lett. 36, 77-84.

White, R.S. and D.P. McKenzie (1989) Magmatism at rift zones: the generation of volcanic continental margins and flood basalts. J. Geophys. Res. 94, 7685-7729.

Wopfner, H., (1991) Permo-Triassic sedimentary basins in Australia and East Africa and their relationship to Gondwanic stress pattern. In: Gondwana Seven Proceedings, eds. H. Ulbrich and A.C. Rocha Campos. Instituto de Geociencias, Universidade de Sao Paulo, 261-271.

Ziegler, A.M., Scotese, C.R.and S.F. Barrett (1983) Mesozoic and Cenozoic paleogeographic maps. In: Tidal Friction and the Earth's Rotation II, eds. Broche and Sundermann. Berlin: Springer-Verlag, 240-252.

RELATIONSHIP BETWEEN HOTSPOTS AND MANTLE STRUCTURE: CORRELATION WITH WHOLE MANTLE SEISMIC TOMOGRAPHY

S. KEDAR, D. L. ANDERSON and D. J. STEVENSON
Division of Geology and Planetary Sciences 170-25
California Institute of Technology
Pasadena, CA 91125 U.S.A.

ABSTRACT. We examine the relation between the locations of hotspots on the surface of the Earth and mantle structure as determined from seismic tomography. In particular, we correlated hotspot locations with Tanimoto's (1989) shear wave velocity structure throughout the mantle. A spherical harmonic representation of both fields enables us to perform a "degree by degree" correlation, and to test possible relationships between features of the same scale. A statistical significance analysis is applied to these results. A similar analysis was performed by Richards et al. (1988) in studying hotspot - geoid relations.

Two major phenomena were observed in the hotspot - shear velocity correlation: 1. Very good correlation between the hotspots and slow (hot) regions in degree 2, and only degree 2, in the bottom half of the lower mantle. The correlation gradually decays to zero in the upper mantle. 2. A good correlation of degree 6 with the deeper upper mantle (200-670 km) which decays rapidly below 670 km. These good correlations are significant both statistically and by the fact that these degrees show peaks in the hotspot amplitude spectrum.

The length scales we are looking at are too large to determine the origin of a single hotspot. However, if we believe that hotspots reflect the general convection pattern in the mantle, we can hope to learn something about the style of this convection. The above observations suggest the possibility of two, not necessarily independent, regions in different depths in the mantle, which control the location of different hotspots. To check whether this is reasonable, we excluded those hotspots which give a negative contribution to the $l=6$ correlation in the upper mantle. The correlation of these excluded hotspots with the lower mantle at degree 2 improves slightly relative to the already significant correlation that the complete set of 47 hotspots demonstrates. This set of observations presents a constraint on mantle convection models.

1. Introduction

Hotspots are areas of extensive and long-lived volcanism. Some hotspots seem to be fixed relative to each other. Even though the phenomenon of hotspots is widely accepted, there is an on-going debate as to the number, composition, mechanism and origin. The following work is an attempt to get a better idea of the regions in the Earth's interior that control the location of hotspots on the Earth's surface.

We choose to study this problem using the most powerful tool for deep Earth exploration, seismic tomography. In particular, we have chosen Tanimoto's (1989) whole mantle tomography and the list of 47 hotspots used by Morgan (1981) and Crough and Jurdy (1980). We correlate the two and discuss the relationship between them.

First we discuss what we might expect to see for various models of mantle convection. For homogeneous whole-mantle convection with strong mantle plumes, such as in Morgan's (1972) model, we expect radial continuity throughout the mantle. Hotspots are expected to correlate well

D. B. Stone and S. K. Runcorn (eds.),
Flow and Creep in the Solar System: Observations, Modeling and Theory, 249–259.

with slow seismic velocity regions in the lower and upper mantles. It was the good correlation of the hotspots with the long-wavelength (*l*=2) geoid and lower mantle velocity structure, that led Richards et al. (1988) to hypothesize that plumes originated in the lower mantle. However, as Richards et al. state, the correlation alone does not carry any cause and effect information, and their test of whether hotspots are responsible for the *l*=2 residual geoid pattern is not conclusive.

Phase changes and possible viscosity and chemical changes near the boundary between the upper and lower mantle may inhibit or prevent material transfer. In this case, the upper and lower mantles may be only weakly coupled and will exhibit less continuity. For example, a high viscosity lower mantle thermally coupled to the upper mantle will control the general locations of hot upwellings, but the scale lengths of the upper mantle and the plates, and plate motions, will control the style and detailed pattern of the upper mantle. Likewise, the locations of past and present subduction zones may control the locations of cold downwelling regions of the lower mantle even if slabs are confined to the upper mantle.

2. Data Sets

We use the whole mantle tomography obtained by Tanimoto (1989). Tanimoto divided the mantle into 11 shells whose depths are listed in Table 1. By fitting wave forms of Love waves and deep turning S, SS, and SSS waves, he retrieved the seismic shear wave velocity variation in each shell. This was represented in spherical harmonics up to degree *l*=6. We note that slower than average areas are probably also hotter than average, whereas the faster regions are colder. The full list of harmonic coefficients for the 11 shells is given in Tanimoto's original paper. Table 2 lists the 47 hotspots that were used for the correlation. This list follows the compilation of Morgan (1981) and Crough and Jurdy (1980) and is the same list used by Richards et al. (1988).

Morgan (1981) used plate reconstructions to identify tracks of hotspots on the moving plates. He then showed that relative to the plates, these hotspots form a fixed frame of reference.

Shell	Depth Range (km)
1	0 - 220
2	220 - 400
3	400 - 670
4	670 - 1022
5	1022 - 1284
6	1284 - 1555
7	1555 - 1816
8	1816 - 2088
9	2088 - 2359
10	2359 - 2630
11	2630 - 2891

Table 1: Shells in mantle model (Tanimoto, 1990)

3. Correlation

Before going into the details of the correlation calculation and analysis, it is important to note that at no point do we presume any relationship between the location of the hotspots on the surface and the deep Earth structure. We simply look for similarities and differences between the two data sets and try to draw some conclusions about their nature.

3.1. METHOD

We start by overlaying the hotspot map on the velocity variation map of each shell. This way of correlating the two fields is valuable to a certain extent. It is clear, however, that in order to examine possible relationships between small scale features in the two fields, such as the scale of a group of hotspots, we need to use a more precise tool, numerical correlation.

Hotspot	Latitude (deg.)	Longitude (deg.)	Hotspot	Latitude (deg.)	Longitude (deg.)
Eifel	50	7	**Easter Island**	-27	-109
Hoggar Mountains	23	6	Galapagos Islands	0	-91
Tibesti, Chad	21	17	San Felix	-27	-80
Jebel Marra, Sudan	13	24	Juan Fernandez	-34	-79
Mount Cameroon	4	9	Jan Mayen	72	-8
Lake Victoria	-3	36	Iceland	64	-20
Afar, Ethiopia	12	42	**Bermuda**	30	-60
Comores Islands	-12	44	Azores	38	-28
Reunion	-21	56	Madeira	33	-17
Crozet	-45	45	Canary Islands	28	-17
Kerguelen	-45	65	New England	29	-29
Christmas Islands	-35	80	Cape Verde	15	-24
Tasmania	-40	150	**Fernando**	-4	-32
Yellowstone	45	-111	**Arnold**	-17	-25
Raton, NM	37	-104	**Trinidade**	-21	-29
Baja California	27	-113	Ascencion	-8	-14
Bowie	53	-135	St. Helena	-16	-6
Juan de Fuca	46	-128	Tristan de Cunha	-37	-12
Hawaii	20	-155	Discovery	-42	0
MacDonald	-29	-140	Bouvet	-54	4
Society Islands, Tahiti	-18	-148	**Vema**	-32	16
Pitcairn	-27	-129	**Mount Erebus**	-78	167
Caroline Islands	3	167	Samoa	-15	-168
Marquesas	-11	-139			

Table 2: Morgan's Hotspots

Since the seismic velocities are described using spherical harmonics, it is only natural to do the same for the hotspots field and to perform the correlation in the harmonic domain. This approach has two advantages: First, the calculation is straightforward as a result of the orthogonality of the harmonic expansion. Secondly, it enables us to treat each harmonic degree l as a separate spherical function and by correlating the two functions "degree by degree," we can compare features of the same scale.

In principal, any function which is defined on a sphere can be written in the form:

$$f(\theta,\phi) = \sum_{l=0}^{\infty} \sum_{m=0}^{l} [C_l^m cos(m\phi) + S_l^m sin(m\phi)]P_l^m(cos\theta)$$ (1)

where θ and ϕ are latitude and longitude, respectively, and P_l^m are Legendre Polynomials.

In particular, we follow Richards et al. (1988) and represent each hotspot as a spike $\delta(\theta,\phi)$, and set every other point on the globe to have a value of zero, we then obtain a set of harmonic coefficients for the hotspots field.

Now, suppose $f(\theta,\phi)$ (hotspots) is expanded into a set of coefficients C_l^m and S_l^m, and $g(\theta,\phi)$ (seismic velocities) is expanded into A_l^m and B_l^m, then for each degree, l, we get a correlation coefficient:

252

Now, suppose $f(\theta,\phi)$ (hotspots) is expanded into a set of coefficients C_l^m and S_l^m, and $g(\theta,\phi)$ (seismic velocities) is expanded into A_l^m and B_l^m, then for each degree, l, we get a correlation coefficient:

$$r_l = \frac{\sum_{m=0}^{l} (C_l^m A_l^m + S_l^m B_l^m)}{\sqrt{\sum_{m=0}^{l} (C_l^{m2} + S_l^{m2})}\sqrt{\sum_{m=0}^{l} (A_l^{m2} + B_l^{m2})}} \tag{2}$$

Figure 1 shows the numerical correlation for the lowermost mantle layer A positive correlation is defined as a correlation between regions of high concentration of hotspots and slow regions in the mantle. A statistical analysis was applied to the results.

Confidence levels (dashed lines) at 10% interval were calculated following Eckhardt (1984). Note that only $l=2$ exhibits a significant correlation. Larger wavelengths, $l=1$, and shorter wavelengths, $l>2$, show no correlation.

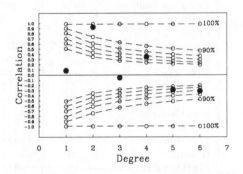

Figure 1: Correlation between hotspots and the bottom of the mantle (shell 11)

3.2. CORRELATION RESULTS

After correlating the hotspots with all the shells in Tanimoto's model, two major phenomena were observed:

1. There is a very good correlation between the hotspots and slow, hot regions in degree 2 in the bottom of the mantle, which decays as we climb up in the mantle (Figure 2). The upper mantle and top of the lower mantle show little correlation in $l=2$.
2. A good correlation in degree 6 in the deeper upper mantle (200-670 km) which drops below 670 km and decreases significantly below shell 4 (Figure 3).
3. There are essentially no significant correlations for other degrees.

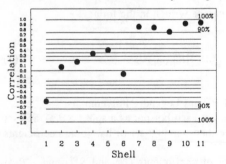

Figure 2: Variation of correlation of $l=2$ throughout the mantle.

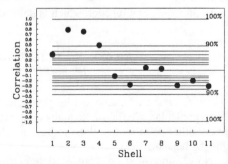

Figure 3: Variation of correlation of $l=6$ throughout the mantle.

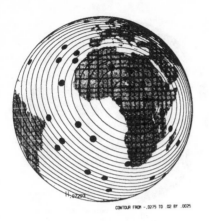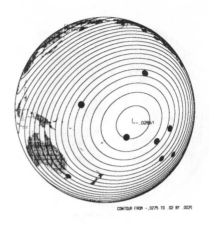

Figure 4a: Shell 11 - Shear wave velocity $l = 2$

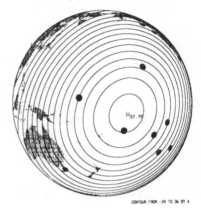

Figure 4b: Hotspots Distribution $l = 2$

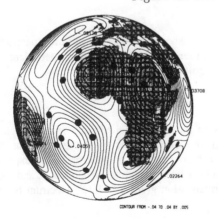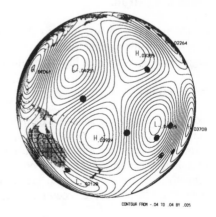

Figure 4c: Shell 3 - Shear wave velocity $l = 6$

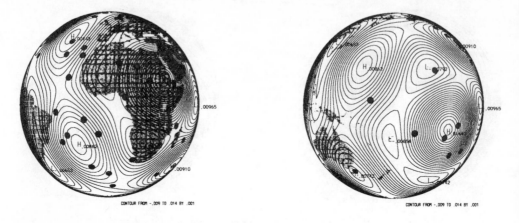

Figure 4d: Hotspots Distribution $l = 6$

These observation are illustrated in Figure 4(a-d). Highs (H) in the hotspots field match lows (L) in the velocity fields of the corresponding shell and harmonic degree. The positive correlations are statistically significant (higher than 95% confidence level). Correlations of other harmonic coefficients were low and statistically insignificant.

We have performed the same analysis for a list of 116 hotspots published by Vogt (1981). The correlation with Vogt's hotspots showed no significant correlation at any degree. The $l=2$ correlation coefficient has decreased to values that range from 0.3 to 0.7 in the lower mantle, and the $l=6$ correlation coefficient has decreased slightly as well. Vogt used a fairly broad definition of a hotspot. His list may be contaminated by volcanically active regions, which are not real hotspots. On the other hand, it should be commented that Morgan's list includes some disputable hotspots (Bermuda) and misses other possible hotspots in regions like China and Siberia.

4. Amplitude Spectrum

We should pay special attention to the harmonic amplitude spectrum. The amplitude spectrum describes what portion of the field a particular harmonic represents. A high correlation coefficient r_l in a degree that is also a peak in the amplitude spectrum is significant.

4.1. CALCULATION

We used fully normalized spherical harmonics. In addition to the full normalization, consistent with Richards et al. (1988), we divide the amplitude spectrum of each harmonic by the square root of $2l+1$. The explanation for this additional normalization can be understood by looking at the expansion of a delta function located at the pole. In that case there is no longitudinal dependency, and the full normalization gives exactly that factor. Thus, this normalization gives a "white" spectrum for a delta function. The amplitude spectrum is therefore:

$$A_l = \sqrt{\frac{\sum_{m=0}^{l} (C_l^{m^2} + S_l^{m^2})}{2l + 1}} \qquad (3)$$

The hotspots' amplitude spectrum is shown in Figure 5.

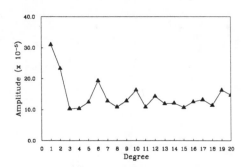

Figure 5: Hotspots - Amplitude Spectrum

4.2. HOTSPOTS' AMPLITUDE SPECTRUM

Three main features dominate the hotspots' amplitude spectrum:

1. The spectrum is peaked in degree $l=1$. This is a reflection of the fact that the hotspots' distribution on the Earth's surface is asymmetric. Most of the hotspots are located in and around Africa and the Atlantic.
2. Degree 2 also has a high value in the amplitude spectrum. The second order characteristic of the hotspots' distribution is the existence of two main groups of hotspots: Africa and the Atlantic; the Pacific.
3. Degree 6 is a peak in the amplitude spectrum. This harmonic reflects the general spacing between groups of hotspots.

We note that degrees 2 and 6 are those that showed significant correlations with different parts of the mantle. Surprisingly, $l=1$ shows little correlation with either the lower or upper mantle.

4.3. AMPLITUDE SPECTRUM - STATISTICAL ANALYSIS

We check the possibility that the above characteristics are a random result. We randomly distributed 47 points on a sphere, representing them as $\delta(\theta,\phi)$, expanded this function in spherical harmonics, and calculated its amplitude spectrum. This test was repeated 6000 times to form histograms of the amplitude spectrum of 47 points randomly distributed on a sphere. This is presented in Figure 6. It is evident that the amplitude at degrees 1, 2 and 6 falls well beyond the mean of the amplitude of the random distribution. At the same time, the degrees which had low amplitude ($l=3,4,5,7,8$) fall within the main body of the histograms. Therefore, we can conclude that the relatively high values we obtained for $l=1,2$ and 6 are not a random effect.

256

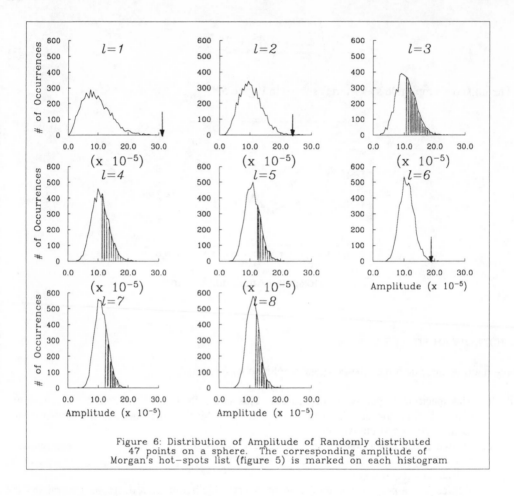

Figure 6: Distribution of Amplitude of Randomly distributed
47 points on a sphere. The corresponding amplitude of
Morgan's hot-spots list (figure 5) is marked on each histogram

4.4. SEISMIC VELOCITIES AMPLITUDE SPECTRUM

According to Tanimoto's (1989) calculations of the tomographic amplitude spectrum, degree 2 dominates throughout the mantle, with the exception of layer 5 in which degree 3 is dominant. Degree 1 also exhibits high power in layers 7-11. We note that the upper mantle (layers 1-3) and the bottom of the mantle (layer 11) have relatively more power than other layers of the mantle, and that there is a better depth resolution for the upper mantle than the lower mantle.

5. Discussion

The key to a better understanding of the resulting correlations lies not only in the good correlations but in the lack of them. We have observed a significant correlation coefficient for $l=2$ in the deep lower mantle, which gradually shifts in a way that results in a negligible correlation by the time we reach the upper mantle. At the same time r_6 is low in the lower mantle and high in the upper mantle, and is changing more abruptly than r_2. Both $l=2$ and $l=6$ are peaks in the hotspots' amplitude spectrum. It seems that the $l=6$ pattern is dominated by the upper mantle. Even though there is a relatively high r_6 at shell 4, we must remember that the resolution in each shell is not independent of the neighboring shells (Tanimoto, 1989). In other words, it is not very likely to have a sharp drop from a very high correlation coefficient, to zero between two neighboring shells.

It is reasonable to ask if there are two populations of hotspots, one controlled by the $l=2$ pattern of the lower mantle, and one by the $l=6$ pattern in the upper mantle. We tested this hypothesis by excluding all the hotspots whose contribution to the high r_6 in the upper mantle was negative. We found 14 hotspots that had this characteristic. These are the 14 hotspots marked by boldface letters in Table 2. We argue that if the excluded hotspots are independent of the other 33, then these 14 points should have no significant r_2 in the lower mantle. However, these 14 points have an even slightly higher r_2 than all the 47. This point is subtle. It is possible that since the complete set of 47 hotspots has such a high r_2, that any randomly chosen 14 points of the 47 will result in a significant correlation coefficient as well. We therefore randomly picked 14 out of the original 47 hotspots and correlated them with the lower mantle. After repeating this 6000 times, we got the histogram shown in Figure 7. In less than 2% of the cases 14 randomly chosen hotspots had as high or higher r_2 as the 14 we excluded. Since the scales we are analyzing are much larger than the scale of a single hotspot, it is hard to draw conclusions about the nature and origin of a single hotspot.

Therefore, this last result does not tell us where hotspots originate. Moreover, based on our observations, we cannot say that we found two sets of hot spots that rise from different depths in the mantle.

The most general statement we can make at this point is that, to the extent that we trust our data sets, if hotspots are a manifestation of the convection pattern in the mantle, then this pattern is affected by the upper mantle, as we can deduce from the $l=6$ pattern of the hotspots field. A possible model that may be examined is the breakup of a major ($l=2$) upwelling into smaller ones across the transition zone. It is possibly significant that the ratio 2/6 is approximately the ratio of thicknesses of the upper mantle and lower mantle.

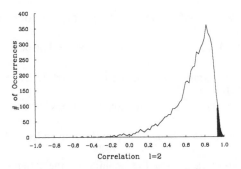

Figure 7: 14 out of 47 original hotspots
6000 random checks

It may be significant that many of these l=2 hotspots are among the strongest or most active (Davies, 1988; Sleep, 1990; Schilling, 1991) (Hawaii, MacDonald, Pitcairn, Reunion, Easter). On the other hand some strong hotspots (Afar, Azores, Iceland, Galapagos, Kerguelen, Tristan, Marquesas, Society) contribute to the l=6 pattern. It appears that strength or long-livedness or temperature excess (Schilling, 1991) are not characteristics that distinguish these groups.

6. Summary

1. There are excellent correlations between the global hotspot distribution and the lowermost mantle shear wave velocity distributions at degree 2. The l=2 geoid also correlates well with the above (Richards et al., 1988). The l=2 component dominates the seismic S wave velocity throughout the mantle, and is a significant component in the hotspots field. Of equal importance is the negligible correlation at l=1 and l>2. The l=2 pattern of the upper mantle does not correlate with the distribution of hotspots or with the l=2 pattern of the lower mantle. This seems to rule out simple models of whole mantle convection. The mismatch in tomographic patterns between upper and lower mantle has been pointed out by Tanimoto (1990).
2. The hotspots field is highly asymmetric. This is reflected by a high amplitude degree 1 which dominates the spherical harmonics expansion of the hotspots field. This harmonic does not correlate with the tomographic results at any depth. It may correlate with prior continental positions (Anderson, 1982).
3. In the upper mantle, the l=6 components of the hotspots and the seismic velocities correlate significantly. l=6 is a peak in the hotspots' amplitude spectrum, representing a scale of average spacing between groups of hotspots.
4. 1-3 is a well established set of observations which should be met by any convection model.

Acknowledgments. This work was supported by the National Science Foundation grants EAR 90-02947 and EAR 90-17893. Contribution number 5066, Division of Geological and Planetary Sciences, California Institute of Technology, Pasadena, California, USA.

7. References

Anderson, D. L. (1982) Hotspots, polar wander, Mesozoic convection, and the geoid. Nature 297, 391-393.

Crough, S. T. and Jurdy, D. M. (1980) Subducted lithosphere, hotspots and the geoid. Earth Planet. Sci. Lett. 48, 15-22.

Davies, G. F. (1988) Ocean bathymetry and mantle convection 1. Large scale flow and hotspots. J. Geophys. Res. 93, 10467-10489.

Eckhardt, D. H. (1984) Correlation between global features of terrestrial fields. Mathematical Geology 16, No. 2.

Morgan, W. J. (1972) Plate motions and deep mantle convection. Mem. Geological Society of America, 132, 7-22.

Morgan, W. J. (1981) Hotspot tracks and the opening of the Atlantic and Indian oceans, in C. Emiliani (ed.), The Sea, Wiley-Interscience, New York, pp. 443-487.

Richards, M. A., Hager, B. H. and Sleep, N. H. (1988) Dynamically supported geoid highs over hotspots: observation and theory. J. Geophys. Res. 93, 7690-7708.

Schilling, J. G. (1991) Fluxes and excess temperature of mantle plumes inferred from their interaction with migrating mid-ocean ridges. Nature 352, 397-403.

Sleep, N. H. (1990) Hotspots and mantle plumes: some phenomenology. J. Geophys. Res. 95, 6715-6730.

Tanimoto, T. (1989) Long wavelength S-wave velocity structure throughout the mantle. Geophys. J. Int. 100, 327-336.

Tanimoto, T. (1990) Predominance of large scale heterogeneity and the shift of velocity anomalies between the upper and lower mantle, I. Phys. Earth 38, 493-509.

Vogt, P. R. (1981) On the applicability of thermal conduction models to mid-plate volcanism: comments on a paper by Gass et al. J. Geophys. Res. 86, 95-96.

POROUS MEDIA FLOW IN GRANITOID MAGMAS: AN ASSESSMENT

NICK PETFORD
Department of Earth Sciences,
University of Liverpool,
Liverpool L69 3BX,
UK

ABSTRACT. Compaction and compositional convection as potential *in-situ* differentiation mechanisms in granitoid intrusions has been investigated numerically for melt fractions of between 10 and 50 percent. The results show that the major factor controlling fluid movement, and hence chemical and mineralogical variation during late stage crystallisation is the viscosity of the interstitial melt. Thus, for anhydrous melts where viscosity increases with crystallisation, fluid migration rates are trivial over the average lifespan of even the largest silicic magma chambers (ca. 10^6 years). Alternatively, if the melt viscosity *decreases* during crystallisation, the relative movement of evolved fluid relative to the solid phase is such that both processes become potentially viable mechanisms of *in-situ* magma chamber differentiation. At initial porosities in excess of 20% and melt viscosities at or less than 10^5 pascal seconds, compositional convection is the dominant process of fluid movement in the crystallising pluton. As crystallisation proceeds and porosities drop to values below ~0.2, convective velocities become subcritical and compaction becomes increasingly dominant. A major consequence of both compaction and compositional convection in basic magmas is the production, through superefficient melt extraction, of texturally equilibrated, layered monomineralic rocks. The absence of similar rock types in granitoid plutons suggests that although compaction and compositional convection may go some way to explain chemical and mineralogical variations in zoned granitoids, neither process is capable of producing the extreme mineralogical variations seen in large basic intrusions.

1. Introduction

Most of the major advances in *conceptual* understanding that have taken place in igneous petrology over the last ten years have come about through the application of numerical and dynamic modelling of fluid processes to igneous systems. However, the great majority of work to date has been concerned with the dynamics of low viscosity (10^1-10^3 Pa s) basic magmatic systems (McBirney and Noyes, 1979; Huppert and Sparks, 1980, 1984; Rice, 1981; Sparks et al., 1984; Marsh, 1989; Hansen and Yuen 1987), where there now exists a wealth of both experimental and theoretical data. In contrast, the fluid behavior of silicic magma systems has not been dealt with in any great detail, although their fossilised remains (now exposed at the earth's surface as granitoid plutons) show clear mineralogical and chemical evidence that they have, in much the same way as their more basic counterparts, undergone a complex fluid history prior to solidification. Where magma differentiation in granitoids has occurred *in-situ,* the gross scale compositional variation within the pluton is generally thought to be controlled by boundary layer fractionation (eg. Sawka et al., 1990). At at melt fractions in excess of 50%, this appears an effective way of producing

261

D. B. Stone and S. K. Runcorn (eds.),
Flow and Creep in the Solar System: Observations, Modeling and Theory, 261–286.
© 1993 *Kluwer Academic Publishers. Printed in the Netherlands.*

compositional variation in zoned plutons. However, at melt fractions (porosities) < ~0.5, the mechanical strength of the magma increases over many orders of magnitude (Arzi, 1978) and the system becomes in effect a rigid porous medium. Under these conditions, large scale boundary layer differentiation will be be strongly inhibited. This contribution is concerned with crystal-liquid fractionation processes that are thought to occur after the main stage of boundary layer differentiation, at melt contents generally less than 50%.

Fluid flow through saturated porous media is an important process in both hydrology and petroleum engineering, and is supported by a large base of technical literature (Bear, 1972; Dullien, 1979; Cushman, 1990). Crystallising magma can also be treated as a porous medium, and recently several macroscopic flow models involving 1) **compaction** and 2) **compositional convection** have been applied to basic intrusions with some success in explaining the origins of certain mineralogical and chemical variations (McKenzie, 1985, 1987; Tait et al., 1984; Tait and Kerr, 1987; Tait and Jaupart, 1989; Martin, 1990). In this study, the equations used to model porous media flow in basic magmas have been applied, in an unmodified form, to higher silica (higher viscosity) granitoid systems using values of grain size, viscosity, density and porosity likely to be found during the crystallisation of granitoid magmas. Although simplistic in approach, it is considered that the results obtained here offer some constraints on the required boundary conditions that must be met for compaction and compositional convection to promote chemical differation in high silica granitoid magma chambers.

Section 2 of this paper reviews the role of thermal convection in silicic magma chambers. Section 3 examines some of the field, mineralogical and chemical evidence for *in-situ* zoning in granitoid plutons, and is followed in section 4 by a review of the compaction process. The results obtained from the compaction equations applied to silicic systems are shown in section 5, while sections 6 and 7 consider the role of compositional convection. In the discussion section (8), the comparative effectiveness of both processes are assessed, along with some of the petrological implications of porous media flow in granitoid plutons.

2. Whole-Body Thermal Convection In Silicic Magmas

Once a hot body of magma has been emplaced within the crust, heat loss through the floor, walls and roof of the newly formed magma chamber will produce thermal gradients that may cause the magma to convect. Induced convection resulting from heat loss is described by the dimensionless thermal Rayleigh number

$$Ra = \frac{g\alpha\Delta Td^3}{\kappa_T\mu} \tag{1}$$

representing the ratio of buoyant to viscous forces in a fluid (see Appendix for a list of symbols). The critical Ra number for convection in magma chambers is generally > 3000 (Sparks et al., 1984). For geological systems, the two most important variables in (1) are the thickness of the fluid layer (d), and the fluid viscosity (μ). The viscosities of silicate melts can be estimated fairly reasonably by the Arrhenis relation (Shaw, 1972),

$$\eta = \eta_o \exp(E / RT) \tag{2}$$

where η_0 is a pre-exponential constant and E is the activation energy. The melt viscosity is temperature dependent, and decreases with increasing temperature. However, once crystallisation begins, the cumulative effect of suspended crystals can lead to an increase in viscosity of more than ten orders of magnitude (Arzi, 1978). Shaw (1965) and Bartlett (1969) have shown that even with the relatively high viscosities calculated for granitic liquids, thermal convection is easily achieved during the early stages of cooling and crystallisation, provided d is large enough, while Spera et al., (1982) have calculated Ra numbers in excess of 10^{11} for large magma bodies. This is shown in figure 1, where magma viscosity (calculated using eqn.2) is plotted against magma chamber height (d) for a suite of typical granitoids. The field defined by the granitoids lies well inside the convecting region of the diagram, indicating that during the initial stages of intrusion, the Ra number for these magmas was supercritical. However, the ability of magma chambers to loose heat through thermal convection has been recently questioned by Marsh (1989, 1991), who has argued that ΔT is rarely large enough to drive vigorous, whole-chamber convection. In the Marsh model, thermal convection is limited to the early stages of (superheated) magma evolution, and rapidly gives way to conductive cooling, although this thesis has been recently criticised by Huppert and Turner (1991).

It is assumed that during thermal convection the granitoid liquids were chemically homogeneous, and behaved as viscous Newtonian fluids. However, Spera et al. (1982) have shown that single phase melts with SiO_2 contents > 65 wt % may deviate from strictly Newtonian behavior. Furthermore, silica-rich liquids may also possess an inherent shear strength. One result of these non-linearities, if present, may be to suppress the onset of thermal convection, even at supercritical Ra numbers. In such instances, the critical Rayleigh number may have to be overstepped before convection can begin (Wickham, 1987). Alternatively, if Marsh (1989) is correct in his assessment of ΔT (very small), any non-linearities in fluid behavior may be large enough to effectively suppress thermal convection.

Magma behavior is also to a large extent density dependent. Magma densities vary with temperature, pressure and composition, the latter being strongly dependent on the extent to which Fe is enriched or removed from the melt during crystallisation (Sparks et al., 1980). The densities of multicomponent silicate melts can be calculated from the partial molar volumes of their oxides from (Bottinga and Weill, 1970; Bottinga et al., 1983),

$$\rho = \sum_i X_i M_i / X_i V_i \tag{3}$$

where X_i is the mole fraction of component i, M_i is the gram formula weight and V_i is the partial molar volume. For most silicic melts, there is a general decrease in density with increasing silica content. Densities calculated using (3) for the suite of granitoids illustrated in figure 1 range from 2700 kg m^{-3} at 50 wt% SiO_2 to 2400 kg m^{-3} at 75 wt% SiO_2. The calculations in the following sections are based on a density contrast ($\Delta\rho$) between the solid and melt phase of 300 kg m^{-3}.

2.1. COOLING TIMES IN SILICIC MAGMA CHAMBERS

Obviously for a fractionation mechanism to be of any relevance in magmatic differentiation, it must operate on timescales well below the solidification times for silicic magma chambers. For a closed system chamber undergoing *in-situ* differentiation, the longevity of the system will depend on several factors including the size and depth of emplacement of the magma

264

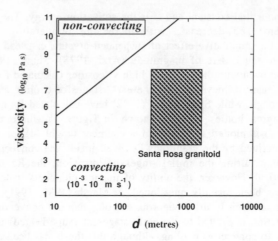

Figure 1. Magma convection as a function of viscosity and magma chamber depth. Numbers in brackets show the approximate range in convective velocities (after Bartlett, 1969 and Wickham, 1987).

body, volatile content and the local geothermal gradient (Shaw, 1965). Heat fluxes for a range of magma bodies have been estimated at between 1×10^{-3} cal cm^{-2} s^{-1} and 1×10^{-5} cal cm^{-2} s^{-1} (Bartlett, 1969 and fig. 1), while some silicic chambers apparently record vertical temperature gradients of the order of 200°C km (Hildreth, 1981). For large (ca 10-15 km sized) intrusions, cooling times are generally between 10^{3}-10^{6} years (Spera, 1980) while studies from long lived volcanic centres such as the Yellowstone Caldera complex suggest magma chamber lifetimes of tens of millions of years. Recent advances in isotope geochronology, in particular the ability to accurately date discrete fractionation events within large silicic magma chambers such as the Long Valley volcanic complex, have been used to imply magma residence times of the order of 0.7 Ma (Halliday et al., 1989). The heat required to to keep the system from solidifying is thought to come from primitive material injected into the base of the chamber (although this has been recently questioned by Sparks et al., 1990).

3. Evidence for Zonation and Chemical Heterogeneities in Granitoid Plutons

Many field based studies from around the world have shown that granitoids are mineralogically and chemically zoned, both vertically and horizontally, on scales ranging from cm to km (Daly, 1933; Vance, 1961; Ragland and Butler, 1972; Stephens and Halliday, 1979; Castro, 1987; Shimizu and Gastil, 1990; Sawka et al., 1990). Some typical examples of zoning patterns are shown in figure 2. The most common form is a concentric zonal geometry defined by a progressive increase in more evolved (higher SiO_2) rock types towards the interiors of the exposed pluton, away from a more 'basic' (lower SiO_2) margin, although this pattern may sometimes be reversed (Speer et al, 1989). It should be stressed that compositional zonation of this kind in the liquid state would be gravitationally unstable. Zoning profiles can be either symmetrical or asymmetric, and individual zones can be either mineralogically and chemically distinct, or diffuse and irregular. Furthermore, some plutons that show no obvious zonation in the field can nevertheless be cryptically zoned with respect to certain trace elements and isotopes (Taylor, 1985; Mohr, 1990).

A number of explanations have been put forward to explain compositional zonation in granitoids, and these are summarised in table 1. Possible mechanisms include: 1) zonation via multiple injection of magmas of differing composition into the same magma body, 2) assimilation (with possible fractionation) of melted wall rock by the invading high temperature magma, and 3) *in-situ* differentiation within a single, homogeneous pulse of magma. The main differences in 1-3 above are that while both 1 and 2 evolve as open systems, with new material being added periodically into the chamber, 3 remains a closed system, effectively isolated from new influxes of heat and mass. Clearly the simplest case is found in 3, and thus only those plutons that are considered to have become zoned *in-situ* are considered further.

3.1. *IN-SITU* ZONING IN GRANITOID PLUTONS

A good example of a horizontally and vertically zoned granitoid is the Santa Rosa unit that forms part of the much larger Coastal Batholith of Peru (Pitcher, 1978). The intrusion shows a range of rock types that are considered to have formed *in-situ* from a single, homogeneous

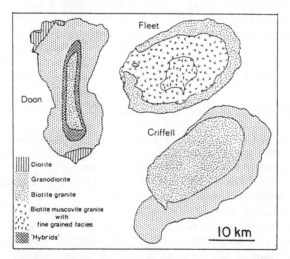

Figure 2. Typical examples of zoned granitoid plutons from southern Scotland (from Stephens and Halliday, 1979). All show a more or less concentric zoning pattern defined by an increase in higher SiO_2 (lower density) rock types towards their cores. Criffell is considered to be a multiple pluton (Stephens and Halliday, 1979) while Doon may have differentiated *in-situ* from a single magma pulse (Tindle and Pearce, 1981).

pulse of magma (Atherton, 1981). Evidence for *in-situ* zoning comes from both field and chemical studies, where detailed sampling has shown a systematic variation in major and trace element concentrations from the margin of the intrusion towards its core (figure 3). In particular, the smooth K/Rb (and Rb/Sr) ratios across individual lithological zones implies that crystallisation occurred unimpeded by repeated influxes of new magma. Similar gradients in trace element ratios are seen in other plutons that are believed to have become zoned *in-situ* (Phillips et al., 1981; Tindle and Pearce, 1981).

3.1.1. In-situ differentiation mechanisms. Most authors invoke some form of boundary layer differentiation to explain the observed zoning profiles seen in granitoids (Bateman and Chappell, 1979; Atherton, 1981; McBirney, 1985; Sawka et al., 1990). In these models, the crystallisation of early formed mineral phases at the side-walls, floor and roof of the cooling magma chamber produces a residual, boundary layer liquid. Depending upon its density, the liquid moves either upwards or downwards, away from the advancing crystallisation front to collect at the floor or the roof of the chamber. Differentiation in the liquid state may also occur through thermogravitational diffusion (Hildreth, 1981).

TABLE 1. Summary of processes that may contribute to compositional zoning in granitoids.

OPEN SYSTEMS	CLOSED SYSTEMS
multiple injection of compositionally discrete magmas (Halliday et al., 1980)	*in-situ* differentiation from a parent magma of uniform composition
assimilation of country rocks during and after emplacement (DePalo, 1981; Stephens et al., 1985)	Possible fractionation mechanisms: 1) crystal settling (Atherton et al., 1979) 2) boundary layer (floor, side-wall and roof) differentiation (Bateman and Chappell, 1979; Sawka et al., 1990)
magma mixing (Wiebe, 1973)	3) filter pressing (Tindle and Pearce, 1981) 4) compaction?
fractional fusion (Prensnall, 1969)	5) compositional convection?

Two processes that may potentially contribute to compositional zoning in granitoid systems at melt fractions < 0.5 are compaction and compositional convection. These melt fractions coincide with the rheological critical melt percentage of Arzi (1978) and the critical melt fraction of Van der Molen and Patterson (1979). Here, at crystal contents of between ca. 40-70%, the mechanical behavior of the system changes from a viscous fluid to an elastic solid. At these crystal contents, compaction at the base of the magma chamber may become an effective differentiation mechanism by removing evolved melt out of the crystalline matrix and into the liquid core of the pluton. Indeed, the analogous process of filter pressing has been cited as an important, late-stage *in-situ* differentiation mechanism in granitoid plutons (Tindle and Pearce, 1981). Finally, there also exists the possibility that, as in basic systems, compositional convection will occur within the crystallising layer. These processes are shown schematically in relation to a crystallising mush in figure 4.

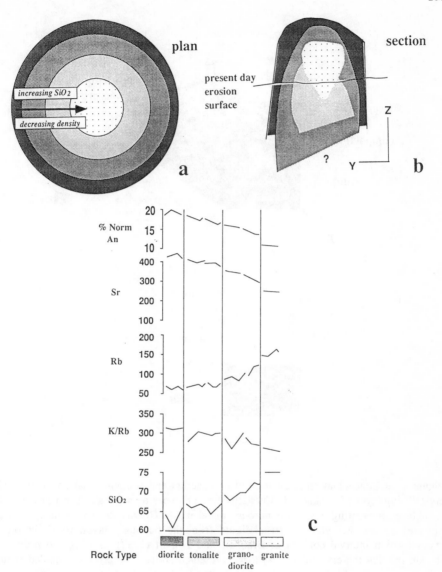

Figure 3. a) Cartoon showing schematically the normally zoned nature of the Santa Rosa granitoid. b) SiO$_2$ (wt%) and trace element (ppm) trends from a traverse at right angles to the contacts of the major rock types that make up the intrusion. The relatively smooth variations in trace elements between each rock type is consistent with closed system, in-situ differentiation. Also shown is the inwards decreasing normative plagioclase content (after Atherton, 1981).

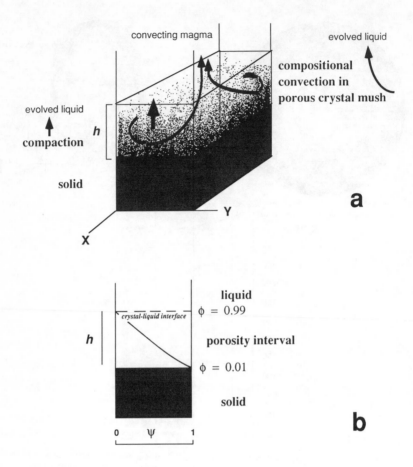

Figure 4. a) Conceptual representation of compaction and compoistional convection within a crystallising layer of thickness h. Crystallisation produces a chemically evolved melt which is capable of convecting out of the porous crystal mush. Gravity driven compaction may also operate to expel evolved interstitial liquid from the boundary layer. b) Variation on the crystallisation interval concept of Brandeis and Jaupart (1987) showing a one-dimensional section through the crystallising layer h represented above. The layer is bounded at the base by fully crystallised solid material and at the top by a crystal-liquid interface, above which is crystal free melt. The crystallising layer forms a porosity interval, with porosity decreasing downwards with increasing crystal content (ψ).

To summarise, there is abundant evidence that at high (1.0-0.5) melt fractions, boundary layer fluid movement has occurred during the *in-situ* crystallisation of granitoid plutons, and that this movement has resulted in mineralogical and chemical variations. The rest of this paper is concerned with the potential role of both compaction and compositional convection in producing chemical and mineralogical variations in granitoid margins at porosities (melt fractions) < 0.5.

4. Compaction and Melt Extraction

McKenzie (1984, 1985) has shown that gravity driven compaction of partially molten rock can be an efficient mechanism of melt extraction and chemical differentiation. The solutions to the differential equations formulated by McKenzie (1984, 1985) governing the the viscous deformation of the compacting crystalline matrix pertinent to this study are set out below. These are a) the compaction length, b) the relative separation velocity between melt and matrix, and c) the compaction time.

In the simplest case, a layer of thickness h with a constant melt fraction (ϕ) is placed on an impermeable surface. If the matrix is denser than the melt, compaction then begins in a layer near to the surface, whose thickness, δ_c, is defined by

$$\delta_c = \left[\mu^{-1}\left(\zeta + \frac{4}{3}\eta\right)k\right]^{\frac{1}{2}} \tag{4}$$

where μ is the shear viscosity of the melt, ζ and η are the bulk and shear viscosities of the matrix, and k is the permeability of the melt network. In layered intrusions, the viscosities of pure olivine matrices range from 2×10^{18} to 10^{17} Pa s (McKenzie, 1985). However, the matrix viscosity in granitoid systems is an unknown variable, and following McKenzie (1985), ($\zeta + 4/3\eta$) has been assigned a value of 10^{18} Pa s. If the resulting upward movement of less dense melt is fast enough to prevent compaction, the relative separation velocity between melt and matrix becomes

$$W_o = K(1-\phi)(\rho_s - \rho_f)g / \mu\phi \tag{5}$$

where ϕ is the porosity (or melt fraction by volume) and g is acceleration due to gravity. As the compacting melt moves upwards, the compacting layer must also increase with time to provide more fluid. The time, t_h, taken to reduce the total amount of buoyant fluid in a layer h by a factor of e (from $h\phi$ to $h\phi/e$), is approximated by

$$t_h = \frac{h}{W_o(1-\phi)} = \frac{\tau_o h}{\delta_c} \tag{6}$$

where the variable τ_o is a scaling factor (Richter and McKenzie, 1984). Eqn. (6) is only valid in cases where $\delta_c \ll h$. Where the compaction length exceeds the layer height, the flow becomes independent of both viscosity and permeability, and is controlled purely by the flow of the matrix. In this case, t_h is expressed by

$$t_h = \frac{\tau_o \delta_c}{h} = \frac{\phi\left(\zeta + \frac{4}{3}\eta\right)}{h(1-\phi)^2 \Delta\rho g} \tag{7}$$

The governing equations and their solutions were originally used to describe the expulsion of small melt fractions form a source region within the mantle (McKenzie, 1984; Richter and McKenzie, 1984). However, compaction may also be important in generating low porosity cumulate rocks in crustal layered basic magma chambers, and in driving fluid movements in sedimentary basins (Irvine, 1980; Bethke, 1985; Kerr and Tait, 1986; McKenzie, 1987; England et al., 1987).

Compaction as a means of segregating granitic melts from a crustal source region undergoing 2-10% partial melting has been considered by Wickham (1987). Using the constants adopted by McKenzie (1984) for mantle melting, Wickhams calculations showed that over timescales of 10^6-10^7 years, only limited melt segregation would occur, and concluded that the compaction process is incapable of producing km sized bodies of granitic magma over reasonable($< 10^9$ years) timescales, due largely to the inhibiting effects of high melt viscosities.

4.1. CRYSTALLISATION, COMPACTION AND MELT SEGREGATION IN SILICIC MAGMA CHAMBERS

Both the approach and the problems addressed here are different from those set out by Wickham (1987). Rather than using the compaction equations to generate a body of granitic magma, they are instead applied to a magma chamber where compaction is occurring within a 100 m thick crystallising layer at the base of the chamber. Similar models have been put forward to explain the origin of some adcumulate layers in basic intrusions (McKenzie, 1987). The analogous process of filter pressing has also been proposed as a mechanism for *in-situ* fractionation in granitoid plutons (Tindle and Pearce, 1981). As a gravity driven process, compaction will not be able to produce large scale horizontal zoning in magmatic systems, and is thus only applicable in the case of vertically zoned plutons (although compaction modelling in sedimentary basins by Bethke (1985) suggests that here the process can lead to horizontal flow).

Although large compositional differences exist between granitoid and basic rock types, McCarthy and Groves (1979) have suggested that granitoids are analogous to basic cumulates, where their textures reflect a mixture of early formed crystals (cumulus phase) and trapped residual liquid (intercumulus melt). Textural and petrological studies of Cordilleran type granitoids suggest that the first major minerals to crystallise are plagioclase ± amphibole ± biotite. Phase relations in the quaternary Ab-Or-An-Qz(-H_2O) system show that for many Cordilleran granitoids, the early stages of crystallisation are dominated by the precipitation of plagioclase (Prensnall and Bateman, 1973; Atherton, 1988). These rocks often contain modal plagioclase in excess of 50% that in three dimensions form an interconnected framework (R. Hunter, pers comm). Thus, during cooling it is possible to envisage a crystallising layer at the margins and floor of the chamber composed essentially of plagioclase ± biotite and hornblende, with the intergranular fluid forming an interconnected porosity of up to 50%. The effects of the crystallisation of biotite and hornblende will lower the density of the intergranular melt by removing iron and magnesium.

4.1.1. Porosity and Permeability. Rapid undercooling at the margins of magma chambers leads to nucleation and crystal growth. The resultant boundary layer between the crystallising and fully crystallised magma forms a crystallisation front that moves inwards with time, away from the solid margins of the intrusion (Brandeis and Jaupart 1987). The nucleation density, and the shape and size of the growing crystals in the crystallisation front is likely to exert a fundamental control on the porosity and permeability, and hence fluid flow, through this region. In this study, where the emphasis is on the volume of melt as opposed to the volume of solid, it is more useful to think of the crystallising magma in terms of melt content, as opposed to crystalline solid. The relationship between porosity (ϕ) and crystal content (ψ) is

$$\phi = (1 - \psi) \qquad (8)$$

where the porosity interval decreases from 1 to 0 as crystallisation proceeds (Fig 4b). A porosity interval of 0.5-0.1 (ie. 50-10%) has been used in the following calculations in order to asses the roles of compaction and compositional convection during the latter stages of crystallisation. Although for the following calculations the porosity interval is kept constant for constant viscosities, in reality the situation may be far more complex, and both viscosity and porosity are likely to vary within the crystallising layer. The major control on porosity is the grain size, itself strongly dependent on the crystallisation kinetics within the layer. Another important variable, controlled to a large extent by ϕ is the permeability (k) of the porous media. Permeability measurements over a range of porosities suggest that the relationship

$$k = \frac{d^2 \phi^{5.5}}{5.6} \qquad (9)$$

where d^2 is the grain diameter, is valid for randomly packed spheres with porosities of between 0.35-0.65 (Dullien, 1979). McKenzie (1984) considered this to be too high an estimate of mantle permeabilities, where porosities range generally between 0.001 and 0.01, and has proposed a slightly modified expression (McKenzie, 1984, eqn. 4.4). However, for the much higher porosities expected during boundary layer crystallisation in crustal magma systems, (9) appears a valid approximation (Kerr and Tait, 1987).

5. Results

Figure 5 shows the variation in compaction length (δ_c) with porosity for the range of magma viscosities appropriate for the Santa Rosa granitoids, calculated using eqns. 4 and 7. The dashed lines in figure 5 show where the compaction length exceeds the compacting layer height (ie $\delta_c \gg h$). The calculations show that for a given viscosity, the compaction length is strongly influenced by the porosity. For example, a melt with a viscosity of 10^7 Pa s and a porosity of 0.3 has a compaction length of 50 m which is half the compacting layer height. But if the porosity is reduced to 0.1, δ_c falls to just 2.4 m. For viscosities higher than ca. 10^7 Pa s, compaction lengths become small even at relatively high (0.3) porosities. However, a favourable combination of high porosities and low viscosities can lead to rapid extraction times. For example, in the most extreme case, where $\mu = 10^5$ Pa s and $\phi = 0.3$, the time (t_h) required to compact the layer h is ~ 7×10^4 years. However, if for the same viscosity the porosity is reduced to 0.1, the required compaction time increases to 2.5×10^5 years.

Figure 6 shows the variation in τ_o (scaled using eqns. (6 and 7)) with porosity for the same range of melt viscosities used in figure 5. Also shown are the upper limits of magma chamber longevity based on cooling rates in silicic plutons (Spera et al., 1982). Taking a minimum

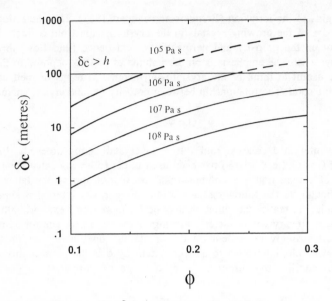

Figure 5. Compaction length (δ_c) versus porosity (ϕ) for a range of melt viscosities appropriate for anhydrous granitoid magmas, calculated using (4) and (7). The dashed lines show where the compaction length exceeds the layer height of 100 m (see text for discussion).

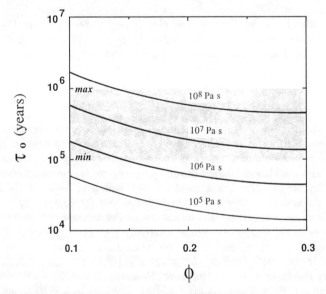

Figure 6. The variation in scaled compaction time (τ_o) with porosity for the range of melt viscosities shown in Fig. 5. The shaded area represents the minimum and maximum estimated time taken for a large silicic magma chamber to cool to its solidus temperature. Compaction is only a viable mechanism of melt extraction at melt viscosities less than ca 10^6 Pa s.

time required for the magma temperature to reach the solidus temperature as 10^5 years, it is clear that compaction will only be important in systems where melt viscosities are > 10^6 Pa s and at porosities of between 30-20%. The implications of prolonged cooling rates are discussed in section 8.

5.1. COMPACTION AND TRACE ELEMENT VARIATIONS

Studies of vertically zoned plutons often show a marked increase in the upwards distribution of certain trace elements such as U, Th, and K (Lachenbruch, 1969; Sawka and Chappell, 1988). The fate of a particular element during magmatic crystallisation is determined by its distribution coefficient (k_d)

$$k_d = \frac{C_{s(i)}}{C_{l(i)}} \tag{10}$$

where $C_{s(i)}$ is the concentration of element i in the solid, and $C_{l(i)}$ is the concentration of element i in the liquid. For $K_ds > 1.0$, the element is concentrated preferentially in the solid phase and is regarded as compatible. Where $K_d < 1.0$, the element is excluded from the crystallising phase, and will remain in the melt. Uranium and Thorium both have $K_ds < 1.0$ for the major mineral phases in granitoid rocks and as a consequence become progressively enriched in the melt during crystallisation and subsequent compaction. An expression that describes the behavior of incompatible trace elements during compaction is given by McKenzie (1985) as

$$W_e / W = \left[(\frac{1}{\phi} - 1)\frac{\rho_s}{\rho_f}k_d + 1 \right]^{-1} \tag{11}$$

where W_e/W is the effective velocity of an incompatible element with respect to the velocity of the melt. McKenzie (1985) first used eqn. (11) to compare the velocities of incompatible trace elements during mantle compaction in matrices of pure olivine and orthopyroxene respectively. A similar approach has been taken here by assuming that the compacting layer is composed entirely of plagioclase. It then becomes possible, using eqn. (11), to assess the effectiveness of compaction in transporting incompatible elements out of the crystalline matrix.

The variation in W_e/W with porosity for the trace elements U, Th, K, Rb and Sr is shown in figure 7. With the exception of Sr, all have $K_ds < 1.0$ for plagioclase. The maximum effective velocity attained by a particular element is clearly a function of its distribution coefficient. Relative velocity increases with increasing porosity, so that for $\phi = 0.5$, U, Th and Rb travel at ~ 95% of the melt velocity. In contrast, the high compatibility ($k_d = 6.0$) of Sr in plagioclase severely inhibits its ability to move through the compacting matrix. Even at porosities in excess of 50% strontium only reaches a maximum velocity 1/10 that of the melt.

274

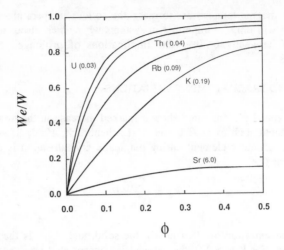

Figure 7. The relative velocities of selected trace elements relative to the compacting melt velocity (W_e/W) for a range of porosities. Bracketted numbers are the distribution coefficients for each element in plagioclase (Henderson, 1982).

6. Compositional Convection

Fluid flow through porous media has been proposed by Tait et al. (1984) and Kerr and Tait (1986) as a means of producing monomineralllic layering in large basic intrusions that have cooled slowly. The assumption is that during crystallisation, the melt phase forms an interconnected three dimensional porosity between the growing crystal grains. The resulting capillary network is capable of transporting chemically fractionated solute away from sites of crystallisation by convective flow, provided the residual liquid is less dense than the surrounding solid phase (Chen and Turner, 1980; Huppert and Sparks, 1984). Although successfully applied to basic systems, the possible effects of compositional convection in high silica liquids has not been dealt with, although compositional convection in viscous melts has been recently investigated by Tait and Jaupart (1989). In this section, the theory developed from the experimental results of Kerr and Tait (1986) describing the physical behavior of a basic melt phase in a crystallising porous medium is applied to granitoid systems. The theory will be used in an attempt to assess firstly under what conditions compositional convection is possible in granitoid magmas and to assess its effectiveness as a differentiation mechanism. As with compaction theory, the most important substituted variables in the following calculations are the melt viscosity and the density contrast between the solid and melt.

Thermal convection can occur in a porous medium in much the same way as in a fluid, provided the solutal Rayleigh number is $> 4\pi^2$ (Lapwood, 1948). The temperature gradient required to initiate convection comes from cooling at the base of the porous media, although the presence of horizontal temperature gradients set up through side wall cooling may also be important (Spera et al., 1982). Compositional convection is defined by the dimensionless solutal Rayleigh number

$$Ra_s = \frac{kg\Delta\rho h}{D\mu} \tag{12}$$

where h is the height of the porous medium, $\Delta\rho g$ is the density contrast between solid and melt, and D is the diffusivity of the chemical components in the melt (Kerr and Tait, 1986). As with compaction, fluid movement through compositional convection will only occur if convective velocities exceed the rate of solidification.

7. Results

The relationship between Ra_s and melt viscosity over a range of porosities calculated using eqn. (12) is shown in figure 8. The shaded area represents the critical Rayleigh number that must be exceeded before convection can begin. The results show that compositional convection is only possible at $\phi > 0.2$, even for the lowest melt viscosity of 10^5 Pa s. Compositional convection occurs in a relatively narrow window at melt viscosities of between 10^7-10^5 Pa s and porosity interval of between 0.5 and ~0.2. The most viscous melt capable of convection ($\mu = 10^7$ Pa s) can only do so at porosities close to 0.5. If the porosity is reduced to 0.3, convection is no longer possible unless accompanied by a systematic drop in melt viscosity to ~ 10^6 Pa s. These results imply that in the simplest case, compositional convection will only occur early on in the crystallisation history of the magma, where initial porosities are likely to be high. The convection velocity (V_c) of the unstable melt has been estimated (after Kerr and Tait, 1986) from

$$V_c = \frac{C\,\phi^{4.5}}{\mu} \tag{13}$$

where ϕ is the porosity and C is a constant relating the density of the fluid to the grain diameter (see Appendix). For eqn. (13) to be valid, the pore Reynolds number, defined by

$$Re_p = \frac{Ud}{\nu} \tag{14}$$

where U is the fluid velocity and n is the kinematic viscosity must be less than 1. At $Re_p > 10$, the flow is likely to be turbulent and Darcy's law is no longer valid (Dullien, 1979). Pore Reynolds numbers for the various melt viscosities used in this study reach a maximum of 1×10^{-4}.

Figure 9 shows the calculated variation in convection velocity with melt viscosity for the granitoid magma. The shaded region correlates with the onset of compositional convection shown in figure 8 (ie. where $Ra_S > $ ~40). Convective velocities within this region range from 400 mm/yr at $\phi = 0.5$ and $\mu = 10^5$, to 4 mm/yr at $\phi = 0.5$ and $\mu = 10^7$ Pa s. Thus, at high porosities, compositional convection appears a viable mechanism for removing *low viscosity* melt from its crystallising matrix, although as crystallisation proceeds and the porosity is

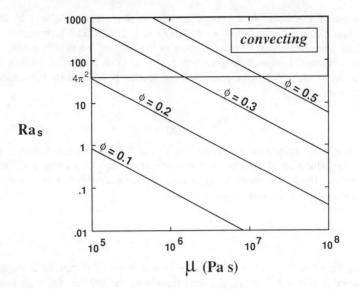

Figure 8. Solutal Rayleigh number (Ra$_S$) as a function of melt viscosity for a range of porosities, calculated using (12). The shaded region shows where Ra$_S$ exceeds $4\pi^2$ (~40) marking the onset of convection within the pore fluid. Convection can only occur at porosities greater than 0.2, even for the lowest initial melt viscosity of 10^5 Pa s.

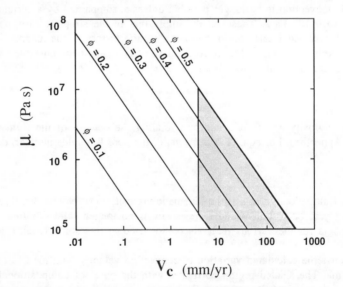

Figure 9. Variation in convective velocity of the interstitial melt as a function of viscosity and porosity. Compositional convection only occurs within the shaded region where Ra$_S$>40 (see Fig. 8). Convective velocities (calculated using (13)) are fastest at low viscosities and high porosities.

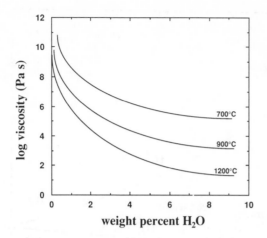

Figure 10. The effects of water (wt%) on melt viscosity over a range of temperatures for granitic liquids. If H_2O and other volatile species become concentrated in the melt phase during crystallisation, then the viscosity can be significantly reduced (after Shaw, 1965).

reduced to values $< \sim 0.2$ the process can no longer operate. Furthermore, for a fixed porosity, a reduction in viscosity by one order of magnitude leads to a ten-fold increase in convection velocity. As a final point, it is interesting to compare the convective velocities calculated by Kerr and Tait (1986) for basic magmas, with those obtained in this study. The far higher melt viscosities in silicic systems at similar porosities inhibit convective velocities to less than one tenth of those in basic systems at similar porosities, even allowing for the compensating effects of a much larger density contrast between the melt and solid used in the above calculations.

8. Discussion

From the previous section it appears that both compaction and compositional convection are viable differentiation mechanisms over a porosity interval of 0.5-0.1. However, the effectiveness of both processes is critically dependent upon the viscosity of the interstitial melt phase during subsequent crystallisation. There is now much evidence to suggest that the viscosity of the interstitial liquid will *decrease* during crystallisation (figure 10), provided the melt phase contains abundant dissolved volatiles such as H_2O, F, Cl and B (Burnham, 1963; Shaw, 1974; Dingwell and Mysen, 1985; Dingwell, 1987; Holtz and Johannes, 1991). Indeed, the presence of tourmaline (B) and fluorite (F) in the evolved rock types of some zoned plutons (Tindle and Pearce, 1981), is clear evidence that volatiles can be enriched in the interstitial liquid during crystallisation.

The effects of decreasing melt viscosity on the effectiveness (as a measure of τ_0) of both compaction and compositional convection as a function of porosity are shown in figure 11. Two hypothetical curves are drawn, one for a dry (volatile-free) melt that becomes

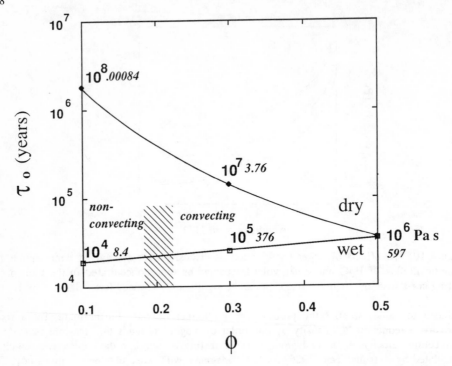

Figure 11. The variation in compaction time (τ_o) porosity for two hypothetical, interstitial melts, one anhydrous (dry) and the other volatile-rich (wet). Bold numbers show the melt viscosity at a particular porosity. Numbers in italics refer to the solutal Rayleigh number appropriate for each viscosity and porosity. The paths taken by the two liquids away from an initial start viscosity of 10^6 Pa s are markedly different. For the dry melt, compaction time increases with crystallisation, and τ_o is unfavourably long, even at relatively high (0.3) porosities. Furthermore, at porosities of ~ <0.4, compositional convection is no longer possible. In such instances, where the increasing viscosity of the melt inhibits intergranular flow, the crystallised pluton is likely to be relatively homogeneous on a macroscopic scale. Only if the melt viscosity decreases during crystallisation can compaction (and compositional convection) be considered as potential differentiation mechanisms over reasonable timescales (See text for further discussion).

increasingly viscous as crystallisation proceeds, and another for a wet (ie. volatile-rich) melt that becomes less viscous with increasing crystallisation. Both melts start at an arbitrarily chosen viscosity of 10^6 Pa s, and an initial porosity of 0.5. Also shown are the associated solutal Rayleigh numbers for the melt during crystallisation. Considering first the dry melt curve, the compaction time clearly increases with increasing melt viscosity. Furthermore, at

$\mu = 10^7$ and $\phi = 0.3$, the solutal Rayleigh number is subcritical, indicating that compositional convection will not occur. At melt viscosities of 10^8 and porosities of 0.1, τ_o exceeds the cooling times of even the largest magma chambers.

However, the path taken by the wet melt is quite different. Here the compaction times are all below $3.5x10^4$ yrs, and at porosities of 0.1, the compaction time for a melt with a viscosity of 10^4 Pa s is reduced to ca 19000 years. Another effect of reduced viscosity is to keep the $Ra_S > \sim40$ until porosities drop to below ca 0.2 (shaded region in figure 10). These results imply that during crystallisation, *decreasing* melt viscosities enable both compaction and compositional convection to operate simultaneously over relatively high (> 0.2) porosities. However, porosity appears to exert a dominating influence on compositional convection, effectively shutting it off at values below 20%.

The relationship between compaction and compositional convection during the crystallisation of both dry and volatile-rich melts can be assessed by comparing the ratio of convective velocity (V_c) to compaction time (τ_o) for a range of porosities. Thus, the time taken for a dry melt to compact our 100 m thick layer at $\mu = 10^5$ and $\phi = 0.3$ is $1.4x10^4$ years. The same parameters give a convective velocity of 40 mm/yr (figure 9). The ratio V_c/τ_o is thus ~6, falling to ca 0.02 at $\mu = 10^7$ and $\phi = 0.1$. For wet crystallisation V_c/τ_o is about a factor of 10 higher. The changing compositional convection/compaction ratio (relative to high melt fraction boundary layer differentiation) is summarised schematically in figure 12 as a function of time. Both the wet and dry curves show that during the early stages of crystallisation, for a given value of μ and ϕ, compositional convection will be the most effective differentiation mechanism. However, as crystallisation proceeds, the reduction in porosity leads to compaction becoming the dominant mechanism of differentiation within the crystal pile. The effectiveness of compaction at low melt fractions is discussed in the following section.

8.1. TEXTURAL DEVELOPMENT DURING POROUS MEDIA FLOW

The experimental studies of Kerr and Tait (1986) on basic systems has shown that at high porosities, vigorous compositional convection will rapidly remove the interstitial melt out of the porous media. However, at low permeabilities, the evolved melt may become trapped in regions of low porosity where it ultimately freezes in chemical isolation. The pegmatitic pods and irregular veins (not to be confused with pegmatite/aplite dykes) commonly found in granitoids may represent areas where dynamically unstable, volatile-rich melt has become frozen within impermeable regions of the crystallising magma. Such a process of 'trapping and freezing' of convecting or compacted melt due to anisotropic permeability relatively late in the crystallisation interval may explain some of the macroscopic (and microscopic) variations in texture and chemistry found in granitoid plutons (eg. Barriere, 1981; Sultan et al., 1985; Speer et al., 1989). Theoretically, the compaction process can continue to operate regardless of the boundary layer permeability if the melt forms an interconnected three-dimensional network. Beere (1975) has shown that such a network can be achieved, even at small (< 0.001) melt fractions, provided the dihedral angle, Θ where two grains and melt meet is < 60°. The geometry of the melt phase with respect to its crystalline matrix has been recently investigated in basic systems, where adcumulate rocks commonly show dihedral angles less than 60° (McKenzie, 1985; Hunter, 1987). Similarly, experimental work on dry granitic systems (Jurewicz and Watson, 1985) has revealed a range of dihedral angles

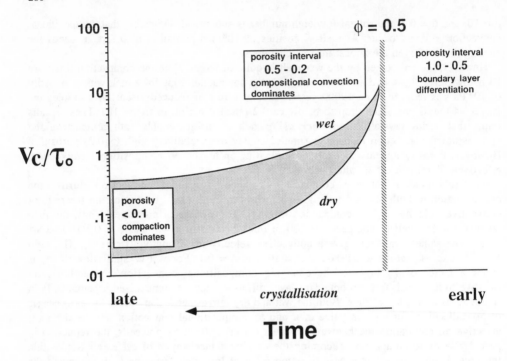

Figure 12. Schematic assessment of the relative roles of compositional convection and compaction, expressed as the ratio Vc/τ_o for the hypothetical anhydrous and volatile-rich melts shown in Fig. 11. Both processes are shown in relation to high melt fraction ($\phi > 0.5$) boundary layer differentiation. The changing ratio suggests that where porosities are still relatively high (0.5-0.2), compositional convection is the dominant process. As crystallisation proceeds (and porosity decreases to ≤ 0.1), compaction processes become increasingly important in removing evolved fluid from the crystallising magma.

of 44 to 60°. However, it is extremely rare to find texturally equilibrated monomineralic granitoid rocks—the silicic counterparts of adcumulate rocks—in silicic plutons. The implication here is that compaction and compositional convection in silicic systems, although viable, are by no means as efficient in expelling melt as in basic magmas, even for $\Theta < 60°$.

9. Conclusions

The aim of this study has been to determine under what boundary conditions the processes of compaction and compositional convection become effective *in-situ* differentiation mechanisms in granitoid magmas. The results show that in order for either process to operate to any significant extent within the cooling times of large silicic magma chambers, the viscosity of the evolved, interstitial melt must *decrease* during crystallisation. Calculations carried out over a porosity interval of 0.1-0.5 and at melt viscosities of 10^4-10^9 Pa s suggest the main factors governing the effectiveness of both processes are: **1)** The initial porosity, **2)** the density contrast between the evolved melt and solid phase and **3)** the viscosity of the interstitial melt. At porosities > 0.2, and melt viscosities < ~10^6 Pa s, compositional convection appears to be a potentially efficient differentiation mechanism. For both processes to operate over reasonable timescales, the required density contrast between melt and solid is ~ 10%. Calculated convective velocities are on the order of 400-4.0 mm/yr and V_c/τ_0 is ~ 6 for a crystallising layer 100 m thick. These velocities are sufficiently high to remove an evolved melt fraction of broadly granitic composition out of the layer over timescales of 250 to 25000 years. At porosities < ~ 0.2, the interstitial melt is no longer able to convect, and compaction becomes the dominant mechanism for melt extraction. For a 'wet' melt viscosity of 10^4 Pa s, and a porosity of 0.1, the compaction time (t_h) is 2.5×10^4 years. Under these conditions, incompatible elements such as U and Th can move through the viscously deforming matrix at between 60-70% of the melt velocity.

Finally, although both mechanisms complement, and may even provide an alternative to *in-situ* fractionation through boundary layer differentiation, textural evidence from granitiod rocks implies that compaction and compositional convection can only remove a finite amount of evolved interstitial fluid before the system 'locks up,' preventing further differentiation. The rate of solidification, and in particular the possible effects of impermeable regions within the crystallising magma acting as fluid traps may explain the locally heterogeneous nature of granitoid rocks.

Acknowledgments. Thanks go to Mike Cheadle, Klaus Mezger and Alex Halliday for helpful reviews and comments, and to Mike Atherton and Bob Hunter for inspiring this work. This paper was written while visiting the University of Michigan, and I would like to take this opportunity to thank the staff and students in the Department of geological sciences and CGTB, Ann Arbor, for technical support during my stay. Financial support from NATO and The Royal Society are gratefully acknowledged.

10. References

Arzi, A.A. (1978) Critical phenomena in the rheology of partially molten rocks. Tectonophysics 44, 173-184.

Atherton, M.P. (1981) Horizontal and vertical zoning in the Peruvian Coastal Batholith. J. Geol. Soc. London 138, 343-349.

Atherton, M.P. (1988) On the lineagecharacter of evolving granites. Fifth international symposium on Sn-W granites in SE Asia and the W Pacific. Shimane University, Japan. pp. 1-6.

Atherton, M.P., McCourt, W.J., Sanderson, L.M. and Taylor, W.P. (1979) The geochemical character of the segmented peruvian batholith and associated volcanics. In Atherton, M.P. and Tarney, J. (eds.), Origin of Granite Batholiths: Geochemical Evidence. Shiva, UK.

Barriere, M. (1981) On curved lamine, graded layers, convection currents and dynamic crystal sorting in the Ploumanac'h (Brittany) subalkaline granite. Contrib. Mineral. Petrol. 77, 214-224.

Bartlett, R.W. (1969) Magma convection, temperature distribution, and differentiation. Amer. J. Sci. 267, 1067-1082.

Bateman, P.C. and Chappell, B.W. (1979) Crystallisation, fractionation and solidification of the Tourmaline intrusive series, Yosemite National park, California. Geol. Soc. Am. Bull. 90, 465-482.

Bear, J. (1972) Dynamics of fluids in porous media. American Elsevier, New York.

Beere, W. (1975) A unifying theory of the stability of penetrating liquid phases and sintering pores. Acta Metallurgica 23, 131-138.

Bethke, C.M. (1985) A numerical model of compaction-derived groundwater flow and heat transfer and its application to the palaeohydrology of intracratonic sedimentary basins. J. Geophys. Res. 90, 6817-6828.

Bottinga, Y. and Weill, D.F. (1970) Densities of silicate systems calculated from partial molar volumes of oxide components. Amer. J. Sci. 269, 169-182.

Bottinga, Y., Richet, P. and Weill, D.F. (1983) Calculation of the density and thermal expansion coefficient of silicate liquids. Bull. Mineral. 106, 129-138.

Brandeis, G. and Jaupart, C. (1987) Characteristic dimensions and times for dynamic crystallisation. In Parsons, I. (ed.), Origins of Igneous Layering. Kluwer Academic Publishers, Dordrecht. pp. 613-639.

Burnham, C.W. (1963) Viscosity of a water-rich pegmatite (abs). Geol. Soc. Am. Spec. Pap. 76, 26.

Castro, A. (1987) On granitoid emplacement and related structures: a review. Geol. Rundschau. 76, 101-124.

Chen, C.F. and Turner, S.J. (1980) Crystallisation in a double diffusive system. J. Geophys. Res. 85, 2573-2593.

Cushman, J.H., ed. (1990) Dynamics of Fluids in Hierarchical Porous Media. Academic Press, London.

Daly, R.A. (1933) Igneous Rocks and the Depths of the Earth. McGraw-Hill, New York.

DePalo, D.J. (1981). Trace element and isotopic effects of combined wall rock assimilation and fractionational crystallisation. Earth Planet. Sci. Lett. 53, 189-202.

Dingwell, D.B. (1987) Melt viscosities in the system $NaAlSi_3O_8-H_2O-F_2O-1$. In Mysen, B.O. (ed.), Magmatic Processes: Physicochemical Principles. The Geochemical Society special publication No. 1. pp. 423-431.

Dingwell, D.D. and Mysen, B.O. (1985) Effects of fluorine and water on the viscosity of albite at high melt pressure: a preliminary investigation. Earth Planet. Sci. Lett. 74, 266-274.

Dullien, F.L.A. (1979) Porous Media Fluid Transport and Pore Structure. Academic Press, New York.

England, W.A., McKenzie, A.S., Mann, D.M. and Quigley, T.M. (1987) The movement and entrapment of petroleum fluids in the subsurface. J. Geol. Soc. London 144, 327-347.

Halliday, A.N., Stephens, W.E and Harmon, R.S. (1980) Rb/Sr and O isotopic relationships in 3 zoned Caledonian granitic plutons, Southern Uplands, Scotland: evidence for varied sources and hybridisation of magmas. J. Geol. Soc. London 137, 329-348.

Halliday, A.N., Mahood, G.A., Holden, P., Metz, J.M., Dempster, T.J. and Davidson, J.P. (1989) Evidence for long residence times of rhyolitic magma in the Long Valley magmatic system: the isotopic record in precaldera lavas of glass mountain. Earth Planet. Sci. Lett. 94, 274-290.

Hansen, U. and Yuen, D.A. (1987) Evolutionary structures in double diffusive convection in magma chambers. Geophys. Res. Lett. 14, 1099-1102.

Henderson, P. (1982) Inorganic Geochemistry. Pergamon Press. Oxford.

Hildreth, W. (1981) Gradients in silicic magma chambers: implications for lithospheric magmatism. J. Geophys. Res. 86, 153-193.

Holtz, F. and Johannes, W. (1991) Effect of tourmaline on melt fraction and composition of first melts in quartzofeldspathic gneiss. Eur. J. Mineral. 3, 527- 536.

Hunter, R.A. (1987) Textural equilibration in layered igneous rocks. In Parsons (ed.), Origins of Igneous Layering. Kluwer Academic Publishers, Dordrecht. pp. 473-503.

Huppert, H.E. and Sparks, R.S.J. (1980) The fluid dynamics of a basaltic magma chamber replenished by influx of hot dense ultrabasic magma. Contrib. Mineral. Petrol. 75, 279-289.

Huppert, H.E. and Sparks, R.S.J. (1984) Double diffusive convection due to crystallisation in magmas. Ann. Rev. Earth Planet. Sci. 12, 11-37.

Huppert, H.E. and Turner, S.J. (1991) Comments on convective style and vigor in sheet-like magma chambers by Bruce D. Marsh. J. Pet. 32, 851-854.

Irvine, T.N. (1980) Magmatic infiltration metasomatism, double diffusive fractionation crystallisation and adcumulus growth in the Muskox and other layered intrusions. In Hargreaves, R.B. (ed.), Physics of Magmatic Processes. Princetown University Press. pp. 325-383.

Jurewicz, S.R. and Watson, E.B. (1985) Distribution of partial melts in a granitic system: the application of liquid phase sintering. Geochem. Cosmochim. Acta 49, 1109, 1121.

Kerr, R.C. and Tait, S.R. (1986) Crystallisation and compositional convection in a porous medium with application to layered igneous intrusions. J. Geophys. Res. 91, 3591-3608.

Lachenbruch, A.H. (1968) Preliminary geochemical model of the Sierra Nevada. J. Geophys. Res. 73, 6977-6990.

Lapwood, E.R. (1948) Convection of a fluid in a porous medium. Proc. Cam. Phil. Soc. 44, 508-521.

Marsh, B.D. (1989) On convective style and vigour in sheet-like magma chambers. J. Pet. 30, 479-530.

Marsh, B.D. (1991) Reply to Huppert H.E. and Turner, S.J. J. Pet. 32, 855-860.

Martin, D. (1990) Crystal settling and in situ crystallisation in aqueous solutions and magma chambers. Earth Planet. Sci. Lett. 96, 336-348.

McBirney, A.R. (1985) Igneous Petrology. Freeman Cooper and Co. San Francisco.

McBirney, A.R. and Noyes, R.M. (1979) Crystallisation and layering of the Skaergaard intrusion. J. Pet. 20, 487, 554.

McCarthy, T.S. and Grooves, D.I. (1979) The Blue Tier batholith, NE Tasmania. Contrib. Mineral. Petrol. 71, 193-209.

McKenzie, D.P. (1984) The generation and compaction of partially molten rock. J. Pet. 25, 713-765.

McKenzie, D.P. (1985) The extraction of magma from the crust and mantle. Earth Planet. Sci. Lett. 74, 81-91.

McKenzie, D.P. (1987) The compaction of sedimentary and igneous rocks. J. Geol. Soc. London 144, 299-307.

Mohr, P. (1991) Cryptic Sr and Nd isotopic variation across the Lenister granite, southern Ireland. Geol. Mag. 128, 251-256.

Phillips, W.J., Fuge, R. and Phillips, N. (1981) Convection and crystallisation in the Criffell-Dalbeattie pluton. J. Geol. Soc. London 138, 351-366.

Pitcher, W.S. (1978) The anatomy of a batholith. J. Geol. Soc. London 135, 157-182.

Prensnall, D.D. (1969) The geometric analysis of partial fusion. Amer. J. Sci., 1178-1194.

284

Prensnall, D.D. and Bateman, P.C. (1973) Fusion relationships in the system Al-An-Or-Qtz-H$_2$O) and the generation of granite magmas in the Sierra Nevada batholith. Bull. Geol. Soc. Amer. 84, 3181-3202.

Ragland, P.C. and Butler, J.R. (1972) Crystallisation of the West Farrington pluton, N. Carolina, USA. J. Petrol. 13, 381-404.

Rice, A. (1981) Convective fractionation: a mechanism to provide cryptic zoning (macrosegregation), banded tuffs and explosive volcanism in igneous processes. J. Geophys. Res. 86, 405-417.

Richter, F.M. and McKenzie, D.P. (1984) Dynamical models for melt segregation from a deformable matrix. J. Geol. 92, 729-740.

Sawka, W.N. and Chappell, B.W. (1988) Fractionation of U, Th and REE in a vertically zoned granodiorite: implications for heat production distributions in the Sierra Nevada batholith, California, USA. Geochim. Cosmochim. Acta 57, 1131-1143.

Sawka, W.N., Chappell, B.W. and Kistler, R.W. (1990) Granitoid compositional zoning by side-wall boundary differentiation: Evidence from the Palisade Crest intrusive suite, central Sierra Nevada, California. J. Pet. 31, 519-553.

Shaw, H.R. (1965) Comments on viscosity, crystal settling, and convection in granitic magmas. Amer. J. Sci, 120-152.

Shaw, H.R. (1972) Viscosities of magmatic silicate liquids: an empirical method of prediction. Amer. J. Sci. 272, 870-893.

Shaw, H.R. (1974) Diffusion of H$_2$O in granitic liquids: Parts I and II. In Hoffman, A.W. (ed.), Geochemical Transport and Kinetics. Carnegie Inst. Washington Publ. 634, 131-170.

Shimizu, M. and Gastil, G. (1990). Recent advances in concepts concerning zoned plutons in Japan and southern and Baja California. The University Museum, The University of Tokyo, Nature and Culture, No. 2, 252 pp.

Sparks, R.S.J., Meyer, P. and Sigurdsson, H. (1980) Density variation amongst mid ocean ridge basalts: implications for magma mixing and the scarcity of primative lavas. Earth Planet. Sci. Lett. 46, 419-430.

Sparks, R.S.J., Huppert, H.E. and Turner, J.S. (1984) The fluid dynamics of evolving magma chambers. Phil. Trans. Roy. Soc. London 310, 511-534.

Sparks, R.S.J., Huppert, H.E. and Wilson, C.J.N. (1990) Reply on Evidence for long residence times for rhyolitic magma in the long valley magmatic system. Earth Planet. Sci. Lett. 99, 390-394.

Speer, J.A., Naeem, A. and Almohandis, A. (1989) Small scale variations and subtle zoning in granitoid plutons: the Liberty hill pluton, South Carolina, USA. Chem. Geol. 75, 153-181.

Spera, F.J. (1980) The thermal evolution of plutons: a parameterised approach. Science 207, 299-301.

Spera, F.J., Yuen, D.A. and Kirschvink, S.J. (1982) Thermal boundary layer convection in silicic magma chambers: effects of temperature dependent rheology and implications for thermogravitational chemical fractionation. J. Geophys. Res. 87, 8755-8767.

Stephens, W.E. and Halliday, A.N. (1979) Compositional variation in the Galloway plutons. In Atherton, M.P. and Tarney, J. (eds.), Origin of Granite Batholiths: Geochemical Evidence. Shiva, UK.

Stephens, W.E., Whitley, J.E., Thirwall, M.F. and Halliday, A.N. (1985) The Criffell zoned pluton: Correlated bahaviour of REE abundances with isotopic systems. Contrib. Mineral. Petrol. 98, 226-238.

Sultan, M., Batiza, R. and Sturchio, N.C. (1986) The origin of small scale geochemical and mineralogical variations in a granite intrusion. Contrib. Mineral. Petrol. 93, 513-523.

Tait, S.R., and Jaupart, C. (1989) Compositional convection in viscous melts. Nature 338, 571-573.

Tait, S.R. and Kerr, R.C. (1987) Modelling of interstitial melt convection in cumulus piles. In Parsons, I. (ed.), Origins of Igneous Layering. Kluwer Academic Publishers, Dordrecht. pp. 569-587.

Tait, S.R., Huppert, H.E. and Sparks, R.S.J. (1984) The role of compositional convection in the formation of adcumulus rocks. Lithos 17, 139-146.

Taylor, W.P. (1985) Three dimensional variation within granite plutons: a model for the crystallisation of the Canas and Puscao plutons. In Pitcher, W.S., Atherton, M.P., Cobbing E.J. and Beckensale, R.D. (eds.), Magmatism at a Plate Edge: The Peruvian Andes. Halsted, New York. pp. 228-234.

Tindle, A.G. and Pearce, J.A. (1981) Petrogenetic modelling of in-situ fractional crystallisation in the zoned Loch Doon pluton, Scotland. Contrib. Mineral. Petrol. 78, 196-207.

Van der molen, I. and Patterson, M.S. (1979) Experimental deformation of partially melted granite. Contrib. Mineral. Petrol. 70, 299-318.

Vance, J.A. (1961) Zoned granitic intrusions—an alternative hypothesis of origin. Geol. Soc. Am. Bull. 72, 1723-1728.

Wickham, S.M. (1987) Segregation and emplacement of granitic magmas. J. Geol. Soc. London 144, 281-297.

Wiebe, R.A. (1973) Relations between coexisting basaltic and granitic magmas in a composite dyke. Amer. J. Sci. 273, 130-151.

APPENDIX Symbols and values used in this study. @(Kerr and Tait, 1986), $(Henderson, 1982), *(Wickham, 1987), #(McKenzie, 1984).

d	grain diameter	1×10^{-2} m
C	constant in eqn. (13) $\left\{ g\Delta\rho d^2 / 10 \right\}^{@}$	0.03 m s^{-1}
d	depth (fluid thickness) of magma chamber	m
D	diffusivity of silicate melt$^{\$}$	2×10^{-10} m^{-2} s^{-1}
g	acceleration due to gravity	10 m s^{-2}
h	height of boundary layer	100 m
K_d	distribution coefficient	none
k	specific permeability	m^2
Ra	thermal Rayleigh number	none
Ra$_s$	solutal Rayleigh number	none
Re$_p$	pore Reynolds number	none
ΔT	temperature difference over d*	10-$100°$C
t	time	s
We/W	transport velocity	m s^{-1}
α	coefficient of thermal expansion*	5×10^{-5} °C
δc	compaction length	m
κ_T	thermal diffusivity*	10^{-6} m^2 s^{-1}
μ	interstitial melt viscosity	10^4-10^9 Pa s
$\Delta\rho$	density difference between solid and melt	300 kg m^{-3}
ρ_f	density of melt	2400 kg m^{-3}
ρ_s	density of solid	2700 kg m^{-3}
τ_0	scaled reference time for compaction	s
ϕ	porosity	0.1-0.5
ν	kinematic viscosity	μ/ρ_m
$\zeta 4/3 + \eta$	viscosity of compacting matrix#	10^{18} Pa s

DYNAMICS OF MAGMA CHAMBERS

ALAN RICE
Dept. of Geology
Rhodes University
Grahamstown 6140
South Africa

ABSTRACT. Discussed herein are three major issues of debate regarding the dynamics of magma chambers:
   I. whether or not crystal settling is important,
  II. whether or not magma densities drop during fractionation, and
  III. whether there is a Soret effect.
On occasion, resolutions to these questions are stated or implied in the literature. Such statements or implications are often premature.

I. CRYSTAL SETTLING

There is a common impression that I am a proponent of crystal settling...the impression has never had any justification.   N.L. Bowen, 1956

Winds carry phenomenal quantities of earthen material, whose densities exceed that of air by about 3000 times, as suspended loads. Examples are the great dust storms that cross the Sahara or Australia and which so devastated the central US in the 1930's. Silicic material of the size of $\approx$ 1mm is often carried by winds from one hemisphere to the next (e.g., Duce, personal communication, 1987; Betzer et al., 1986; Duce, 1986; Uematsu and others, 1985; Dauphin, 1983; Carder et al., 1986) which negates some theoretical efforts that would conclude volcanic debris is never transported far from its source (e.g., Sparks et al., 1978). Mined ore particulates of sizes up to ten centimeters have been transported as air/rock slurries for distances up to 25km between pumping stations by forced draft. Air driven slurries have lofted ore vertically through mine shafts over heights in excess of a kilometer (e.g., Wohlbier, 1986). Hailstones are kept aloft in the atmosphere long enough to sometimes grow to the size of grapefruits. Although the density differences are not nearly so great, being only a factor of two or three, water as a transport agent of sedimentary loads is equally efficacious. The importance of these transport processes to the environment, to economic considerations, etc. has led to the establishment of a great store of technical literature in the chemical, civil, mechanical and nuclear engineering sciences in terms of sewage treatment, fluidized bed reactors, etc.

    In view of the relatively diminutive density difference between magmas and crystals solidifying from them, the relatively diminutive settling velocities of crystals imposed by relatively much higher viscosities, and the comparable vigor of the convection attending

D. B. Stone and S. K. Runcorn (eds.),
Flow and Creep in the Solar System: Observations, Modeling and Theory, 287–305.
© *1993 Kluwer Academic Publishers. Printed in the Netherlands.*

edifices the size of magma chambers, it is puzzling that in the face of such plainly manifest and awesome abilities of fluids to keep material aloft, such as is experienced in ice fogs in Canada and Alaska (again, these ice crystals are suspended in quiescent air in spite of the fact their densities are about a thousand times greater), that the carrying capacity of fluids has been ignored for so long in theoretical igneous petrology. Particularly so in view of the fact that much of these mechanisms may be quantified. The concentration has, instead, been on the theoretical effects of crystal settling, not crystal carrying. Whether or not crystal settling obtains such an importance has not been adequately determined and in essence, has simply been assumed. Hence a vast literature that may be quite pertinent to the very foundations of igneous petrology has largely been ignored since the concept of crystal settling was advanced over a hundred years ago.

Incidentally, crystal settling is almost never reported in the industrial literature of solidifying melts (see Rice, 1981, 1985, for a review and references to this literature).

The appeal of crystal settling in magma appears to be "the want of an alternative explanation" (McBirney and Noyes, 1976) for the field data. McBirney and Noyes (1976) have objected to these interpretations of field data, pointing out that:

1. The olivine crystals in Kilauea Iki lava lake have not sunk since the lake was emplaced. They should have done so within months. The olivine phenocryst distribution in Kilauea Iki lava lake is concentrated near the top. This latter point is discussed in Rice (1990).

2. Plagioclase does accumulate on the bottom of iron-rich intrusions but cannot possibly sink in ferrogabbros.

3. Yield strength considerations dictate crystals cannot sink (this because crystal growth is matched by an increase in yield strength. See discussion further on).

4. Skaergaard rocks have textures indicating mineral growth in situ (e.g., McBirney and Noyes, 1976, 1979).

5. Trough bands of the Skaergaard are completely concordant, lack braided structure, do not migrate laterally, do not cut across earlier layers and outlying rocks cannot be associated with turbidity flows. Therefore, the trough bands are not sedimentary features (e.g., McBirney and Noyes, 1976, 1979).

6. Size grading is uncommon and layers are distinguished by color and not grain size, therefore the grading is not sedimentary.

7. Heaviest layers can be on top as in the Jimberlana Complex in Australia (e.g., McBirney and Noyes, 1976, 1979).

8. So-called "sedimentary" features can be seen on the walls.

9. Sedimentation would require cooling from below which, as is discussed later, is impossible.

10. Jackson (1961) and Wilshire (1969) have pointed out that crystals are not hydraulically equivalent.

In addition:

11. Rhyolite is thought not likely to be able to fractionate from basalt (D.K. Bailey, personal communication, 1978).

I.A. SUSPENDED LOAD FLOW

A sufficient condition for retaining particulate matter in suspension is that the convective velocities exceed the stagnant fluid settling rates. Magma chambers can be expected to convect on the order of at least a cm/s. The settling velocity is given by $w = D^2(\rho_s - \rho)g/18\mu$

where D is crystal size, ρ_s is the density of the crystal, ρ is the density of the melt, g is the acceleration of gravity and μ is the viscosity (e.g., Middleton and Southard, 1985). For D on the order of a millimeter, $\rho_s = 3.3\text{gm/cm}^3$, $\rho = 2.57\text{gm/cm}^3$, $\mu = 100\text{poise}$ (as expected for a basalt melt near the liquidus), then $w \approx 4 \times 10^{-3}\text{cm/s}$. The above criterion is readily met. It should be emphasized that the value calculated here for w is conservative. Use of the "hindered velocity", $w' = Kw$, $K = 1.0 + 1.56 C_s^{1/3}$ where C_s is the absolute volume concentration, indicates that a 1% volume concentration of suspended matter will reduce the fall velocity w by 25%, a concentration of 10w/o (w/o = weight percent) will cut the fall velocity by a factor of 10. The hindered velocity is by no means connected to the Einstein relationship for the dependence of viscosity on particle content, but instead reflects the continuity condition that as particulate matter falls, fluid must stream upwards, hindering settling. Shape factors may lower the value of w several times even further (flat particles will fall slower than round). A necessary condition that crystals remain in suspension is that the critical shear velocity $U*_c$ exceeds the settling velocity w. The critical shear velocity is given by $U*_c = \sqrt{\tau * \rho}$ and reflects the critical shear $\tau*$ necessary to mobilize and keep particulate matter suspended in the flow. The critical shear velocity may also be given by $U*_c = [\beta(\rho_s - \rho)gD/\rho]^{1/2}$ where β is the Shields parameter which reflects effects of lift, bursts (turbulent gusts that mobilize bottom particulate matter, see, for instance, Middleton and Southard, 1985), etc. For conditions here β takes on a value of 0.8 (see, for instance, Middleton and Southard, 1985). For a mafic magma chamber, $U*_c \approx 11$. Since $U*_c$ is a constant for the corresponding crystal size for w, the criterion for suspension is met for millimeter size crystals by four orders of magnitude. For either criteria, the particulate matter will not settle. Note that viscosity does not appear in the relationship for $U*_c$, indicating this criterion for maintaining suspension holds for all viscosities of the magma as it cools. $U*_c$ remains a constant of the magma chamber throughout its cooling history. That is, as the melt becomes more viscous on cooling, the easier it is to keep particles in suspension as w is inversely proportional to viscosity. This further implies that the crystal content of a magma will remain in suspension even as the crystal content of the melt passes through 65%, when the exponential dependency of viscosity on crystal content renders the melt too viscous to continue convecting. The response of fluids to an increase in particulate load up to approximately 65% is universal for all flow types carrying a suspended load, be they slurries, rivers, whatever. To adapt the terminology from the sedimentation literature, the flow "congeals" when the suspended load approaches $\approx 65\%$, i.e., the flow stops. This is because at these concentrations, the particles begin to touch one another. Therefore, from initiation of crystallization until crystal content forces the magma to congeal, the melt will carry an increasing suspended load. A variation on the critical shear velocity is the critical velocity U_{cr} which is the main stream velocity of flow sufficient to "scour" the floor, i.e., mobilize particles on the bottom. This velocity has the relationship $U_{cr}/\sqrt{g(\rho_s - \rho)D/\rho} = 0.5\ln(d/D)^{-1} + 1.63$ where d is the depth of flow (e.g., Garde, 1985). This criterion indicates crystals in a magma boundary layer at the bottom can be mobilized into the flow if the main stream velocity is of the order of 1cm/s, hence this criterion is consistent with the others. Such velocities are well in excess of that required to provide sufficient "lift" to entrain crystals in main stream flow as may be inferred from $F_L = C_L \rho A U^2/2$ where F_L is the lift force on an immersed sphere, C_L is the lift coefficient and A is the cross-sectional area of the sphere projected into the flow (see, for instance, Morisawa, 1985). Lift is implicit in the use of the Shields parameter above.

Scaling through the Reynolds numbers applicable to settling particles yields the size of magma chamber crystals predicted by the experiments of Martin (1990) which purport to establish crystal settling in magma chambers. Taking a settling velocity of $V = 1$cm/s in the experimental tank, crystal size d of ≈ 1cm as reported, assuming a density ρ for the solution of 1gm/cm^3 and the viscosity μ of the solution as 10^{-2} poise, then the Reynolds number of the crystal is Re $= \rho Vd/\mu \approx 1 \times 10^2$. The crystal in the magma chamber must have the same Reynolds number. For a density of 2.5gm/cm^3, a velocity of 4×10^{-3}cm/s (as obtained previously) and a viscosity of 10^4 poise, this Reynolds number is given as $1 \times 10^{-6}d_m$ where d_m is the size of the crystal in the magma chamber. Solving for d_m by equating Reynolds numbers predicts crystal sizes in the magma chamber of $\approx 10^8$cm. This is on the order of a lunar diameter. These numbers shouldn't be that surprising. The crystals grown in the experimental apparatus are approximately 1/20 the depth of the tank. So to start, a corresponding magma chamber could only contain a handful of crystals. That the viscosities employed are six orders of magnitude apart further compounds the problem.

I.B. YIELD STRENGTH

It has been suggested in the petrological literature that yield strength (or stress) in any type of fluid does not really exist (Kerr and Lister, 1991). From a more mature and extensive literature, "Common materials with yield stresses include catsup, toothpaste, and mayonnaise. If toothpaste did not have yield stress, it would flow out whenever the cap was off the tube. Many solid suspensions exhibit yield stresses; in addition, human blood has a yield stress, which plays an important role in a number of circulatory disorders" (Fahien, 1983). Yield strengths are a topic of interest in the engineering of cosmetics production, food and candy production, medicine and drug production, plastics, polymers, concrete, etc.

Measured yield strengths for magmas forbid crystal settling. Crystals nucleate and grow during cooling of the magma. So does yield strength (e.g., McBirney and Noyes, 1979). Although yield strengths are smallest as the magma cools through the liquidus, crystals are also smallest at this stage and yield strength matches their growth as the fluid temperature falls to the solidus near which only crystals tens of meters in diameter could overcome the yield strength attending this stage of evolution of the fluid and sink. Writing on the sinking of crystals in magmas, Kerr and Lister (1991) imply that yield strength is a fictitious concept which arises from incorrectly extrapolating as a straight line the stress/strain rate curve to the stress coordinate when in reality this curve should swing in to become tangent to the strain rate coordinate at the origin (Fig. 5b, Kerr and Lister, 1991). These authors provide no experimental justification as to why this curve should swing in to the origin. No experimental points were indicated between the origin and the point from which a yield strength was extrapolated. The material under discussion might even be a lower yield Houwink fluid (e.g., Sherman, 1968). However, if their Figure 5b is inverted such that stress is plotted against strain rate, then the slope of that curve is, by definition, the viscosity of the fluid. If the curve then actually swings in to the origin rather than intersecting the stress coordinate as they claim it should, then the viscosity approaches infinity as the strain rate approaches zero. If Kerr and Lister are correct, the implication still remains that crystal settling in magmas cannot occur.

All fluids exhibiting yield stress are "Bingham" fluids. Suspensions and polymers are Bingham fluids (e.g., Bird et al., 1960; Lane et al., 1958). The concept of a "yield strength" for a fluid carries over from that of a solid: the curves are very much the same except that for a solid, strain takes place of strain rate. The portion of the curve Kerr and Lister claim missing for Bingham fluids is that region analogous to elastic deformation for a solid. Plastic (permanent) deformation occurs only for stress greater than the yield strength (e.g.,

Askeland, 1990). Kerr and Lister infer that only suspension flows exhibit yield strength and at that only when the particulate content rises to 50%. This leaves unexplained Bingham or yield stress in polymers which include plastics (Bingham fluids are often referred to as Bingham plastics). There are no suspensions in these Bingham fluids. PIB's (polyisobutylenes) have very definite yield strengths wherein the stress versus strain rate curve does not extrapolate to the stress coordinate but actually swings up to parallel itself with the strain rate coordinate and continues as a horizontal line to intersect the stress coordinate at a much higher yield stress than would have been obtained by extrapolation from a section of the curve representing higher strain rates. All silicate melts are polymers first, meaning they should display yield strength even if all the crystals are removed from the melt. And as polymers, the increasing complexity of melt structure with decreasing temperature will be reflected in the yield strength. McBirney and Noyes apparently employed an undercooled melt in which particulate content was not yet in equilibrium, i.e., there was no significant crystal content. In this case, the increase in yield strength needs to relate to the degree of polymerization and not suspension content. As shown below in the section on magma densities, magmas are also suspension flows. Yield strengths have been assigned to lavas (e.g., Shaw et al., 1968; Hulme, 1974; Wilson, 1980; Wolff and Wright, 1981; Fisher and Schmincke, 1984; Cas and Wright, 1987). Since an activation energy of flow must apparently be exceeded in any fluid to initiate flow, one might argue that all fluids in a sense possess a yield strength albeit infinitesimal for some. Suspensions apparently have yield strengths because of the additional boundary layers attending each suspended particle for which static friction must be exceeded to set flow in motion. The snap through increase of the viscosity of suspension flows when the particulate content exceeds 50%, which Kerr and Lister claim is "yield strength," is called "congealing" in other disciplines wherein the literature is vast (e.g., Middleton and Southard, 1984). Congealing may have been noted in some experiments replicating freezing magma chambers. These experiments employed paraffin as the model fluid and cooling was from the top. An overlying crust was formed which grew downward into the fluid. No crystals grew on the floor nor was there bottom growth. When the convecting paraffin had cooled halfway between its liquidus and solidus, i.e., contained about 50% solidified particulate, the convection precipitously stopped (Brandies and Marsh, 1989). This is in keeping with the history of magma chambers as proposed by Rice (e.g., 1981, 1985 and earlier articles referenced therein).

The work of Sharpe et al. (1983) on the viscosity of mafic melts has been referred to as evidence that yield strengths were not observable in magmas (Sharpe, personal communication, 1989). These experiments were done above the liquidus where even those reporting yield strengths indicate that none is to be found (e.g., McBirney and Murase, 1984). As there are no crystals above the liquidus, it is meaningless to infer anything concerning crystal settling and yield strength from the work of Sharpe et al. (1983).

I.C. THE QUESTION OF FREEZING FROM THE BOTTOM

There have been a number of attempts to promote the notion of freezing from below in magma chambers. If the crystals simply grew from the floor, this would serve to overcome the stringencies imposed on crystal settling in a convecting magma chamber. Experimental examination of this possibility has yielded "compositional convection" (e.g., Tait and Jaupart, 1989), i.e., evolved melt streaming up from the solidification front on the floor. Streaming from the floor upward, however, again spells complication for settling mechanisms in that it "hinders" fall velocities as discussed above. This applies even in the face of laboratory experiments that would argue for crystal settling in magma chambers such as those performed by Martin (1990). In these experiments, KNO_3 crystals grown in a solution of water and sodium carboxymethal cellulose reached sizes on the order of a centimeter but scale to igneous crystals with dimensions on the order of lunar diameters as discussed below, i.e., there is no dynamic similarity in these experiments. It has been recognized for some time that the melt in a magma chamber must be cooler than the floor. Lovering (1935) was apparently the first to demonstrate that the temperature of the country rocks below a crystallizing magma will eventually exceed that of the magma itself (this is implicitly noted by Morse, 1986, 1988, for instance). Freezing cannot take place from the floor of a magma chamber when the floor becomes warmer than the magma itself regardless the rise in melting point with pressure. Pressure would have little effect anyway if the temperature drop from bottom to top of the chamber exceeded about 10°C. This also applies to the effect of the adiabatic lapse rate which would account for only a few degrees in a chamber 10km in depth. The reason the floor of a magma chamber becomes warmer than the magma itself is simply demonstrated. An initial charge of magma into the crust of the earth will form a magma chamber with chill or quench zones of original melt at the boundaries of the chamber. The components of the magma with the highest melting point then freeze out upon the quench margin, this material laid down from subcooled magma at its melting point (which is higher than the nucleation temperature) due to release of latent heat. Having deposited the highest melting point material at its melting point on the walls, the magma must then be at a lower temperature than the immediate wall let alone the need to subcool magma to the nucleation temperature to assist further freezing. The new solid material on the wall at close to its freezing temperature (or melting point) transfers heat into its surroundings by conduction only which is orders of magnitude less effective than that of convection. Hence once in place, this most refractory material never cools as fast as the magma which experiences most of the convective heat transfer out the top of the chamber. Material quenched to the ceiling would be resorbed and assimilation or melting to some degree of overlying country rock would characterize processes at the ceiling. This scenario attends the Bushveld Complex. Beneath the marginal zone of ≈ 100m thickness lies a metamorphic aureole penetrating into the floor rock beneath the Bushveld Complex to distances in excess of 4000m. Highest grade metamorphism occurs at the marginal zone contacts but there is no evidence of melting (e.g., Engelbrecht, 1988; Nell, 1985). On the other hand, estimated thicknesses of the melting of overlying roof rock ranges from 800m to about 3km. This due to the input of mafic magma (e.g., von Gruenewaldt, 1972; Walraven, 1987). Heat transfer out the top of the magma chamber is certainly enhanced over that at the bottom due to the geothermal gradient alone which reflects a 20°C to 30°C temperature rise per km of depth. This would account for an initial temperature difference of about 200°C to 300°C between the ceiling and the floor of the Bushveld Complex magma chamber, simply from the geothermal gradient alone. Heat transfer out the top of the Bushveld is aided over that out the bottom by about 300°C. In addition, the thermal impedance of the material beneath a magma chamber is infinitely more greater that above it.

I.D. THE QUESTION OF STAGNANT BOUNDARY LAYERS AT THE BOTTOM OF MAGMA CHAMBERS

In order to secure a stagnant boundary layer at the bottom of a magma chamber through which crystals could settle unimpeded by the vigor of the convection, some authors have proposed cooling through the bottom of the magma chamber. This purportedly would lead to an increase in the viscosity next to the chamber floor such that convection would have to cease there (McBirney and Noyes, 1979) or establish reversed temperature gradients stable against convection (Brandeis and Jaupart, 1986; Jaupart and Brandeis, 1986). A temperature drop at most of ten degrees would not secure sufficient viscosity change to generate a stagnant boundary layer. Nor do the scour criteria given in a previous section allow one to develop. The suggestion has no application, however, in light of the discussion above. A stagnant boundary layer cannot exist while the floor remains hotter than the overlying magma.

I.E. FREEZING FROM THE TOP

It has been put forward that magmas may freeze from the top down and sides inward as occurs in many industrial situations. As most of the heat transfer from a magma chamber is out the top, top-down solidification would seem favored. Appeal to the industrial literature has led to successful explanations of layering in border groups as manifestations of solute banding or microsegregation. This is particularly so in the marginal border groups where the layering is vertical and can in no way be attributed to crystal settling. Such layering arises purely from convective processes (e.g., see Rice, 1981, 1985 for detail). A few workers still maintain that the border groups are the manifestation of Liesegang banding, a purely diffusional mechanism. Even these workers note that if the magma is not stagnant, "the magma would tend to mix and homogenize, and oscillating plagioclase-rich and plagioclase-poor magma compositions could not be maintained long enough to build up substantial thicknesses of rhythmic layers" (Komor and Elthon, 1990). That magma chambers must convect vigorously may be ascertained from the ratio of streaming terms to viscous terms in the momentum equation of the Navier-Stokes equations. This ratio must hold in the large, i.e.,

$$\left| v \bullet \nabla v \right| \, / \, \left| v \nabla^2 v \right| \; \approx \; VL \, / \, v \; \approx \; Gr^{1/2}$$

where Gr is the Grashof number, V velocity, v kinematic viscosity and L the depth of the convecting fluid (e.g., Tritton, 1977). Gr = Ra/Pr where Ra is the Rayleigh number and Pr the Prandtl number, hence for Pr = 10^4 and Ra = 10^{12}, L= 10^5cm, v = 10^2cm^2/s, all reasonable values for a mafic magma chamber, then V $\approx$ 10cm/s.

Further, it has been pointed out (e.g., Rice, 1981, 1985) that the first material to go into border groups are the high temperature components of the magma which should leave evolved magma overlying magma of composition close to that of the primary melt. This will stratify the chamber. The compositional variation with depth coupled with the temperature gradient now leads to double-diffusive layering and convection. This "cores" (an industrial term relating to castings) the magma chamber, i.e., provides it with cryptic variation (macrosegregation). This cryptic variation runs from primitive material at the roof through evolved material further down and then back towards primary melt as one proceeds to the floor of the chamber. Increasing crystal content in the magma chamber eventually increases the viscosity to the point that the chamber "congeals" much like the paraffin experiments noted above. The chemical variation in the vertical direction would then be frozen in within the unit. Such a picture attends the Skaergaard Unit (e.g., Rice, 1981, 1985). Downward directed solidification, i.e., top-down solidification or solidification from the boundary in, is now reported for some igneous units (e.g., Stephenson, 1990; Le Roex et al., 1990; Sawka et al., 1990).

I.F. EXPERIMENTS PURPORTING TO DEMONSTRATE CRYSTAL SETTLING BUT WHICH ESTABLISH SOLIDIFICATION FROM THE TOP DOWN

Martin (1990) writes of his experimental efforts to model magma chambers, "The heat exchanger was prevented from coming into contact with the experimental fluid by floating...kerosene on top of the fluid between it and the heat exchanger. Allowing the copper surface of the heat exchanger to come into contact with the fluid would have provided a site to which the crystals could have adhered." That is, if the crystals had a competent roof to adhere to, the growth would have been at the start from the top down. Further, "During the initial period of crystallization...elongate crystals of KNO_3 nucleated and grew on the fluid/kerosene interface...but fell away from the interface when their length was still only a fraction of a centimeter." Further yet, "Eventually...Eutectic solid formed across the top of the fluid and proceeded to grow into the interior of the fluid." That is, down. Increasing the viscosity increased the propensity of crystals to grow at the top. At the largest viscosity, "crystals were able to grow to the size of a few centimeters long before falling away from the fluid/kerosene interface. Some crystals managed to adhere to the sides of the tank near the top of the fluid, eventually enabling falling crystals to be trapped and causing a network of crystals to form across the top of the fluid....Eventually, as in the initial experiment, in all these experiments eutectic solid formed across the top of the fluid and proceeded to grow downward into the tank." Crystal growth did not initiate on the bottom in these experiments until crystals falling from above provided nucleation sites for this undercooled fluid that was at much the same temperature throughout (KNO_3/water solution 20cm deep with viscosity enhancer added). In short, these experiments demonstrate the likelihood of top-down solidification. And would have done so completely if the fluid had been allowed to come into contact with the copper roof as occurs in the industrial analog of freeze concentration. Freeze concentration (FC) is the industrial equivalent of fractional crystallization. FC separates substances by crystallization and is employed in the concentration of juices, milk, and is a process method for beer, vinegar, pulp, paper, pharmaceuticals, chemicals and petroleum products (e.g., Rosen, 1990). If the freezing is from the top, that is the surface to which the concentrate or fractionate will adhere.

Weinstein et al. (1988) conducted both an experimental and computational program to demonstrate that crystal settling is the fractionation mechanism in magma chambers. The computational analysis was limited to laminar flow. Noting that laminar flow must attend for Rayleigh numbers less than about 6×10^4 (the Rayleigh number Ra = $g\alpha\Theta L^3/\nu\kappa$ where g is gravity, α is the coefficient of thermal expansion, Θ is the temperature difference across the convecting layer, L is the depth of the layer, ν is kinematic viscosity and κ is thermal diffusivity) and taking physical values appropriate for magma chambers and solving for L when Ra $\approx 6 \times 10^4$ yields L $\leq$ 1m for the depth of the magma chamber. These are not very deep magma chambers. A logic decision, not physics, dictates for these authors whether or not particles settle. An "if" statement inquires as to the position of a particle: if it's near the floor, the program "sticks" it to the floor and calls it settled. In addition, it is noted in the accompanying experimental effort that all particles sink "after (my underlining) the heat source was switched off and the convective flow spinned down." That is, the particles sank only after the convection stopped. The increase in viscosity of magmas as they solidify prevents any sinking even after the convection stops. The particles under discussion in these experiments were glass spheres with a density 2 to 3 times that of the working fluid which was a sucrose solution of viscosity of about 10^3 poise. Under quiescent conditions, i.e., no convection, these particles sank through 20cm in 2 hours. Correcting for the density differences associated with crystals in a magma, this implies settling rates in a magma chamber of $\approx 7 \times 10^{-4}$ cm/s, in the same order of magnitude as inferred above. Convective velocities in the tank were on the order of 10^{-3} cm/s, four orders of magnitude less than that expected for a magma chamber. The only points where crystals gathered on the floor were at the stagnation points of the upwelling, i.e., where opposing flows along the floor of the tank met, stopped their horizontal motion, and turned upward. Otherwise, up to 60% of the glass spheres remained in the flow! It must be kept in mind that this was not a freezing system that allowed attachment of crystals to the ceiling. These experiences, however, corroborated some of our own wherein a 10-cm-deep beaker filled with Lyll syrup and laced with aluminum flakes was heated from below, yielding a Rayleigh number of about 5000, i.e., barely convecting. The only places wherein "crystals" gathered were in the corners where the bottom of the beaker joined its walls and where the surface of the syrup joined the walls, i.e., only at the stagnation points where the flow necessitated a right angle turn. At that, the flakes filled these corners only up to the point of matching the streamlines of the turning flow. These flakes comprised a negligible portion of those in the syrup which remained aloft over the four month period the beaker sat on the heat source (unpublished material, University of Pretoria, 1989). For myself, I reproduced the calculational program of Weinstein et al. but reversed the logic statement. If the particles were next to the ceiling, the logic statement stuck them to the ceiling. In this fashion, I was computationally able to solidify a magma chamber from the ceiling downward.

The usage of the term "homogenous nucleation" in Weinstein et al. (1988) may cause some confusion. These authors mean by this term that the particles are initially evenly distributed throughout the melt as opposed to "inhomogeneous nucleation" wherein they mean the particles are initially distributed near the top of the melt. The term "homogeneous nucleation" refers to the initiation of solidification in a melt without advantage of a nucleation site such as occurs in "heterogeneous nucleation." There is a complete misunderstanding by Martin (1990) of the concept of homogenous nucleation which he confuses with the undercooling of a melt necessary to instigate solidification, i.e., the nucleation temperature. More will be said on this later, but the theoretical homogenous nucleation temperature is given by

$$32\gamma^3(MT_m/(T_m - T)L_m)^2 = 50kT$$

where k is Boltzmann's constant, T is the nucleation temperature, T_m is the melting point, L_m is the latent heat of fusion, M is the molar volume and γ is the surface interfacial energy between the crystal and its melt (e.g., Cottrell, 1975). These theoretical temperatures have never been seen. Homogenous nucleation generally initiates well above them and undercoolings necessary to secure homogenous nucleation are generally about 15 to 25% of the melting point. As an example, a magma with a melting point of 1200°C (1473 K) may have to see an undercooling of 140°C to 350°C to initiate homogenous nucleation if the fluid is not convecting. Homogenous nucleation can be initiated at higher temperatures depending on other phases in the melt, pressure gradients induced by convection, etc. (e.g., Kingery et al., 1976). The solutions employed by Martin are, of course, not solidifying melts.

I.G. SOME UNCERTAINTIES AND THEIR RESOLUTION

While the model of freezing from the top down has been successful in yielding the features of the small and highly (near the surface of the crust) emplaced Skaergaard intrusion, there are aspects of a number of large gabbroic bodies that do not immediately fit the top-down evolution. The Bushveld Complex will be taken as an example. The compositional variation from top to bottom of the Bushveld only increases from evolved to primitive material (in the gross sense) downward. There is no reversal upward from evolved to primitive material as is seen on moving upward into the upper border groups of the Skaergaard. In short, where are the Mg rich olivines that should be in the upper zone of the Bushveld if it is an analog to the Skaergaard? It is clear, however, that extensive melting occurred at the top of the Bushveld as cited previously in this paper. Assimilation of the upper border group analog of the Bushveld Complex into melted pelitic sediments at the top of the unit incorporates additional silica to push the magma out of the olivine field (e.g., Ghiorso and Kelemen, 1987) and this is where the olivine went (Rice, 1992, work in preparation).

The cryptic variation in the Bushveld generally trends downward from evolved to primitive melt which has been the major reason for interpretating this unit as a manifestation of crystal settling. However, similar trends can be expected from the convective fractionation model applied to the Skaergaard and which involves solidification from the top down. Again, swinging from primitive to evolved and back to a composition close to that of the original melt as one descends into the chamber. The problem in applying this to the Bushveld has been the observation (e.g., Eales, personal communication, 1992) that the material at the bottom of the Bushveld is so replete with dunites that the original melt had to have had an Mg content well above that normally observed. This would seem to call for crystal settling of olivines into the bottom of the chamber.

An examination of the core logs of the Lower Zone of the Bushveld Complex (Teigler, 1990) shows, however, that the so-called dunites are actually layering in the sense of a border group. That is, cyclic alternations between two different phase fields—olivine and pyroxene— which cannot occur by crystal settling. In fact, the band widths of this microsegregation increase as one proceeds into the chamber. This is as one expects for microsegregation since the Rayleigh number, hence the frequency of convective oscillations, painting the two phases alternately to the walls, decreases during the cooling and inward solidification of the chamber (e.g., see Rice, 1981, 1985). In short, the so-called dunites—olivines interleaved with pyroxene layers—were frozen to the margins of the Bushveld Complex. They did not settle there. This also requires that the most refractory material go to the margins first as is observed. This has led to the conjecture that the Lower Zone may really be a sloping wall.

However, even the convective fractionation model calls for crystal transport downward into the chamber simply because this transport is driven by a concentration gradient of suspended crystals from top to bottom which does not involve crystal settling at all. Freezing from the top down simply keeps pushing an increasing concentration of suspended crystals to the bottom (e.g., Rice and von Gruenewaldt, 1991a, 1991b; Rice, work in preparation, 1992).

That it is easy to keep particles in suspension should be kept in mind when attempting to provide a model for the field evidence. Figure 1 displays the change in the nature of a lava flow or infilling of another chamber. This might initially be considered proof of crystal settling in that the last material out of the originating chamber is charged with the most phenocrysts. Figure 1 shows, however, that another interpretation is as equally valid. In this case, no crystal sinking occurred at all in the original chamber. It was simply tapped at various stages as a suspended load built up.

COOLING MAGMATIC POT AND EVOLUTION OF PHENOCRYST CONTENT OF DISCHARGE

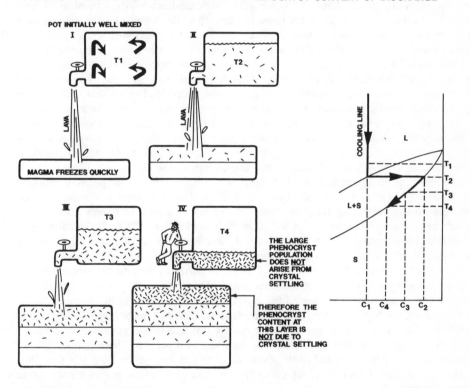

Figure 1. This diagram depicts the variation in composition of the discharge of an evolving magma chamber in which there is a developing suspended load. As the outpouring will show increasing phenocryst content and evolved melt, it would give the erroneous appearance that the variation of composition in time was due to crystal settling wherein the bottom of the chamber was the last material exuded out. In fact, the ranging compositions of the outpourings could reflect a chamber that has settled nothing but is simply gathering a suspended load, the sequential tapping of which reflects stages of crystal content supported throughout the entire chamber by convection.

II. ON THE EVOLVING DENSITIES OF MAGMAS

Sparks and Huppert (1984) offer a model of magma mixing in a stratified magma chamber based on the evolution of magma wherein the density of a basaltic melt becomes lighter as the melt evolves. See for instance their Figure 4 wherein density is plotted against magnesium number. This model presumes a melt devoid of crystals via crystal settling. Analyses of multiphase flows (water and steam; air and sand, air and salt, air and water or other particulates; slurries; turbidity currents; avalanches of snow, soil, rock; sewage transport, etc.) always treat the density of the total flow as being a function of the densities of both the transporting fluid and its load (i.e., the particulate matter). That is, the flow density is taken to be the sum of the products of volume fractions and densities of each material component of the flow. For instance, the density of water-laden hurricane-force winds is equal to the volume fraction of water carried by the wind times the density of the water, plus the volume fraction of the remaining air times the density of the air. Since the density of water is about 1000 times greater than that of air at STP, a 1% (by volume) water load of a wind implies a flow density some 10 times greater than that of dry air. This correct practice of including the particulate load in determining the density of multiphase flows carries over into most areas of the geologic sciences, e.g., sediment transport as suspended load in streams, nuées ardentes, etc. Although similar processes are recognized in some theories of magmatic evolution, it is not in some recent work or the original insights have been abandoned. It is assumed that there is no suspension load of crystals at all. As a more general example, Naldrett et al. (1987) report density changes as a crystal free melt evolves through the olivine field, then orthopyroxene to finally include plagioclase and clinopyroxene as it cools. There is an initial density decrease in the melt from $2.58g/cm^3$ to $2.57g/cm^3$ as the temperature drops from approximately 1330°C to 1300°C. This rather small density change is the foundation for some theories of mixing of magmas. Note the density of crystal free melt rises on further cooling but remains less than that of the initial melt until about 1175°C. Included in Figure 2 is the effective magma density to be expected if there is crystal (solid particulate) retention in the melt. Note that a 5% crystal suspension load yields an effective fluid density of approximately $2.67g/cm^3$ as opposed to $2.57g/cm^3$ corresponding to the minimum density inferred for a crystal free melt at the same stage of evolution. This density difference is a factor of ten greater than that inferred for crystal free dynamics and in the opposite direction, i.e., is heavier, not lighter, and reflects the much larger density change involved in going from liquid to solid phase. This effect still overwhelms that due to the evolution of crystal free liquid even if the suspended load is only of the order of a percent. There is little doubt that magmas carry suspended crystal loads. Evidence of this is seen in the Bushveld Complex by 1) the abundance of inclusions of partially resorbed plagioclase grains within cumulus bronzite crystals, 2) similar inclusions hosted in cumulus orthopyroxene, 3) the hosting grains being both sub-hedral and anhedral, 4) inclusion free margins and cores of host grains, 5) reversed zoning in the inclusions, 6) Sr-isotopic disequilibrium between hosts and inclusions, 7) pods and schlieren of anorthosite entangled with pyroxene cumulates, and 8) TiO_2 vs Al_2O_3 plots characteristic of magma mixing events. These features establish the existence of feldspar crystals suspended at an early stage within the melts before they were encapsulated within orthopyroxene or olivine and lead to the conclusion that these textures and characteristics arose from the mixing of partially crystalline magma (Eales et al., 1991; Eales, 1991; de Klerk, 1991; Eales et al., 1990). Hydrodynamic criteria for retention of particulate content in convecting fluids as discussed previously provides physical affirmation of these field observations. Hence determinations of the density of the melt must include the contribution of its particulate content throughout the evolution of the melt. As most replenishing magmas are close to or on their liquidus, the same will generally apply to them also. Models of magmatic evolution that rest on the notion that plumes may rise within magma chambers to form a liquid layer at some distance above the crystalline floor may then be quite incorrect. It is possible that this could occur only when the plume possesses superheat, i.e., is wholly liquid, without crystals. This assumption that magmas newly intrusive into existing magma chambers can be treated as single-phase, wholly liquid systems is questionable. The same applies to models wherein the evolution of melts in stratified systems is thought to lead to 'rollover' and mixing of the stratification due to a density decrease in underlying melts (e.g., Eales and Rice, 1991).

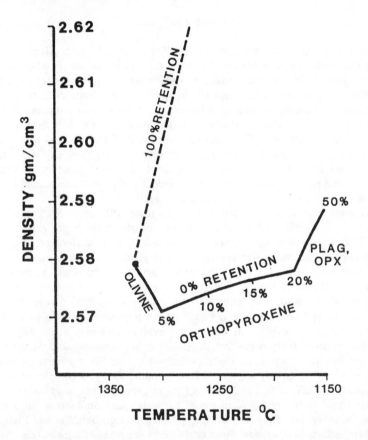

Figure 2. The variation of the density of an evolving melt retaining 100% of its crystal content is shown as the dashed line. Even if only a few percent of crystals precipitated out were retained in the melt, the effective melt density changes would overwhelm that due to evolution of the fluid alone which is depicted by the solid line (adapted from Eales and Rice, 1991).

III. Is There a Soret Effect?

The solute Rayleigh number which plays such an important role in describing double diffusive convection may be written as

$$Ra_s = S_T C(\partial\rho/\partial C) / (\partial\rho/\partial T)Ra$$

where C is concentration, ρ is density, T is temperature, Ra is the Rayleigh number and S_T is the Soret coefficient. Ra_s may also be written as $Ra_s = ScS_TRa/Pr$ where Sc is the Schmidt number and Pr is the Prandtl number. For sea water, $Ra_s \approx 80S_TRa$. The thermal diffusion coefficient is given by

$$TS_T = \sigma_T = (\beta/C\alpha)R$$

where R is the double diffusion parameter $R = \beta\Delta C/\alpha\Delta T$, and β is the coefficient of change in density with concentration, α is the coefficient of thermal expansion. Further, the transport equations describing double diffusive processes (which are a coupled system written for each layer of double diffusive convection) turn out to approximate a finite difference form of the continuum equation depicting a thermal diffusion (i.e., Soret) swing separation cascade as employed in industry. The transport equations are written in terms of both heat and mass film coefficients which are experimentally determined functions of both Ra and Ra_s. The Soret or thermal diffusion coefficients do not appear explicitly nor are they explicitly determined experimentally (see Rice, 1981, 1985 for discussions). They are buried in the functions for Ra and Ra_s. To set $S_T = 0$ apparently zeros out any double diffusive convection. So to say the Soret effect is negligible implies double diffusion is negligible.

Most fractionation processes are proportional to mass difference. Soret fractionation is more complicated in liquids and therefore not correctable in the usual mass spectrometer procedures. The Soret effect notes only that there is a mass difference. The degree of fractionation depends on molecular configuration (e.g., degree of polymerization). It's an "either-nor" mechanism in liquids. The Soret effect fractionates chemicals, elements, isotopes. Its complexity is reflected in the fact that given three isotopes, two of them may be equally fractionated from the third. The Soret coefficient $S_T \propto \ln(T'/T)$ where T' is the high temperature of the system and T is the cold end temperature. It is also a function of concentration. Hence S_T evolves as the system evolves and may eventually reverse the direction of fractionation. If S_T is positive, the lighter components move to the warm end. If it is negative, the heavier components move to the warm end. Soret separation factors ε are given by $\ln\varepsilon = S_T\ln(T'/T)$. Steady state mass flux including the Soret component is given by J $= D[\nabla C + S_TC(C-1)\nabla\ln T] = 0$. The Soret coefficient generally has magnitudes in the range of 10^{-2} to 10^{-3} and in the purely diffusional mode is quite diminutive. In a convecting configuration, however, the effect may be very prominent (e.g., Benedict and Pigford, 1957). There have been efforts to determine Soret coefficients for magmatic melts (e.g., Walker and DeLong, 1982; Rice, 1985).

The importance of the efficacy of convection on Soret separation is lost to some workers. Cygan and Carrigan (1992) conclude that the short residence time of about 100hr for the thermal boundary layer between magma and the magma chamber wall (to which they restrict all their Soret separation) allows a fractionation of only 0.004%, hence the Soret effect is ostensibly negligible. The magma chamber, however, will likely be convecting on the order of

10^5yrs, hence the boundary layer will be replaced about 10^7 times, each time gathering more fractionate. This would increase the 0.004% by a considerable amount. Further, the amount fractionated in any fractionation column is proportional to the area of the reactive surfaces. The surface area of the container of a magma chamber is large. A point continually missed, however, is that the boundary layers between layers of double diffusive convection can also be employed as Soret separators, in which case the magma chamber then behaves as multi-staged separation cascade, the total separation being the single layer separation factor multiplied by the number of layers (whose total surface area quite exceeds that of the walls of the chamber). If there are a hundred double diffusive layers in the magma chamber, then the separation is 100 times that of a single layer (e.g., Rice, 1985). The Soret coefficient, of course, evolves throughout the life of a magma chamber, perhaps eventually becoming negligible or reversing itself to destroy all its previous work. Only further investigation can resolve this. Cygan and Carrigan (1992) employ a constant Soret coefficient, hence they have not truly analyzed "time-dependent Soret transport."

It has never been the intention of those interested in Soret processes in magma chambers to supplant the importance of crystal fractionation. Without fractional crystallization, marginal border groups could not be formed (we apply fractional crystallization from the top down), hence magma chambers could not be stratified and double-diffusive assisted Soret fractionation could not come into play. There are peculiar distributions to be found in nature, however, that have no explanation in terms of classic fractionation concepts (e.g., Hildreth, 1981, in which the unreferenced Clusius-Dickel column—an early Soret separation device— may be found in Benedict and Pigford, 1957). Perhaps here Soret fractionation plays a role.

The reason I became interested in the possibility of Soret processes in magmas arose from observations that the silicic volcanism appearing at the end of the life of a bimodal center often displayed Sr ratios indicative of a primitive source. The interpretation then was that this silicic material was the end product of an evolved magma originally from depth, i.e., around 100km. That may be, but this is very complicated plumbing if one could call upon upper crustal assimilation as the source of the silicic material. Let alone side stepping the difficulty of actually evolving a rhyolite out of such material. The maximum theoretical amount of assimilation of silicic material by a mafic magma in the upper crust is about 40% of the original volume of the mafic magma (e.g., Ghiorso and Kelemen, 1987). This, of course, would not be realized in the real world. Nonetheless, some assimilation does occur (e.g., perhaps 10% of the volume of the Bushveld as discussed in earlier sections). Where does this material go? Sr ratios for Long Valley silicic volcanism did indicate a crustal origin until the last activity whereas the silicic material of the explosive end event looked primitive. Except for the momentary end event, most of the history of the activity was characterized by silicic venting of a crustal nature. It is common wisdom that so-called silicic centers are silicic only to the extent that the silicic melt occupies 5-25% of the top of an otherwise mafic chamber. In an initially stratified system of pure silicic magma over pure mafic, double-diffusive layering will not develop until the last 10% or so of the history of the chamber before rollover (see Rice, 1981). Significant Soret processes would not come into play until this period when a large number of boundary layers attending the double-diffusive layering would become available to bring into play fractionation cascades that would greatly enhance Soret separation. If the Soret coefficient for ^{87}Sr became negative at this time, the ^{87}Sr of the upper silicic melt would stream into the lower chamber whose greater volume would absorb it with little impact on its own strontium ratio signatures. The history of such a configuration would be crustal silicic venting until the last 10% of the lifetime of the system which would then see increasingly more primitive silicic melt: all without the necessity of getting the primitive silicic material up from 100km depth somehow. The Soret coefficient in magmatic melts is unknown for strontium isotopes, so all the above remains a conjecture neither proved nor disproved with certainty.

IV. References

Askeland, D.R. (1990) The Science and Engineering of Materials. Chapman and Hall, 880 pp.

Barnes, S.J. and Naldrett, A.J. (1986) Geochemistry of the J-M reef of the Stillwater Complex, Minneapolis adit area. II. Silicate mineral chemistry and petrogenesis. J. Petrol. 27, 791-825.

Benedict, M. and Pigford, T.H. (1957) Nuclear Chemical Engineering. McGraw-Hill, New York.

Betzer, P.R., Bernstein, R.E., Carder, K.L., Breland, J.B., Duce, R.A., Uematsu, M. and Freely, R.A. (1986) Particle fluxes in the North Pacific Ocean: responses to major atmospheric dust storms. EOS 67, 899.

Bird, R., Stewart, W. and Lightfoot, E. (1960) Transport Phenomena. John Wiley and Sons, New York.

Brandeis, G. and Jaupart, C. (1986) On the interaction between convection and crystallization in cooling magma chambers. Earth Planet. Sci. Lett. 80, 345-361.

Brandies, G. and Marsh, B.D. (1989) The convective liquidus in a solidifying magma chamber: a fluid dynamic investigation. Nature 339, 613-616.

Carder, K.L., Steward, R.G., Betzer, P.R., Johnson, D.L. and Prospero, J.M. (1986) Dynamics and composition of particles from an aeolian input event to the Sargasso Sea. J. Geophys. Res. 91, 1055-1066.

Cas, R.A.F. and Wright, J.V. (1987) Volcanic Successions. Allen and Unwin, London, 528 pp.

Cottrell, A. (1975) An Introduction to Metallurgy. Edward Arnold, Ltd., London, 548 pp.

Cygan, R.T. and Carrigan, C.R. (1992) Time-dependent Soret transport: applications to brine and magma. Chem. Geol. 95, 201-212.

Dauphin, J.P. (1983) Eolian quartz granulometry as a paleowind indicator in the northeast equatorial Atlantic, north Pacific and southeast equatorial Pacific. Ph.D. Thesis, University of Rhode Island. Kingston, RI. 335 pp.

De Klerk, W.J. (1991) Textures exhibited by feldspars in the Giant Mottled Anorthosite (GMA) of the Bastard Unit in the Upper Critical Zone, Western Bushveld Complex. Abstracts, Conference on Medium and Small-Scale Structures in Mafic and Ultramafic Rocks. Dept. of Geology, University of Natal, Pietermaritzburg, 9-10 Dec., p. 9.

Duce, R.A. (1986) Aeolian mineral particles; effects of atmosphere and marine processes. EOS 44, 898.

Eales, H.V. (1991) Inclusions of partially resorbed plagioclase, and mineral chemistry of pyroxenes, as indicators of magma-mixing events. Abstracts, Conference on Medium and Small-Scale Structures in Mafic and Ultramafic Rocks. Pietermaritzburg, 9-10 Dec., p. 5.

Eales, H.V. and Rice, A. (1991) On the densities of evolving igneous melts. Conference on Medium and Small-Scale Structures in Mafic and Ultramafic Rocks. Pietermaritzburg, 9-10 Dec., p. 10.

Eales, H.V., de Klerk, W.J. and Teigler, B. (1990) Evidence for magma mixing processes within the Critical and Lower Zones of the northwestern Bushveld Complex, South Africa. Chem. Geol. 88, 261-278.

Eales, H.V., Maier, W.D. and Teigler, B. (1991) Corroded plagioclase feldspar inclusions in orthopyroxene and olivine of the Lower and Critical Zones, Western Bushveld Complex. Mineralogical Magazine 55, 479-486.

Engelbrecht, J.P. (1988) The metamorphic aureole of the Bushveld Complex in the Marico District, western Transvaal, South Africa. Institute for Geological Research on the Bushveld Complex Report No. 77, University of Pretoria, South Africa, 22 pp.

Fahien, R.W. (1983) Fundamentals of Transport Phenomena. McGraw-Hill Book Co., New York, 613 pp.

Fisher, R.V. and Schmincke, H.-U. (1984) Pyroclastic Rocks. Springer-Verlag, Berlin, 472 pp.

Garde, R.J. (1985) Mechanics of Sediment Transportation. Wiley, New York, 618 pp.

Ghiorso, M.S. and Kelemen, P.B. (1987) Evaluating reaction stoichiometry in magmatic systems evolving under generalized thermodynamic constraints: examples comparing isothermal and isenthalpic assimilation. In Myson, B.O, ed., Magmatic Processes: Physiochemical Principles. The Geochemical Society, Special Publication No. 1, University Park, PA.

Hatton, C.J. and Sharpe, M.R. (1989) Significance and origin of boninite-like rocks associated with the Bushveld Complex. In Crawford, A.J., ed., Boninites and Related Rocks. Unwin Hymand, London, pp. 174-207.

Hildreth, W. (1981) Gradients in silicic magma chambers: implications for lithospheric magmatism. J. Geophys. Res. 36, 10153-10192.

Hulme, G. (1974) The interpretation of lava flow morphology. Geophys. J. Royal Astr. Soc. 39, 361-383.

Jackson, E.D. (1961) Primary textures and mineral associations in the ultramafic zone of the Stillwater Complex. U.S. Geological Survey Professional Paper 358.

Jaupart, C. and Brandeis, G. (1986) The stagnant bottom layer of convecting magma chambers. Earth Planet. Sci. Lett. 80, 183-199.

Kaufman, H.S. and Falcetta, J.J. (1977) Introduction to Polymer Science and Technology. John Wiley and Sons, New York, 613 pp.

Kerr, R.C. and Lister, J.R. (1991) The effects of shape on crystal settling and on the rheology of magmas. J. Geol. 99, 457-467.

Kingery, W.D., Bowen, H.K. and Uhlmann, D.R. (1976) Introduction to Ceramics. John Wiley and Sons, New York, 1032 pp.

Komor, S.C. and Elthon, D. (1990) Formation of anorthosite-gabbro rhythmic phase layering; an example at North Arm Mountain, Bay of Islands ophiolite. J. Petrol. 31, 1-50.

Kreith, F. (1967) Principles of Heat Transfer. International Textbook Co., Scranton.

Lane, J.A., MacPherson, H.G. and Maslan, F. (1958) Fluid Fueled Reactors. Addison-Wesley, Reading, MA.

Le Roex, A.P., Watkins, R.T. and Milner, S.C. (1990) Chemical stratification of the gabbros of the 'differentiated suite': Okenyenya igneous complex. Conference on Medium and Small-Scale Structures in Mafic and Ultramafic Rocks, Pietermaritzburg, Abstracts volume, p. 30.

Lovering, T.S. (1935) Theory of heat conduction applied to geological problems. Geol. Soc. Amer. Bull. 46, 69-94.

Martin, D. (1990) Crystal settling and in situ crystallization in aqueous solutions and magma chambers. Earth Planet. Sci. Lett. 96, 336-348.

McBirney, A.R. and Murase, T. (1984) Rheological properties of magmas. Ann. Rev. Earth Planet. Sci. 12, 337-357.

McBirney, A.R. and Noyes, R.M. (1976) Factors governing crystal settling and layering in igneous intrusives. Report, Center for Volcanology and Department of Chemistry, University of Oregon, Eugene.

McBirney, A.R. and Noyes, R.M. (1979) Crystallization and layering of the Skaergaard intrusion. J. Petrol. 20, 487-554.

Middleton, G.V. and Southard, J.B. (1985) Mechanics of Sediment Movement. Society of Economic Paleontologists and Mineralogists (Eastern Section), Providence, 297 pp.

Morisawa, M. (1985) Rivers. Longman, Inc., New York.

Morse, S.A. (1986) Convection in aid of adcumulus growth. J. Petrol. 27, 1183-1214.

Morse, S.A. (1988) Motion of crystals, solute and heat in layered intrusions. Canadian Mineralogist 26, 209-244.

Naldrett, A.J., Cameron, G., von Gruenewaldt, G. and Sharpe, M.R. (1987) The formation of stratiform platinum-group element deposits in layered intrusions. In Parsons, I., ed., Origin of Igneous Layering. NATO ASI Series, D. Reidel Publishing Co., Dordrecht.

Nell, J. (1985) The Bushveld metamorphic aureole in the Potgietersrus area: evidence for a two-stage metamorphic event. Institute for Geological Research on the Bushveld Complex Research Report No. 53, University of Pretoria, South Africa, 58 pp.

Rice, A. (1981) Convective fractionation: a mechanism to provide cryptic zoning (macrosegregation), layering, crescumulates, banded tuffs, and explosive volcanism in igneous processes. J. Geophys. Res. 86, 405-417.

Rice, A. (1985) The mechanism of the Mt St Helens eruption and speculations regarding Soret effects in planetary dynamics. Geophys. Surveys 7, 303-384.

Rice, A. (1992) Could the Lower Zone really be a wall? The South African Geocongress, Bloemfontain, 1-3 July.

Rice, A. and von Gruenewaldt, G. (1991a) Convective scavenging in Bushveld melts: a proposed mechanism for the concentration of PGE and chromite mineralized layers. Institute for Geological Research on the Bushveld Complex Research Report 95, University of Pretoria.

Rice, A. and von Gruenewaldt, G. (1991b) Convective scavenging of PGE's in Bushveld magmas. Conference on Medium and Small-Scale Structures in Mafic and Ultramafic Rocks. Pietermaritzburg, 9-10 Dec., p. 13.

Rosen, J. (1990) Freeze concentration beats the heat. Mechanical Engineering, 48f.

Shaw, H.R., Peck, D.L. and Okamura, A.R. (1968) The viscosity of basaltic magmas: an analysis of field measurements in Makaopuhi lava lake, Hawaii. Amer. J. Sci. 266, 225-264.

Sharpe, M.R., Irvine, T.N., Mysen, B.O. and Hazen, R.M. (1983) Density and viscosity characteristics of melts of Bushveld chilled margin rocks. Carnegie Institution of Washington Year Book 82, pp. 300-305.

Sherman, P. (1968) Emulsion Science. Academic Press, 496 pp.

Sparks, R.S.J., Wilson, L. and Hulme, G. (1978) Theoretical modelling of the generation, movement and emplacement of pyroclastic flows by column collapse. J. Geophys. Res. 83, 1727-1739.

Sparks, R.S.J. and Huppert, H.E. (1984) Density changes during the fractional crystallization of basaltic magmas: fluid dynamic implications. Contrib. Mineral. Petrol. 85, 300-309.

Stephenson, P.J. (1990) Layering in felsic granites in the main East pluton, Hinchinbrook Island, North Queensland, Australia. Geol. J. 25, 325-336.

Swaka, W.N., Chappell, B.W. and Kistler, R.W. (1990) Granitoid compositional zoning by side-wall boundary layer differentiation: evidence from the Palisade Crest intrusive suite, central Sierra Nevada, California. J. Petrol. 31, 519-553.

Tait, S. and Jaupart, C. (1989) Compositional convection in solidifying viscous melts. Nature 338, 571-574.

Teigler, B. (1990) Petrology and geochemistry of the Lower Zone and the Lower Critical Zone of the Western Bushveld Complex. Doctoral Dissertation, Rhodes University, Grahamstown, 293 pp.

Tritton, D.J. (1977) Physical Fluid Dynamics. Van Nostrand Reinhold, New York, 362 pp.

Uematsu, M., Duce, R.A. and Prospero, J.M. (1985) Deposition of atmospheric mineral particles in the North Pacific Ocean. J. Atmos. Chem. 3, 123-138.

von Gruenewaldt, G. (1972) The origin of the roof-rocks of the Bushveld Complex between Tauteshoogte and Paardekop in the eastern Transvaal. Geol. Soc. South Africa Trans. 75, 121-134.

Walker, D. and DeLong (1982) Soret separation of mid-ocean ridge basalt magma. Contrib. Mineral. Petrol. 79, 231-240.

Walraven, F. (1987) Textural, geochemical and genetic aspects of the granophyric rocks of the Bushveld Complex. Geological Survey of South Africa. Memoir 72, 145 pp.

Weinstein, S.A., Yuen, D.A. and Olson, P.L. (1988) Evolution of crystal-settling in magma-chamber convection. Earth Planet. Sci. Lett. 87, 237-248.

Wilshire, H.G. (1969) Mineral layering in the Twin Lakes granodiorite, Colorado. Geological Society of America Mem. 115, 235-261.

Wolff, J.A. and Wright, J.V. (1981) Rheomorphism of welded tuffs. J. Volcanol. Geotherm. Res. 10, 13-34.

A MECHANISM FOR SPONTANEOUS SELF-PERPETUATING VOLCANISM ON THE TERRESTRIAL PLANETS

P. J. TACKLEY
Seismological Laboratory
Division of Geological and Planetary Sciences
California Institute of Technology
Pasadena, California 91125, USA

D. J. STEVENSON
Division of Geological and Planetary Sciences
California Institute of Technology
Pasadena, California 91125, USA

ABSTRACT. We present a model for self-perpetuating magmatism resulting from Rayleigh-Taylor like instabilities developing spontaneously in regions that are partially molten, or at the solidus. The mechanism is capable of generating large volumes of magma without the need of a plume, or other deep source. Numerical models have been used to determine characteristic timescales, spacings and eruption rates in terms of nondimensional parameters. Scaled to realistic parameter space, the results correspond to timescales and eruption rates compatible with observations of small-scale intra-plate volcanism. Applications to oceanic seamount production, volcano spacing and rapid lithospheric erosion are discussed.

1. Introduction

Volcanoes occur in diverse forms on the Earth's surface. However, they are usually attributed to one of two distinct causes: plate-boundary effects (spreading centers, back-arcs), and deep mantle plumes. Several separate arguments lead to the conclusion that deep mantle plumes reaching the base of the lithosphere must exceed a certain minimum size; these are, diapirs arising from D" exhibit a characteristic size (Griffiths, 1986; Olson et al., 1987), weak plumes may get swept up by the large-scale flow (Boss and Sacks, 1985; Richards and Griffiths, 1988), and weak plumes cannot penetrate phase changes (Liu et al., 1991) and possible chemical boundaries (Kellogg, 1991) in the mantle. However, intra-plate volcanism is observed to have a broad spectrum of sizes, ranging down to small (100s of meters high) seamounts. Hence an alternative mechanism for intra-plate volcanism is required. This argument is reinforced by recent observations of Venus by the Magellan spacecraft. Hundreds of mainly small (<100km diameter) features, probably volcanic in origin, and probably too small to be caused by deep mantle plumes, have been observed.

In this paper we present a novel mechanism, previously proposed by Stevenson (1988), for the production of such intra-plate volcanism. The mechanism involves Rayleigh-Taylor instabilities developing spontaneously in regions of the asthenosphere which are partially molten, or at the onset of melting. The emphasis in this paper is on an understanding of the

307

D. B. Stone and S. K. Runcorn (eds.),
Flow and Creep in the Solar System: Observations, Modeling and Theory, 307–321.
© 1993 *Kluwer Academic Publishers. Printed in the Netherlands.*

basic concepts and physics, and an identification of the important parameters and their effects, by means of numerical models and physical argument.

2. Description of the Instability

2.1. GEOMETRY AND INSTABILITY

Figure 1 shows schematically the unstable initial condition which may be representative of, for example, young oceanic lithosphere and asthenosphere. By 'lithosphere' we refer to rock which is over three orders of magnitude more viscous than the asthenosphere, rendering it effectively immobile to asthenospheric flow. With realistic temperature-dependent viscosity the base of this viscous lithosphere corresponds to a geotherm about 300°C lower than the asthenospheric temperature, i.e. typically 1000°C.

As a first approximation, a simple univariant solidus is assumed. The geotherm, which is adiabatic at great depth, intersects the solidus at some lesser depth. Phase equilibria between solid and melt buffer the temperature at the solidus temperature or 'wet adiabat' above this depth, until the influence of the cold upper boundary causes the temperature profile to depart from this wet adiabat and drop steeply to the surface (following approximately an error function in the case of oceanic lithosphere). D, the distance between the depth of onset of melting and the base of the viscous lithosphere, is the most important length scale.

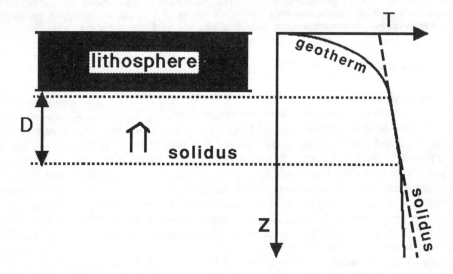

Figure 1. Initial condition. On the left is a cross-section through the lithosphere and asthenosphere, and on the right is the corresponding temperature profile, with the solidus indicated (dashed lines). D is the most important length scale.

The layer of rock at the solidus, which may be either partially molten or on the onset of melting, is unstable to a Rayleigh-Taylor like instability as follows: If an element of rock is given an infinitesimal velocity upwards, it will experience pressure-release partial melting, and hence a decrease in density since the melt (and possibly the solid residuum - see section 2.3) is less dense than the solid. The resulting buoyancy causes increased upwelling velocity and a higher rate of melting, a positive feedback situation which can lead to an episode of circulation, magma production and possibly surface magmatism. Since the degree of partial melting and hence buoyancy is proportional to the vertical distance moved, the growth of velocity is exponential, as in a conventional Rayleigh-Taylor instability at early times (Turcotte and Schubert, 1982).

2.2. EFFECT OF MELT MIGRATION

The situation is made more complicated by migration of melt by percolation through the solid matrix. Since melt buoyancy drives the bulk flow, percolation of melt up and out of the system diminishes the buoyancy and vigor of the flow. Melt percolation can reasonably be described by Darcy's law (McKenzie, 1984; Stevenson and Scott, 1991); arguably the most realistic form in this application is one in which the permeability, k, is proportional to porosity squared. This predicts that the melt velocity relative to solid is proportional to the porosity. The volume flux of melt per unit area of solid is therefore proportional to porosity squared.

Thus, the bulk average upwelling velocity of solid and liquid is linear in melt fraction, and the flux of melt through the solid is quadratic in melt fraction. Since the melting rate is proportional to upwelling velocity, at low melt fractions melting will exceed percolation. Therefore the melt fraction and hence flow velocity will increase, reaching a nearly steady-state at which the melting rate is balanced by percolation.

To get a clear picture of how melt migration affects the dynamics it is instructive to consider limiting cases. In the limit of infinite permeability the melt is removed instantaneously, hence there is no buoyancy, and no bulk flow occurs. In the limit of zero permeability (k=0), the melt fraction, hence buoyancy and circulation velocity, is at its maximum possible value determined by the vertical dimension of the melting region D_{melt}, and the rate of melting of an adiabatically upwelling element, $\partial f/\partial z$:

$$f_{max} = D_{melt} \frac{\partial f}{\partial z} \Big|_{k=0} \tag{1}$$

However, in this case no magmatism is observed at the surface.

2.3. COMPOSITIONAL CHANGES

Since it is the denser components, garnet and clinopyroxene, that melt first, the residual solid component is also less dense than the unmelted rock, and provides an additional contribution to the buoyancy. Even in the case of infinite permeability, buoyancy would be present. This effect has been used as a buoyancy source in some models (Mutter et al., 1988). However, the situation is made complicated on Earth by garnet-spinel-plagioclase phase transitions occurring at depths of interest, and by the possible existence of a stably stratified depleted layer at the top of the asthenosphere as a result of mid-ocean ridge melting processes. On other planets the presence of this buoyancy source is less certain. All these effects have been numerically modelled, but since the aim of this paper is to describe the basic physics in a way which is scalable to different dimensional parameters and different planets, compositional buoyancy is ignored here.

3. Equations and Numerical Model

Three sets of equations are necessary, describing the bulk flow, melt percolation and advection/diffusion.

It is appropriate to nondimensionalise the equations to the vertical dimension of the melting region (D, shown in figure 1), the diffusive timescale (D^2/κ, where κ=thermal diffusivity), the maximum melt fraction (f_{max}, (1)), and the superadiabatic temperature drop across the mobile part of the upper boundary layer (ΔT), which is about 300°C in this case (section 2.1). Nondimensional variables are denoted by tildas.

The bulk flow v of the local center of mass of solid and liquid, assumed to be incompressible and of infinite Prandtl number, is described by mass conservation and the Stokes equation with an extra term describing the buoyancy due to melting:

$$\nabla \bullet \tilde{v} = 0 \tag{2}$$

$$\nabla \tilde{p} - \nabla^2 \tilde{v} = \left(Ra.\tilde{T} + Rm.\tilde{f} \right).\hat{z} \tag{3}$$

where $\tilde{T}$ is the nondimensional temperature, $\tilde{f}$ is the normalized melt fraction and $\tilde{p}$ is the nondimensional pressure. Ra and Rm are the Rayleigh numbers for thermal buoyancy-driven and melt buoyancy-driven flow respectively:

$$Ra = \frac{\rho g \alpha \Delta T D^3}{\eta \kappa} \tag{4}$$

$$Rm = -\frac{g f_{max} D^3}{\eta \kappa} \frac{\partial \rho}{\partial f} \tag{5}$$

where α is the thermal expansivity, η is the asthenospheric viscosity at the depth of onset of melting, κ is the thermal diffusivity, and f_{max} is defined in (1).

A realistic temperature dependent viscosity law is used, so that the lithosphere arises naturally from the equations rather than being artificially imposed.

$$\eta(T) = \eta_o \exp\left(\frac{E_{act}}{RT} \right) \tag{6}$$

R is the gas constant. We use a value of E_{act}, the activation energy, of 420kJ/mole.

The segregation of solid and melt **u**, is described by Darcy's law. We assume melt percolation is in the vertical direction, and porosity is equal to melt fraction.

$$\tilde{u} = \tilde{f}\left(\tilde{v}_{liq} - \tilde{v}_{sol}\right) = \frac{Rm}{M}(1-f).\tilde{f}^2.\hat{\tilde{z}}$$ (7)

where M is the melt retention number:

$$M = \frac{\eta_{liq}D^2}{\eta\, k_0}$$ (8)

η_{liq} is the viscosity of the melt, and k_0 is the permeability constant. High values of M correspond to very slow percolation, resulting in vigorous bulk circulation, and low values of M correspond to fast percolation, resulting in slow bulk circulation, as discussed earlier.

The full form of Darcy's law for melt migration problems includes a term related to compaction of the solid matrix. This term is only important on lengths of the order of the compaction length, which is around 0.1-1km for reasonable mantle properties in this case (Stevenson and Scott, 1991). The length scales of interest here are tens of kilometers, so we neglect this term. It is also possible that the permeability is zero below a certain melt fraction threshold of order 1% (Nakano et al., 1989). Since this does not fundamentally alter the physics of the mechanism, in fact slightly favoring the development of these instabilities, we neglect this possibility.

Finally, advection/diffusion equations are required for temperature and melt fraction, f:

$$\frac{\partial \tilde{T}}{\partial \tilde{t}} = \nabla^2\tilde{T} - \tilde{v}_{sol} \bullet \nabla\tilde{T} - L.\tilde{m}$$ (9)

$$\frac{\partial \tilde{f}}{\partial \tilde{t}} = -(1-f)\frac{\partial \tilde{u}}{\partial \tilde{z}} - \tilde{v}_{sol} \bullet \nabla\tilde{f} + \tilde{m}$$ (10)

where L is the latent heat, m is the rate of melting, and $v_{sol} = v-u$ is the velocity of the solid component.

The equations are solved using a Petrov-Galerkin finite element scheme (Hughes, 1987; Hughes et al., 1989) with a varying element size, designed to give maximum resolution at the top of the asthenosphere where it is most needed.

In the lithosphere the mechanism for melt migration changes to one of rapid propagation through cracks. Since this is beyond the scope of the numerical model, melt percolating by Darcy flow in the upper part of the partially molten layer is simply removed, and measured as an 'eruption rate'. In reality only a fraction of this material would erupt at the surface.

4. Results

In order to isolate and study the effects of melt buoyancy, the thermally-driven component to the flow was eliminated by setting α, hence Ra, to zero, and imposing zero heat flux at the lower boundary. Values of Ra appropriate to this situation are probably close to critical (Haxby and Weissel, 1986).

The velocity boundary conditions are impermeable, stress-freeoon the base and outside of the box and rigid at the top. The temperature boundary conditions are isothermal at the top, and zero heat-flux at the base. Calculations in larger boxes verified that box is sufficiently large that the limited box size and velocity boundary conditions have negligible effect on the flow, and calculations at higher resolutions verified that the resolution is adequate. The initial push is provided by a temperature perturbation of 0.4% of the total temperature drop (scaling to about 5°C), exponentially decreasing away from the symmetry axis. This temperature perturbation is immediately converted into partial melt.

4.1. TYPICAL SIMULATIONS

Figure 2 show a typical simulation for the values $Rm=10^5(=100k)$, $M=150$. Time is normalized to the diffusive timescale D^2/κ, which is about 50Ma for D=40km. The four frames show the growth, peak, decline and death of the circulation. The circulation dies out for two reasons. Firstly, some of the depleted and cooled material that has already upwelled once gets recirculated into the upwelling. This material cannot contribute to the buoyancy, and so has the effect of progressively pinching off the upwelling. Secondly, the system as a whole is undergoing cooling, particularly near the base of the viscous lithosphere. Thus the viscous lithosphere thickens, reducing D and hence M and Rm. Cooling also causes increased viscosity in dry areas of the asthenosphere, further diminishing the flow.

The time evolution of bulk flow velocity and eruption rate are shown in figure 3 for three cases. Velocity, time and eruption rate are nondimensionalised to thermal quantities, κ/D, D^2/κ and κD respectively (about 0.1cm/yr, 50Ma and 1300km^3/Ma for D=40km, $\kappa=10^{-6}$m^2/s). In each case the rapid rise, leading to a long, gradually decreasing plateau, and a fairly rapid dying out, are apparent. The growth of the velocity with time looks linear on this semi-logarithmic plot, confirming that the growth is indeed exponential. The time lag between the velocity curve and the eruption curve is indicative of the melt percolation time through the layer.

The three cases shown are a reference case ($Rm=50k=5.10^4$, $M=150$), one with Rm doubled, which corresponds to figure 2 ($Rm=100k$, $M=150$), and one with M increased by an order of magnitude ($Rm=50k$, $M=1500$). The effect of doubling Rm is, to first order, a doubling of nondimensional velocity with the nondimensional timescale remaining constant. Increasing M increases both the flow velocity and timescale.

4.2. EFFECT OF PARAMETERS

The effect of the main parameters Rm, M and D on the length scale, timescale and eruption rates are now investigated more thoroughly. All results are nondimensionalised to thermal quantities, as described above.

time = 0.0000

time = 0.0359

time = 0.0685

time = 0.1123

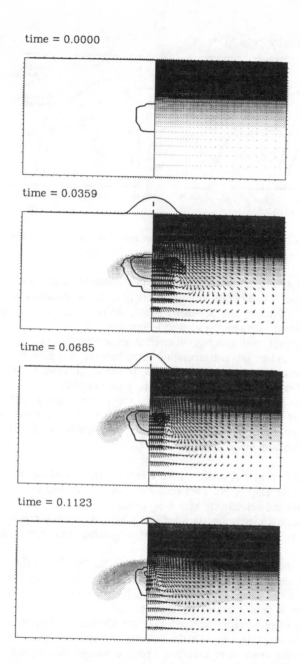

Figure 2. Simulation for Rm=100k, M=150, Ra=0, in axisymmetric geometry, at four times showing start, peak, decline and death of the instability. Each frame shows temperature (right, shaded), velocity (right, arrows), melt fraction (left, contours at 0 and intervals of 2.5%), composition (left, shaded means depleted), and eruption rate per unit area (graph on top). Dimensions of box are 3.25D and 2.75D, initial lithospheric thickness is D.

314

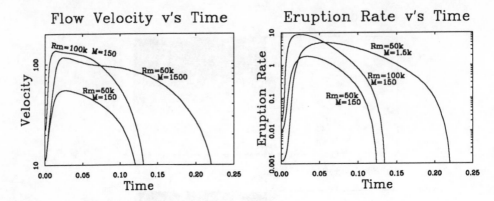

Figure 3. Time evolution of velocity and eruption rate for three cases. Time, velocity and eruption rate are nondimensionalized to D^2/κ, κ/D and κD respectively.

4.2.1. *Length scale*. Drawing an analogy between this instability and the classical Rayleigh-Taylor instability between two highly viscous fluids of different densities bounded by upper and lower boundaries (Turcotte and Schubert, 1984), the preferred spacing of upwellings is expected to be determined solely by the vertical length scale, D.

To determine this preferred spacing, numerical experiments were performed in long Cartesian boxes with random initial temperature perturbations. A typical result, in an aspect ratio 4 box with impermeable boundary conditions, is shown in figure 4. The initial random temperature perturbations are of amplitude 0.4% the total temperature drop, as before. After a short time, five upwellings have formed, two of them being strong, and three weak. In general, experiments at this aspect ratio lead to between three and five upwellings, indicating that the preferred number is about four, corresponding to a spacing of 2.5xD. This preferred spacing is also observed with other aspect ratios.

4.2.2 *Timescale*. Figure 5 a) and b) shows the effect of Rm and M on the nondimensional timescale, defined as the elapsed time between the start of the simulation and the point at which the melt fraction and velocity are zero everywhere.

The timescale is approximately constant for Rm > 50k, decreasing rapidly at low values. However it increases significantly with increasing M, beginning to saturate at about $M=10^5$.

4.2.3. *Peak Velocity and Eruption Rates*. The effect of Rm and M on the peak velocity is shown in figure 5 c) and d). The relationship between peak velocity and Rm appears linear, except at low Rm values where thermal diffusion becomes important. The peak velocity increases steadily with M until it saturates at about $M=10^4$.

Similar curves for eruption rate, averaged over the lifetime of the system, are shown in figure 5e) and f). The eruption rate, like the velocity, appears approximately linear in Rm. However, in M-space the eruption rate reaches a peak at around $M=10^3$, and declines rapidly either side of this.

time = 0.0000

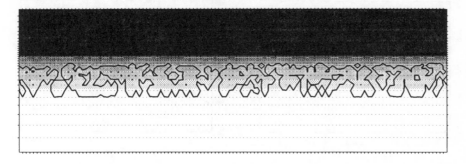

time = 0.0162

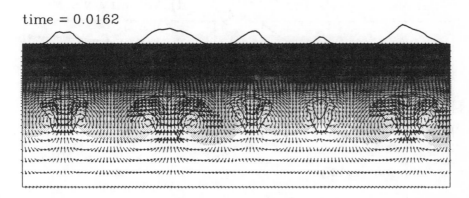

Figure 4. Simulation in a 4x1 cartesian box for Rm=50k, M=150, Ra=0. Upper plot shows random initial condition, lower plot shows subsequent development of upwellings. Temperature (shaded), velocity (arrows), melt fraction (contours) and eruption rate (graph), as figure 2.

5. Discussion

5.1. INTERPRETATION OF PARAMETERS

The results in the previous section lead to a qualitative interpretation of the main parameters.

Rm may be interpreted as $V_{stokes}/V_{thermal}$, where V_{stokes} is the characteristic velocity induced by melt buoyancy, derived by dimensional analysis from the Stokes equation, and $V_{thermal}$ is the same from the energy equation. Although one can interpret the thermal Rayleigh number Ra in the same way (except with thermal buoyancy), there is an important difference between the buoyancy forces in each case: in melt-driven flow, the buoyancy of an element increases as it rises, but in thermally-driven flow within an otherwise adiabatic region, the buoyancy remains constant.

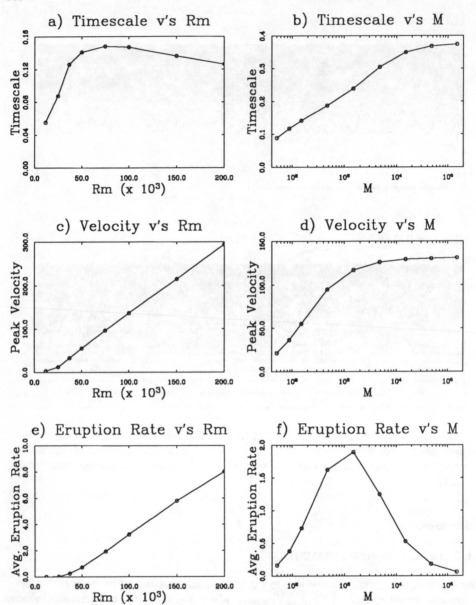

Figure 5. Effect of Rm and M on timescale, velocity and eruption rate. In a) c) and e) M is fixed at 150, in b), d) and f) Rm is fixed at 50k. Time, velocity and eruption rate. nondimensionalized to $D2/\kappa$, κ/D and κD respectively.

Rm primarily affects the nondimensional peak velocity and average eruption rate in an approximate proportionality, and has little effect on the nondimensional timescale except at low values. At these low values, thermal cooling becomes comparable with advection of heat, particularly during the early growth of the instability when the thermal velocity may exceed the flow velocity, and so the size of the initial perturbation becomes important. In the low Rm limit, thermal cooling dominates advection of heat, and the observed activity is simply due to relaxation of the initial condition.

M may be interpreted as V_{stokes}/V_{melt}, and in a sense is closer to the thermal Rayleigh number in physical meaning than Rm is, since both M and Ra are the ratio of flow velocity to the velocity at which the buoyancy force is dissipating. It affects the peak velocity and the timescale in a similar way, both increasing steadily with M and then saturating at high M. However, the eruption rate increases with M, then decreases. The effect on velocity can be understood as follows: In the limit of zero permeability (M=infinity) the maximum possible peak flow velocity, determined by Rm, is obtained; at finite M the buoyancy is reduced due to melt percolation and only some fraction of this velocity is reached. The effect on the timescale is due to the presence of melt buffering the system against cooling: melt fraction and thus timescale increase with increasing M. The eruption rate is low for high M because it is related to the percolative flux of melt at the top of the asthenosphere, and at high M the permeability and hence flux is small. At low M the eruption rate follows the trend of the velocity and other indicators of activity.

5.2. INITIAL STATE

In a partially molten layer where the temperature is buffered at the solidus, infinitesimal perturbations will grow exponentially into significant flow, so that these instabilities will develop spontaneously, as in the classical Rayleigh-Taylor instability (Turcotte and Schubert, 1984). In this case the instability greatly amplifies the amount of melt available for magmatism.

If the temperature is merely close to the solidus, then a vertical "push" is required to initiate melting and melt-driven flow. The required "push" could be provided by local small-scale flow, for example thermally-driven flow (Buck and Parmentier, 1986; Haxby and Weissel, 1986; Richter and Parsons, 1975), flow induced by fracture zones (Robinson et al., 1988), and passive corner-flow circulation induced at spreading centers. Hence a pervasively partially-molten asthenosphere is not required for these instabilities to develop.

5.3. INTERACTION WITH OTHER SMALL-SCALE FLOW

Simulations which include thermal buoyancy indicate that the melt-driven flow, once induced, can dominate thermally-driven flow. Even at lower values of M and Rm, melt buoyancy is observed to have a significant effect on the thermally-driven flow. The interaction between the two driving forces can result in a richness of phenomena not observed by treating each force separately. Thus, to correctly model flow in the asthenosphere it is necessary to include melt buoyancy.

5.4. VOLCANO SPACING

Characteristic spacing of major volcanic centers is observed in many situations, for example along hotspot chains. One current theory explains this by invoking stresses in the lithosphere (ten Brink, 1991). However, in the results presented in section 4.2.1, upwellings exhibit a characteristic spacing related to D, the vertical distance between the onset of melting and the base of the viscous lithosphere. Thus, fluid dynamics in the asthenosphere provides an alternative explanation for the spacing observed in nature. One upwelling may correspond to one volcano or a cluster of volcanoes, depending on the details of crack dynamics in the lithosphere.

5.5. LITHOSPHERIC EROSION

Melting of the rock as it rises adiabatically is accomplished by the absorption of latent heat by the melt. The subsequent vertical percolation of the melt to the top of the partially molten layer is effectively transporting this latent heat energy upwards, enhancing the vertical heat transport. In the calculations presented in this paper, much of this melt was removed from the top of the layer, removing the energy from the system. Calculations in which all the melt is allowed to freeze at the top of the asthenosphere, releasing its latent heat, indicate considerable thinning or erosion of the viscous lithosphere on timescales much shorter than predicted by diffusion alone.

The exact mechanism for this is as follows: The presence of partial melt at the top of the asthenosphere effectively buffers the temperature to the solidus temperature. The entire partially molten layer is being advected upwards by the bulk circulation described by Stokes equation. Thermal diffusion is important on small length scales near the top of the partially molten region, warming and hence softening the base of the viscous lithosphere, allowing further upward movement by Stokes flow. While melt is being supplied to the top of the layer by Darcy flow, Stokes flow and pressure-release melting at a faster rate than it is freezing due to thermal conduction, the lithosphere will thin. This combination of diffusion on small length scales and advection causes much faster erosion than thermal diffusion alone. Since the erosion timescale is much less than the corresponding diffusion timescale, a non-equilibrium lithospheric temperature profile results.

Preliminary results suggest that lithospheric thinning by a factor of up to two in a few million years may be possible. Due to difficulty in resolving the high thermal, viscosity and compositional gradients resulting from the non-equilibrium temperature profile, further investigation is needed to validate these numerical results.The situation is further complicated by the possibility of a high melt-fraction compaction layer at the base of the lithosphere (Sparks and Parmentier, 1991).

5.6. OCEANIC SEAMOUNTS

There are a large number of oceanic seamounts, which have been particularly well studied in the Pacific Ocean. Batiza (1982) estimated that there are 22-55,000 >1km high seamounts on the Pacific floor, and Smith and Jordan (1988) estimate roughly one million of all sizes. Smith and Jordan's analysis suggests that 78% of the large (>1km high) non-hotspot, non-fracture zone seamounts were formed away from the axis of the East Pacific spreading center, on lithosphere that was already up to several tens of Ma old. Examples of seamounts produced off-axis are described by Honda et al. (1987).

There is much uncertainty in the values of parameters appropriate to young oceanic asthenosphere, particularly in η and k_0. However, Rm and M are most sensitive to D. The depth of onset of melting for n-type Mid-Ocean Ridge Basalt (MORB) has been calculated to be at around 80km (McKenzie and O'Nions, 1991). Taking reasonable parameters: $\eta=10^{18}$Pa.s, $\eta_{liq}=10$Pa.s, $d\rho/df=-500$kg/m^3, $k_0=3\times10^{-10}$m^2, D=30-50km, we obtain Rm=20-74k and M=31-90. Scaling the numerical results to this dimensional space, we obtain flow velocities of up to ~10cm/yr, timescales of ~10Ma, spacings of ~100km, and magma production rates of up to ~1500 km^3/Myr. The total magma volume can be several times that required to build the largest seamounts, and the observed seamount volume per unit area in the Pacific (Smith and Jordan, 1989). Thus, the mechanism provides a plausible explanation for oceanic seamounts. The fertile asthenospheric material would be provided by rock which is drawn up by the mid-ocean ridge corner-flow but misses the proposed focussed upwelling directly under the ridge (Buck and Su, 1989; Scott and Stevenson, 1989). Some distance from the ridge, there is still a vertical component to the velocity, leading to the temperature profile in figure 1. Each upwelling may feed several volcanoes, the number being determined by cracks in the lithosphere, and the height determined by the depth of the feeding magma chamber, related perhaps to the thickness of the lithosphere.

5.7. LARGER VOLCANOES

Where anomalously high asthenospheric temperatures exist, for example where material has been emplaced by a plume, the depth of onset of melting may be much deeper than 80km; McKenzie and O'Nions (1991) calculate 120km for e-type MORB. Using D=50-90km and otherwise the same parameters as in section 5.6, we obtain Rm=74-430k, M=90-290. The numerical results scale to flow velocities of ~10-20 cm/yr, timescales of 10-40 Ma, eruption rates of ~1000-40000 km^3/Ma, and spacings of 125-200km. Thus, this mechanism may be important in the dynamics of such regions.

6. Conclusions

We have presented a model for self-perpetuating, melt-buoyancy-driven asthenospheric flow and volcanism arising from Rayleigh-Taylor instabilities in the asthenosphere. The mechanism can generate surface volcanism without a deep source, such as a plume. In the presence of pre-existing small-scale circulation, the instability can be triggered without a pervasively partially-molten asthenosphere. The important parameters are the melt-buoyancy Rayleigh number Rm, the melt retention number M, and the vertical distance between the onset of melting and the base of the viscous lithosphere, D: the time scale is determined mainly by M and the amount of magmatism by M and Rm.

The melt-driven upwellings exhibit a characteristic spacing proportional to D. This may provide an explanation for the frequently observed characteristic spacing of volcanic centers.

If a large fraction of the melt generated freezes at the base of the viscous lithosphere rather than erupting, then significant lithospheric thinning is possible.

When scaled to parameters realistic to young oceanic lithosphere, the magma volumes produced and timescales are compatible with observations of intra-plate volcanism, in particular the tens of thousands of large seamounts in the Pacific which appear unrelated to major hotspot chains.

Melt buoyancy can have a major effect on asthenospheric dynamics. Flow resulting from the interaction of thermal buoyancy and melt buoyancy exhibits greater complexity than flow driven by either of these individually. Hence, it is important to consider partial melting when studying the dynamics of the asthenosphere.

Acknowledgments. We are grateful to David Scott for furnishing us with his finite element code. We thank Don Anderson for a thoughtful review of the manuscript and helpful suggestions, and Craig Scrivner and Helen Qian for improving the quality of the text. The work benefitted from discussions with Geoff Davies. Supported by NSF grant EAR9017893. Contribution Number 5147, Division of Geological and Planetary Sciences, California Institute of Technology, Pasadena, CA 91125.

7. References

Batiza, R. (1982) Abundances, distribution and sizes of volcanos in the Pacific Ocean and implications for the origin of non-hotspot volcanoes, Earth Plan. Sci. Lett., 60, 195-206.

Boss, A.P. and Sacks, I.S. (1985) Formation and growth of deep mantle plumes, Geophys. J.R. Astr. Soc., 80, 241-255.

ten Brink, U. (1991) Volcano spacing and plate rigidity, Geology, 19, 397-400.

Buck, W.R. and Parmentier, E.M. (1986) Convection beneath young oceanic lithosphere: Implications for thermal structure and gravity, J. Geophys. Res., 49, 1961-1974.

Buck, W.R. and Su, W. (1989) Focussed mantle upwelling below mid-ocean ridges due to feedback between viscosity and Melting, Geophys. Res. Lett., 16, 641-644.

Griffiths, R.W. (1986) Dynamics of mantle thermals with constant buoyancy or anomalous internal heating, Earth Plan. Sci. Lett., 78, 435-446.

Haxby, W.F. and Weissel, J.K. (1986) Evidence for small-scale mantle convection from seasat altimeter data, J. Geophys. Res., 91, 3507-3520.

Honda, M., Bernatowicz, T., Podosek, F.A., Batiza, R. and Taylor, P.T. (1987) Age determinations of eastern pacific seamounts (Henderson, 6 and 7) - implications for near-ridge and intraplate volcanism, Marine Geology, 74, 79-84.

Hughes, T.J.R. (1987) The Finite Element Method, Prentice-Hall, Englewood Cliffs NJ, 631pp.

Hughes, T.J.R., Liu, W.K. and Brooks, A. (1979) Finite element analysis of incompressible viscous flows by the penalty function formulation, J. Comput. Phys., 30, 19-35.

Kellogg, L.H. (1991) Interaction of plumes with a compositional boundary at 670km, Geophys. Res. Lett., 18, 865-868.

Liu, M., Yuen, D.A. and Honda, S. (1991) Development of diapiric structures in the upper mantle due to phase transitions, Science 252, 1836-1839.

McKenzie, D.P. (1984) The generation and compaction of partially molten rock, J.Petrol., 25, 713-765.

McKenzie, D.P. and O'Nions, R.K. (1991) Partial melt distributions from inversion of rare earth element concentrations, in press, J. Petrol.

Nakano, T. and Fuji, N. (1989) The multiphase grain control percolation: Its implication for a partially molten rock, J. Geophys. Res., 94, 15653-15661.

Olson, P., Schubert, G. and Anderson, C. (1987) Plume formation in the D" layer and the roughness of the core-mantle boundary, Nature, 327, 409-413.

Richards, M.A. and Griffiths, R.W. (1988) Deflection of plumes by mantle shear flow: experimental results and a simple theory, Geophys. J., 94, 367-376.

Richter, F.M. and Parsons, B. (1975) On the interaction of two scales of convection in the mantle, J. Geophys. Res., 80, 2529-2541.

Robinson, E.M., Parsons, B. and Driscoll, M. (1988) The effect of a shallow low-viscosity zone on the mantle flow, the geoid anomalies and depth-age relationships at fracture zones, Geophys. J., 93, 25-43.

Scott, D.R. and Stevenson, D.J. (1989) A self-consistent model of melting, magma migration and buoyancy-driven circulation beneath mid-ocean ridges, J. Geophys. Res., 94, 2973-2988.

Smith, D.K. and Jordan, T.H. (1988) Seamount statistics in the pacific ocean, J. Geophys. Res., 93, 2899-2918.

Sparks, D.W. and Parmentier, E.M. (1991) Melt extraction from the mantle beneath spreading centers, in press, Earth Plan. Sci. Lett.

Stevenson, D.J. (1988) Rayleigh-Taylor instabilities in partially molten rock, EOS, 69, 1404.

Stevenson, D.J. and Scott, D.R. (1991) Mechanics of Fluid-Rock systems, Annual Reviews of Fluid Mechanics, 23, 305-339.

Turcotte, D.L. and Schubert, G. (1982) Geodynamics, John Wiley and Sons, New York, pp. 251-257.

HIGH LATITUDE OCEAN CONVECTION

BERT RUDELS

Institut für Meereskunde der Universität Hamburg
Troplowitzstr. 7,
D-2000 Hamburg 54 Germany

ABSTRACT. The main part of the world ocean deep waters is formed by heat loss at high latitudes. In contrast to the thermal winter convection occurring over most of the oceans the thermohaline forcing is, in the polar areas, dominated by freezing and melting. The density changes are due to variations in salinity rather than in temperature. The characteristics of the created waters depend upon topographic conditions, the nature of the ice cover and the ambient water masses. Four areas of deep and bottom water formation are identified: the shelves of the Arctic Ocean; the shelf areas around the Antarctic continent; the Weddell Sea; and the Greenland Sea. These areas represent two types: boundary and open ocean convection. The different areas are presented in order of increasing complexity of the possible active processes. The inter-actions between the convecting waters and the ambient water masses and the communication between the source areas and the World ocean and the renewal rates of the World ocean deep waters are briefly described.

Introduction

All energy of the solar radiation reaching the earth is returned to space. Part is reflected, part is radiated back at longer wavelengths. The earth's fluid envelopes - the atmosphere and the oceans - redistribute part of the heat before it is lost. The resulting meridional heat transport leads to an area of net, annual heat gain around the equator and areas of net annual heat loss at high latitudes and to weaker meridional temperature gradients (Figure 1). At mid latitudes the atmospheric and oceanic heat transports are comparable, but at higher latitudes the atmospheric transport dominates.

The atmosphere absorbs little radiation directly and is mainly heated from below. Hence the troposphere is well mixed by convection. The atmospheric motions are swift and predominantly driven by the latent heat released by the condensation of water evaporated from the oceans. The equator to pole heat transport is at low latitudes primarily through the Hadley cells - warm air rising near the equator and sinking at higher latitudes. Poleward of 30 degrees the principal meridional exchange is due to eddies generated by frontal instabilities.

The oceans by contrast are heated from above and stratified. They gain energy at low and lose it at higher latitudes. This could, in principle, give rise to a Hadley type meridional circulation with warm water being advected towards the poles and gradually becoming cooler because of heat loss to the atmosphere and to space. The water would eventually sink into the deep and return towards the equator, where it slowly becomes heated and rises towards the surface (Figure 2).

D. B. Stone and S. K. Runcorn (eds.),
Flow and Creep in the Solar System: Observations, Modeling and Theory, 323–356.
© *1993 Kluwer Academic Publishers. Printed in the Netherlands.*

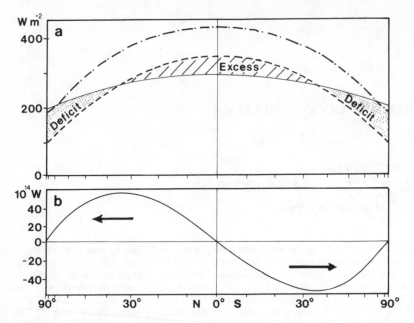

Figure 1. a) Radiation balance of the earth as a function of latitude. Short wave radiation reaching the top of the atmosphere (line-dot line), short wave radiation absorbed at the earth's surface (broken line), long wave radiation to space (full line). b) Meridional heat balance needed to maintain radiation balance. Positive transport towards the north. The horizontal scale is such that the spacing between latitudes is equal to the area between them (from Wells, 1986).

Such simple, thermally driven meridional flow is not predominant in the oceans. The surface circulation is mainly wind driven and consists of several gyres, anticyclonic in the subtropics, cyclonic in the sub-polar areas (Figure 3). The continents provide the meridional boundaries, which allow for organized north-south flows that are not possible in the atmosphere.

The oceanic circulation is much slower than the atmospheric and the water spends several years to pass through the gyres. The seasonal heating and cooling cycle stores heat and stratifies the water in summer and removes the heat and homogenizes the water by convection in winter.

Because the horizontal advection water crosses latitudes and experiences stronger (or weaker) cooling from one year to the next, more heat is lost and the convection reaches deeper at higher latitudes. The seasonal heating and cooling thus overlies the meridional circulation (Figure 4).

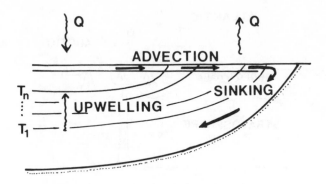

Figure 2. Idealized, thermally driven meridional ocean circulation.

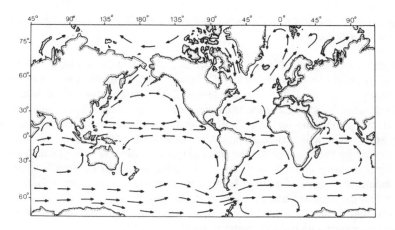

Figure 3. Surface currents of the world oceans in northern summer (from Apel, 1987).

The atmospheric meridional heat transport is associated with a fresh water flux. Water evaporated at low latitudes is returned to the oceans either by direct precipitation or by continental run-off and glacial melting. The density of sea water is determined by its salinity as well as by its temperature and the density decrease caused by the fresh water input at high latitudes compensates the density increase due to stronger cooling. The densities of the deep waters cannot therefore be attained by cooling of the surface water. Thermal convection is limited to the poleward extensions of the gyres - the subtropical in the southern, the subpolar in the northern hemisphere - and does not reach deeper than 1000 m. These are the horizontal and vertical limits of the oceanic troposphere (Defant, 1961).

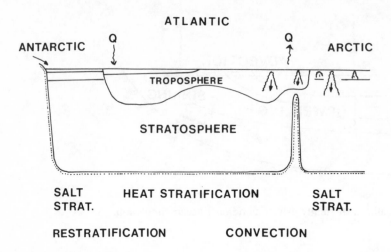

Figure 4. Seasonal convection and restratification in the Atlantic Ocean. Southern hemisphere summer, northern hemisphere winter.

At higher latitudes the surface salinity is too low to allow for deep reaching convection even if the water is cooled to the freezing point. Fresh water has to be removed to create waters dense enough to convect into the deeper layers. Such a mechanism exists. When sea water freezes, fresh water is extracted as ice while most of the salt remains in the liquid phase. Freezing thus creates conditions for deep "haline" convection driven by salinity increase (Figure 4). The areas where haline stratification dominates thermal stratification can, at high latitudes, be identified by the extent of the ice cover (Figure 5). However, the stability due to the freshwater input is strong and only at a few favoured locations will the stratification, the heat loss, the amount of ice formed and the bathymetry combine to create water sufficiently dense to convect into the deep and bottom layers of the oceans.

Our theme is to follow the chain: fresh water input - freezing - salt ejection - dense water formation - deep convection - global deep water renewal.

The equation of state, freezing and ice formation are reviewed in section 1. Different aspects of dense water formation are brought out in connection with discussions of the active areas - the Arctic Ocean (section 2.1), the rim of the Antarctic continent (section 2.2), the Weddell Sea (section 2.3) and the Greenland Sea (section 2.4).

The world ocean is almost everywhere, also in areas where deep, haline convection occurs, thermally stratified in the deeper layer. This points to the importance of the interactions between the convecting dense surface water and the ambient water column, which is discussed in section 3. Apart from a few sources at lower latitudes: the Mediterranean Sea, the Labrador Sea, the Red Sea, the deep waters of the world oceans derive from high latitudes. Section 4 considers the importance of this deep water renewal on the global scale.

Excellent general reviews of ocean convection and mixing processes are found in Killworth (1983) and Carmack (1986, 1990). The books edited by Smith (1990) and Untersteiner (1986) also contain much useful information.

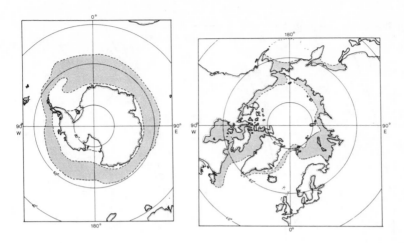

Figure 5. The distribution of sea ice in winter and summer (from Fairbridge, 1966).

1. Forcing

Convective motions are driven by a gravitationally unstable distribution of density in the water column (heavy water above lighter). This situation arises from diurnal cooling, by winter heat loss, by evaporation and by freezing.

Sea water differs from fresh water in several important aspects. Its density depends on salinity as well as on temperature and pressure. The temperature of freezing and the temperature of the maximum density are both lowered by increasing salinity. The rate of decrease of the maximum density temperature is faster and for salinities above 24.7 psu (24.7g salt per kilogram of water) the coldest water is also the densest.

The effects of temperature and salinity on the density and on the freezing of sea water at atmospheric pressure are seen on a Temperature - Salinity (T-S) diagram (Figure 6). The characteristics of polar waters are limited by the open rectangle and the polar deep waters are indicated by the circle. If the temperature and salinity of the water column are plotted on the T-S diagram they will form a line (going from lower to higher densities). T-S diagrams or rather θ-S diagrams, θ being the potential temperature, will be used extensively below to describe mixing processes and deep water formation.

The fact that in the oceans the coldest water is also the densest cannot be over emphasized. In contrast to lakes the entire water column has to be cooled to freezing before ice can form. The seasonal cooling is, in general, only strong enough to bring a few hundred meters thick layer to freezing. In this section we briefly describe some features of freezing and the formation of saline water (brine) leaving the question, how the freezing temperature is attained, for section 2. For thorough discussions of sea-ice and ice formation we refer to Weeks (1986) and Gow and Tucker (1990).

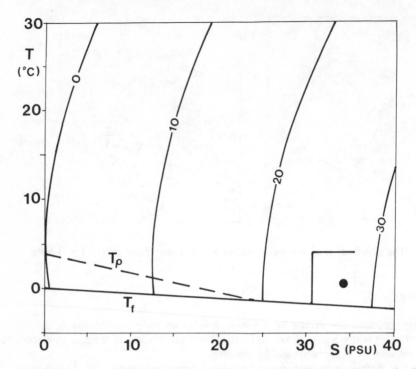

Figure 6. Temperature - Salinity diagram. The curved lines indicate isopycnals. T_f is the freezing temperature, T_ρ the temperature of maximum density, all at atmospheric pressure. The open rectangle limits the T-S range for the Polar waters and the circle shows the range of the Polar deep waters (Adapted from Apel, 1987).

1.1. FREEZING

With open water at freezing temperature a continued heat loss to the atmosphere will lead to supercooling of the topmost layer. If no strong, mechanically generated, turbulence is present, the thickness of the supercooled layer will be limited by the conduction of heat. Only within this layer can ice formation take place. The newly formed ice consists of small 0.1 mm spheres, which quickly grow to discs with a diameter of 1 mm. This "frazil" ice rises and accumulates at the surface. The ice is pure water and the salinity of the liquid phase increases. This unstable situation leads to the sinking and replacement of the top, now saline, layer. The heat loss only penetrates a few mm, before the water is replaced (Figure 7a). Strong turbulence may, however, redistribute the supercooled water and the frazil ice formation over a much deeper layer. With intense cooling the ice formation is so rapid that large volumes of sea water are trapped between the ice crystals and give sea-ice a salinity different from zero.

Once an ice cover is established, the ice crystals begin to grow in the only possible direction, vertically downward. Water may be trapped between the crystals and further contribute to the salinity of the sea ice (Figure 7b). In addition to an organized crystal structure the sea ice may also grow larger platelets. In the Arctic the platelets have vertical extents of a few mm, while in the Antarctic they may reach 10 meters down (Gulliksen pers. comm.). This difference is probably due to the Antarctic iceshelves which provide a subsurface heat sink and maintain a much deeper layer at the in situ freezing point (see also section 2.2). The growth of platelets could result from water freezing and losing fresh water onto a crystal while it, due to its increased salinity, sinks downward (Figure 7b). This effect, not present in lakes where water at freezing is less dense and no instability occurs, might also account for the different crystal orientations in sea ice and fresh water ice.

In older sea ice (1 year) the trapped salt (brine) slowly drains into the underlying water. The difference between the air temperature and the water temperature beneath the ice creates a temperature gradient within the ice. Water will freeze in the upper, colder part and the brine becomes more saline. When the high salinity brine is redistributed to the lower part of the brine pocket its salinity will be too high with respect to the surrounding temperature and it will melt part of the ice. This allows the brine to migrate through the ice.

The brine channels might converge to localized, intense drainage points. The brine has, when it leaves the ice, a high salinity and a temperature below the freezing temperature of the underlying sea water. This may then freeze around the sinking brine and form stalactites. The salinity of the surrounding water thus increases and a secondary, haline convection is induced (Figure 7c). These features, as the platelets, are more frequent and of larger dimensions in the Antarctic than in the Arctic. They have also been studied in the laboratory (Martin, 1974).

1.2. COOLING VERSUS FREEZING

The density of sea water ρ (T, S, P) is a non-linear function of temperature, salinity and pressure. The temperature and salinity dependence at atmospheric pressure is indicated by the isopycnals in the T-S diagram (Figure 6). It is clear that changes in temperature become less important at lower temperature and close to freezing salinity variations strongly dominate the density changes.

The salinity change ΔS due to freezing corresponding to a heat loss E can be found from

$$\int_0^t Qdt = E = C\rho H\Delta T = \rho Ld \tag{1.1}$$

where Q is rate of the heat loss, C, the heat capacity, ρ, density, H, the depth of the water, d, the amount (thickness) of water frozen to ice, L, the latent heat of freezing. If the salinity of the ice is put to zero we have with the initial salinity S the salt balance

$$(H-d)(S+\Delta S) = HS \tag{1.2}$$

which gives

$$\Delta S = \frac{S}{(L/C\Delta T)-1} \tag{1.3}$$

where ΔT is the temperature change. For the salinity of 38 psu a temperature change of 1°C corresponds to a salinity change of 0.4 psu. The effects on the density can be seen from the T-S diagram (Figure 6).

Freezing should be compared to evaporation. The latent heat of evaporation is 8 times as large as that of freezing and the salinity change is correspondingly smaller. Moreover, the influence of salinity on density is less at higher temperatures. The larger density increase during evaporation is therefore due to cooling and not due to the salinity increase resulting from loss of fresh water.

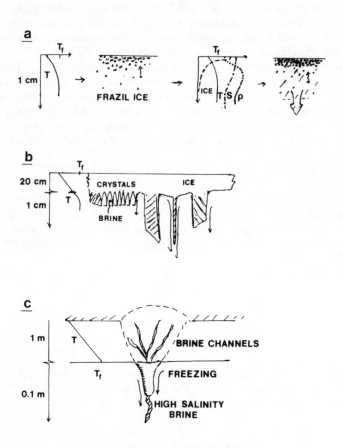

Figure 7. Freezing of the sea water (conceptual). a) Formation of frazil ice and separation of ice and brine in open water. Profiles of temperature, salinity, ice concentration and total (ice and water) density are indicated. b) Crystal and platelet growth beneath solid ice. c) Brine drainage from old ice. The high salinity and the corresponding lower temperature lead to freezing and the formation of the stalactites of the surrounding water and induce a secondary, haline convection.

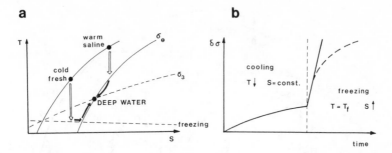

Figure 8. a) Evolution of temperature and salinity in the surface layer during heat loss, when the initial temperature and salinity are higher respectively lower than those of the deep water. Isopycnal at atmospheric pressure, σ_θ, Isopycnal referred to 3000 db, $\sigma_3 \cdot$ b) Density increase $d\rho$ at constant heat loss. During the cooling phase the salinity is constant and the density increase is slow. During freezing the density increase is due to the ejection of salt and more rapid. When the thickness of the ice cover grows, the heat loss is reduced and the density increase becomes slower (from Rudels and Quadfasel, 1991).

The cold, dry polar air also induces evaporation, especially over open leads. However, the temperature is at freezing and the only heat source is latent heat released by freezing. The effect of evaporation is thus primarily a salinity increase due to larger ice production, not due to loss of freshwater to the atmosphere.

The relative importance of the salinity and temperature of a water parcel is brought out by considering two situations. Cold, fairly saline deep water is in one case overlaid by a warmer, more saline and in a second case by a colder, fresher surface layer (Figure 8a). The initial density difference is the same. The density increase with time at constant heat loss is shown in Figure 8b. If the isolating effects of increased ice thickness are ignored, freezing is by far the most efficient way to create dense water. This is even more obvious when the pressure effect on density is considered. Cold water is more compressible than warm and the slope of the isopycnals change (Figure 8a). When the surface water is warmer and saltier, moving the water vertically will make it less dense relative to the deep water. By contrast cold water will, when it sinks, experience a relative density increase due to the higher pressure (Figure 8a, see also section 2.2).

2. The Convection Areas

Ocean convection can be separated into two principal types (Killworth 1983) - boundary or shelf convection and open ocean convection. The two types are discussed in order of increasing complexity. The shelf convection in the Arctic Ocean and at the rim of the Antarctic continent is considered first and the open ocean convection in the Weddell Sea and in the Greenland Sea later.

Figure 9. Surface currents in the Arctic Ocean. Warm currents, full arrows, cold currents, broken arrows.

2.1. THE ARCTIC OCEAN: STRONG STRATIFICATION AND SHALLOW SHELVES

2.1.1. *The oceanographic setting.* The Arctic Ocean is an enclosed sea. One third of its area consists of the wide and shallow Siberian shelves, while the remainder is made up by two deep basins; The Eurasian and the Canadian Basins, separated by the 1400 m deep Lomonosov Ridge.

The Eurasian Basin receives warm, saline Atlantic Water over the Barents Sea and through the Fram Strait. The principal outflow of Polar Surface Water is through Fram Strait. Some water, mainly consisting of Pacific Water entering the Canadian Basin through Bering Strait, leaves through the Canadian Arctic Archipelago. All deep water exchanges between the Arctic Ocean and the surrounding seas occur through Fram Strait (Figure 9).

In the upper part of the Arctic Ocean water column are found the locally homogenized 50 m deep Polar Mixed Layer, at freezing temperature and with a salinity of 32-33 psu, and the

100 m thick halocline with water close to freezing down to 100 m but with increasing salinity. Below the halocline lies the thermocline above the warm, 250-400 m thick Atlantic Layer. The temperature is destabilizing and the stable stratification is also here due to increasing salinity. Only below the Atlantic Layer is, as in other oceans, a stabilizing thermocline encountered. The deep water is thermally stratified in its upper part. However, in the deepest layers the temperature is constant while the salinity increases drastically (Figure 10).

2.1.2. *Dense water formation on the shelves.* In summer the river run-off interacts, on the shelves, with Atlantic and Pacific Waters entering over the Barents Sea, through Bering Strait and across the shelf break to create waters later to supply the Polar Mixed Layer. In winter the water on the shelves may, because of the shallow depth, be cooled to freezing and its density is increased by the ejection of salt (Figure 10). Depending upon initial salinity, bathymetry, ice cover and wind conditions a wide range of densities (salinities) are produced, all at freezing temperature. The dense water leaves the shelves as entraining density flows sinking down the continental slope. The higher the initial density the deeper into the water column the density flow will penetrate. Much of the halocline is supplied from the shelves and the θ-S characteristics of the Atlantic and deeper layers are altered by deep reaching density flows (Figure 10 and section 3).

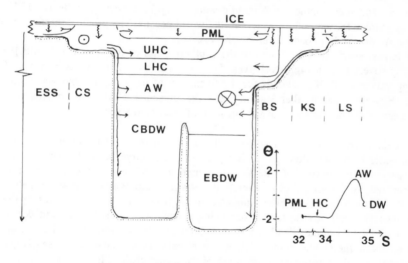

Figure 10. Stratification and water masses in the Arctic Ocean. The view is along the Lomonosov Ridge towards the Eurasian continent and the shelf is split between the Laptev and the East Siberian Sea. Barents Sea (BS), Kara Sea (KS), Laptev Sea (LS), East Siberian Sea (ESS), Chukchi Sea (CS). The main water masses are the Polar Mixed Layer (PML), upper Halocline (uHC), lower Halocline (lHC), Atlantic Layer (AL), Canadian Basin Deep Water (CBDW), Eurasian Basin Deep Water (EBDW). Fram Strait inflow, ⊗, Bering Strait inflow, ⊙.

Observations of shelf waters reaching different depths of the Arctic Ocean have been reported by Melling and Lewis (1982) (halocline) Rudels (1986) (Atlantic Layer) and Quadfasel et al. (1988) (bottom water). Of special interest is the characteristic θ-S signal of the deep and bottom water, which with its increasing salinity and constant temperature shows the added contributions from the high salinity shelf water and the warmer entrained water (section 3). Areas of persistent open water (polynyas) produce large amounts of ice and are likely sources for most of the dense shelf waters. The river run-off, while contributing to the strong stratification, is not crucial for its maintenance. Important is the shallow depth of the shelves which allows the water to cool to freezing and ice to form during winter. The seasonal cooling and heating cycle makes the ice melt and refreeze. A stable stratification, the prerequisite for a permanent ice cover, is created. The fresh water content of the Bering Strait inflow adds to the separator effect provided by the shelves. If the river run-off were less, the shelves would still produce the same amount of ice in winter, which later melts and supplies the mixed layer. They would, however, produce waters with higher salinity, and the deep water renewal would be more vigorous (Rudels, 1989).

2.2. THE SHELVES OF THE ANTARCTIC CONTINENT

In the southern hemisphere the oceans surround the ice covered Antarctic continent (Figure 11). Glacial melt and net precipitation are the fresh water sources for the Southern Ocean. The ice cover is less restricted than in the Arctic and its seasonal variation is much larger (Figure 5). The seas around the Antarctic were recognized early (Brennecke, 1921; Mosby, 1934) as sources for the bottom waters of the world oceans. The production of dense water is, as in the Arctic, likely to occur over the shelves.

The largest shelf areas are found in the southern parts of the Ross and Weddell Seas and are partly covered by floating iceshelves. The stratification is much weaker than in the Arctic and importance of shelf processes for the formation of deep and bottom water was realized earlier. However, the weaker stability and the presence of the ice shelves increase the importance of the non-linearity of the equation of state of sea water.

2.2.1. Weddell Sea Shelf Area. The Weddell Sea shelf is the most important area for deep and bottom water formation although similar processes are active in the Ross Sea. The Weddell Sea water column consists of: the Winter Water, a 150-200 m thick, fresher surface layer homogenized by freezing during winter; the Warm Deep Water, below ultimately derived from the North Atlantic; and the Antarctic and Weddell Sea Bottom Water in the deepest layers. The Weddell Sea Bottom Water being somewhat colder, fresher and denser (Figure 12). The circulation in the Weddell Sea is cyclonic and brings water from the east up onto the broad southern shelf area from which it flows, permanently covered by sea ice, along the Antarctic Peninsula, turning eastward south of the island chain (Figure 11).

Close to the iceshelf tidal motions keep the sea surface free from ice during winter, which leads to large heat loss and extensive ice production. The ejected salt creates the saline High Salinity Shelf Water. Part of this water drains towards the north, part drains southward, filling the depression below the Filchner Iceshelf.

The shelf water crossing the shelf break will mix with the waters of the Weddell Sea Water column. Fofonoff (1956) realized that the mixing between waters of similar densities but with different θ-S characteristics may form mixing products denser than both the original waters. This process, caballing or contraction on mixing, could then add to the density of the waters sinking from the shelves (Figure 12). A mixing between shelf waters and the deep basin waters across the shelf break was also proposed by Foster and Carmack (1976) for the formation of Antarctic Bottom Water.

The High Salinity Shelf Water entering the Filchner depression gets into contact with the bottom of the floating ice shelf. The freezing point of sea water is lowered by increasing pressure and the temperature of the water is above the in situ freezing temperature. The bottom of the ice shelf melts and fresh water is added to the High Salinity Shelf Water to create the colder, fresher and less dense Ice Shelf Water. This flows northward under the ice then across the shelf to the shelf break, where it sinks into the deep basin. This scenario was first suggested by Carmack and Foster (1975) and was later elaborated by Foldvik et al. (1985), who observed cold water (-2°C) sinking down the continental slope.

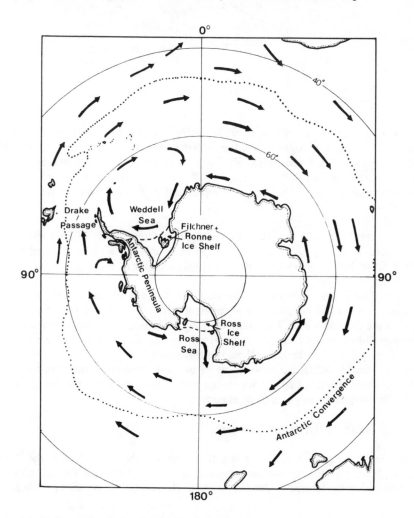

Figure 11. Surface Circulation in the Southern Ocean.

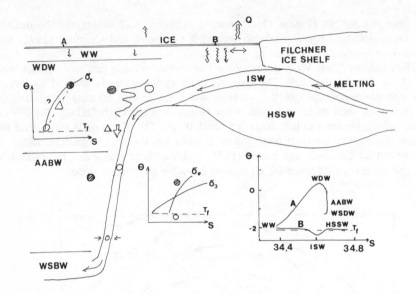

Figure 12. Stratification and water masses in the southern Weddell Sea. Winter Water (WW), Warm Deep Water (WDW), Antarctic Bottom Water (AABW), Weddell Sea Bottom Water (WSBW), High Salinity Shelf Water (HSSW), Ice Shelf Water (ISW). The positions of the T-S curves are indicated by A and B. The schematic T-S diagrams show the effects of contraction on mixing (caballing) and of the higher compressibility of cold water, (Adapted from Carmack, 1986).

The descent into the deep is facilitated by the higher compressibility of cold water, which increases the relative density of the Ice Shelf Water as it sinks. This effect, discussed by Gill (1973) and worked into a model by Killworth (1977) appears to be necessary to make the shelf water reach the bottom.

The colder, fresher signature of the Weddell Sea Bottom Water as compared to the Antarctic Bottom Water indicates that the Ice Shelf Water, not the denser High Salinity Shelf Water reaches the deepest levels. This could be a further indication of the importance of the higher compressibility at cold temperatures. However, the temperature difference is only 0.5°C and it is surprising that the effect should be this strong.

There is a qualitative difference between the variation of compressibility with temperature and caballing, which requires that the waters are mixed down to molecular level. This is a much slower process, which might limit its importance.

2.3. OPEN OCEAN DEEP CONVECTION I: THE WEDDELL SEA

Open ocean deep convection occurs only at a few locations, the convection in the Western Mediterranean being perhaps the one most studied. At high latitudes the Greenland Sea has long been assumed to be an area of deep convection. However, the first direct evidence of a deep reaching convection event was reported from the Weddell Sea (Gordon, 1978). A narrow (30 km), cold, low salinity funnel or chimney (Killworth, 1979) reaching from the surface layer to 4000 m was observed close to Maud Rise in the Weddell Sea gyre and was interpreted as the remnant of deep convection from the preceding winter (Figure 13). The anomaly was also reported in oxygen and density and the "chimney" was associated with strong cyclonic circulation.

This observation has been related to the occurrence of the Weddell Polynya, coupling deep convection to the overturning of water masses, the melting of sea ice and a large heat loss to the atmosphere. The process has mostly been considered as the evolution of a mixed layer influenced by cooling, freezing and ice melt.

2.3.1. *Weddell Sea Mixed Layer.* The evolution of an upper layer with the mixing (entrainment) across the pycnocline limited by the density increase caused by heat loss at the sea surface has recently been modelled by Martinson (1990). The energy loss, either through the ice or at open leads, increases the salinity (and density) by removing freshwater. The mixed layer deepens and incorporates more and more of the pycnocline. However, the deepening also brings warmer water into the mixed layer, which retards the ice formation and might even cause some melting beneath thicker ice floes (Figure 14). The erosion of the pycnocline increases the temperature gradient and diffusive heat transfer through the pycnocline into the mixed layer will add to the heat flux due to entrainment. The diffusive heat flux might become quite significant, if the temperature step is large and the net stability weak.

Martinson concludes that only in areas where the stability and the temperature gradients are weak, as they are close to Maud Rise (Figure 15), can a density increase of the mixed layer to and above that of the underlying water occur. This is in the area of the Weddell Polynya and close to the location of the chimney reported by Gordon (1978).

A similar configuration was studied by Walin (1991). He primarily considers mechanically (wind) generated mixing and entrainment. In early winter the stratification is strong and essentially no entrainment occurs ($H = H_0$, $W_e = 0$). The heat loss E to the atmosphere is supplied by the release of latent heat (ρ, L, Q, d as in section 1).

$$ E = \int_0^t Q dt = \rho L d \qquad (2.1) $$

and leads to a growing ice cover and increased salinity in the upper layer. When the heat loss becomes larger than a critical value E_1, the removal of fresh water has decreased the stability drastically and entrainment becomes important. However, this implies that warmer water is incorporated into the mixed layer. The ice starts to melt, which inhibits the entrainment. Ice can again form and decrease the stability and cause new entrainment. The system will attain a state with slow entrainment and deepening of the mixed layer and a slow reduction of the ice cover (Figure 16).

The added fresh water is just enough to lower the salinity of the entrained water to that of the mixed layer and keep the salinity step at the base of the mixed layer constant. In the asymptotic case the density steps due to temperature and salinity balance at the base of the mixed layer

338

$$-\alpha\Delta T = \beta\Delta S \tag{2.2}$$

α and β are the coefficient of heat expansion and salt contraction, respectively. The entrained water supplies the heat necessary to melt the ice as well as the heat loss to the atmosphere. We may write

$$Q_e = \frac{Q_e}{R} + \frac{1 - R}{R} Q_e \tag{2.3}$$

where the first term accounts for the melting and the second for the atmospheric heat loss and

$$R = \frac{C\beta S}{\alpha L} \tag{2.4}$$

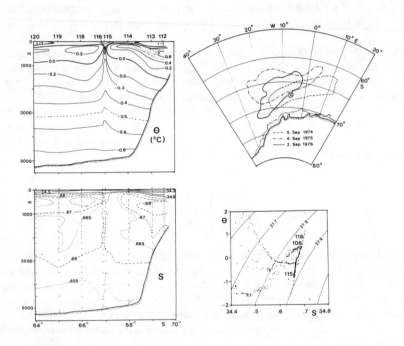

Figure 13. a) temperature and b) salinity sections observed at Maud Rise, 1977. c) shows the θ-S diagram of stations, 108, 115, 118, d) indicates the positions of the Weddell Polynya (from Gordon, 1978).

is a parameter introduced by Walin, which essentially gives the ratio of density variations due to salinity changes to those due to temperature changes (compare equation 1.3 above). The density buffer supplied by the ice cover will be removed when

$$E_2 = \int_{t_1}^{t_2} Q_e dt = RE_1 \tag{2.5}$$

Since $R \approx 10$ (Walin, 1991) $t_2 - t_1 \gg t_1$ and a removal of the ice will normally not occur within one winter. If it does happen, deep overturning and the development of a polynya are expected by the model.

2.3.2. Convection and the Weddell Polynya. The convective explanation for the Weddell Polynya was first elaborated by Killworth (1979) and later by Martinson et al. (1981). The approach is based on one-dimensional mixed layer models with constant depth of the mixed layer and constant salinity and temperature in the lower layer. The fluxes between the mixed layer and the ice and between the mixed and the lower layer were described by the "Newtonian" flux laws $K(T - T_f)$, $K_T(T_2 - T_1)$ and $K_s(S_2 - S_1)$. In the ideal case $K_T = K_s = 0$ and K put equal to infinity, which implies no entrainment between the layers and immediate melting or freezing of the ice. The configurations and the transitions between different possible states are shown in Figure 17 and an ideal cyclic evolution of the T-S characteristics of the mixed layer is given in Figure 18.

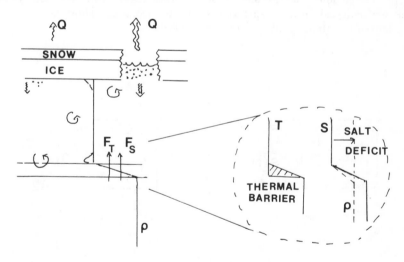

Figure 14. Deepening of the mixed layer by increased density and buoyancy generated turbulence. F_T and F_S are diffusive fluxes of heat and salt through the pycnocline. The blow-up shows the two features inhibiting overturning: The thermal barrier, and the salt deficit which must be removed to allow for deep convection (from Martinson, 1990).

340

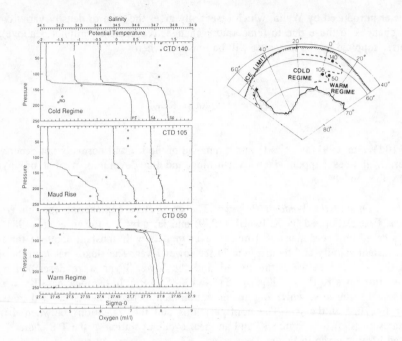

Figure 15. Temperature, salinity and density profiles taken in the Southern Ocean. The positions are indicated on the map together with the front separating the warm and cold regimes, broken line (from Gordon and Huber, 1990).

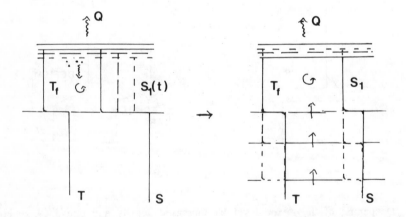

Figure 16. The freeze melting concept. a) Early winter, strong stability, freezing and salinity increase, no entrainment. b) High and late winter, weak stability, melting, entrainment and deepening of the mixed layer. The salinity of the mixed layer is constant (from Walin, 1991).

When the mixed layer reaches freezing temperature, the salinity increases and the T-S characteristics of the mixed layer move along the freezing point line until the density of the lower layer is reached. The column overturns and the lower layer gets in contact with the ice. Ice melts and an upper layer is reestablished. Depending upon the stability and on the temperature of the lower layer the overturning might lead to a partial melting of the ice or a complete removal of the ice cover. In the first case a cyclic deep convection might be established during the winter. In the second case the mixed layer will experience further cooling during the winter, but the density of the mixed layer is too low to permit deep convection by cooling alone. The temperature is likely to be high enough to prevent new ice from forming in the same winter, and not until next autumn will the upper layer again reach freezing temperature and the cycle repeats itself.

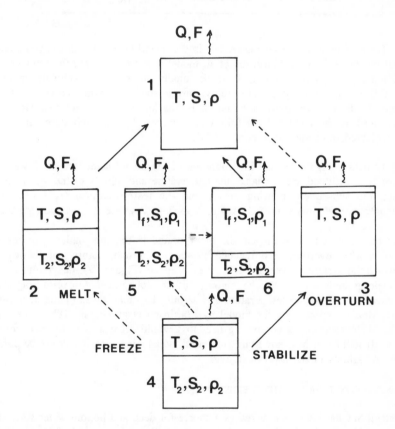

Figure 17. Possible states of the Weddell Sea water column. 1 to 4 represent the asymptotic case with constant mixed layer depth and no entrainment discussed by Martinson et al. (1981). 5 and 6 show the freezemelting for marginal stability (Walin, 1991) (Adapted from Martinson et al., 1981).

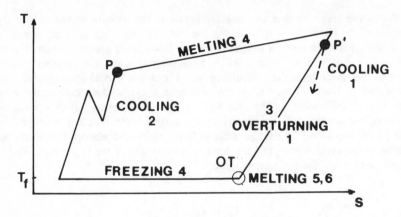

Figure 18. The evolution of T-S characteristics in the mixed layer through the year. When the mixed layer reaches OT in the Martinson et al. model it overturns along the sequence $4 \to 3 \to 4 \to 2$ and a polynya is formed at P. In the model proposed by Walin the mixed layer remains at OT, evolving through 5 and 6, to overturn to state 1 to form a polynya at P^1. In the second route the deep convection is not shut off by the forming of the polynya. However, the mixed layer will not be recreated the following summer. The referred states are shown in Figure 17 (Adapted from Martinson et al., 1981).

In the schematics (Figure 18) the incorporation of the mixed layer into the lower layer at the overturning is assumed not to change the temperature and salinity of the lower layer. The overturning, the subsequent melting of ice and the restratification are taken to occur instantaneously by the routes $4 \to 3 \to 4$ or $4 \to 3 \to 2$ where 2, 3, 4 indicate the states shown in Figure 17.

The concept proposed by Walin can be incorporated in the schematics (Figures 17 and 18). However, when overturning is reached by state 4 the temperature and salinity remain constant while the system moves through states 5 and 6 (Figure 17), deepening the mixed layer and melting the ice. When the ice is gone, further cooling leads to overturning in state 1 and a polynya is formed with temperature and salinity less than those of the deep layer but higher than what is expected by the model by Martinson et al. (Figure 18). Since all ice is removed before overturning a continuing heat loss would drive a deep thermal convection and no shut-off will occur as in the Martinson et al. model. However, it will not be possible to reform the low salinity mixed layer next year.

2.4. OPEN OCEAN CONVECTION II: THE GREENLAND SEA

The Greenland Sea has been recognized as a source for deep and bottom water since the turn of the century (Nansen, 1906). However, to date no observations but one of convection to the bottom have been reported (Bogorodsky et al., 1987). This has led to proposals of several different mechanisms as the causes for the renewal of the Greenland Sea deep and bottom waters. However, recent observations indicate that no renewal of the bottom water has occurred in the last 10 years, suggesting that every possible mechanism by which the deep water formation occurs only operates intermittently (Rhein, 1991).

2.4.1. *The oceanographic setting.* The circulation of the Greenland Sea is bounded to the east by the warm, saline northward flowing West Spitsbergen Current and to the west by the cold, low salinity East Greenland Current exiting through Fram Strait from the Arctic Ocean. The circulation is cyclonic and mainly wind driven. The dense deep waters are forced towards the surface and lighter waters such as the outflow from the Arctic Ocean are confined to the boundaries. This results in an upward doming of the isopycnals in the center of the Greenland Sea Gyre (Figure 19). The stratification in the gyre is weak and the θ-S characteristics of the waters display a hook-like shape (Figure 19). The fresh and cold signature of the bottom water is not found elsewhere in the northern seas and was one early indication that the Greenland sea was one, perhaps the most important, source for the deep and bottom waters north of the Greenland-Scotland Ridge (Nansen, 1906; Kiilerich, 1945).

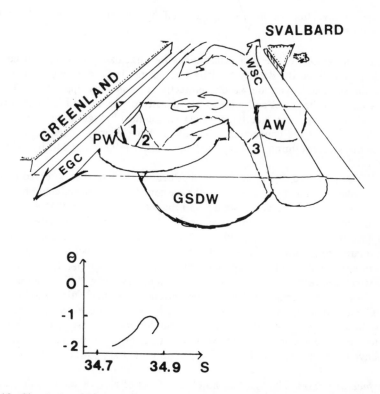

Figure 19. Circulation and watermasses in the Greenland Sea. East Greenland Current (EGC), West Spitsbergen Current (WSC), Polar Water (PW), Atlantic Water (AW), Greenland Sea Deep Water (GSDW), upper Polar Deep Water (1), Canadian Basin Deep Water (2), Norwegian Sea Deep Water (3). The T-S diagram indicates the waters of the central Greenland Sea Gyre (Adapted from Carmack, 1986).

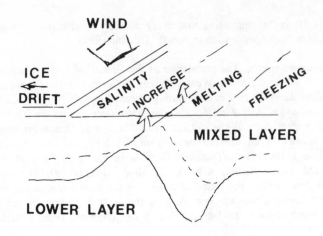

Figure 20. Deep convection at an ice edge caused by upwelling, increased mechanical entrainment and cooling.

2.4.2. *Preconditioning.* One of the first explanations for Greenland Sea deep convection was that the winds precondition the density field by forcing the dense deep water of the central gyre closer to the surface, leaving only a thin protective low stability upper layer. The strong heat loss in winter can then cool and increase the density of the surface layer to that of the slightly more saline and warmer underlying water. This is then brought to the surface and the cooling and deepening of the convection proceeds until the density of the bottom water is reached. Such massive overturnings, suggested by Nansen (1906) and Mosby (1959) have not been observed.

Killworth (1979) proposed that more localized structures, akin to that observed by Gordon (1978) in the Weddell Sea and those found in the western Mediterranean, could be generated by baroclinic instability. Saline water would be brought still closer to the sea surface and exchanges across the sloping density interface would add to the density increase.

Another approach, put forward by Häkkinen (1987) focuses on the ice edge. Wind makes the ice move relative to the water. If the wind is blowing almost parallel to the ice (Figure 20) the induced relative motion of the ice will cause upwelling, similar to an atmospheric mountain lee wave, seaward of the ice. In a two layer situation this will bring the lower, more saline water closer to the surface and it will be entrained, by mechanical mixing, into the upper layer. Combined with the density increase due to cooling this eventually leads, in the model, to instability and overturning. The salinity (and density) increase in the upper layer is mainly due to the salt entrained from below, not to freezing as in the mixed layer models considered above (section 2.3) (Martinson, 1990; Martinson et al., 1981).

2.4.3. *Subsurface processes.* The fact that no observations of surface water with deep water density and θ-S characteristics exist led Carmack and Aagaard (1973) and later McDougall (1983) to conclude that the formation of the Greenland Sea deep water essentially is a subsurface process.

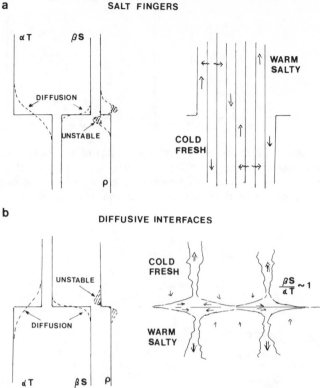

a SALT FINGERS

b DIFFUSIVE INTERFACES

Figure 21. Double-diffusive convection. a) Saltfingers; b) Diffusive interfaces. The profiles αT and βS indicate the density contributions due to heat and salt.

Sea water is a two component solute with one component, heat, diffusing much faster than the other component, salt. This implies that, if the distribution of one component is unstable, potential energy can be released and drive convective motions in spite of a stable net stratification.

Two such modes of convection are possible. If the salt distribution is unstable, "saltfingers" are formed. These transport mass, heat and salt vertically. They are highly viscous and are, in principle, not limited vertically by the stratification (Figure 21a). However, they do go unstable (Kunze, 1987). The opposite situation, when cold, fresher overlies warmer more saline water, leads to the formation of diffusive interfaces. No vertical mass transport takes place across the interfaces and the heat and salt transfer occur by a combination of molecular diffusion and laminar plumes or thermals. The convection is limited vertically by the stratification (Figure 21b). The ratio of heat to salt transfer is different for the two modes and different from that of mechanical mixing. For further details see Turner (1973) (chapter 8).

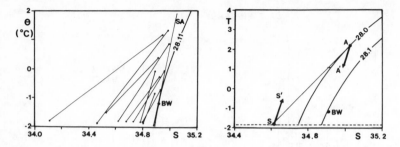

Figure 22. a) The relations between the T-S characteristics of the surface water and the Atlantic Water in the Greenland Sea. The difference is largest at the edge and smallest at the center of the gyre. b) Expected changes of temperature and salinity of the surface and the Atlantic Water due to diffusive interfaces (from Carmack and Aagaard, 1973).

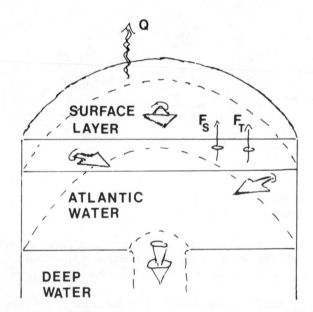

Figure 23. Subsurface transformation of Atlantic Water to Greenland Sea Deep Water by double diffusive convection.

Carmack and Aagaard (1973) and McDougall (1983) assumed that Atlantic Water was advected from the perimeter towards the center of the Greenland Sea Gyre while losing heat through diffusive interfaces to the cold surface layer (Figure 22a). The heat loss at the sea surface keeps the surface layer at freezing temperature and its salinity increases by the diffusive flux. The Atlantic water loses both heat and salt and observations indicate that the ratio of salinity step to temperature step remains constant along the assumed advection path from the perimeter towards the center (Figure 22a). The ratio of the heat flux to the salt flux then also remains constant and by using flux ratios observed in the laboratory (Turner, 1965) (Figure 22b), Carmack and Aagaard (1973) concluded that the Atlantic Water loses heat and salt at such a rate and in such a ratio that its θ-S characteristics approach that of the bottom water. When the Atlantic Water reaches the center of the gyre it has attained the bottom water temperature and salinity and convects into deep (Figure 23).

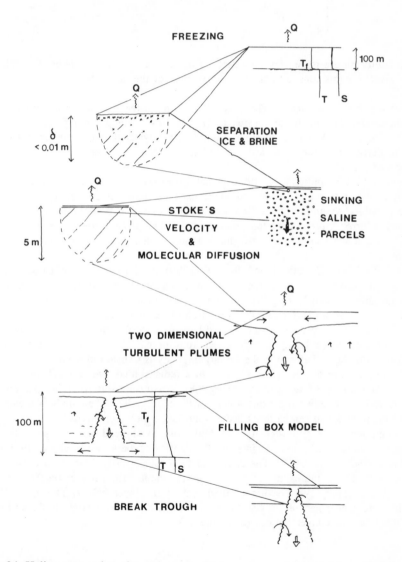

Figure 24. Haline convection pictured as a multistage process (Adapted from Rudels, 1990).

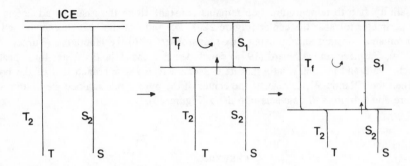

Figure 25. Reforming a mixed layer by melting of sea ice on warmer water. The salinity of the upper layer remains constant during the deepening of the mixed layer.

2.4.4. *Freezing at the sea surface.* The idea that freezing at the sea surface might trigger deep convection entered late in the discussion of Greenland Sea deep water formation. The salt needed to increase the density of the low salinity surface water to that of the deep waters was assumed to derive from the Atlantic layer and several possible processes for this exchange were proposed (see above).

Clarke et al. (1990) applied a one-dimensional mixed layer model with prescribed heat loss to observed temperature and salinity distributions to examine if convection to the bottom would be possible. Assuming gradual homogenization of the water column and allowing ice to melt and reform as the convection penetrated downwards they found that the Greenland Sea Gyre could convect to the bottom within a winter season with a net ice formation of only tens of cm.

Rudels (1990) and Quadfasel and Rudels (1990) tried to avoid the parameterization implied in convective adjustment (the instant merging of two layers once the stratification becomes unstable) by considering the freezing and salt ejection process in detail. The saline parcels, formed at the sea surface by the separation of ice and water, will sink with a velocity determined by a balance between viscosity and buoyancy (Stoke's resistance law). Because of their small dimensions ($<$ 1 cm) the salt is lost from the parcels by molecular diffusion. This implies that a time lag exists before the salinity anomaly is transferred from the parcels to the ambient water. During this time the salt becomes redistributed over a much larger volume because of the sinking of the parcels. Once the diffusion of salt from the parcels to the ambient water is accomplished a finite volume with a finite density anomaly is created. This volume will then sink and can be described as a turbulent plume or thermal.

The entrainment lowers the density and convection is stopped at the base of the upper layer. The convection will gradually increase the density of the upper layer by filling it from below (Baines and Turner, 1969). The density of the upper layer will finally become so large that the density of the plume in spite of the entrainment will, when it reaches the lower boundary of the upper layer, be higher than that of the layer below. The upper layer is emptied into the deeper part of the water column. Warmer water is brought to the surface, the freezing stops and the convection ceases (Figure 24).

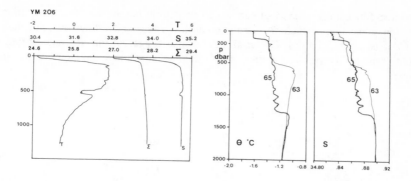

Figure 26. a) Intrusion of cold shelf water into the Arctic Ocean Water Column (from Rudels, 1986); b) Interleaving of convectively homogenized water with the ambient Greenland Sea water column (from Rudels et al., 1989).

Such a process could explain the sudden deepening of the convection and the persistence of the low salinity surface layer (Figure 26b), which is recreated after each convection event by the melting of the ice.

The mixed layer temperature is at the freezing point and the fresh water input is due to the melting of sea ice. The description of the mixed layer evolution would be simple if the ratio of the heat used for melting to the heat lost to the atmosphere were constant. This strong requirement, which is also implied in the work by Walin, leads to constant salinity of the mixed layer and a deepening proportional to √t (Figure 25). This assumes that at least in the initial phase, when the lower layer is close to the sea-surface, the entrainment, not the atmospheric forcing, dominates the oceanic heat loss (Rudels, in prep.).

When a sufficiently deep upper layer is reestablished, the entrained heat becomes so small that the atmospheric heat loss takes over. Ice will reform, and the convection cycle starts. The density of the intermediate layers has increased and the next convection event penetrates deeper. After several events the convection perhaps reaches the bottom.

Appearance and disappearance of the ice cover are frequently observed in the Greenland Sea. In fact, were it not for the warm Atlantic Water on the eastern side, the Greenland Sea gyre would appear as an opening and refreezing polynya throughout the winter.

The convection is driven by released salt. However, because the cold, low salinity upper water is entrained into the plume as it convects the θ-S characteristics of the deep (and bottom) water become colder and fresher. This is in spite of the initially high salinity at the surface. This effect is contrary to what would be expected if the salt source were the Atlantic Water as in the case of double-diffusive convection.

2.4.5. *Preconditioning revisited.* The convection increases the density of the waters in the central gyre and leads to a doming of the isopycnals. This allows for speculations about the importance of and the causes for the preconditioning of the Greenland Sea density field. The convection may, through the doming, generate a cyclonic circulation which reinforces that driven by the wind, and the convection thus preconditions itself.

Recent three-dimensional numerical convection modelling has shown structures, driven by the convection, consisting of cyclonic circulation at the surface and anticyclonic circulation in the deep (Marshall pers. comm.). Gill et al. (1979) also suggested that the deep water, as it spreads out from the convection center, may establish an anticyclonic circulation in the deeper layers.

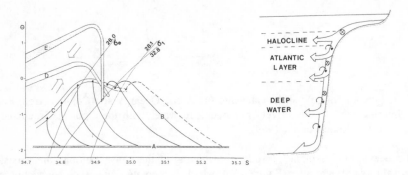

Figure 27. The merging of dense shelf water with the Canadian Basin water column. The θ-S diagram indicates the evolution of the θ-S characteristics, B, of the brine enriched shelf water, A, as it sinks down the continental slope as an entraining density flow. C connects the θ-S characteristics of different density flows at their terminal density level. D is the ambient water column and E is the advective component needed to maintain the θ-S characteristics of the basin (from Rudels and Quadfasel, 1991).

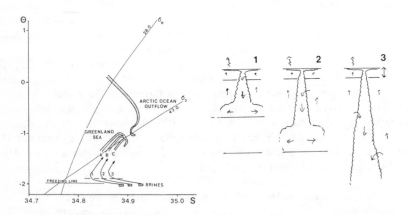

Figure 28. Evolution of the Greenland Sea water column, A, B, C, during successive convective events, 1, 2, 3. The θ-S characteristics of the convective contribution are indicated by the broken line. The balancing Arctic Ocean outflow is also shown (from Rudels and Quadfasel, 1991).

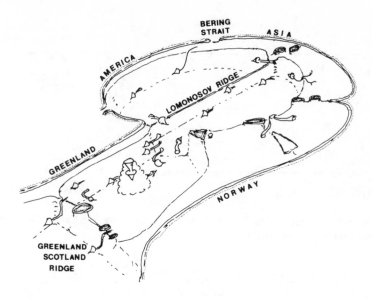

Figure 29. The deep circulation in the Arctic Ocean - Greenland Sea system. Convection areas are indicated.

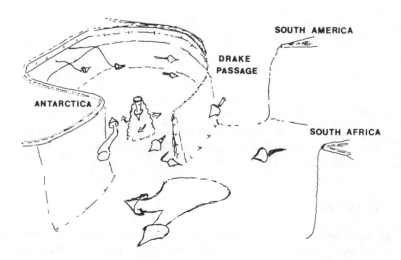

Figure 30. The deep circulation in the Weddell Sea convection areas are indicated.

ATLANTIC

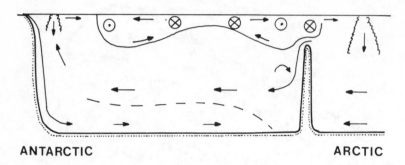

ANTARCTIC   ARCTIC

Figure 31. The meridional circulation in the Atlantic Ocean. North Atlantic Deep Water (NADW), Antarctic Bottom Water (AABW).

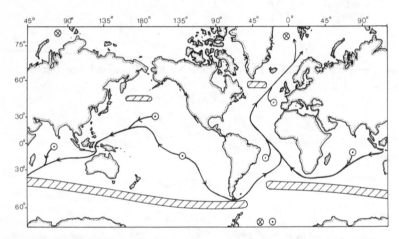

Figure 32. The buoyancy driven circulation of the world oceans, deep convection, ⊗, upwelling, ☉, convection in the troposphere (shaded). Only in the Greenland Sea and Arctic Ocean is there a surface connection between the troposphere and the stratosphere.

3. Merging With Ambient Water

Convection, driven by whatever process, will introduce waters with different θ-S characteristics into the water column. Injections of dense shelf water show up clearly (Figure 26a). The resulting large vertical gradients and inversions of temperature and salinity induce double diffusive processes, which act to remove the anomalies.

Open ocean convection changes the θ-S structure of the water column and leads to large horizontal gradients. These may be so strong that slight disturbances can result in vertical gradients capable of establishing double diffusive convection, which may drive horizontal interleaving and mixing of the waters (Figure 26b).

Haline convection is characterized initially by large density anomalies and small volumes. No well defined water mass is created off the surface as compared to the 1000 deep homogenous layers formed by thermal winter convection at lower latitudes. The "penetrative" nature of the convection also implies that the subsurface entrainment of lighter ambient water is important (dominant) in forming the "convective" water mass.

To maintain the θ-S characteristics of the basin an advective component must be present to balance the convective contribution. In the Canadian Basin, where the residence time of the waters ranges from 50 years in the upper layer to 500 years in the deepest layer the dense, sinking shelf water entrains and redistributes waters of the existing column. The shelf waters are injected at their density levels and form a distinct curve in the θ-S diagram. If only the shelf process supplied water to the deep Canadian Basin water column, it would gradually approach the characteristics of the shelf contributions. However, water with other characteristics crosses the Lomonosov Ridge and balances the convective water mass and acts to preserve the Canadian Basin deep water θ-S characteristics (Figure 27). The situation is similar in the Eurasian Basin.

However, in the smaller and more rapidly ventilated Greenland Sea the picture is different. The water column cannot be considered a constant environment throughout the winter. The convection changes the water column and the entrained water will be different at each convection event (Figure 28).

Nevertheless, here also is an advected component necessary to balance the convection and restore the long term characteristics of the water column. In the Greenland Sea the advective contribution is supplied by the deep outflow from the Arctic Ocean.

4. Global Abyssal Circulation

The deep circulation in the Arctic Ocean - Greenland Sea system is mainly confined to the area north of the Greenland-Scotland Ridge (Aagaard et al., 1985) (Figure 29). The shelf processes of the Arctic Ocean stratify the water column by forming lighter upper water from river run-off and ice melt and by creating saline and dense water from freezing on the shelves. The temperature of the sinking waters increases because of entrainment and the Arctic Ocean supplies heat and salt to the deep waters. By contrast the open ocean convection in the Greenland Sea removes the stratification and causes upward fluxes of heat and salt. The two regions act together to keep the system in balance (Rudels and Quadfasel, 1991).

Only less dense deep water spills over the Greenland-Scotland Ridge into the deep North Atlantic. This overflow, which also includes the waters formed by thermal convection in the Icelandic Sea, has a total strength of about 6×10 m^3 s^{-1} and is augmented by entrained water south of the ridge to a volume of 11×10 m^3 s^{-1} (Dickson et al., 1990). Together with a comparable amount formed by thermal convection south of the ridge, especially in the Labrador Sea, the outflow forms the North Atlantic Deep Water, which flows as a boundary current along the western side of the basins and ventilates most of the deep water of the world oceans (Warren, 1981).

The high density water formed in the south can easily penetrate into the different oceans basins (Figure 30). With the densest water of the Arctic remaining north of the Greenland-Scotland Ridge the waters formed at the rim of the Antarctic continent supply the bottom

354

waters of the world oceans (Figure 31). The total Antarctic contribution amounts to about 20 x 10 m^3 s^{-1}. However, in the south, waters from the cold stratosphere, mainly North Atlantic Deep Water, are transformed into the bottom water. The warm troposphere waters are ventilated north of the polar front and isolated from deep water formation.

The residence time for the waters of the stratosphere is about 1000 years. The interactions between the troposphere and the stratosphere occur in most parts of the ocean by a slow upwelling of deep water through the oceanic thermocline into the troposphere. Only in the Nordic Seas and the Arctic Ocean does a horizontal communication exist which permits warm troposphere water to penetrate into the cold stratosphere (Figure 32).

Apart from forming a window, which allows water from the upper layers to sink and renew the deep water, the North Atlantic also provides a conduit for warm water into high northern latitudes. This accounts for the warm and temperate climate of the European continent. In spite of its small extent and the slow rate of overturning the North Atlantic "window" is an essential link in the global climate system. The importance of this link and how stable it is to disturbances is far from understood at this time.

Acknowledgment. I wish to thank Professor Gösta Walin for his critique and advice while preparing this manuscript. Thanks are also due to M.-C. Beaupré, H. Friedrich, P. Jones, D. Quadfasel, N. Verch and H. Wohlert. The work was financed by the Deutsche Forschungsgemeinschaft (SFB 318).

5. References

Aagaard, K., Swift, J.H. and Carmack, E.C. (1985) Thermohaline circulation in the Arctic Mediterranean Seas. J. Geophys. Res. 90, 4833-4846.

Apel, J.R. (1987) Principles of Ocean Physics. Academic Press, New York, pp. 634.

Baines, W.D. and Turner, J.S. (1969) Turbulent buoyant convection from a source in a confined region. J. Fluid Mech. 37, 51-80.

Bogorodsky, R.V., Makshtas, A.P., Nagurny, A.P., Sovchenko, V.G. and Ivanov, R.V. (1987) Features of the ocean-atmosphere mesoscale interaction in the Greenland Sea area. Meteor. Gidsol. 10, 69-74 (in Russian).

Brennecke, W. (1921) Die Ozeanographischen Arbeiten der deutschen antarktischen Expedition 1911-1912. Arkt. Dsch. Seewarte 39, 1-214.

Carmack, E.C. (1986) Circulation and mixing in ice covered waters. In N. Untersteiner (ed), The Geophysics of Sea Ice, Plenum Press, New York, pp. 641-712.

Carmack, E.C. (1990) Large-scale physical oceanography of polar oceans. In W.O. Smith, Jr. (ed), Polar Oceanography, Part A, Physical Sciences, Academic Press, New York, pp. 171-222.

Carmack, E.C. and Aagaard, K. (1973) On the deep water of the Greenland Sea. Deep-Sea Res. 20, 687-715.

Carmack, E.C. and Foster, T.D. (1975) Circulation and distribution of oceanographic properties near the Filchner Iceshelf. Deep-Sea Res. 22, 711-724.

Clarke, R.A., Swift, J.H., Reid, J.L. and Koltermann, K.P. (1990) The formation of Greenland Sea Deep Water: double diffusion or deep convection? Deep-Sea Res. 37, 1385-1424.

Coachman, L.K. and Aagaard, K. (1974) Physical oceanography of Arctic and subarctic seas. In Y. Hermann (ed), Marine Geology and Oceanography of the Arctic Seas, Springer Verlag, Berlin, pp. 1-72.

Defant, A. (1961) Physical Oceanography. Volume I. Pergamon Press, New York, pp. 729.

Dickson, R.R., Gmitrowicz, E.M. and Watson, A.J. (1990) Deep-water renewal in the northern North Atlantic. Nature 344, 848-850.

Fairbridge, R.W. (1966) The Encyclopedia of Oceanography. Reinolds Publishing Corp., pp. 1021.

Fofonoff, N.P. (1956) Some properties of sea water influencing the formation of Antarctic bottom water. Deep-Sea Res. 4, 32-35.

Foldvik, A., Gammelsrød, T. and Tørresen, T. (1985a) Hydrographic observations from the Weddell Sea during the Norwegian Antarctic Research Expedition 1976/1977. Polar Res. 3, 177-193.

Foldvik, A., Kvinge, T.K. and Tørresen, T. (1985b) Bottom currents near the continental shelf break in the Weddell Sea. Antarct. Res. Ser. 43, 21-24.

Foster, T.D. and Carmack, E.C. (1976a) Temperature and salinity structure in the Weddell Sea. J. Phys. Oceanogr. 6, 36-44.

Foster, T.D. and Carmack, E.C. (1976b) Frontal zone mixing and Antarctic bottom water formation in the southern Weddell Sea. Deep-Sea Res. 23, 301-317.

Foster, T.D., Foldvik, A. and Middleton, J.H. (1987) Mixing and bottom water formation in the southern Weddell Sea. Deep-Sea Res. 39, 1771-1794.

Gill, A.E. (1973) Circulation and bottom water production in the Weddell Sea. Deep-Sea Res. 28, 111-140.

Gill, A.E., Smith, J.M., Cleaver, R.P., Hirche, R. and Jonas, P.R. (1979) The vortex created by mass transfer between layers of rotating fluid. Geophys. Astrophys. Fluid Dyn. 12, 195-220.

Gordon, A.L. (1978) Deep Antarctic convection West of Maud Rise. J. Phys. Oceanogr. 10, 600-612.

Gordon, A.L. and Huber, B.A. (1990) Southern Ocean mixed layer. J. Geophys. Res. 95, 11655-11673.

Gorshkov, S.G. (1983) World Ocean Atlas 3; Arctic Ocean. Pergamon, Oxford.

Gow, A.J. and Tucker III, W.B. (1990) Sea ice in the Polar Regions. In W.O. Smith, Jr. (ed), Polar Oceanography. Part A, Physical Science. Academic Press, New York, 47-122.

Häkkinen, S. (1987) A coupled dynamic-thermodynamic model of an ice-ocean system in the marginal ice zone. J. Geophys. Res. 92, 9469-9478.

Kiilerich, A.B. (1945) On the hydrography of the Greenland Sea. Medd. Grönl. 144, 1-63.

Killworth, P.D. (1977) Mixing on the Weddell Sea continental slope. Deep-Sea Res. 24, 427-448.

Killworth, P.D. (1979) On chimney formations in the oceans. J. Phys. Oceanogr. 9, 531-554.

Killworth, P.D. (1983) Deep convection in the world ocean. Rev. Geophys. Space Phys. 21, 1-26.

Kunze, E. (1987) Limits of growing, finite-length saltfingers: A Richardson number constraint. J. Marine Res. 45, 533-556.

Martin, S. (1974) Ice stalactites: comparison of a laminar flow theory with experiment. J. Fluid Mech. 63, 51-79.

Martinson, D.G. (1990) Evolution of the southern Ocean winter mixed layer and sea ice: Open ocean deep water formation and ventilation. J. Geophys. Res. 95, 11641-11654.

Martinson, D.G., Killworth, P.D. and Gordon, A.L. (1981) A convective model of the Weddell Polynya. J. Phys. Oceanogr. 11, 466-488.

McDougall, T.J. (1983) Greenland Sea bottom water formation: a balance between advection and double-diffusion. Deep-Sea Res. 30, 1109-1117.

Melling, H. and Lewis, E.L. (1982) Shelf drainage flows in the Beaufort Sea and their effect on the Arctic Ocean Pycnocline. Deep-Sea Res. 29, 967-985.

Midttun, L. (1985) Formation of dense bottom water in the Barents Sea. Deep-Sea Res. 32, 1233-1241.

Mosby, H. (1934) The Waters of the Atlantic Antarctic Ocean. Sci. Results Norw. Antarct. Exped. 1927-1928, 1, 1-131.

Mosby, H. (1959) Deep Water in the Norwegian Sea. Geophys. Publ. 21, 1-62.

Nansen, F. (1906) Northern Waters: Captain Roald Amundsen's oceanographic observations in the Arctic seas in 1901. Videnskabs Selskabets Skrifter I. Mathematisk-Natur Christiania. J. Dybwad. Klasse, 1-145.

Quadfasel, D., Rudels B. and Kurz, K. (1988) Outflow of dense water from a Svalbard fjord into the Fram Strait. Deep-Sea Res. 35, 1143-1150.

Quadfasel, D. and Rudels, B. (1990) Some new observational evidence for salt induced convection in the Greenland Sea. Inst. f. Meereskunde, Hamburg, Techn. Rep. 4-90, pp. 30, (unpublished manuscript).

Rhein, M. (1991) Ventilation rates of the Greenland and Norwegian Seas derived from distributions of the chlorofluormethanes F11 and F12. Deep-Sea Res. 38, 485-503.

Rudels, B. (1986) On the θ-S structure in the Northern Seas: Implications for the deep circulation. Polar Res. 4, 133-159.

Rudels, B. (1989) The formation of Polar Surface Water, the ice export and the exchanges through the Fram Strait. Progr. Oceanogr. 22, 205-248.

Rudels, B. (1990) Haline convection in the Greenland Sea. Deep-Sea Res. 37, 1491-1511.

Rudels, B. and Quadfasel, D. (1991) Convection and deep water formation in the Arctic Ocean - Greenland Sea System. J. Mar. Systems 2, 435-450.

Rudels, B., Quadfasel, D., Friedrich H. and Houssais, M.N. (1989) Greenland Sea convection in the winter of 1987-1988. J. Geophys. Res. 94, 3223-3227.

Smith, W.O., Jr. (ed) (1990) Polar Oceanography. Part A, Physical Science. Academic Press, New York, pp. 406.

Turner, J.S. (1965) The coupled transports of salt and heat across a sharp density interface. Int. J. of Heat and Mass Transfers 8, 759-767.

Turner, J.S. (1973) Buoyancy effects in Fluids. Cambridge Univ. Press. Cambridge, pp. 367.

Untersteiner, N. (ed) (1986) The Geophysics of Sea-Ice. Plenum Press, New York, pp. 1196.

Walin, G. (1991) On the formation of ice on deep weakly stratified water. Tellus (in press).

Warren, B.A. (1981) Deep circulation of the World ocean. In B.A. Warren and C. Wunsch (eds), The Evolution of Physical Oceanography. MIT press, Cambridge, Mass., pp. 6-41.

Weeks, W.F. and Ackley, S.F. (1986) The growth structure and properties of Sea Ice. In N. Untersteiner (ed), The Geophysics of Sea-Ice. Plenum Press, New York, 9-164.

Wells, N. (1986) The Atmosphere and Ocean, A physical introduction. Taylor and Francis, London, pp. 347.

ANALOGOUS MODES OF CONVECTION IN THE ATMOSPHERE AND OCEAN

S. A. CONDIE* and P. B. RHINES
School of Oceanography
WB- 10
University of Washington
Seattle, WA 98195
U.S.A.

ABSTRACT. Various modes of large scale convective overturning relevant to the atmosphere and ocean will be briefly reviewed, with emphasis on similarities and fundamental differences between the two systems. Analogous dynamics on both the mesoscale (convective vortices) and the planetary scale (Hadley cells) can be identified. A new convective mode, generated by strong interaction between overturning motions and topography, is also suggested as relevant to circulation on continental shelves.

1. Introduction

Convection is directly or indirectly responsible for all large scale motions in the atmosphere and ocean and is thus a principal determinant of the earth's climatic conditions. Both systems play a critical role, carrying approximately equal quantities of heat between equatorial and polar regions. In the atmosphere, cumulus convection near the equator and radiative cooling elsewhere, drive strong zonal flows such as the jet stream. These become unstable and form eddies which carry the meridional heat flux. When zonally averaged, this flow field takes the form of the atmospheric Hadley cell. Ocean convection on the other hand, is principally driven by cooling and brine formation at high latitudes. The resultant downwelling is concentrated within rotating vortices and flows down continental slopes. Most of the meridional heat transfer between high and low latitudes is carried by boundary currents flowing along meridional coastlines. The zonally averaged flow again takes the form of a planetary scale Hadley cell, although the overturning time-scale is much longer than that of the atmosphere.

The convective motions outlined above will be reviewed in more detail, particularly those larger scale modes which are strongly influenced by the earth's rotation. Section 2 begins by considering mesoscale convective structures, which take the form of small coherent vortices. These contribute to the global scale Hadley cell circulations which will be examined in Section 3. A new convection mode dependent on topography is analyzed in Section 4. Basic dynamical aspects will be emphasized throughout. More complex factors, such as the nonlinear interaction between the various modes, will not be considered here.

---

* Present address: Research School of Earth Sciences, Australian National University, G.P.O. Box 4, Canberra, A.C.T. 2600, Australia.

D. B. Stone and S. K. Runcorn (eds.),
Flow and Creep in the Solar System: Observations, Modeling and Theory, 357–370.
© 1993 *Kluwer Academic Publishers. Printed in the Netherlands.*

358

Figure 1. A photograph of the sea surface (courtesy of NASA). The cyclonic vortices (southern hemisphere) are of order 10 km across.

2. Localized Modes: Convective Vortices

In both the atmosphere and the ocean, strong convection tends to be concentrated within small localized structures. In the atmosphere these most often take the form of cumulus convection towers. Because these features are formed in hot equatorial regions and generally have short lifetimes, the earth's rotation has little direct effect on their dynamics. However, other features, referred to as mesoscale convective complexes (Maddox, 1980; Raymond and Jiang, 1990) have longer lifetimes and have been observed at mid-latitudes (Fritsch and Maddox, 1981; Menard and Fritsch, 1989). Analogous structures, often referred to as convective chimneys, appear to be ubiquitous in regions of the ocean forced by surface cooling or evaporation. Although they are often difficult to locate and identify in situ, the characteristic cyclonic sea surface convergence zones can easily be identified in photographs of the sea surface taken from space (Figure 1).

Isolated rotating convective structures are characterized by the horizontal convergence of hydrostatically unstable fluid, which then rises (atmosphere) or sinks (ocean) to a new level where it diverges again. In order to conserve angular momentum, cyclonic motion is generated in the convergence zone and anticyclonic motion in the divergence zone (Haynes and McIntyre, 1987). In the case of the atmosphere, rising thermals entrain surrounding air, then diverge at their level of neutral buoyancy. A similar process is also thought to operate in the oceanic chimneys.

If entrainment of surrounding fluid is sufficiently large, then convection does not extend beyond a well mixed boundary layer. In this case penetrative convection can lead to deepening of the boundary layer. Many one-dimensional models of this phenomenon have been proposed, following the early work of Kraus and Turner (1967). Rotational effects have also been included in both atmospheric boundary layer models (Therry and Lacarrere, 1983) and ocean mixed layer models (Garwood, 1977; Gaspar, 1988). However, the simple parameterizations used in such studies reveal nothing of the detailed processes involved. The dynamics may be further complicated by interactions with wind driven Langmuir cells (Weller and Price, 1988) which are known to align themselves with the wind direction.

In shallow or weakly stratified seas, chimney vortices extend to the sea floor, thus constituting a form of deep convection. Oceanographic observations of this phenomenon are very infrequent (Killworth, 1983). However, theoretical and laboratory studies of processes such as rotating Bénard convection (Nakagawa and Frenzen, 1955; Rossby, 1969; Boubnov and Golitsyn, 1986; Fernando, 1991), in which a fluid is heated from below and/or cooled from above, may also be relevant to this problem. While heating from below is inappropriate for modeling ocean convection, recent laboratory experiments reveal that the general features of time-dependent flow forced only by surface cooling are very similar those of rotating Bénard convection. Figure 2 shows an example of a surface streak photograph taken during the purely surface cooled experiments. The flow is dominated by intense cyclonic downwelling vortices, surrounded by less visible regions of anticyclonic upwelling.

Bénard convection in a fluid of depth H, rotating with angular velocity $f/2$ and driven by a heat flux Q, is usually described in terms of three independent non-dimensional parameters:

Figure 2. Surface flow in a 30 cm diameter cylinder of water cooled from above. The flow was visualized with fine aluminum powder. The water depth was 12 cm, the rotation rate $0.7s^{-1}$ and the temperature difference between the air and water 13°C.

$$Ra = \frac{BH^4}{\nu\kappa^2} \qquad \text{(flux Rayleigh number),} \qquad (1a)$$

$$Pr = \frac{\nu}{\kappa} \qquad \text{(Prandtl number),} \qquad (1b)$$

$$Ta = \frac{f^2 H^4}{\nu^2} \qquad \text{(Taylor number),} \qquad (1c)$$

where the buoyancy flux is

$$B = \frac{g\alpha Q}{\rho_o c_p}. \qquad (2)$$

Here ν is the kinematic viscosity, κ is the thermal diffusivity, g is the gravitational acceleration, α is the coefficient of thermal expansion, ρ_o is the mean density and c_p is the specific heat.

Linear stability theory (Nakagawa and Frenzen, 1955; Chandrasekhar, 1961) predicts that the distance between downwelling centers λ is given by $\lambda/H = 5.6Ta^{-1/6}$. This is confirmed by laboratory experiments for slightly supercritical flows. Theoretical progress on the nonlinear problem is more difficult because the flow is unsteady and non-hydrostatic. However, laboratory studies of larger Rayleigh number flows have yielded $\lambda/H = 7.2Ra^{1/12}Ta^{-1/4}$ for Ra $> 0.04Ta$ (Boubnov and Golitsyn, 1986) and $\lambda/H = 2.1\times10^2 Pr^{-1/2}Ta^{-1/4}$ (Chen et al., 1989). The dependency $\lambda \propto f^{1/2}$ therefore seems to be well established, while dependencies on other variables is less certain. The turbulent characteristics of the convective motions have been reported in more detail by Fernando et al. (1989) and Fernando et al. (1991). A number of two and three dimensional numerical models of these processes have also recently been developed and should provide an interesting comparison with the laboratory results.

Convective vortices in both the atmosphere and ocean may be regarded as the convective arm of much larger scale Hadley circulations. The zonally averaged component of these planetary scale phenomena will now be described in Section 3.

3. Global Modes: Hadley Cells

Equatorial cumulus convection, with a smaller contribution from the mesoscale convective complexes described above, drives a mean upwelling in the low latitude atmosphere. Radiative cooling over the remainder of the atmosphere causes air to sink again. These motions combine to form the planetary scale Hadley cell component of the mean circulation. The analogous oceanic motions consist of convective chimneys and dense flow down continental boundaries which provide downwelling in high latitude regions. The mid-latitude and equatorial oceans respond with slow broad-scale upwelling. The resulting zonally averaged flow forms the oceanic analog of the Hadley cell. In both these cases, a relatively small geographical region of active convection is balanced by broad regions of slower vertical motion.

361

3.1. THE ATMOSPHERIC HADLEY CELL

There have been a number of linear viscous theories proposed for the atmospheric Hadley circulation (Leovy, 1964; Schneider and Lindzen, 1976). However, a later study by Held and Hou (1980) considered a more realistic inviscid solution. A simplified version of this theory is outlined below. The steady, zonally averaged, meridional (θ) and vertical (z) momentum equations for hydrostatic Boussinesq flow on a spherical earth are,

$$fu + \frac{u^2 \tan\theta}{a} = -\frac{1}{\rho_o a}\frac{\partial p}{\partial \theta},$$ (3)

$$0 = -\frac{1}{\rho}\frac{\partial p}{\partial z} + g.$$ (4)

Here, u is the zonal velocity, a is the radius of the earth rotating at a rate Ω and $f = 2\Omega\sin\theta$ is the planetary vorticity (Figure 3). The remaining notation is standard. Because we are considering planetary scale motions, the nonlinear advection terms have been neglected, while a nonlinear term associated with the earth's curvature has been retained.

In the inviscid limit, the upper boundary condition on the zonal velocity is set by conservation of angular momentum of fluid particles moving poleward from the equator. The angular momentum is the sum of planetary and local contributions,

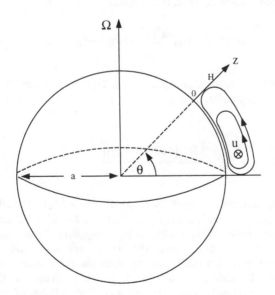

Figure 3. Schematic view of the atmospheric Hadley circulation.

$$M = \Omega a^2 \cos^2\theta + ua \cos\theta, \tag{5}$$

so that at the top of the Hadley cell ($z = H$) fluid diverges from the equator with $M = \Omega a^2$. Conservation of M then implies,

$$u(\theta,H) = \Omega a \sin\theta \tan\theta. \tag{6}$$

The only restriction at ground level is that

$$u(\theta,0) \ll u(\theta,H). \tag{7}$$

Eliminating the pressure term using (3) and (4), then integrating over the depth of the cell yields

$$f\{u(\theta,H) - u(\theta,0)\} + \frac{\tan\theta}{a}\{u^2(\theta,H) - u^2(\theta,0)\} = -\frac{gH}{a}\frac{\partial\sigma}{\partial\theta}, \tag{8}$$

where

$$\sigma(\theta) = \frac{1}{\rho_o H}\int_0^H \rho\,dz \tag{9}$$

is the non-dimensional depth averaged potential density distribution. Utilizing the boundary conditions (6) and (7), then integrating from 0 to θ gives

$$\sigma(\theta) = \sigma(0)\frac{a^2\Omega^2}{2gH}\sin^2\theta \tan^2\theta. \tag{10}$$

If the Hadley cell extends to a latitude denoted by θ_H, then its size can be calculated from (10) if $\sigma(\theta_H)$ and $\sigma(0)$ are known. For example, if the small angle limit is applicable then

$$\theta_H \approx \frac{\left(2gH[\sigma(0) - \sigma(\theta_H)]\right)^{1/4}}{(a\Omega)^{1/2}}. \tag{11}$$

By specifying a density distribution $\sigma_R(\theta)$ corresponding to a purely radiative equilibrium, $\sigma(\theta_H)$ and $\sigma(0)$ can be calculated explicitly. In particular, if a radiative balance holds outside the Hadley cell, then continuity of σ implies that $\sigma(\theta_H) = \sigma_R(\theta_H)$. Conservation of potential density further implies that σ integrated over the Hadley cell must equal the integrated radiative distribution. This gives an additional equation for $\sigma(0)$, which allows θ_H to be calculated. For realistic $\sigma_R(\theta)$, Held and Hou (1980) predicted a Hadley Cell extending to a latitude around 24°. Numerical solutions indicated that the addition of friction tends to increase the size of the cell by balancing part of the planetary vorticity gradient and thus breaking the angular momentum constraint.

3.2. OCEANIC HADLEY CELLS

It is more difficult to formulate simple models which describe the zonally averaged oceanic Hadley cell. In particular, inviscid formulations are inappropriate because of the very long convective overturning timescale. Most progress has therefore been made with the use of numerical simulations (Bryan, 1987; Colin de Verdière, 1988; Cox, 1989; Suginohara and Aoki, 1991) and laboratory experiments (Rossby, 1965; Speer and Whitehead, 1988; Condie and Griffiths, 1989). With meridional boundaries, most of the heat transport is concentrated in boundary currents and the resulting convection cell generally fills an entire basin. The meridional heat transfer is then independent of zonal motions (i.e. independent of f). This has led to interest in two dimensional models of global oceanic Hadley cells (Stocker and Wright, 1991) which may be applicable to long term climate problems.

There are a number of essential characteristics common to most models of oceanic convection which include a planetary vorticity gradient. Convective downwelling is always concentrated in small regions at high latitudes and balanced by slow upwelling over the rest of the ocean. At abyssal depths, the vortex stretching associated with upwelling is off-set by horizontal flow over the planetary vorticity gradient (Stommel and Arons, 1960). Local topography may alter this equilibrium (Straub and Rhines, 1990), but should contribute little to the zonally averaged Hadley cell. Nearer the surface, upwelling is thermodynamically balanced by downward diffusion, which maintains the ocean thermocline. The Hadley cell itself is closed by upper ocean advection toward the convective downwelling zones.

The general features of these models are consistent with what is known about the convectively driven component of the large scale ocean circulation. In particular, deep convection into the abyssal ocean has been identified at only a few high latitude locations. Surface cooling of high salinity water in the Norwegian-Greenland Sea (Rudels, 1990) results in dense flow through the Denmark Strait into the deep ocean, while intense cooling and brine formation due to ice formation produces very dense water at a number of locations around the antarctic coast (Killworth, 1983). Limited observations (Warren, 1981) suggest that the abyssal circulation is generally weak outside the deep western boundary currents as predicted by the Stommel and Arons (1960) formulation. The predicted broad scale upwelling is too slow to measure, however the existence of the thermocline in the upper 1 km of the ocean provides strong evidence for a balance between upwelling and downward diffusion.

There are two fundamental differences between the atmospheric and oceanic Hadley cells. Firstly, meridional boundaries allow the oceanic cell to fill an entire basin via boundary current transport. Whereas the atmospheric cell must rely on eddies for meridional transport and is thus restricted to lower latitudes. The second difference lies in the efficiency of the vertical heat transfer processes. In the atmosphere cumulus convection near the equator is balanced by radiative cooling throughout the atmosphere. There is no counterpart to radiative cooling in the ocean. High latitude convection can only be balanced by the far less efficient mechanism of vertical turbulent diffusion. This results in the large asymmetry between the size of the upwelling and downwelling zones and also accounts for the very long overturning time of the oceanic Hadley cell.

4. Intermediate Modes: Topographic Cells

It is well known that the effects of sloping topography on a rotating fluid are analogous to those of the planetary vorticity gradient, particularly where stratification is weak. Laboratory experiments suggest that convection cells dynamically similar to the atmosphere Hadley cell can be produced by surface cooling over topography. Figure 4 shows a streak photograph of flow in a hemispherical bowl forced by surface cooling. While chimney vortices similar to those in Figure 1 are clearly evident, the prominent zonal jets are indicative of overturning cells. This is illustrated schematically in Figure 5 for topography which might be representative of a continental shelf. The surface jets flow along isobaths with the shallow region on their left (in the northern hemisphere). This is the opposite direction to both Kelvin waves and topographic Rossby waves and thus can be easily distinguished from these phenomena. The topographic mode should be most prominent in weakly stratified seas forced by surface cooling, brine formation or evaporation. While surface winds could modify or destroy their structure, cells should prevail during low winds and in ice covered regions.

An inviscid linear theory of a topographic convection cell, in many respects analogous to the Hadley cell theory in Section 3, will now be presented. Neglecting the earth's curvature and assuming that along isobath variations are small, the inviscid cross isobath (r) momentum balance is geostrophic,

$$f u = - \frac{1}{\rho_o} \frac{\partial p}{\partial r}, \tag{12}$$

where f is now taken to be constant. Within the cell, the hydrostatic balance (4) applies in the vertical. The angular momentum of the surface flow $M = f r^2 / 2 + r u$, is assumed to be conserved in the absence of wind stress. If the convection cell extends across isobaths from $r_o - \Delta r$ to r_o, then angular momentum conservation implies

$$u(r,0) = \frac{f r_o}{2} \left(\frac{r_o}{r} - \frac{r}{r_o} \right), \tag{13}$$

while bottom friction should ensure that

$$u(r,-H) \ll u(r,0). \tag{14}$$

Eliminating the pressure term using (4) and (12), then integrating over the depth of the cell yields an equation analogous to (8),

$$f \{ u(r,0) - u(r,-H) \} = - g H \frac{\partial \sigma}{\partial r}. \tag{15}$$

The non-dimensional depth averaged density distribution σ is again defined by (9), except that the integration is now from the seafloor z=-H(r), to the sea surface z=0. Also the reference density ρ_o is defined as the mean density of the overturning cell, such that

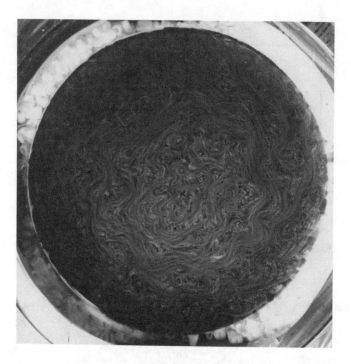

Figure 4. Surface flow in a 30 cm radius hemispherical bowl of water cooled from above. The flow was visualized with fine aluminum powder. The maximum water depth was 11 cm, the rotation rate $0.7s^{-1}$ and the temperature difference between the air and water 12°C.

$$\int_{r_o - \Delta r}^{r_o} (\sigma - 1)\, dr = 0. \tag{16}$$

Substituting boundary conditions (13) and (14) into (15), then integrating from r to r_o yields an expression for the depth averaged density distribution,

$$\sigma(r) = \frac{1}{H(r)} \left(\sigma(r_o) H(r_o) - \frac{f^2 r_o^2}{4g} \left(2 \ln\left(\frac{r}{r_o}\right) - \frac{r^2}{r_o^2} + 1 \right) \right). \tag{17}$$

To progress further, we must now prescribe the form of the topography. A simple analytically tractable choice is

$$H(r) = kr^2. \tag{18}$$

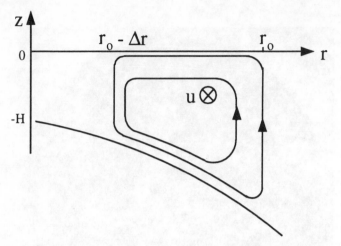

Figure 5. Schematic view of a topographic convection cell for continental shelf like topography.

where k is a constant. This is more general than it might first appear, since (18) need only approximate the topography over the width of the convection cell. Substituting (17) and (18) into (16) yields an expression for the density on the off-shore side of the cell,

$$\sigma(r_o) = \gamma + \frac{\chi}{4}\left(\frac{2\ln\gamma + \gamma^2 - 4\gamma + 3}{(1 - \gamma)}\right), \tag{19}$$

where $\gamma = 1 - \Delta r/r_o$ and $\chi = f^2/gk$. It then follows from (17), (18) and (19), that the density distribution within the cell is

$$\sigma(r) = \gamma\left(\frac{r_o}{r}\right)^2 + \frac{\chi}{4}\left(\left(\frac{r_o}{r}\right)^2\left(\frac{2\gamma\ln(r/r_o) - 2\ln(r/\gamma r_o)}{1 - \gamma} + (2 - \gamma)\right) + 1\right). \tag{20}$$

For realistic oceanic density gradients, it turns out that the terms in (19) and (20) proportional to χ can generally be neglected. In this case we are left with the very simple distribution,

$$\sigma(r) = \frac{r_o(r_o - \Delta r)}{r^2}. \tag{21}$$

It follows directly from (18) that

$$r_o = \frac{2H(r_o)}{\partial H / \partial r(r_o)}. \tag{22}$$

So evaluating (21) at $r=r_0$ and then using (22) to eliminate r_0, yields an expression for the cell aspect ratio,

$$\frac{\Delta r}{H} = \frac{\Delta \sigma}{\partial H / \partial r}. \tag{23}$$

where $\Delta \sigma = 2(1 - \sigma(r_0))$ is the change in σ across the cell. Strictly speaking, H and $\partial H / \partial r$ in (23) should be calculated at r_0. However, for realistic oceanic parameters, these quantities will probably vary little over the cell width.

Equation (23) indicates that the cell width increases with the depth and density contrast across the cell, whereas the bottom slope acts in an analogous way to the planetary vorticity gradient by restricting the width of the cell. On a continental shelf with typical parameters of $\partial H / \partial r \sim 10^{-4}$ and $\Delta \sigma \sim 10^{-3}$, we obtain a cell aspect ratio of $\Delta r / H \sim 10$. Friction in the system would tend to relax the angular momentum constraint, thereby increasing the width of the cell. This should certainly be true of the laboratory experiments, which are quite viscous compared to oceanic flows. For the example shown in Figure 4, $H \approx 9$cm, $\partial H / \partial r \approx 0.3$ and $\Delta \sigma \approx 10^{-3}$, giving an inviscid width of 0.03 cm. Since the measured width is at least an order of magnitude greater, we might conclude that friction is a first order effect in the laboratory model. On the other hand, the cell width does tend to decrease toward the outer edge of the bowl, consistent with the slope dependence in relation (23).

Although the size of an inviscid convection cell can now easily be determined from (23), its location cannot be predicted. This did not present any difficulties in the atmospheric theory in Section 3, since the Hadley cell could reasonably be assumed to start near the equator. Although there is no corresponding reference location for ocean convection, the additional degree of freedom allows the interesting possibility of multiple cells across the shelf with their axes aligned along isobaths. Qualitative support is again supplied by the experiment in Figure 4 which reveals a number of interacting concentric cells.

5. Summary

The basic convection modes of the atmosphere-ocean system are summarized schematically in Figure 6. The atmospheric Hadley cell has two major components: near equatorial cumulus convection and broad scale sinking due to radiative cooling. The oceanic Hadley cell is somewhat more complex. Convection at high latitudes is a combination of dense flow down continental slopes and open ocean chimney features. The cell is closed by slow upwelling over the remainder of the ocean. Smaller scale topographic convection modes have also been included in Figure 5, as a potentially significant new mode of ocean convection.

It should be remembered that the scenario outlined in Figure 6 is extremely simplistic. Geophysical convection is invariably unsteady and highly nonlinear. The basic convective modes discussed here also interact non-linearly with each other. For example, atmospheric winds drive much of the upper ocean circulation, which in turn interacts with the convectively driven component of the ocean circulation. Understanding such interactions is one of the principal thrusts of contemporary research in atmosphere-ocean dynamics.

Acknowledgments. Our thanks to Dr. D. Ohlsen for helpful comments on the manuscript.

368

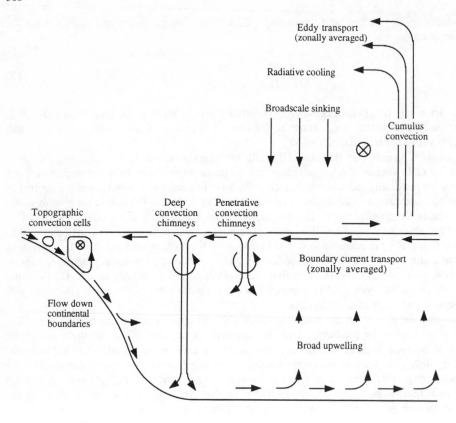

Figure 6. Schematic illustration of various modes of convection in the atmosphere and ocean.

6. References

Boubnov, B.M. and G.S. Golitsyn (1986) Experimental study of convective structures in rotating fluids. J. Fluid Mech. 167, 503-531.

Bryan, K. (1987) Parameter sensitivity of primitive equation general circulation models. J. Phys. Oceanogr. 14, 970-985.

Chandrasekhar, S. (1961) Hydrodynamic and Hydromagnetic Stability, Oxford University Press, London, 652 pp.

Chen, R., H.J.S. Fernando and D.L. Boyer (1989) Formation of isolated vortices in a rotating convecting fluid. J. Geophys. Res. 94, 18445-18453.

Colin de Verdière, A. (1988) Buoyancy driven planetary flows. J. Mar. Res. 46, 215-265.

Condie, S.A. and R.W. Griffiths (1989) Convection in a rotating cavity: modelling ocean circulation. J. Fluid Mech. 207, 453-474.

Cox, M.D. (1987) An idealized model of the world ocean. Part 1: The global-scale water masses. J. Phys. Oceanogr. 19, 1730-1752.

Fernando, H.J.S., D.L. Boyer and R. Chen (1989) Turbulent thermal convection in rotating stratified fluids. Dyn. Atmos. Ocean. 13, 95-121.

Fernando, H.J.S., R. Chen and D.L. Boyer (1991) Effects of rotation on convective turbulence. J. Fluid Mech. 228, 513-547.

Fritsch, J.M. and R.A. Maddox (1981) Convectively driven mesoscale weather systems aloft. Part 1: Observations. J. Appl. Meteor. 20, 9-19.

Garwood, R.W. (1977) An oceanic mixed layer model capable of simulating cyclic states. J. Phys. Oceanogr. 7, 455-468.

Gaspar, P. (1988) Modeling the seasonal cycle of the upper ocean. J. Phys. Oceanogr. 18, 161-180.

Griffiths, R.W. (1987) Effects of earth's rotation on convection in magma chambers. Earth Planet. Sci. 85, 525-536.

Haynes, P.H. and M.E. McIntyre (1987) On the evolution of vorticity and potential vorticity in the presence of diabatic heating and frictional or other forces. J. Atmos. Sci. 44, 828-841.

Held, I.M. and A.Y. Hou (1980) Nonlinear axially symmetric circulations in a nearly inviscid atmosphere. J. Atmos. Sci. 37, 515-533.

Killworth, P.D. (1983) Deep convection in the world ocean. Rev. Geophys. Space Phys. 21, 1-26.

Kraus, E.B. and J.S. Turner (1967) A one-dimensional model of the seasonal thermocline. II. The general theory and its consequences. Tellus 19, 98-105.

Leovy, C. (1964) Simple models of thermally driven mesospheric circulation. J. Atmos. Sci. 26, 841-853.

Maddox, R.A. (1980) Mesoscale convective complexes. Bull. Amer. Meteor. Soc. 61, 1374-1387.

Menard, R.D. and J.M. Fritsch (1989) A mesoscale convective complex-generated inertially stable warm core vortex. Mon. Wea. Rev. 117, 1237-1261.

Nakagawa, Y. and P. Frenzen (1955) A theoretical and experimental study of cellular convection in rotating fluids. Tellus 7, 1-21.

Raymond, D.J. and H. Jiang (1990) A theory for long-lived mesoscale convective systems. J. Atmos. Sci. 47, 3067-3077.

Rossby, H.T. (1969) A study of Bénard convection with and without rotation. J. Fluid Mech. 36, 309-335.

Rudels, B. (1990) Haline convection in the Greenland Sea. Deep-Sea Res. 37, 1491-1511.

Schneider, E.K. and R.S. Lindzen (1976) The influence of stable stratification on the thermally driven tropical boundary layer. J. Atmos. Sci. 33, 1301-1307.

Speer, K.G. and J.A. Whitehead (1988) A gyre in a non-uniformly heated rotating fluid. Deep-Sea Res. 35, 1069-1077.

Stigebrandt, A. (1985) A model for the seasonal pycnocline in rotating systems with application to the Baltic proper. J. Phys. Oceanogr. 15, 1392-1404.

Stocker, T.F. and D.G. Wright (1991) Rapid transitions of the oceans deep circulation induced by changes in surface water fluxes. Nature 351, 729-732.

Stommel, H. and A.B. Arons (1960) On the abyssal circulation of the world ocean II. An idealized model of the circulation pattern and amplitude in oceanic basins. Deep Sea Res. 6, 217-233.

Straub, D.N. and P.B. Rhines (1990) Effects of large-scale topography on abyssal circulation. J. Mar. Res. 48, 223-253.

Suginohara, N. and S. Aoki (1991) Buoyancy-driven circulation as horizontal convection on β-plane. J. Mar. Sci. 49, 295-320.

Therry, G. and P. Lacarrere (1983) Improving the eddy kinetic energy model for planetary boundary layer description. Bound.-Layer Meteor. 25, 63-88.

Warren, B.A. (1981) Deep circulation in the world ocean. In Warren, B.A. and C. Wunsch (eds.) Evolution of Physical Oceanography, MIT Press, Cambridge MA, pp 6-41.

Weller, R.A. and J.F. Price (1988) Langmuir circulations within the ocean mixed layer. Deep-Sea Res. 35, 711-747.

THE DYNAMICS OF SUBCRITICAL DOUBLE-DIFFUSIVE CONVECTION IN THE SOUTHERN OCEAN: AN APPLICATION TO POLYNYAS

J. SCHMALZL and U. HANSEN
Institute of Earth Science
Department of Theoretical Geophysics
Budapestlaan 4
3508 TA Utrecht
The Netherlands

ABSTRACT. We investigated the nature of subcritical double diffusive convection in the southern ocean with a time dependent two-dimensional finite-element method based on stream-function, compositional and temperature fields. The initial and boundary conditions are chosen with special respect to open-water polynyas which play an important role in the heat budget and in the gas exchange of the antarctic ocean and the polar atmosphere. The temperature and salinity profiles from the *Mikhail Somov* measurements in the Weddell gyre (1981) have been used in our calculations. They showed cold fresh water overlaying a warm layer with a higher salt concentration. Input of freshwater has been neglected in our model. We have investigated the evolution of overturning convection from an initially layered state and its dependence on the thermal and chemical Rayleigh number. Our results indicate that the initially layered period is important for the transport of heat and salt from the lower to the upper, cold and fresh layer. Even under conditions which are stable in the static sense, we observed overturning convection with a high heat transport rate.

1. Introduction

Since large ice-free areas called "polynyas" have first been observed in the vicinity of the Weddell Sea, interest has focused on the dynamical mechanism which is capable of maintaining a sufficient heat flux from the deep water to the surface. The occurrence of polynyas is unpredictable. It seems that, once a polynya has evolved, it is maintained over years, thus leading to a strong increase in the gas exchange between the southern ocean and the polar atmosphere (Comiso and Gordon, 1987). In the antarctic summer of 1981 a joint U.S.S.R. - U.S.A. expedition was carried out in the southern ocean on the soviet ship "Mikhail Somov." The hydrographic dataset resolves the thermohaline stratification during the transition phase from waxing to waning of the southern ocean sea-ice cover. These measurements showed an approx. 100 m deep cold and fresh water layer overlaying the warmer and salty "Weddell Deep Water" (W.D.W.). The combined effect of heat and salt on the density results in a stable stratification. Deep reaching convection, as has been proposed to be the mechanism responsible for polynyas (Martinson et al., 1981), can occur during the antarctic winter when the stratification becomes unstable due to the cooling of the upper layer by the polar atmosphere. A sufficiently unstable layering can cause the warm and salty water to rise from the deep to the surface, thus transporting enough heat to prevent ice-formation at the surface. What are the conditions that allow for deep reaching "overturning" convective

371

D. B. Stone and S. K. Runcorn (eds.),
Flow and Creep in the Solar System: Observations, Modeling and Theory, 371–383.
© 1993 *Kluwer Academic Publishers. Printed in the Netherlands.*

motions? Most work devoted to the polynya phenomenon has considered only thermal convection as a possible dynamical mechanism. However, during the last years it became clear that fluids whose density is influenced by two components with different molecular diffusivities can evolve very differently. Therefore the transport properties of such *"double-diffusive convection"* (D.D.C. in the following) need to be investigated. For thermal convection the onset of instabilities can be calculated from linear theory and expressed in terms of a non-dimensional stability parameter, the Rayleigh number **Ra** (e.g. Chandrasekhar, 1961). Above a critical threshold value Ra_c which depends on the boundary conditions and on the mode of the initial perturbation but is always on the order of 10^3, motion takes place. In the most general case of two components affecting the density, the effects can generally be linearly superimposed as long as both ingredients (i.e. temperature and salt in this case) are contributing to the density in the same way (i.e. both are stabilizing or both are destabilizing). If temperature and salinity are oppositely stratified, however, nonlinear effects can occur which change drastically the stability properties of the system (Huppert and Moore, 1976; Hansen and Yuen, 1989).

Two scenarios can be envisaged:

a) the "finger-regime"
If a cold and fresh water layer underlies warm and salty surface water, resembling the situation that occurs in tropical latitudes due to evaporation processes, then *"finger-like"* instabilities are observed even if the overall density stratification is stable (Huppert and Turner, 1981). From this type of instability the name *"finger-regime"* was chosen (Stern, 1960).

b) the "diffusive-regime"
The opposite situation, i.e. a heat distribution, acting as the driving force, in a stably stratified salt field can be observed in polar regions. It is this regime which has relevance for the the polynya phenomenon. Due to different diffusivities of heat and salt (heat diffuses about 100 times faster than salt) the potential energy which is stored in the unstably stratified field can be converted into driving force, despite the presence of a stabilizing component. In other words: the onset of motion can take place, even in a statically stably stratified medium (Turner, 1973; Hansen and Yuen, 1989).

Models of the polynya phenomenon which have been developed so far considered an unstable stratification of the seawater as a necessary condition for the occurrence of overturning convection (Martinson et al., 1981; Martinson, 1990). Overturning convection can lead to a strong increase of heat transport from the depth to the surface, thus being a potential candidate for the dynamical mechanism which maintains polynyas.

The main purpose of this research was to investigate the stability of two seawater layers against double-diffusive instabilities. In particular we wanted to explore the parameter space in which an overturn of the two layers is possible for initially weakly unstable or even stable density stratifications. We have developed a time-dependent two-dimensional finite element model that is explained in section 2. The heat and salt transport mechanism and the parameter conditions under which overturning convection with a high heat transport rate can occur are explained in section 3. We finish in section 4 with a discussion of the results and the possible implications for the dynamics in the southern ocean.

2. Formulation of Method and Solution

In two dimensions double-diffusive convection for a Boussinesq fluid with constant properties is described by the following partial differential equations:

$$\frac{1}{Pr}\frac{\partial}{\partial t}\nabla^2\psi + \frac{1}{Pr}\left(\frac{\partial\psi}{\partial z}\frac{\partial^3\psi}{\partial x\partial z^2} - \frac{\partial\psi}{\partial x}\frac{\partial^3\psi}{\partial z^3} + \frac{\partial\psi}{\partial z}\frac{\partial^3\psi}{\partial x^3} - \frac{\partial\psi}{\partial x}\frac{\partial^3\psi}{\partial x^2\partial z}\right) =$$

$$\frac{\partial^4\psi}{\partial x^4} + 2\frac{\partial^4\psi}{\partial x^2\partial z^2} + \frac{\partial^4\psi}{\partial z^4} - Ra_T\frac{\partial T}{\partial x} + Ra_S\frac{\partial C}{\partial x} \tag{2.1}$$

(the equation of momentum transport)

$$\frac{\partial T}{\partial t} - \frac{\partial\psi}{\partial x}\frac{\partial T}{\partial z} + \frac{\partial\psi}{\partial z}\frac{\partial T}{\partial x} = \nabla^2 T \tag{2.2}$$

(the equation of heat transport)

$$\frac{\partial C}{\partial t} - \frac{\partial\psi}{\partial z}\frac{\partial C}{\partial x} + \frac{\partial\psi}{\partial x}\frac{\partial C}{\partial z} = \frac{1}{Le}\nabla^2 C \tag{2.3}$$

(the equation of salt transport)

The equations have been scaled according to Turcotte et al. (1973). Ra_T and Ra_S are respectively the thermal and compositional Rayleigh-numbers and Le is the Lewis-number which is the ratio between the thermal κ_T and the haline diffusivity κ_S. Pr is the ratio between the kinematic viscosity v and the thermal diffusivity κ_T called Prandtl-number which is 7 for water. The buoyancy ratio R_ρ is given by $Ra_S/Ra_T = \beta\cdot\Delta C / \alpha\cdot\Delta T$. The quantities ψ, T and C are the streamfunction, temperature and concentration fields. The dimensionless time is given by t, non-dimensionalized by the thermal diffusion time across the layer depth d. The horizontal and vertical coordinates are denoted by x and z, with gravity g aligned opposite to z. The assumption of a linear equation of state (Fig. 2.1) is in this case justified by the small variations in the temperature and salinity as they occur in the southern ocean.

The linearized equation of state for the density is thus given by:

$$\rho = \rho_0\left[1 - \alpha(T - T_o) + \beta(C - C_o)\right] \tag{2.4}$$

where the coefficients of thermal expansion and its analogous compositional counterpart are given respectively by α and β. Reference values are denoted by the subscript o.

In characterizing D.D.C. flow with specified temperature difference over the layerdepth d and salinity, we defined the two Rayleigh-numbers by

$$Ra_T = \frac{\alpha g \Delta T d^3}{\kappa_T v}$$

and

$$Ra_s = \frac{\beta g \Delta C d^3}{\kappa_T v}.$$

We adopt the following boundary conditions:

$$\psi = \frac{\partial^2 \psi}{\partial z^2} = 0 \text{ on } z = 0,1;$$

$$\psi = \frac{\partial^2 \psi}{\partial x^2} = \frac{\partial T}{\partial x} = \frac{\partial C}{\partial x} = 0 \text{ on } x = 0, \lambda;$$

thus resembling stress-free conditions all around the box. (λ corresponds to the aspect-ratio of the box: $\lambda = x/z$) We fixed T to unity at $z=0$ (bottom) and to zero at $z=1$. For the salinity C we fix the bottom $z=0$ to unity but assume no flux on the top $\partial C / \partial z = 0$. The fixed salinity at the bottom resembles the up to 2000-meter-deep warm and salty water reservoir. Since we only consider short time ranges we can neglect the influence of the freshwater input due to precipitation. This is achieved by insulating the upper boundary. We have solved eqs. (1), (2) and (3) using a finite-element method. A non-confirming type of quartic elements for the stream functions in the momentum equation (1) and a bilinear type for the energy and compositional equation (2) and (3) have been used. A locally sensitive upwind scheme has been employed (Heinrich et al., 1977). Additional details of the spatial discretisation can be found in Hansen and Ebel (1984). The eqs. (1), (2) and (3) are advanced in time by a multi-predictor-corrector method (Hansen and Ebel, 1988). The predictor step is a fully implicit one, while the corrector step is accomplished by the Cranck-Nicholson scheme. The overall

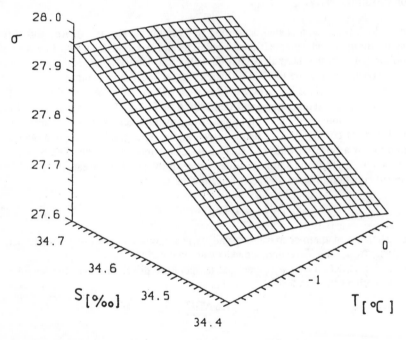

Fig. 2.1: The density $\sigma_{s,\theta,p} = (\rho_{s,\theta,p} - 1)) * 1000$ as a function of the temperature (T) and the salinity (s).

accuracy of the time-stepping is second-order in t. The sub-iteration that is generally necessary to solve the Navier-Stokes Equation can be dropped if a sufficiently small time step is used (Heinrich, 1984). In this study we limited the time-step Δt to half of the maximum of the Courant-Friedrichs-Levy (CFL) criterion. The overall numerical accuracy is second order correct in space and time for the temperature, velocity and compositional fields. In order to check the code, cases of thermal convection have been successfully compared with the work of Davis et al. (1983). The infinite Prandtl version of this code has been benchmarked extensively and compares well in the study of Blankenbach et al. (1989).

For runs in boxes of $\lambda = 1$. usually a mesh of 25 * 25 elements has been employed. Some results have been verified on finer grids up to 51 * 51 elements.

3. Heat and Salt Transport Mechanism

A convective overturn of the material is of particular interest since it can increase the heat transport substantially, thus preventing ice-formation at the top. In order to determine at least a few conditions which are essential for the occurrence of overturning convection, we have carried out numerical experiments in a Rayleigh number range of $10^4 \leq Ra_T \leq 10^7$ and $0.4 \cdot 10^4 \leq Ra_S \leq 1.4 \cdot 10^7$. The onset of overturning convection is accompanied by a sharp increase in the heat transport efficiency. Usually the heat transport efficiency is expressed in terms of the Nusselt-number Nu which is defined as:

$$Nu = \frac{convective + conductive}{conductive} \, heat \; transport$$

$Nu = 1$ means that the heat is solely transported by conduction while in the convective case Nu is always greater than unity. By identifying the onset of overturning convection from the time history of the Nusselt-number we are able to measure the time until overturning type of convection sets in.

In what follows we will discuss three typical cases in detail. A summary of the cases is given in table 1.

Table 1: The non-dimensional time t until convection with $Nu > 1$ occurred. At higher thermal Rayleigh-numbers (Ra_T) overturning convection occurred after a short period of time even if the system is stable in the static sense $(R\rho \geq 1)$. The underlined parameter-combinations are the characteristic cases discussed in the text.

time	$Ra_T = 10^4$	$Ra_T = 10^5$	$Ra_T = 10^8$	$Ra_T = 10^7$
$R\rho = 0.4$	0.01	< 0.01	< 0.01	< 0.01
$R\rho = 0.6$	0.35	< 0.01	< 0.01	< 0.01
$R\rho = 0.8$	2	0.6	≤ 0.01	< 0.01
$R\rho = 1.0$	7	1	0.15	< 0.01
$R\rho = 1.2$	> 10	1.6	0.3	< 0.01
$R\rho = 1.4$	> 10	2.2	0.6	< 0.01

376

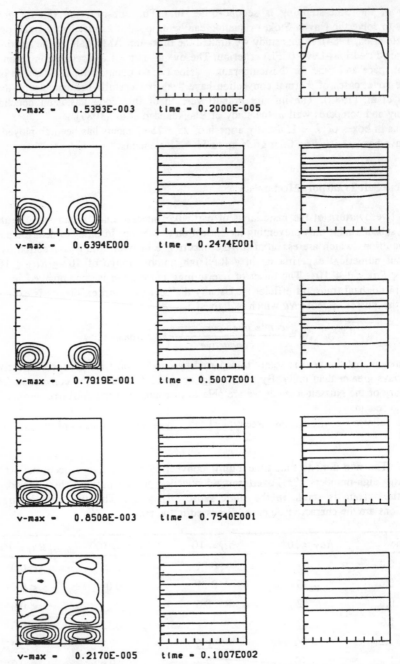

v-max = 0.5393E-003   time = 0.2000E-005

v-max = 0.6394E000   time = 0.2474E001

v-max = 0.7919E-001   time = 0.5007E001

v-max = 0.8508E-003   time = 0.7540E001

v-max = 0.2170E-005   time = 0.1007E002

Fig. 3.2: From the left to the right: streamfunction, temperature and compositional field for the first case.

Case 1. $Ra_T = 10^4$, $R_\rho = 1.2$

The system is stable even after a long period of time. The evolution of the streamfunction (left row), the temperature-field (middle row) and the compositional-field (right row) can be seen in Fig. 3.2.

The upper three plots show the initial condition resembling the two water layers separated by a sharp interface. The streamfunction as shown in the first line is calculated with one model-step and is a result of the temperature and salinity distribution. The correspondent evolution of the Nusselt-number is shown in Fig. 3.3.

After the decay of the initial perturbation the system remains at rest. The isohalines and isothermals are spreading due to diffusion. Because of the higher thermal diffusivity the isothermals are spreading much faster than the isohalines as shown in the middle row and the right row of Fig. 3.2. After this initial period the Nusselt-number remains at 1 (Fig. 3.3).

It seems unlikely that such a low heat transport can prevent ice-formation. A significant dynamical contribution of D.D.C. to the polynya phenomenon can thus be excluded in this parameter range.

Case 2. $Ra_T = 10^6$, $R_\rho = 0.8$

In this case the two Rayleigh-numbers are chosen in such a way that the system is statically instable $(R_\rho < 1)$. It is not trivial that this configuration causes convection to occur since the buoyancy-forces have to overcome not only the stabilizing influence of the concentration field but also that of the viscous forces. For pure thermal convection the critical value (Ra_{TC}) can be determined from linear theory but simple perturbation theory is not available for

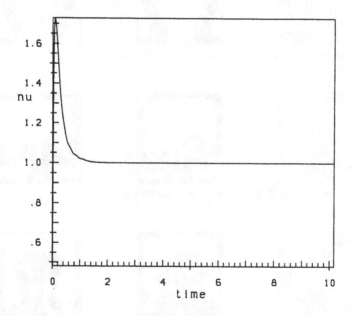

Fig. 3.3: Time history plot of the Nusselt-number for case 1. Parameters are $Ra_T = 10^4$ and $R_\rho = 1.2$. On the x-axis we plotted the time t in thermal diffusion times and on the y-axis we ploted the Nusselt-number as defined in the text.

378

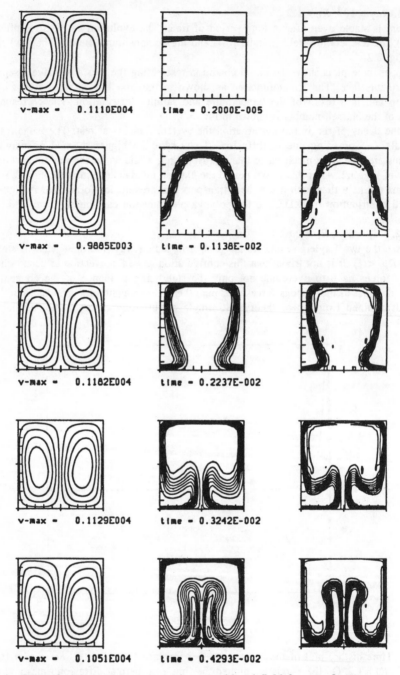

Fig. 3.4: Streamfunction, temperature and compositional field for case 2.

double-diffusive convection (Huppert and Moore, 1976). Thus, numerical experiments are required to cover this range. The first row of Fig. 3.4 displays again the initial conditions for this parameter combination. A sudden increase in the heat transport is clearly reflected in the time-history plot of the Nusselt-number (Fig. 3.5).

The decaying oscillation in the evolution of the Nusselt-number is caused by parts of the initial upper cold and fresh layer which are entrained and transported by the mean-stream. After some overturns they vanish due to mixing. As can be seen in Fig. 3.4 the initial cellular pattern with one upstream in the middle of the cell and two downstreams on the sides has subsequently disappeared, giving rise to a one-cell pattern.

Case 3. $Ra_T = 10^5$, $R_\rho = 1.2$
This parameter configuration resembles a subcritical situation where the restoring force exceeds the thermal driving force. Within the framework of thermal convection motion can not develop under such circumstances. As clearly reflected in Fig. 3.6 the isotherms spread much faster than the isohalines. This is not only due to the Lewis number being 100, i.e. a thermal diffusivity being significantly larger than the compositional one, but also by advective heat transport, caused by an oscillatory motion. This oscillatory motion which is triggered by the initial condition persists, despite the presence of the restoring force. The difference in the molecular diffusivities cause a phase-lag in the T- and C-field.

After a certain time an unstable temperature stratification builds up in the lower part of the box while this part is still isohaline. Thus there is no restoring force caused by salinity in the

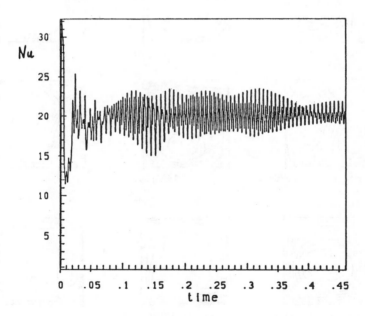

Fig. 3.5: The time-evolution of the Nusselt-number for case 2.

380

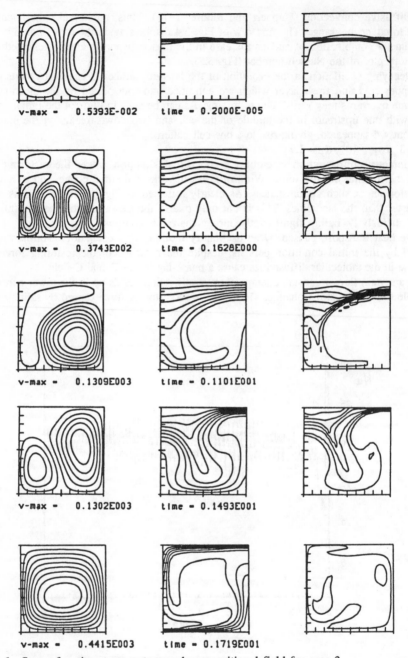

Fig. 3.6: Streamfunction, temperature and compositional field for case 3.

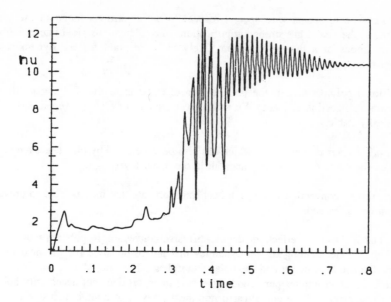

Fig. 3.7: The time-evolution of the Nusselt-number for case 3.

lower part of the box $(\partial C / \partial z = \partial C / \partial x = 0)$, and convection starts in this area. The occurrence of such a type of layered convection basically can account for profiles as obtained by profiles taken by "Mikhail Somov" in the Weddell Sea and resemble somehow the phenomenon of warm W.D.W.-cells as described in Gordon and Huber (1984). The heat transport to the surface in this state is still small. The evolution of the Nusselt-number for this case is shown in Fig. 3.7.

The amount of heat transported to the surface in the period of layered convection is about two times higher than in the purely conductive state. During the time of layered convection cold and fresh water of the upper water layer is entrained within the main flow. Since we do not consider a freshwater input from above the upper layer vanishes after some time. Now overturning convection extends through the entire layer. The amount of heat transported to the surface increases strongly as reflected in Fig. 3.7. Such a high heat-flow may be able to melt the ice or prevent the ice from being formed. If this type of convection is once established even an eventually limited strong freshwater input would probably not be able to stop the convective motion because of the efficient mixing processes caused by the high velocities.

4. Conclusions

In this work we have investigated the nature of subcritical, double-diffusive convection in a two-layer system in view of the phenomenon of polynyas. The initial and boundary conditions have been chosen with special respect to the conditions found in the southern ocean. Although the model has strong simplifications, the calculations yield interesting results about the dynamics of a two-layer system likely to be relevant for the dynamics of the southern ocean:

1.) In traditional polynya models the presumption of static instability for the occurrence of overturning convection is made. We found overturning convection even in systems that are statically stable.

2.) The period of layered convection, which has also been detected by direct measurements, is important for the transport of heat and salt in the upper layer.

3.) The overturning convection has a high heat transport rate that may be able to prevent the ice from being formed.

We expect that inclusion of effects of double-diffusive convection in polynya modelling will lead to a better understanding of the dynamics that are responsible for the heat exchange between the the southern ocean and the polar atmosphere. The often-drawn conclusion that a stably stratified system implies pure conductive heat transport does not necessarily hold. This is a direct consequence of the complicated nonlinear physics of double-diffusive convection under subcritical conditions.

5. References

Blankenbach, B., F. Busse, U. Christensen, L. Cserepes, D. Gunkel, U. Hansen, H. Harder, G. Jarvis, M. Koch, G. Marquart, D. Moore, P. Olson, H. Schmeling, T. Schnaubelt (1989) A benchmark comparison for mantle convection codes. Geophys. J. Int. 98, 23-38.

Busse, F.H. (1981) Transition to turbulence in Rayleigh-Benard convection. Topics in Applied Physics 45, 97-137, Springer Verlag, Berlin, Heidelberg, New York.

Chandrasekhar, S. (1961) Hydrodynamic and Hydromagnetic Stability. Clarendon Press, Oxford.

Comiso, J.C. and A.L. Gordon (1987) Recurring polynyas over the Cosmonaut Sea and the Maud Rise. J. Geophys. Res. 92, 2189-2833.

Davis, G. de Vahl and I.P. Jones (1983) Natural convection in a square cavity: A comparison exercise. Int. Journal for Num. Meth. in Fluids 3, 227-248.

Gordon, A.L. and J. Comiso (1988) Polynyas in Sudpolarmeer, Spektrum der Wissenschaft 8.

Gordon, A.L. and B. Huber (1984) Thermohaline stratification below the Southern Ocean sea ice. J. Geophys. Res. 89, 641-648.

Gordon, A.L., Chen, C.T.A. and W.G. Metcalf (1984) Winter mixed layer entrainment of Weddell Deep Water. J. Geophys. Res. 89, 637-640.

Hansen U. and A. Ebel (1984) Numerical and dynamical stability of convection rolls in the Rayleigh number range 10^3 - $8 \cdot 10^5$. Annales Geophysicae 2, 3, 291-302.

Hansen, U. and A. Ebel (1988) Time dependent thermal convection - a possible explanation for a multi-scale flow in the Earth's mantle. Geophys. J. 94, 181-191.

Hansen, U. and D.A. Yuen (1989) Subcritical double-diffusion convection at infinite Prandtl number. Geophys. Astrophys. Fluid Dynamics 47, 199-224.

Heinrich, J.C. (1984) A finite element model for double diffusive convection. Int. J. for Numerical Methods in Engineering 20, 447-464.

Heinrich, J.C., Huyacorn, P.S. and O.C. Zienkiewicz (1977) An upwind finite element scheme for the two-dimensional convective equation. Int. J. Num. Meth. Eng. 11, 131-143.

Hughes, T.J.R., Wing Kam Liu and A. Brooks (1978) Finite Element Analysis of Incompressible Viscous Flows by the Penalty Function Formulation. Division of Engineering and Applied Science, California Institute of Technology, Pasadena, California 91125.

Huppert, H.E. and D.R. Moore (1976) Nonlinear double-diffusive convection. J. Fluid Mech. 78, 821-854.

Huppert, H.E. and R.S.J. Sparks (1984) Double diffusive convection due to crystallization in magmas. Ann. Rev. Earth Planet. Sci. 12, 11-37.

Huppert, H. and D. Turner (1981) Double-diffusive convection. J. Fluid Mech. 106, 299-329.

Martinson, D.G. (1990) Evolution of the Southern Ocean winter mixed layer and sea ice; open ocean deep water formation and ventilation. J. Geophys. Res. 95, 1741-1755.

Martinson, D.M., Killworth, P.D. and A.L. Gordon (1981) A convective model for the Weddell polynya. J. Phys. Oceano. 11, 466-487.

Rayleigh, Lord (1916) On convection currents in a horizontal layer of fluid when the higher temperature is on the under side. Phil. Mag. 32, 529-546.

Stern, M.E. (1960) The salt fountain and thermohaline convection. Tellus 12, 172-175.

Stommel, H., Arons, A. and D. Blanchard (1956) An oceanographical curiosity: The perpetual salt fountain. Deep-Sea Res. 3, 152-153.

Tritton, D.J. (1988) Physical Fluid Dynamics. Clarendon Press, Oxford.

STRATEGIES FOR MODELLING CLIMATE CHANGES

L. A. LLIBOUTRY
Laboratory of Glaciology and Environmental Geophysics
B. P. 96
38402 Saint-Martin d'Hères
France

ABSTRACT. An impressive amount of hard facts about ice ages has been obtained from oceanic and polar ice cores. They contradict some popular ideas, and raise new puzzles. At the climatic time scale (3 years - 3 ka), and at the ice ages time scale (3 ka - 3 Ma), the atmosphere (with its greenhouse gases), oceans, ice, and vegetation form a complex nonlinear oscillator. At the latter time scale, it is subject to periodic changes in the solar energy input, of astronomical origin. Several lines of approach are reviewed: spectral analysis, logical analysis of delayed feedbacks, response of a slow non-linear system to fast stochastic inputs and search of regularities (empirical, or drawn from atmospheric-oceanic GCM simulations) to be used for ice sheet modelling. Ice sheet models used so far are rather naive, hindering any progress in ice ages simulations.

1. Introduction

1.1. EMERGENCE OF PALEOCLIMATOLOGY

Most actual investigations in meteorology, oceanography, and glaciology deal with climate changes. The increasing content of greenhouse gases in the atmosphere is a matter of concern that has been fully realized by political authorities and research funding agencies. They have accepted the argument that to study climate changes in the past is the only way to understand the complex atmosphere-oceans-ice-vegetation system, in order to build up predictive models for the near future.

Therefore, as a result of an international endeavour, a new discipline has emerged: paleoclimatology. Its results appear in more than 30 international journals. In recent review papers on paleoclimatology the most quoted journals are, in alphabetical order: Advances in Geophysics, Climate Dynamics, J. Atmosph. Sci., J. Geophys. Res. (C, D), Nature, Palaeogeog. Palaeoclim. Palaeoecol., Paleooceanography, Quat. Res., and Science. Moreover, in the last eight years, the proceedings of at least ten international symposia or workshops have been published (cf. references, 1st. part).

Obviously, this paper cannot be a review of such a wide subject. It tries only to give some landmarks for the beginner in this field, and to help him avoid locking himself in the first theory that he has read.

D. B. Stone and S. K. Runcorn (eds.),
Flow and Creep in the Solar System: Observations, Modeling and Theory, 385–398.
© 1993 *Kluwer Academic Publishers. Printed in the Netherlands.*

1.2. TIME SCALES

It is convenient to consider four time scales (rather than three as proposed by Saltzman, 1990), each one ranging over three orders of magnitude.

(1) The weather time scale (1 day to 3 years). At this time scale are studied the atmospheric circulation, the seasonal snow and sea ice covers, and the oceanic circulation (excluding bottom water). The external factors acting on this system are the diurnal and annual periodicity of solar energy input and of albedo (seasonal snow, deciduous trees), large fires or big volcanic eruptions.

(2) The climatic time scale (3 years to 3 ka). At this time scale changes in vegetation and in oceanic bottom water are relevant. The only external factor that might exist at this time scale is changes in the solar constant linked with solar activity, a controversial topic.

(3) The ice ages time scale (3 ka to 3 Ma). The waxing and waning of northern ice sheets, with the corresponding variations of sea level and isostatic adjustments, occurred at this time scale. Glacial erosion and drift, weathering of exposed rocks, absorption of CO_2 by coral reefs, burial of organic matter under sediments, etc. may also be considered. The external factors operating at this time scale are the well-known periodicities in the Earth's orbital parameters: precession of equinoxes (19 and 23 ka), variation of the tilt of the ecliptic plane (41 ka), and variation of the eccentricity of the Earth's orbit (100 ka), the so-called Milankovich variations.

(4) The geologic time scale (3 Ma to 3 Ga). At this time scale, continental drift, orogeny, uplift and erosion become important. They should explain why extensive glacierization happened several times during the Earth's history, or why ocean surface temperatures cooled progressively by 8°C during the Eocene (54.9 to 38.0 Ma B.P.), before any Antarctic glacierization occurred (Kennett, 1977). The only external factor that has been suggested is variations in the solar constant, due to dust clouds that the solar system might cross during its motion through the Galaxy.

The geologic time scale will not be considered in this paper.

2. Hard Facts Drawn from Oceanic Sediments, Polar Ice, and Loess

Twenty years ago, quantitative data on past climates came only from chronicles (back to 1 ka B.P.), tree rings (6 ka), and pollen analysis in cores from peat bogs (130 ka). Peat cores were dated by C^{14} (back to 35 ka), and next by U/Th decay. The advent of magnetostratigraphy has allowed the dating of loess cores (Tungsheng et al., 1988), and oceanic cores (Kennett, 1977; Le Treut et al., 1988) back to 2,500 ka B.P.

The dramatic development of paleoclimatology has been possible thanks to the data provided by oceanic cores and polar ice cores. In the former, microfossils, O^{18}/O^{16} and D/H ratios lead to the knowledge of past ocean temperatures in the biotic layer, and on the global ice volume on Earth. Granulometry reveals drift carried by icebergs, and ash layers indicate major volcanic activity. In polar ice cores (Jouzel et al., 1987, 1989; Legrand et al., 1988a, 1988b; Raynaud et al., 1988), the O^{18} content yields the mean temperature above the atmospheric inversion layer, where snow crystals form. The air content yields the altitude of the surface, because the porosity when the firn closes off is a known mild function of the temperature. The content of carbon dioxide, methane, etc. in past atmospheres is drawn from the composition of the air bubbles trapped at the close-off. Dust and aluminium are indicators of the aerosol input with continental origin, that depends on the extent of bare land and on the meridian circulation. Acidic layers correspond to big volcanic eruptions that injected SO_2 into the stratosphere. Last, since the production rate of cosmogenic Be^{10} should be constant (excepting some spurious events), its concentration in ice cores yields relative values of past accumulation.

Data from oceanic, polar ice, and loess cores have definitively contradicted several traditional hypotheses. Let us point out these wrong theories, since they are still found in most textbooks.

(a) If an ice age is defined by the waxing and subsequent waning of large ice sheets over North America and northern Eurasia, there were only three or four ice ages during the Pleistocene. (There might have been about twenty ice ages, each ice sheet erasing the surface evidence of previous smaller ones.)

(b) The glacierization of the antarctic continent caused the formation of northern ice caps. (Since the Oligocene, during 25 Ma, an ice cap existed over East Antarctica, without providing glaciations in the northern hemisphere.)

(c) Ice ages follow intense volcanic activity. (Not true at the onset of Wisconsinian, the only documented case: Legrand *et al.*, 1988a.)

(d) During the growth of northern ice sheets, the Arctic Ocean was ice-free, a speculation of Ewing and Donn in 1956. (Oceanic cores have shown that it was not so. Moisture came from the Atlantic, not from the Arctic Ocean: Ruddiman and McIntyre, 1979.)

(e) During the last ice age the Sahara was wet. (Datings show that it was wet later, during the Hypsithermal, about 6 ka B.P.)

(f) There have been generalized surges of the whole antarctic ice cap, a speculation by Wilson and Hollin that has been taken up by popular writings. (No signal of such an event is found in circumantarctic oceanic cores.)

(g) An ice age began by the formation of a large ice sheet over Tibet, a recent speculation by Kuhle. (Careful, still unpublished, Chinese field studies show that there has never been a large ice cap over the whole of Tibet.)

On the other hand, among the large amount of hard facts available today, there are several puzzling ones, which deserve special attention. In my opinion, they are the five following ones:

(1) The appearance of large ice caps in the northern hemisphere about 25 Ma later than in the Antarctic (Kennett, 1977).

From about 30 Ma B.P. (Oligocene) to 14.4 Ma B.P. (middle Miocene) East Antarctica had calving glaciers. (Note that the main uplift of the Transantarctic Range was 19 Ma ago, and thus these glaciers should have started from the central high mountains, today hidden under the ice sheet.) Meanwhile, West Antarctica had a mild climate, without glaciers. From 11.3 to 5.1 Ma B.P., West Antarctica had calving glaciers, but no large ice sheet yet. It appeared about 4.8 Ma B.P., during the maximum glaciation for the southern hemisphere (the Ross glaciation).

On the other hand, the first ice cap in the northern hemisphere appeared, it seems, in Iceland, 3.1 Ma ago, and the first glacial drift on Rockall Plateau (North Atlantic) is dated 2.7 Ma B.P. There is no sign of simultaneous cooling in Antarctica. On the contrary, wood and leaves of southern beech 2-3 Ma old have been found near the Beardmore Glacier, in the Transantarctic Range, testifying that temperatures there were not always as cold as today.

(2) Long lasting changes of the polar front. Large paleotemperature changes (± 5°C), lasting several centuries, are found in cores from the North Atlantic every 1,000-10,000 years,

independently of the existence of ice caps in the northern hemisphere.

(3) The fast waxing of ice sheets. At the beginning of Wisconsinian, 120,000 years ago, the sea level dropped by 20-70 m in 5,000 years (Andrews and Mahaffy, 1976). This drop corresponds to the formation of $(8-28) \times 10^6$ km^3 of land ice, the latter value equalling about 60% of the estimated volume of northern ice sheets. Over an area of 10^7 km^2, ice sheets should have thickened each year by 0.16-0.56 m. Assuming annual precipitations of the same order as today, this large positive balance demands a long-lasting drop of snow melting in summer.

(4) The fast waning of ice sheets. Around 9,000 B.P., the edge of the Scandinavian ice sheet receded by more than 300 km in 1,000 years. This retreat by 300 m per year, that faint moraines in the Stockholm area confirm, has never been observed in actual ice cold caps ending on land. (It happened however during the 20th century in temperate Patagonian ice fields.)

(5) The variations of atmospheric CO_2. The CO_2 content of the atmosphere increased simultaneously with the mean temperature at the end of the two last ice ages (Wisconsinian, Illinoian), whereas it lagged by 10-20 ka after the temperature drop at the beginning of Wisconsinian (Barnola et al., 1987; Jouzel et al., 1990). The first author to offer a tentative explanation of this asymmetric behaviour is Shaffer (1990).

3. The Old Naïve Concept: Linear Forcing by Periodic External Factor

The activity of man is punctuated by periodic phenomena allowing prediction: day and night, seasons, tides, feminine rhythm. Therefore, he tries to find out periodicities in any phenomenon. When, in the 19th century, the variations of the Alpine glaciers began to be monitored, the commission in charge of it called itself "International commission on the periodic variations of glaciers," until the prediction of an advance every 35 years failed. Given the traditional curricula in physics, that insist on linear operators, and the widespread codes for spectral analysis, the search for periodicities in any time series is more flourishing than ever.

Two wrong beliefs justify this search: (1) strict periodicities do exist. For instance, in the seventies, Dansgaard, having found some oscillations of the O^{18} content with a periodicity of 350 years in the upper part of Greenland ice cores (where annual layering can still be recognized), used this assumed permanent periodicity at depth to date old ice. He surmised that the Sun was a variable star having this periodicity.

(2) When a climatic variable displays the same period as an external factor, this factor is the cause of the climatic variation. This is only true for a system giving a linear response. Otherwise, the external factor may just act as a trigger, a pacemaker that fixes the period of an already self-oscillating system.

Let us give some details on spectral analysis and its difficulties. Given a stochastic function X(t) of the continuous variable t ($- \infty < t < + \infty$), according to the Wiener-Khintchin theorem, the self-correlation function:

$$\Gamma(t) = \int_{-\infty}^{+\infty} X(t') X(t'-t) \, dt' \tag{1}$$

and the spectral power density:

$$P(v) = |\text{ F. T. of } X(t)|^2 \tag{2}$$

are Fourier transforms (F.T.) of each other, i.e.:

$$\Gamma(t) = \int_{-\infty}^{+\infty} P(v)\, e^{i2\pi vt}\, dv$$

$$P(v) = \int_{-\infty}^{+\infty} \Gamma(t)\, e^{-i2\pi vt}\, dt \tag{3}$$

In fact t is digitized with equal time steps h (t = kh). Consequently the frequencies higher than Shannon's, $v_s = 1/2h$, are ignored.

The essential point is that the spectral analysis of a limited sample $0 \leq k \leq N$ is <u>not</u> a well-posed problem. Its solution requires the assumption of some signal structure.

(a) In the <u>Fast Fourier Transform</u> method, the assumption is $\Gamma(kh) = 0$ for $|k| > N$. It follows:

$$P(v) = \sum_{-N}^{+N} \Gamma(kh)\, e^{-i2\pi vkh} \tag{4}$$

This method leads to a broadening of existing spectral lines, and yields secondary lobes that are artifacts.

b) In the <u>maximum entropy</u> method, it is assumed that (putting X_k for X(kh):

$$X_{k+1} = \sum_{0}^{N} b_j\, X_{k-j} + \text{white noise} \tag{5}$$

The adopted values of b_j are computed from all the data thanks to the best linear unbiased estimator, i.e. by a least square fit method.

(c) In the <u>Pisarenko</u> method it is assumed that the signal X_k is the sum of M pure frequencies ($M \leq N/2$), and white noise.

When the time series is drawn from a core, the practical difficulties are, according to Benoist (unpublished):

- The dating is inaccurate.
- The ignored high frequencies $v > v_s$, that normally vanish with depth by diffusion, may still exist. They are folded on the low frequency side of Shannon's cut-off. Therefore, high frequencies must be previously suppressed by some elaborate smoothing (spline function adjusted by the Reinsch method, filtering with a Blackmann window, etc.)
- Sampling is irregular. The sections of core that are analysed are equidistant along the core. They do not correspond to equal intervals in time, because of the variability of accumulation. Sampling at equal intervals in time is impossible, since dating is obtained only after analysis of the selected sections.
- The time series is too short, excluding Fast Fourier Transforms. The maximum entropy method is the best, since Pisarenko's is too optimistic about the existence of exact periodicities.

In Table 1 are listed the periodicities found in polar ice cores and in oceanic ones, within the climatic time scale. They are probably short-lived, if they exist at all. Consider in particular the 2100-2600 "period." Terminal moraines in the northern hemisphere are well dated by varves (annual sequence of sand and silt layers, deposited in turn by glacial streams), and by C^{14}. They confirm the dates obtained by pollen analysis. As shown in Table 2, although the mean interval between two glacier advances was 2325 years, within the range 2100-2600, this lapse of time ranged in fact between 1400 and 3900 years.

TABLE 1. Short-lived periodicities within the climatic time scale (years)

Tree rings:					405		2400
Oceanic cores							
(Pisias *et al.*, 1973):					380	1300?	2600
Greenland ice cores							
(Dansgaard *et al.*, 1970):		80	180	350			2100
Dome C ice core							
(Benoist, unpublished):	42	55	75	140	380	1300	2450

TABLE 2. Climatic changes during late Wisconsinian and Holocene

Moraines		Pollens		Age	Interval
North America	Europe	cold	warm	(B.P.)	
Holocene					
Valders	Stockholm-Salpausselka	Younger Dryas		10700	
			Allerød		1400
Mankato	Scania-St.- Petersburg	Older Dryas		12100	
			Bølling		3900
Wisconsinian=Weichselian					
Cary	Pomerania			16000	
			Lascaux optimum		2000
Tazewell	Poznan			18000	
					2000
Iowa	Brandenburg			20000	

At the ice ages time scale, the astronomical periods appear clearly in the spectral analysis,especially the 100 ka period of Earth's orbit eccentricity. A more elaborate study shows that only 30% of the variance can be ascribed to linear forcing (Kominz et al., 1979). Clearly, there are in the climatic system positive feedbacks that magnify the response.

Wegmann (1940) explained the formation of the Greenlandic ice sheet by such a positive feedback, due to altitude: when an ice sheet thickens, the mass balance at its surface (snow accumulation - melting) increases, because summer rain is replaced by snow, and colder air reduces melting. (This altitude feedback was emphasized by Cailleux in 1952, and expressed in an English paper by Bodvarson in 1955). In fact, under cold arctic climates, the very existence of this feedback may be questioned. Rain is unknown at any altitude, and melting is almost entirely due to the solar energy.

4. Non-Linear Oscillators

The idea that the climatic system (atmosphere, oceans, different snow and ice covers) may present self-sustained oscillations appeared around 1960. To oscillate, the system must include delayed negative feedbacks. They do exist in the climatic system (Oerlemans, 1982; LeTreut et al., 1988; Shaffer, 1990). That climatic oscillators should be non-linear was realized later. Non-linear oscillators had already been studied with analytical models, specially by van der Pol.

A linear oscillator has one or several eigenfrequencies $v_1, v_2, ..., v_k$, with k denoting its number of degrees of freedom. In self-sustained oscillations, only these k frequencies and their harmonics can be present. With a non-linear oscillator, these fundamental frequencies can add or substract, i.e. the frequency may be $v = n_1 v_1 + n_2 v_2 + ..., n_1, n_2, ...$ being positive or negative integers. Moreover, when the system is subject to some external periodic forcing factor, one among these periods may shift slightly to adjust itself to the period of the external "pacemaker," or to a subharmonic of it.

With the development of computers, it has become possible to simulate the transient behaviour of non-linear dynamical systems, and not only their asymptotic periodic responses. In the phase space, the point having as coordinates the values of different state variables of the system moves towards attractors, that are points in case of a steady state, and closed loops in case of a periodic response. The system may reach bifurcations, where the slightest disturbance (the mere truncation errors in running computation) makes the representative point move towards one or another attractor. It has been discovered that with only three degrees of freedom there may be fractal attractors, that infill totally some region of the phase space.

Therefore, a complex non-linear system may generate practically aperiodic noise: either because it has a large number of degrees of freedom with unrelated eigenfrequencies, that add or substract producing many more frequencies; or because there is a fractal attractor. With an analytical model, i.e. a set of coupled non-linear evolution equations, there are methods for detecting a fractal attractor, as the computation of Poincaré sections, or of Lyapunov exponents. But, in the real world, with a set of time series at hand, it is impossible to distinguish both cases. It is easy to assume a simple model that leads to Lorenz equations between three variables, a system possessing a fractal attractor, as done by Salazar and Nicolis (1988), but their simplistic model may be questioned.

Another approach of large complex systems is possible, albeit qualitative. First, a logical analysis of feedbacks and their delays is made (an example will be given later). Next, the state variables are replaced by Boolean variables, having only two possible values, say 0 and 1. Then the highly idealized model consists of a set of Boolean delay equations (Ghil et al., 1987). This method has been well-tried in genetics, but in climatology a flip-flop change of the variables is often unrealistic (see Koerner, 1980, in case of albedo). With Boolean delay equations, processes with a cumulative effect are precluded. Moreover, the qualitative results can strongly depend on the arbitrary initial conditions (Wright et al., 1990).

5. Output of Fast Processes as an Input for Slower Ones

For modelling complex natural phenomena, some space scale and some time scale must be chosen. Most inputs, i.e. the boundary conditions in the space-time frame of reference, are the outputs of processes at smaller scales, that must be studied separately. In the case of oceanic circulation, the sharing into different space scales is well explained by Holland and McWilliams (1987). Let us consider rather the different time scales.

Since the climatic system is not linear, the response of slow processes to fast inputs does not depend on the mere averaged value of these inputs. For instance, assuming roughly that the snow cover melts only when the air temperature is positive, given the diurnal variations of the latter, the mean annual or mean monthly temperatures cannot be used as an input. (Even if for T > 0°C the melting rate is proportional to the air temperature, since for T < 0°C it is zero, the law is not linear.) Therefore, some statistical properties of the instantaneous input must be known.

Ignoring forcing factors with long periods, we may consider how the slow system responds to fast inputs that are stochastic, i.e. random and unpredictable, although having some permanent statistics. General results, when there is a conspicuous gap between the time constants of the fast processes and those of the slow ones, are given in a seminal paper by Hasselmann (1976). The slow system deviates progressively from the linear response (from the state corresponding to the mean values of the fast inputs). At first, the variance of the response increases linearly with time t. The most probable deviation from the mean, although unpredictable, increases as $\sqrt{t}$. The case is similar to the Brownian random walk of a small particle, hit by molecules that, on the average, afford no momentum to the particle. (Mathematically, the position of the representative point in the phase space is governed by a Fokker-Planck equation, with an essential diffusive term.)

For large deviations, some internal negative feedback should operate, that limits the variance of the slow system. Up to this limiting very low frequency, the response of the slow system to a random, fast, white noise, is a red noise, with a spectral power density proportional to v^{-2} (i.e. an amplitude of each Fourier component proportional to its period).

This approach has been used sometimes when studying the response of the ocean to atmospheric variability. The variations of climate at the time scale 3-3000 years might be explained in this way. At the ice ages time scale, the periodical variations in the Earth's orbital parameters lessen the interest of this approach.

In the purely deterministic approach, there are two ways of using the output of fast processes as an input for slow ones: empirical, or numerical.

Field measurements lead to empirical regularities, i.e. correlations (improperly called "laws" in natural sciences). The danger is that they may not be valid at large time scales, in a totally modified context. Therefore, the physical grounds of these regularities must be investigated first.

For instance, the O^{18} content of surface snow in Greenland or Antarctica is, with good accuracy, a linear function of the temperature at 10 m deep, where the seasonal fluctuations are smoothed out. The physical basis of this relationship is that the O^{18} content depends on the air temperature when and where the snow crystals form, above the inversion layer, and is smoothed out after deposition by diffusion, whereas the snow temperature follows the mean air temperature close to the surface. During an ice age, the thickness of the inversion layer should not have changed very much, but the annual distribution of snowfalls might have been totally different. If snowfalls were concentrated in winter, instead of being distributed all the year around as today, the O^{18} paleotemperatures would be lower without change in the mean air temperature.

It is generally possible to distinguish in a particular area a few weather types. In the same way, meteorological patterns over a whole hemisphere are sometimes classified. Such idealizations are more fruitful than the standard measure of variability. Of particular interest are the infrequent extreme cases, as large droughts, El Niño events, etc. Unfortunately, although the return periods of such extreme cases are more or less known, nothing is known about the statistics of the sequences. Which is, for instance, the return period of ten or more consecutive years with large winter snowfalls and cloudy cool summers, that might initiate a glacierization over a broad area?

Numerical computation with an atmospheric GCM allows synoptic forecasts for a few days only, with non-negligible errors. Nevertheless calculated mean values and variability are surer. An atmospheric GCM, with time steps of 1 hour, may simulate the daily weather, to be used in an oceanic GCM that predicts the ocean circulation, sea surface temperatures, etc. during several years. In the same way, a coupled atmospheric-oceanic GCM might yield the input for an ice sheet model, running with time steps of 100 years or more. Two procedures are possible:

(a) The GCM results may be parameterized. Not only mean values must be computed, but also distribution laws or moments, because the climatic system is not linear. This point is generally missed. It would be also possible to determine the probability of sequences of extreme cases, as requested above.

(b) A simpler solution, although more expensive in computer time, is to run asynchronously an atmospheric GCM and an oceanic GCM, coupling both at each time step of the latter (Chalikov and Verbitsky, 1990). To tackle longer time scales, it becomes impossible to couple an ice sheet model with this atmospheric-oceanic GCM. It should simulate the evolution during 1 million years, that means 1 billion time steps for the GCM!

Therefore, the results of an atmospheric-oceanic GCM have to be parameterized first, before tackling the ice sheet model. Up to this date, in the absence of such computed regularities, empirical regularities only have been used.

Note that there might be bifurcations. Therefore, a single simulation might miss another possible evolution.

6. Ice Sheet Modelling

6.1. ICE SHEET MODELS USED BY METEOROLOGISTS

For the time being, when GCM specialists tackle ice sheets, they use simplistic models that glaciologists developed thirty years ago. They underestimate the complexity and difficulties of modelling an ice sheet. Ice rheology depends strongly on its temperature (a rise by 5.4°C halves the viscosity). Therefore, it is compulsory to compute the coupled velocity and temperature fields. Both evolve at time scales on the order of 10,000 years. Moreover, ice rheology depends strongly on its fabric (the statistical distribution of hexagonal axes in the different crystals). Polar ice (specially ice from the last ice age) is often very anisotropic, and this fact may enhance flow by a full order of magnitude. Neither the rheology of anisotropic ice for a given fabric is well known, nor do we have a predictive model for the formation of a fabric, and its subsequent destruction by recrystallization.

I address these difficulties in another contribution to this volume. Only the usual misconceptions about the boundary conditions at the surface of an ice sheet (temperature and balance) will be pointed out here.

(a) The "surface" temperature (in fact the temperature at 10 or 30 m deep, where annual oscillations are damped) is higher than the mean air temperature in areas with some snow melting in summer, because of internal refreezing at depth.

(b) For the same reason, when the energy balance indicates that snow melts at the surface of an ice sheet, the melted layer must not be debited from the mass budget of the ice sheet, contrarily to the case of seasonal snow over unglaciated land, or of temperate glaciers of middle latitudes, entirely at melting point.

(c) The variation of balance (annual mass budget per unit area) with altitude, near the equilibrium line (where the balace is zero), a parameter called the activity coefficient, varies at least by a factor of three according to whether the climate is continental or maritime. To model climate change by a mere change in the altitude of the equilibrium line is misleading. When, between 16,000 and 13,000 BP, western European climate shifted from continental and cold to maritime and cool, the activity coefficient rose. The Scandinavian ice sheet may have shrunk without any rise of the equilibrium line.

6.2. FEEDBACKS IN ICE SHEET BEHAVIOUR

We shall end with a logical analysis of internal feedbacks that modify the response of an ice sheet, at the climatic time scale. First, a graph of the effects is drawn (fig. 1). A straight arrow indicates that the effect is immediate at this time scale. A winding arrow denotes a delayed effect. Signs + or - indicate that the effect is positive or negative.

Many feedback loops are noticed (fig. 2). Feedbacks are negative or positive according to whether the number of minus signs is odd or even. Loops (1), (2), (4), and (6) cannot be activated by external factors, that can modify T_S or b only (or both). Nevertheless, they should be taken into account if changes in the discharge Q due to a progressive change in the ice fabrics were considered. Otherwise, the main delayed internal feedback is the balance-altitude feedback (3), that is either positive or negative. With a further delay, it is reduced by the isostatic adjustment (feedback loop (5)), and, since the discharge depends on the bottom temperature, by the change in altitude due to the change in discharge (loop (7)). With a third delay, feedback (7) is somewhat reduced by isostasy (loop (8)). Do not forget that we are considering internal feedbacks only. There are also crucial external feedbacks, that involve greenhouse gases, sea level, vegetation, etc.

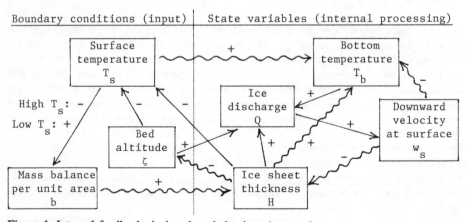

Figure 1. Internal feedbacks in ice sheet behaviour (see text).

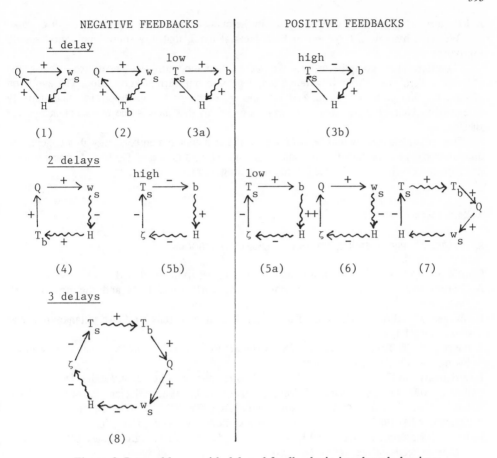

Figure 2. Internal loops with delayed feedbacks in ice sheet behaviour

7. Final Remarks and Conclusion

Science progresses thanks to a permanent dialectic between observation and modelling. In experimental sciences, the predictive value of models can be tested by new experiments. In geophysics, since experimentation is often impossible, models are not properly predictive. Often the only goal of modelling is to show that some suggested process can operate. It is a valuable contribution, but not a proof that this process operates indeed. In other cases, the model is claimed to allow a realistic simulation of the phenomena. Then, the model has no value when it cannot reproduce, qualitatively at least, all the known facts in the same space and time scales, even those that have not been considered for setting up the model. This condition is essential for modelling to be Popperian science, not mere play with a computer.

This stage of sound simulation has been reached at the weather and climatic time scales, but not yet at the ice ages time scale. The models put forward argue for some process or other, but never deal with all the pertinent aspects, as coupled temperature and velocity fields in ice sheets, changes in greenhouse gases in the atmosphere and in vegetation cover on land, etc. We have by now a large set of hard facts at hand, that any sound simulation should reproduce.

Therefore, more complex models are needed. For modelling ice sheets realistically, the analysis of feedbacks above has shown major gaps in our current knowledge. The evolution of the ice fabric must be well understood, and the rheology of anisotropic ice quantitatively determined. Another factor that is poorly modelled to this date is the mass balance of ice sheets.

This critical review would be useful even if the reader remembers only this conclusion: don't trust parts of previously published models to set up a new one. Go back to basic studies and hard facts, still largely ignored in the modelling literature!

8. References

8.1. SYMPOSIUM PROCEEDINGS, WORKSHOP REPORTS, TEXTBOOKS

I. Allison (ed.) (1981) Sea level, ice and climate change, IAHS Publ. 131.

A. Berger (ed.) (1981) Climatic variations and variability: facts and theories, Reidel, Dordrecht.

A. Berger, J. Imbrie, J. Hays, G. Kukla and B. Saltzman (eds.) (1984) Milankovitch and climate, Reidel, Dordrecht.

J. Hansen and T. Takahashi (eds.) (1984) Climate processes and climate sensitivity, Maurice Ewing series 5, AGU, Washington.

J. Oerlemans and C.J. van der Veen (eds.) (1984) Ice sheets and climate, Reidel, Dordrecht.

Carbon dioxide Res. Div. of the US Dept. of Energy (ed.) (1985) Glaciers, ice sheets and sea level: effect of a CO_2-induced climatic change, DOE/EV/60235-1, Washington.

A.P. Hecht (ed.) (1985) Paleoclimate analysis and modeling, Wiley, New York.

B. Bolin, B.R. Doos, J. Jaeger and R.A. Warrick (eds.) (1986) The greenhouse effect climatic change and ecosystems, Wiley, New York.

E.D. Waddington and J.S. Walder (eds.) (1987) The physical basis of ice sheet modelling, IAHS Publ. 170.

W.H. Berger and L.D. Labeyrie (eds.) (1987) Abrupt climate change, Reidel, Dordrecht.

H. Wanner and U. Siegenthaler (eds.) (1988) Long and short term variability of climate, Springer Verlag, Berlin.

A. Berger, S.H. Schneider and J.C. Duplessy (eds.) (1989) Climate and geosciences, Kluwer Acad. Publ., Dordrecht.

H. Oeschger and C.C. Langway (eds.) (1989) The environmental record in glaciers and ice sheets, Dahlem Workshop Reports, Wiley, Chichester.

M.E. Schlesinger (ed.) (1990) Climate-ocean interaction, NATO A.R.W., Kluwer Acad. Publ., Dordrecht.

T.J. Crowley and G.R. North (1991) Paleoclimatology, Oxford Univ. Press, Oxford.

J. Jouzel (ed.) (in press) Paleoclimate modeling, NATO A.R.W., Kluwer Acad. Publ.

8.2. SELECTED PAPERS (WITH SOME COMMENTS)

Andrews, J.T. and M.A.W. Mahaffy (1976) Growth rate of the Laurentide ice sheet and sea level lowering. Quatern. Res. 6, 167-183.

Benoist, J.-P. (1986) Analyse spectrale de signaux glaciologiques: étude des glaces sédimentaires déposées à Dome C, morphologie d'un lit de glacier (Thèse d'Etat, Univ. de Grenoble I). Lab. de Glaciol. et Géophys. de l'Environnement.

Berger, A., H. Gallee, T. Fichefet, I. Marsiat and C. Tricot (1990) Testing the astronomical theory with a coupled climate-ice-sheet model. Paleogeog. Paleoclim. Paleoecol. (Global change section) 89, 125-141. (Ice-sheet mechanics assumed to yield a parabolic profile.)

Chalikov, D.V. and M.Ya. Verbitsky (1990) Modeling the Pleistocene Ice Ages. Adv. in Geophys. 32, 75-131. (Assume mean ice sheet thickness proportional to (area)$^{1/4}$ Snow mass budget computed from air moisture transport.)

Ghil, M., A. Mulhaupt and P. Pestiaux (1987) Deep water formation and Quaternary glaciations. Clim. Dyn. 2, 1-10.

Hasselmann, K. (1976) Stochastic climate models. Tellus 28, 473-484.

Holland, W.R. and J.C. McWilliams (1987) Computer modeling in physical oceanography from the global circulation to turbulence. Physics Today 40 (10), 51-57.

Ives, J.D., J.T. Andrews and R.G. Barry (1975) Growth and decay of the Laurentide ice sheet and comparisons with Fenno-Scandinavia. Naturwissenschaften 62, 118-125.

Jouzel, J. et al.; Barnola, J.-M. et al.; Genthon, C. et al. (1987) Vostok ice core. Nature 329, 403-418.

Jouzel, J., N.I. Barkov, Y.S. Korotkevitch, V.M. Kotlyakov, M. Legrand, C. Lorius, J.-R. Petit, V.N. Petrov, G. Raisbeck, D. Raynaud, C. Ritz and F. Yiou (1989) Global change over the last climatic cycle from the Vostok ice core record (Antarctica). Quat. Intern. 2, 15-24.

Kennett, J.P. (1977) Cenozoic evolution of Antarctic glaciation, the Circum-Antarctic ocean, and their impact on global paleoceanography. J. Geophys. Res. 82 (C27), 3843-3860.

Koerner, R.M. (1980) Instantaneous glacierization, the rate of albedo change, and feedback effects at the beginning of an Ice Age. Quat. Res. 13, 153-159.

Kominz, M.A., G.R. Heath, T.-L. Ku and N.G. Pisias (1979) Brunhes time scales and the interpretation of climatic changes. Earth Planet. Sci. Letters 45, 394-410.

Legrand, M.R., C. Lorius, N.I. Barkov and Y.N. Petrov (1988a) Vostok (Antarctica) ice core: atmospheric chemistry changes over the last climatic cycle (160,000 years). Atmosph. Envir. 22, 317-331.

Legrand, M.R., R.J. Delmas and R.J. Charlson (1988b) Climate forcing implications from Vostok ice-core sulphate data. Nature 334, 418-419.

Le Treut, H., J. Portes, J. Jouzel and M. Ghil (1988) Isotopic modeling of climate oscillations: implications for a comparative study of marine and ice core records. J. Geophys. Res. 93 (D8), 9365-9383. (Simple climate model yields self-sustained oscillations with periods of order 10 ka. Observed longer ones are interpreted as subharmonics due to non-linearity.)

Lin, R.Q. and G.R. North (1990) A study of abrupt climate change in a simple nonlinear climate model. Clim. Dyn. 4, 253-261. (They find bifurcations in the snow line seasonal behaviour, but this line has been unduly assumed to follow the instantaneous 0°C isotherm through the seasons.)

Lloyd, C.R. (1984) Pre-Pleistocene palaeoclimates: the geological and paleontological evidence, modeling strategies, boundary conditions and some preliminary results. Advances in Geophys. 26, 35-140.

Meehl, G.H. (1990) Development of global coupled ocean-atmosphere general circulation models. Clim. Dyn. 5, 19-33.

Mitrovica, J.X. and W.R. Peltier (1991) Radial resolution in the inference of mantle viscosity from observations of glacial isostatic adjustment. In R. Sabadini, K. Lambeck and E. Boschi (eds.), Glacial isostasy, sea level and mantle rheology, Kluwer Academic Publ., Dordrecht, pp. 63-78.

Oerlemans, J. (1982) Glacial cycles and ice-sheet modeling. Climatic Change 4, 353-374. (The only climatologist to consider bottom temperatures and the role of subglacial water, although with questionable equations.)

Peltier, W.R. (1991) The ICE-3G model of late Pleistocene deglaciation: construction, verification, and applications. In R. Sabadini, K. Lambeck and E. Boschi (eds.), Glacial isostasy, sea level and mantle rheology, NATO ASI series C, Kluwer Academ. Publ., Dordrecht, pp. 95-119.

Pollard, D. (1983) A coupled climate-ice sheet model applied to the Quaternary Ice Ages. J. Geophys. Res. 88 (C12), 7705-7718. (Ice flow is modelled assuming a uniform and constant temperature.)

Ramanathan, V., B.R. Barkstrom and E.F. Harrison (1989) Climate and the Earth's radiation budget. Physics Today 42 (5), 22-32.

Raynaud, D., J. Chappellaz, J.-M. Barnola, Y.S. Korotkevich and C. Lorius (1988) Climatic and CH_4 cycle implications of glacial-interglacial CH_4 change in the Vostok ice core. Nature 333, 655-657.

Ruddiman, W.F. and A. McIntyre (1979) Warmth of the subpolar North Atlantic ocean during northern hemisphere ice-sheet growth. Science 204, 173-175.

Salazar, J.M. and C. Nicolis (1988) Self-generated aperiodic behaviour in a simple climatic model. Clim. Dyn. 3, 105-114.

Saltzman, B. (1990) Three basic problems of paleoclimatic modeling: a personal perspective and review. Clim. Dyn. 5, 67-78.

Shaffer, G. (1990) A non-linear climate oscillator controlled by biogeochemical cycling in the ocean: an alternative model of Quaternary Ice Age cycles. Clim. Dyn. 4, 127-143.

Tungsheng, L., Z. Shouxin and H. Jiaomao (1985) Stratigraphy and paleoenvironmental changes in the loess of central China. Quatern. Sci. Rev., 489-495.

Wegmann, E. (1940) Einleitung zur Vortragereihe über die Geologie von Grönland. Mitt. Nat. Ges. Schaffhausen 15, 29-66.

Wright, D.G., T.F. Stocker and L.A. Mysak (1990) A note on Quaternary climate modelling using Boolean delay equations. Clim. Dyn. 4, 263-267.

ICE SHEET DYNAMICS

LOUIS A. LLIBOUTRY
Laboratoire de Glaciologie et Géophysique de l'Environnement
B.P. 96, St-Martin d'Hères, 38402, France

ABSTRACT. When modelling large ice sheets, local equilibrium of a vertical column with the size of the horizontal mesh (~ 100 km) can be assumed. The transit time of temperature disturbances from surface to bottom is ~ 60 ka. Thus, in general, actual and ice ages ice sheets are not in a steady state. If they were, a rule of thumb for the thickness, $H \sim (2b/\pi\kappa)(T_s/G_0)^2$, would be more realistic than the usual assumption of a constant bottom shear stress. Less crude calculations assume uniform vertical strain rates. Still better ones consider a bottom boundary layer (BBL) whose temperatures get locally steady within a time step. In general, sliding demands that the temperature reach melting point not only at the ice-bed interface, but within a temperate BBL. The slight permeability of rocks at large scale should impede the accumulation of subglacial water, put forward by some authors to predict ice sheet surges. Polar ices are often very anisotropic, and we lack quantitative models for predicting their fabrics and behavior. The full rheological law of anisotropic ice with rotational symmetry, to be used in 3-D modelling, is given. A theory allows us to simplify it in the case of c-axes clustered near the z-axis, with two unknown parameters only, instead of ten.

1. INTRODUCTION

Numerical computation of the dynamics of an ice-sheet (as for a glacier, an ocean, or the whole atmosphere) first demands adoption of pertinent space and time scales. Only large ice-sheets, as those existing today over Greenland (the Indlandsis) and over Antarctica, will be considered here.

Eastern Antarctica is a little larger than the USA, and western Antarctica is larger than Mexico. The Indlandsis is a little smaller than Mexico. Therefore, the horizontal mesh size may be on the order of 100 km. This coarse spatial scale allows us to consider the isostatic equilibrium as local, and in most cases to assume a local equilibrium of driving forces acting on a vertical column of ice.

On the other hand, the thickness of the ice cover (H) is on the order of 3 km. Since the driving shear stress increases from zero at the surface to its maximum value at the bottom, and since the temperature may differ by up to 55°C between surface and bottom, the vertical mesh size must be in the range 20-100 m. At 20 m from the surface the annual fluctuation of snow temperature is damped out. In this context, this smoothed temperature is termed the "surface" temperature (T_s). At the bottom, the melting temperature (T_m) may be reached.

The pertinent time scale may be drawn from the transit time of a temperature perturbation from surface to bottom, when T_s changes abruptly. In a motionless ice-sheet this time is $H^2/4\kappa$, with κ denoting the thermal diffusivity ($\kappa \simeq 36.5$ m^2 a^{-1}). When H ~ 3000 m, it is on the order of 60,000 years. The ratio of the thickness H to the annual snow accumulation at

399

D. B. Stone and S. K. Runcorn (eds.),
Flow and Creep in the Solar System: Observations, Modeling and Theory, 399–416.
© 1993 *Kluwer Academic Publishers. Printed in the Netherlands.*

the surface (b, in meters of equivalent ice per year), is of the same order. It gives a characteristic time for advection. We shall see that, since both diffusion and advection intervene, the characteristic transit time is the geometrical mean of both transit times. Therefore, time steps will be on the order of 1,000 years. Since most of the isostatic adjustment is performed within a single time step, it may be considered as instantaneous, as a first approximation. Moreover, the elastic adjustment (that is truly instantaneous) is neglected.

The governing equations, and the usual approximations that are done to handle them, will be given for the 2-D plane problem only, with the z-axis vertical upwards, and the x-axis in the flow direction. The 3-D problem affords more complicated equations (stream and stress functions cannot be defined), but does not lead to new conceptual difficulties.

We shall consider successively the kinematics, the driving forces, and the temperature distribution, with the case of a base at the melting point. In this case high sliding velocities may occur, making invalid the assumption of local equilibrium at the horizontal mesh size scale. Last, the exact rheology of polar ice, that depends on a slowly evolving fabric, will be examined.

2. GOVERNING EQUATIONS

The upper snow and firn layers (that may reach 100 m thick) will be replaced by an ice layer of the same weight. (In glaciology, "ice" means the impermeable rock, after close-off of communicating porosity.) Ice density (ρ) is considered as a constant, in spite of the contraction of air bubbles with depth, and of the slight sensitivity of bubble-free ice density to pressure and to temperature. (The latter might induce thermal convection. Although the Rayleigh number might be high enough to allow it, ice has gone to the ocean before convection could develop.)

In the 2-D problem, the continuity condition (i.e., for an incompressible body, the conservation of volume) yields the existence of a stream function $q(x,z,t)$ such that the two components of velocity are:

$$u = \partial q/\partial z, \quad w = -\partial q/\partial x \tag{1}$$

The altitude above sea level of the bed is denoted $\zeta(x,t)$. (It is a function of time because of changes in sea level and isostasy.) In general, we may adopt $q(x,\zeta,t) = 0$ because bottom melting is non-existent, or one order of magnitude less than the mass balance at the surface. The mass balance $b(x,t)$ equals the snow accumulation because any meltwater refreezes at depth. Putting $q(x,\zeta+H,t) = Q(x,t)$, the continuity condition yields the equation governing the changes in the profile:

$$\partial Q/\partial x + \partial H/\partial t = b \tag{2}$$

The problem is to determine, at any x and any instant $Q = \int_{\zeta}^{\zeta+H} u \, dz$. Pure kinematics is insufficient. We must consider the stress, for which von Karman's notation will be adopted:

$$\underline{\underline{\Sigma}} = \begin{vmatrix} \sigma_x & 0 & \tau_{xz} \\ 0 & \sigma_y & 0 \\ \tau_{xz} & 0 & \sigma_z \end{vmatrix} \tag{3}$$

Since inertia forces and Coriolis forces are negligible, the momentum does not appear in the momentum equations. They should be called rather the equilibrium conditions. Let $\alpha = -\partial (H+\zeta)/\partial x$ be the surface slope. On large ice sheets it is very small, on the order of 10^{-2} or 10^{-3}. Tilting the axes forward by α does not significantly change the components of stress, and allows us to write as stress boundary conditions at the surface $\sigma_z = \tau_{xz} = 0$. With tilted axes the equilibrium conditions for the y-independent problem read:

$$\partial\sigma_x / \partial x + \partial\tau_{xz} / \partial z + \rho g \alpha = 0$$
$$\partial\tau_{xz} / \partial x + \partial\sigma_z / \partial z - \rho g = 0 \tag{4}$$

With tilted axes, $\partial(H+\zeta)/\partial x = 0$. From (4) the existence of an Airy's stress function $\chi(x,z,t)$ may be inferred, such that:

$$\sigma_x = -\rho g(H + \zeta - z) + \partial^2\chi / \partial z^2$$
$$\sigma_z = -\rho g(H + \zeta - z) + \partial^2\chi / \partial x^2 \tag{5}$$
$$\tau_{xz} = \rho g \alpha(H + \zeta - z) - \partial^2\chi / \partial x\partial z$$

We have three scalar functions of x, z, and t to determine: q, χ, and T. They are governed and linked by three partial derivative equations of second order:
- two rheological equations giving the strain rates $\dot{\varepsilon}_{xx} = \partial^2 q/\partial x\partial z$, and $2\,\dot{\varepsilon}_{xz} = \partial^2 q/\partial z^2 - \partial^2 q/\partial x^2$ as instantaneous functions of stress and temperature;
-the heat equation, giving the rate of temperature change as a function of the temperature and velocity fields.

3. THE LOCAL EQUILIBRIUM ASSUMPTION

Ice-sheet dynamics can be simplified by assuming local equilibrium of a vertical column, i.e. x-independent stresses. Then, when the surface slope is more or less uniform, the expression (5) of σ_x implies χ independent of x. When $\partial H/\partial x$ is very small, the bottom drag is approximately equal to τ_{xz} for $z = \zeta$. The traditional value of the bottom drag follows:

$$\tau_b = \rho\,g\,H\,\alpha \tag{6}$$

Is the local equilibrium assumption justified? Consider flow over a rigid bedrock having a sine profile, namely, with tilted coordinates, $\zeta = a \cos \omega x$, without ice-bed separation. It has been shown (Nye, 1969) that, if ice were an isoviscous medium, when $H \gg \lambda$ the stress and strain rate fluctuations would damp upwards as $e^{-\omega z}$. They would be divided by 20 at $\omega z = 3$, $z = \lambda/2$. With finite H, small surface oscillations would be found when the wavelength equals about 3 H. Curiously, such oscillations often exist at the periphery of the Indlandsis or of eastern Antarctica, but they should come from fluctuations in the accumulation. These fluctuations have been discovered and attributed to internal gravity waves in the downslope katabatic wind, by Pettre and others (1986).

In fact ice-sheets are not at all isoviscous. The upper part is much more viscous than the lower one, because it is much colder. Therefore, it acts as a quasi-rigid lid, that transmits horizontal stresses over long distances. Internal reflecting horizons found in VHF soundings, that are of sedimentary origin and closely follow the stream lines, prove that oscillations of the streamlines shorter than 50 km are confined in the bottom layers (Robin, 1977). Since in numerical computations the horizontal mesh size is equal to or larger than 50 km, the assumption of local equilibrium is generally justified.

For the same reason, in 3-D models, the assumption that flow at any depth is in the direction of the steepest surface slope is justified. This assumption allows us to determine, when the topography of the surface and of the bed are known, and the accumulation has been determined by shallow corings during traverses, which would be the ice fluxes if the ice-sheet were in a steady state. They are termed "balance fluxes" and the corresponding mean velocities along a vertical profile are termed "balance velocities" (Budd and others, 1971; Radok and others, 1987). Note that, even if measured velocities fit the computed balance velocities, the ice sheet might not be in a steady state because temperatures might be evolving with time.

Local equilibrium cannot be assumed when ice streams with high velocities appear. It is the case for outlets whose bottom is below sea level, probably because there are subglacial water cavities connected with the ocean. With horizontal mesh sizes of 100 km, most local ice streams can be ignored, but not the largest ones. In "marine" ice-sheets, i.e., ice sheets whose bed is entirely below sea level, such large ice streams may appear anywhere. We shall come back later on this case.

In 1951 Nye introduced the assumption τ_b = a constant, as a consequence of his perfectly plastic model. Implicit prerequisites for this model are no sliding, and a steady flow. With sliding, the yield stress at the bottom may be reached on a plane other than the bed. Without sliding, there may be no flow at all in the central area.

On a flat horizontal bed, assuming τ_b as given by (6) to be constant leads to a parabolic surface profile. This assumption has been widely used by modellers. For instance the CLIMAP group (1976) offered to the climatologists calculated thicknesses of the northern hemisphere ice sheets 18,000 years ago assuming that τ_b had some constant values, different according to the subglacial terrane. This assumed relationship has no sound grounding.

Unfortunately, the first surface profile accurately determined, the EGIG profile in central Greenland, favored the constant bottom shear stress assumption (τ_b = 0.87 bar from the west coast to very near the ice divide). It was a red herring. In general τ_b increases more or less linearly from the ice divide to the periphery. In this case, an elliptic profile is obtained with semi-axes H_0 and L:

$$H^2 + \frac{p}{L} x^2 = H_0{}^2 = pL \quad (p = \tau_{max} / \rho g) \tag{7}$$

Since $\tau_{max} \sim 1$ bar, $p \sim 11$ m.

4. THE HEAT EQUATION

In the Eulerian frame of reference used so far, the 2-D heat equation reads (taking into account that $\partial u/\partial x + \partial w/\partial z = 0$):

$$\frac{\partial}{\partial x}\left(K\frac{\partial T}{\partial x}\right) + \frac{\partial}{\partial z}\left(K\frac{\partial T}{\partial z}\right) + \dot{W} = \rho C\frac{\partial T}{\partial t} + \left(u\frac{\partial(\rho CT)}{\partial x} + w\frac{\partial(\rho CT)}{\partial z}\right) \qquad (8)$$

$$\text{(a)} \qquad\qquad \text{(b)} \qquad\quad \text{(c)} \qquad \text{(d)} \qquad\qquad \text{(e)} \qquad\qquad \text{(f)}$$

where $\dot{W}$ is the viscous dissipation of heat per unit volume and unit time.

At 0°C, K = thermal conductivity = 67 MPa m^2 a^{-1} K^{-1},

ρC = thermal capacity per unit volume = 1.83 MPa K^{-1}.

Term (a) is often negligible, but (e) is never so, as the use of stretched coordinates x/L, z/H can show. The viscous dissipation of heat (c) is negligible outside the bottom layers.

In most calculations Nye's flow is assumed, i.e. the vertical strain rate $\dot{\varepsilon}_{zz} = \partial w/\partial z = \partial^2 q/\partial x\partial z$ is assumed to be constant with depth. This assumption requires, as a necessary but not sufficient condition, sliding on the bed, or at least "pseudo-sliding," i.e., a concentration of shear in a bottom boundary layer (BBL). Then, h denoting the thickness of the BBL (assumed to be more or less uniform), u_s denoting the surface velocity, u_1 denoting the forward velocity at the top of the BBL, and $\beta = -\partial\zeta/\partial x$ denoting the bed slope, the vertical velocity at any z is:

$$w = -u_1\beta\left(1 - \frac{z-\zeta-h}{H-h}\right) - \left(b + u_s\alpha + \frac{\partial H}{\partial t}\right)\frac{z-\zeta-h}{H-h} + \frac{\partial\zeta}{\partial t} \qquad (9)$$

The terms $\partial H/\partial t$ and $\partial\zeta/\partial t$ may be neglected. In most cases $u_1\beta$ and $u_s\alpha$ are much smaller than b, and (9) may be replaced by the approximation:

$$w \approx -b\frac{z-\zeta-h}{H-h} \qquad (10)$$

Then, assuming h $\ll$ H, and the viscous dissipation of heat to be negligible outside the BBL, the computation of the temperature field can be done without calculating the forward velocities. Consider the steady state only. Assume H and the boundary conditions to be independent of x. Consequently, T is a function of z only, and the horizontal advection term (e) vanishes. K and C are considered as temperature independent (a rough approximation). Then the heat equation reads (with $K/\rho C = \kappa$):

$$K\frac{d^2T}{dz^2} = -b\frac{z-\zeta}{H}\frac{dT}{dz} \qquad (11)$$

At the top of the BBL the temperature is T_1, and the thermal gradient is $-G_1$. It follows (Robin, 1955):

$$T_s = T_1 - \sqrt{\frac{\pi \kappa H}{2 b}} \, G_1 \, \text{erf} \sqrt{\frac{bH}{2\kappa}} \tag{12}$$

Confusing T_1 with T_b, the temperature at the ice-bed interface, Table I gives the values of G_1 inferred from this formula at the three Antarctic sites of deep drilling. At all three the bottom temperature T_b is the melting temperature T_m corresponding to the ice pressure, and the viscous dissipation of heat is negligible.

Table I: Numerical values in (12) for three deep drilling sites

	H	b	$\text{erf}\sqrt{\frac{bH}{2\kappa}}$	T_s	$T_b = T_m$	$1/G_1$
	(m)	(m a^{-1})		(°C)	(°C)	(m K^{-1})
Byrd station	2164	0.16	0.998	-28.8	-1.8	32.4
Dome C	3440	0.03	0.954	-54.3	-2.8	49.7
Vostok station	3700	0.025	0.890	-56.4	-3.0	54.4

With the further rough approximations $T_1 = 0°C$, erf $= 1$, and $G_1 = G_0$, the geothermal gradient, formula (12) yields the following rule of thumb giving the thickness of a steady ice sheet over flat land, in its central area:

$$H \cong \frac{2 b}{\pi \kappa} \left(T_s / G_0\right)^2 \tag{13}$$

This rough estimation, which is totally different from the one obtained by assuming some bottom shear stress, shows the key roles of accumulation, surface temperature, and geothermal flux. Nevertheless, it is too crude to allow any computation of the surface profile. Anyway, both rules of thumb cannot be used for modelling of past ice sheets in the northern hemisphere. They waxed and waned during lapses of time on the same order as the characteristic evolution time (~ 60,000 years), and thus probably they never reached a steady state.

5. COMPUTATION OF BALANCE TEMPERATURES

Given measured actual values of $H(x)$, $\zeta(x)$, $b(x)$, and $T_s(x)$, assuming that the ice sheet is in a steady state, and adopting some value for the geothermal heat flux (say KG_0), it is possible to determine the corresponding "balance temperatures" within the ice-sheet.

In the sixties, Jenssen and Radok computed balance temperatures without introducing ice rheology at all. They assumed Nye's flow (i.e., a constant vertical strain rate $\dot{\varepsilon}_{zz}$), and a depth-independent forward velocity: $u(x,z) = U(x)$. Then it was convenient to use a frame of reference moving forward at velocity U. In the heat equation (9), term (e) disappears, but (d) has to be kept, even if the ice sheet is in steady state. All the viscous dissipation of heat ($U\tau_b$ per unit area and unit time) happens in the thin BBL. Thus, KG denoting the geothermal flux entering the ice (it differs from KG_0 when there is either melting or refreezing at the glacier base), the heat flux at the top of the BBL is:

$$KG_1 = KG + U\tau_b \qquad (14)$$

Later, one of the two rheological equations was considered, namely:

$$2\,\dot{\epsilon}_{xz} \simeq \frac{\partial u}{\partial z} = B_{\parallel}^0 \exp\left[\frac{E}{R}\left(\frac{1}{T_m} - \frac{1}{T}\right)\right]\tau_{xz}^3 \qquad (15)$$

where $B_{\parallel}^0$ is an unknown parameter, assumed to be constant. It allowed determination of the vertical profile u(z) to the factor $B_{\parallel}^0$, that was determined next by the continuity condition. A loop is necessary, because $\partial u/\partial z$ depends on the temperature profile. Nye's assumption was dropped. Budd and others (1973) calculated the vertical velocity w(z) by an ungrounded relationship that they assumed to hold between w(z) and u(z), whereas Philberth and Federer (1971) used the correct formula:

$$w(z) = -\int_{\zeta}^{z} \frac{\partial u}{\partial x}\,dz \qquad (16)$$

When $B_{\parallel}^0$ is assumed to be constant (ignoring any anisotropy more or less favorable to shear), the problem can be solved without prior knowledge of this rheological parameter because the actual thickness, a given, is assumed to correspond to the steady state. As already stated, this crucial assumption is probably not true. Therefore, balance temperatures and velocities can only be used to plan deep drilling work. Note, however, that drilling sites are chosen at ice divides, because the ice there is of local origin and the results of core analysis are more easily interpreted. At ice divides horizontal deviatoric stresses are not negligible in front of the vanishing shear stress τ_{xz}, and (15) cannot be used. A correct calculation of balance velocities and temperatures at an ice divide (assuming always, unduly, isotropic rheology) is found in Ritz and others (1982).

6. RESPONSE OF AN ICE SHEET TO CHANGES IN THE SURFACE TEMPERATURE OR ACCUMULATION

In the aforementioned paper, assuming a constant value of thickness H, the temperature changes that would follow changes in the surface temperature T_S are computed. First the response to a unit step (the Heaviside function) is calculated. Its derivative is the response of the system to a Dirac impulse. Since for small fluctuations of $T_S(t)$ the system is linear, its response to T_S is obtained by convolving T_S with this impulsive response. This method might be used for any site. It is also possible to compute the periodic response of an ice sheet to an oscillating T_S (Ritz, 1987). For a 100 ka period the temperatures oscillate in the bedrock below an ice sheet 3 km thick down to about 5 km further. Such calculations may be considered as sensitivity studies.

Similar studies of the reaction of the local temperature field to changes in b are not possible. Then, temperatures change because the velocity field is modified, as a consequence of changes in the thickness and the surface slope. Therefore, the reaction of the whole ice sheet has to be computed.

For numerical computation, variables x, z, and t are discretized. Assume that the surface profile of an ice sheet, the velocity and temperature fields have been computed at time t. Then, the volumetric discharge per unit width Q(x) is known. By using Eq. (2) we may compute locally, at two consecutive values of x, the two values of H, hence the surface slope α, at the next instant of time (t+Δt). The new value of τ_b is known, as is the new value of T_s. We need the new value of Q, but, since it depends on the vertical temperature profile, the heat equation (8) has to be introduced. Nevertheless, we may assume that the horizontal advection term (e) has not been significantly modified. If it is known, (8) becomes an equation in z and t only.

A further simplification is to consider separately the BBL and the bulk of the ice sheet. With the large time steps that should be used to cope with the ice ages, we may consider that the BBL has always reached a steady state (for the boundary conditions at its limits at time t+Δt), whereas the bulk of the ice sheet is in an unsteady state. In the bulk of the ice sheet Nye's flow may be assumed, i.e., the vertical velocity w entering (8) is given by (9), with only u_1 as unknown. The only other unknown quantities at time t+Δt that we need to solve (8) are the thickness h of the BBL, and either the thermal gradient G_1 or the temperature T_1 at its top. As we shall see, when the glacier base is below melting point these three quantities can be approximated by simple formulas. When bottom melting occurs, there may be sliding and the problem becomes more complex.

7. THE BOTTOM BOUNDARY LAYER MODEL

The BBL is assumed to be in a steady state, and subject to simple shear. The x-axis is now taken along the ice-bed interface, which is tilted by β from the horizontal. The creep law (15) may be simplified, since the temperature range in the BBL is moderate. Putting:

$$T - T_m = \theta, \qquad \frac{E}{R\,T_m^2} = k = 0.125 \text{ K}^{-1},$$
$$\dot{\gamma} = \frac{\partial u}{\partial z} = B_{\parallel}^{0}\,\tau_{xz}^{3}\,e^{k\theta} \tag{17}$$

Values at the very bottom (z = 0) will be indicated by subscript b. Eqs. (17) and (8) read:

$$\frac{\partial u}{\partial z} = \dot{\gamma}_b \left(1 - \frac{z}{H}\right)^3 \exp\left[k(\theta - \theta_b)\right]$$
$$\frac{\partial^2 \theta}{\partial z^2} = \frac{1}{\kappa} u \frac{\partial T}{\partial x} - \frac{\dot{\gamma}_b \tau_b}{K}\left(1 - \frac{z}{H}\right)^4 \exp\left[k(\theta - \theta_b)\right] \tag{18}$$

They allow us to write Taylor expansions in z about z = 0 of u and T. Note that $u_b\,\partial T_b/\partial x$ is zero, because either T_b equals the melting point and does not vary with x, or sliding velocity u_b is zero. With the notations:

$$kGH = \Gamma, \quad \frac{\dot{\gamma}_b \tau_b H^2}{K} = \mu, \quad \frac{H^3}{\kappa}\left[\frac{\partial T_b}{\partial x}\dot{\gamma}_b - \frac{\partial G}{\partial x}u_b\right] = \upsilon \quad (19)$$

it is found:

$$\theta = \theta_b - Gz - \mu\frac{z^2}{2H^2} + [\mu(\Gamma+4) + \upsilon]\frac{z^3}{6H^3} + \cdots$$

$$u = u_b + \dot{\gamma}_b H\left\{\frac{z}{H} - (\Gamma+3)\frac{z^2}{2H^2} + \left(\Gamma^2 + 6\Gamma + 6 - k\mu\right)\frac{z^3}{6H^3} - \cdots\right\} \quad (20)$$

The top of the BBL, z = h, is the level where heat advection and viscous dissipation of heat are of equal importance. To estimate heat advection in the BBL the simple shear approximation above cannot be used, as I have unduly done in the past (Lliboutry, 1987a). The vertical advection (term (f) in Eqn. (8)) must be considered. To terms in h^2, the equality $\dot{\gamma}\tau = \rho C \lvert u\partial T/\partial x + w\partial T/\partial z\rvert$ reads:

$$\dot{\gamma}_b\tau_b\left[1 - (\Gamma+4)\frac{h}{H}\right] = \rho C\left|(u_b + \dot{\gamma}_b h)\left(\frac{dT_b}{dx} - \frac{dG}{dx}h\right) + \left(-\frac{du_b}{dx}h\right)(-G)\right|$$

$$= \rho Ch\left|\dot{\gamma}_b\frac{dT_b}{dx} - u_b\frac{dG}{dx} + \frac{du_b}{dx}G\right| \quad (21)$$

In case of no sliding ($u_b = 0$):

$$\frac{H}{h} = \Gamma + 4 + \frac{\rho CH}{\tau_b}\left|\frac{dT_b}{dx}\right| \quad (22)$$

In case of sliding (that implies $T_b = T_m$):

$$\frac{H}{h} = \Gamma + 4 + \frac{\rho CH}{\tau_b\dot{\gamma}_b}\left|\frac{du_b}{dx}G - u_b\frac{dG}{dx}\right| \quad (23)$$

Numerically, when all the geothermal flux enters the ice, $\Gamma + 4 = 8$ to 14. The last term is of the same order of magnitude or larger.

Once h has been calculated, the velocity u_1 and the temperature T_1 at level z = h are given by (20). Since derivatives in x are needed for calculating h, a loop is necessary, but the best algorithm has not yet been sought.

8. ICE SHEETS WITH THE BOTTOM AT MELTING POINT

When the base is at melting point, several cases are theoretically possible. My statements in the sixties (Lliboutry, 1964, p. 413) can be improved as follows.

(1) <u>Part of the geothermal flux enters the BBL</u>. There may be some sliding velocity u_b. Sliding is defined as the extra velocity due to processes at a smaller scale than the smoothing length that has to be introduced for defining the bed profile, with the BBL dynamics in view. Over hard beds, there are two sliding processes: enhancement of stresses and strain rates to overcome the microrelief, and melting on the stoss side of any minute obstacle with simultaneous refreezing on the lee side. The former provides extra heat within the ice, the latter provides heat within the water film of micrometric thickness that carries water from the melting areas to the refreezing ones. Thus, the Newtonian energy that is lost by sliding, $u_b \tau_b$ per unit area and unit time, is divided into two parts, say $u_v \tau_b$ and $u_m \tau_b$, with $u_v + u_m = u_b$. Only $u_m \tau_b$ is dissipated at the interface. Eq. (14) reads:

$$G_1 = G + (u_1 - u_b) \tau_b / K \qquad (24)$$

With ρL denoting the melting heat per unit volume, and $K_b G_b$ the heat flux in the bedrock (it may differ from the steady geothermal flux considered so far), the melting rate per unit area is:

$$M = K_b G_b - K G + u_m \tau_b \qquad (25)$$

In this case, G may be very small, or even zero. Consequently, Γ may vanish, and the thickness h of the BBL be very large. The strain in the BBL differs from simple shear, vertical advection of heat cannot be neglected, and the BBL model becomes very inaccurate.

(2) <u>Some water is steadily refreezing at the ice-bed interface</u>, at the macroscopic scale. The released heat increases the heat flux entering the BBL. Eqs. (24-25) remain valid, with a negative value of M.

(3) <u>There is a temperate bottom boundary layer</u> (TBBL), i.e. a bottom layer of ice entirely at melting point. All the geothermal flux melts ice. The viscous dissipation of heat within the TBBL continuously produces water inclusions, that sooner or later should drain off to the ice-bed interface.

On the TBBL there is a cold (below melting point) BBL. The formulas above are valid, with G = 0 and u_b = the true sliding velocity + the velocity due to the shearing of the TBBL. However, as has been said, they become quite unaccurate.

The existence of a TBBL is documented by radar soundings in Spitsberg, at the bottom of large and relatively thin tidal glaciers (Dowdeswell and others, 1984).

Two new problems have appeared. In case (2), is water available for refreezing? In all three cases, how can the processes at the scale of the microrelief be parameterized, to yield a macroscopic sliding law?

In the opinion of most authors, refreezing is possible because water has been produced upstream, in areas where the sliding velocity is surely $u_b < 10$ m/a, and the bottom shear stress $\tau_b < 1$ bar. The corresponding water production is less than 0.3 cm/a. On the other hand, when all the geothermal flux is used to melt ice, the corresponding water production is on the order of 1 cm per year. Water flow does not depend on the bed slope β, but on the horizontal gradient of the hydraulic head, that is $(\rho/\rho_w)g \, \alpha(\rho_w$ denoting the water density). It may be shown that, even without a layer of porous sediments, the mere permeability of rocks at a large scale (joints are always present) is enough to ensure drainage. Locally, subglacial lakes may exist (they have been detected in eastern Antarctica), and there may be some subglacial streamlets. But a general water sheet at the ice-bedrock interface in the interior of ice sheets, as assumed by many authors (following Weertman, 1972), is unrealistic. Therefore, case (2) seems only possible when the bed consists of water-soaked porous sediments. Then, a frozen ground forms, that would be dragged along by the ice sheet, producing the well-known shear moraines at its edge.

9. SLIDING, AND SHEAR OF UNFROZEN SOFT SEDIMENTS

As for sliding, rocky beds and unfrozen soft beds must be considered separately.

(1) Over a rough bedrock, in general, sliding should require a temperate bottom boundary layer. Otherwise, there are large knobs and hillocks protruding into cold ice that act as pinning points. They are covered with minute obstacles that can be overcome by a melting-refreezing process only. In general, the overpressure on their stoss face is not high enough for the melting temperature to be reached locally.

Nevertheless, whether a TBBL exists at the bottom of fast tidal ice streams is a matter of conjecture. We speculate here that, without TBBL, but with G small enough, the melting-refreezing process around minute obstacles should operate on the stoss sides of any hillock, whereas a smooth fault should exist over the minute obstacles on the lee side, making this side smooth.

The sliding law of a temperate ice layer over a bumpy bedrock has been a matter of debate for a long time. Conflicting results came from the fact that the sliding law depends on the model of microrelief and the subglacial water connections that have been adopted. (A claimed "experimental law" used by Radok and others, 1987, is criticized in Lliboutry, 1987a, p. 184-185.)

Two quite distinct, extreme cases, may be recognized, with all the intermediate situations (Lliboutry, 1987b). They correspond respectively to viscous friction without ice-bedrock separation at slow sliding velocities, and solid friction with ice-bedrock separation at high sliding velocities. Separation is possible because the cavities that appear at the lee of the bumps are filled with water under pressure. Moreover, it is assumed that all these cavities are interconnected, with about the same hydraulic head (thereafter called the subglacial water pressure).

Let N denote the ice lithostatic pressure on the bed, at the macroscopic scale, minus the subglacial water pressure. Some friction coefficient f intervenes. It is of the same order as the mean quadratic slope of the microrelief, ignoring the largest and smallest wavelengths.

When $\tau_b < fN$, the sliding law is of the viscous kind:

$$u_b = C\,\tau_b^2 \quad \text{(Weertman, 1957)}$$

$$\text{or:} \quad u_b = C_1\,\tau_b + C_3\,\tau_b^3 \quad \text{(Lliboutry, 1987a, p. 157-159)}$$

When N is small enough, $\tau_b = fN$, and u_b is no longer locally determined. The surrounding areas where viscous friction operates control the sliding velocity, and the transmission of horizontal stresses becomes essential. There is no local equilibrium at the 10 km scale. Since 10 km is less than the mesh size used to deal with the entire ice sheet, at the ice edge some mixture of viscous and solid friction may be adopted without considering the precise geometry of fast ice streams and slow ice ramps.

(2) When the bed consists of soft unfrozen sediments, it has been smoothed by the ice sheet, and sliding should be possible without TBBL. Anyway, most of the sliding may come from shear within the water-soaked sediments (Boulton and Jones, 1979; Alley, 1991).

In this case, fast velocities should appear when the pore pressure in the sediments is high enough, as is well known and studied in soil mechanics. Large ice streams exist in western Antarctica that drain the ice of Marie Byrd Land into the Ross Ice Shelf. Seismic studies have shown that sliding there is due to the shearing of a water-soaked layer of porous sediments about 6 m thick (Blankenship and others, 1986; Alley and others, 1986).

It seems that the discharge of ice by the different ice streams has considerably changed at the time scale of a few centuries (Shabtaie and others, 1988). Surges, i.e. catastrophic changes (in the mathematical sense, meaning a jump from a steady state to another one), might occur in West Antarctica. Nevertheless, the triggering of an Antarctic surge would differ from the case of popularized surges in mountains of middle latitudes. In the latter, each summer, one or several meters of melt water appear at the surface and find their way to the glacier base. Therefore, to use for surges empirical formulas fitted to middle latitude glaciers (Radok and others, 1987) makes little sense.

The trouble with modelling the West Antarctica ice-sheet is that the size of its ice streams is of the same order as its mesh size. Therefore, ice streams cannot be ignored, and 3-D models must be used. Results obtained with sophisticated codes and supercomputers will be questionable, however, as long as anisotropic ice rheology is not introduced.

10. GRAIN GROWTH AND FABRICS

Up to now, the shear strain rate has been assumed to vary as the third power of the shear stress (Eq. 15). When the ice is macroscopically isotropic, symmetry considerations show that the strain rate tensor (matrix $\dot{\underline{E}}$, with elements $\dot{\varepsilon}_{ij}$) must be proportional to the deviatoric stress (matrix $\underline{S}$, with elements s_{ij}):

$$2\,\dot{\underline{E}} = \frac{1}{\eta}\,\underline{S} \tag{27}$$

The fluidity $1/\eta$ (the inverse of viscosity η) must be a function of temperature and of the three stress invariants only. Experiments on isotropic ice have shown that only the second invariant, $\mathrm{Tr}(\underline{S}^2) = s_{ij}\,s_{ij} = 2\,\tau^2$, intervenes, not the third one, $\mathrm{Tr}(\underline{S}^3) = 3\,\det\underline{S}$. (In plane flow $\det\dot{\underline{E}} = 0$, and thus, if ice behaves isotropically, $\det\underline{S} = 0$). In general, laboratory data are analyzed by fitting a power law:

$$1/\eta = B\,(T)\,\tau^{n-1} \tag{28}$$

When $n = 1$, the material is said to be Newtonian. Accurate creep tests, long enough to eliminate transient creep, give $n = 3$ in the range $0.02\ \mathrm{MPa} < \tau < 0.7\ \mathrm{MPa}$ (Duval, 1981). At lower stresses, n is less than 3. It seems more satisfactory to fit the data by:

$$1 / \eta = \psi + B \tau^2 \qquad (29)$$

so avoiding singularities at points where $\tau = 0$, that make any computational scheme unstable.

Ice sheet coring has shown that the crystal (= grain) size increases with depth, owing to interfacial energy, until the elastic energy due to linear defects in the lattice (dislocations) becomes predominant. Rare processes (superplasticity by grain growth, Harper-Dorn creep with a constant dislocation density) should provide a Newtonian viscosity in the upper layers (Lliboutry and Duval, 1985). Since $\tau_{xz} \cong \tau$ increases linearly with depth, the precise law in these upper layers is unimportant. More at depth, standard dislocation creep leads to (29). Strain in each crystal is due to the motion of numerous screw dislocations in slip planes [0001], perpendicular to the hexagonal axis of symmetry (the c-axis) (Higashi, 1988).

Simultaneously, strain causes the c-axes to cluster near the vertical direction, or, sometimes, in a vertical plane perpendicular to the flow (Lipenkov and others, 1989). More or less vertical c-axes favor the shear strain rate $\partial u/\partial z$, and hinder the extensive strain rate $\partial u/\partial x$. In the creep law relative to simple shear:

$$2 \dot{\varepsilon}_{xz} = \left(\psi_{\|} + B_{\|} \, \tau_{xz}^2 \right) \tau_{xz} \qquad (30)$$

$B_{\|}$, and maybe $\psi_{\|}$, are larger than B and ψ for macroscopically isotropic ice at the same temperature. Ignoring the linear term, the ratio $B_{\|}/B$ has been termed the enhancement factor (Russell-Head and Budd, 1979).

We have not, at this time, a trustworthy law relating $B_{\|}$ with the fabrics. Models relating the fabrics with past strain have been suggested (Alley, 1988), but we have no model predicting when this fabric is destroyed by recrystallization, as observed near the bottom (Budd and Jacka, 1989). Higher temperatures should have a key role, but a state of stress differing from simple shear should also intervene (Duval, 1981). Much more experimental work must be done to model the evolution of ice fabrics during the flow.

An intriguing fact is the abrupt decrease in grain size, and corresponding enhancement of the fabric, when entering Wisconsin ice, i.e. ice from snow deposited during the last ice age (Gow, 1970; Gundestrup and Hansen, 1984). The same has been found with the Vostok ice core, that reaches ice from the Illinoian ice age (Lipenkov and others, 1989). Explanation by the higher dust content (Alley and Bentley, 1988) does not fit detailed stratigraphy. A fuzzy explanation by persistent defects in the lattice, as given by Petit and Duval, seems contradicted by the fact that fine-grained Wisconsin ice is found in Greenland, with surface temperatures of about -30°C, as well as in East Antarctica, with surface temperatures of about -60°C. This fine-grained ice is more anisotropic than the ice just above it, a hard fact that must be included in any accurate predictive model.

Recrystallization must not be confused with the mere migration of grain boundaries, which allows the strains of individual grains to be compatible, and grains to remain equisized. It consists of the appearance of new grains, that are not yet work-hardened. A new stage of steady creep is reached when the birth rate of new crystals balances their work-hardening. This "recrystallization creep" leads to a peculiar fabric, with the c-axes clustered into 3 to 5 (generally 4) directions, that makes ice macroscopically isotropic (Duval, 1981). Then $n = 3$ and, at melting point, $B = 220 \pm 40$ MPa^{-3} a^{-1} when ice is dry (as in the BBL). When it is wet (as in the TBBL) it may rise up to 1000 MPa^{-3} a^{-1} (Lliboutry and Duval, 1985). These values are higher than for isotropic ice without recrystallization (B = 63 ± 7 MPa^{-3} a^{-1}), but smaller than B$_{\parallel}$ for very anisotropic ice, with almost vertical c-axes.

11. CREEP LAW FOR ANISOTROPIC, TRANSVERSELY ISOTROPIC ICE

As long as 3-D models are not tackled, this subject is of academic interest only. Some hints will be given, however, because it raises important problems in materials science. More details may be found in Lliboutry (1987a).

Consider any viscous material, and assume that the mean normal stress does not enter its constitutive law (its creep law in this case). Then, at a given temperature, $\dot{\underline{\underline{E}}}$ is a function of $\underline{\underline{S}}$ only. Since Tr $\dot{\underline{\underline{E}}}$ = Tr $\underline{\underline{S}}$ = 0, only five independent scalar components for $\dot{\underline{\underline{E}}}$ or $\underline{\underline{S}}$ must be considered, say $\dot{\gamma}_i$ and s_i (i = 2 to 6) respectively. They may be chosen such that the viscous dissipation of heat per unit volume and unit time be:

$$\dot{W} = \dot{\varepsilon}_{ij} \, s_{ij} = \dot{\gamma}_i \, s_i \tag{31}$$

Thermodynamics shows that there is a dissipation function $\Phi(s_i)$ such that:

$$\dot{\gamma}_i = \partial\Phi \, / \, \partial s_i \tag{32}$$

Consider now a viscous material that is orthotropic, transversely isotropic, i.e. that has a rotational symmetry about some material axis, that will be chosen as z-axis, the x-y plane being a plane of symmetry. In this case, convenient $\dot{\gamma}_i$ and s_i are as follows:

$$\underline{\dot{\gamma}} = \begin{bmatrix} \dot{\gamma}_d \\ \dot{\gamma}_{ax} \\ \dot{\gamma}_4 \\ \dot{\gamma}_5 \\ \dot{\gamma}_6 \end{bmatrix} = \begin{bmatrix} \dot{\varepsilon}_{xx} - \dot{\varepsilon}_{yy} \\ \sqrt{3} \, \dot{\varepsilon}_{zz} \\ 2 \, \dot{\varepsilon}_{yz} \\ 2 \, \dot{\varepsilon}_{zx} \\ 2 \, \dot{\varepsilon}_{xy} \end{bmatrix}, \quad \underline{s} = \begin{bmatrix} s_d \\ s_{ax} \\ s_4 \\ s_5 \\ s_6 \end{bmatrix} = \begin{bmatrix} (\sigma_x - \sigma_y)/2 \\ \{\sigma_z - (\sigma_x + \sigma_y)/2\}/\sqrt{3} \\ \tau_{yz} \\ \tau_{zx} \\ \tau_{xy} \end{bmatrix} \tag{33}$$

To avoid singularities for $\underline{s} = 0$, Φ is assumed to be the sum of two homogeneous polynomials of degree 2 and 4 respectively. They must be invariant for any rotation about the z-axis. Thus they are functions of the following quantities, that are invariant for such a rotation, are independent from each other, and are polynomials in the s_i of the lowest degrees possible:

$$s_{ax}, \, \tau_{\perp}^2 = s_d^2 + s_6^2, \, \tau_{\parallel}^2 = s_4^2 + s_5^2, \, K_3 = s_d \left(s_5^2 - s_4^2\right) + 2 \, s_4 \, s_5 \, s_6 \tag{34}$$

There are no other independent invariants. For instance:

$$\tau^2 = s_{ax}{}^2 + \tau_\perp{}^2 + \tau_\|{}^2, \quad \det \underline{\underline{S}} = K_3 + \frac{2\,s_{ax}}{\sqrt{3}}\left[\frac{s_{ax}{}^2}{3} - \tau_\perp{}^2 + \frac{\tau_\|{}^2}{2}\right] \tag{35}$$

The final result is more easily written by introducing three fluidities, where 9 rheological coefficients enter:

$$\begin{bmatrix} \phi_{ax} \\ \phi_\perp \\ \phi_\| \end{bmatrix} = \begin{bmatrix} \psi_{ax} \\ \psi_\perp \\ \psi_\| \end{bmatrix} + \begin{bmatrix} B_{ax} & A_{ax\perp} & A_{ax\|} \\ A_{ax\perp} & B_\perp & A_{\|\perp} \\ A_{ax\|} & A_{\|\perp} & B_\| \end{bmatrix} \begin{bmatrix} s_{ax}{}^2 \\ \tau_\perp{}^2 \\ \tau_\|{}^2 \end{bmatrix} \tag{36}$$

The general strain rate - stress relationship is:

$$\dot{\underline{\gamma}} = \underline{\underline{M}}\,\underline{s}$$

$$\underline{\underline{M}} = \begin{bmatrix} \phi_\perp & \dfrac{D}{2}\left(s_5{}^2 - s_4{}^2\right) & 0 & 0 & 0 \\[2ex] \dfrac{D}{2}\left(s_5{}^2 - s_4{}^2\right) & \phi_{ax} & D\,s_5\,s_6 & 0 & 0 \\[2ex] 0 & D\,s_5\,s_6 & \phi_\| - D\,s_d\,s_{ax} & 0 & 0 \\[2ex] 0 & 0 & 0 & \phi_\| - D\,s_d\,s_{ax} & D\,s_{ax}\,s_4 \\[2ex] 0 & 0 & 0 & D\,s_{ax}\,s_4 & \phi_\perp \end{bmatrix} \tag{37}$$

The tenth parameter D, that makes the matrix tridiagonal, comes from a term $(D/2)\,s_{ax}\,K_3$ in the expression of Φ. Thus anisotropy makes the third stress invariant to enter the rheological law.

Neither the ten parameters could all be easily determined by creep tests, nor a problem in continuum mechanics managed. Therefore, the author has inferred these parameters from homogenization. Two assumptions are made: (1) negligible heat dissipation at the grain boundaries; (2) stresses in every grain more or less equal to the macroscopic stress. For a single crystal embedded in polycrystalline ice, the dissipation function is $\psi_B\,\tau_B{}^2/2 + B_B\,\tau_B{}^4/4$, where τ_B denotes the shear stress resolved on the shear plane of the crystal, that is perpendicular to its c-axis. ψ_B and B_B are not the same as for an isolated crystal. They depend on the fabric, and thus cannot be drawn from tests on macroscopically isotropic ice. Summing up the dissipation functions of every grain, the dissipation function of polycrystalline ice can be calculated. The ten rheological parameters are, to the unknown factors ψ_B and B_B, linear functions of the four first even moments of the c-axes distribution.

The creep law gets still much simpler when all the c-axes are clustered near the z-axis. Then, we have:

$$\Psi_{ax} = 3\,\Psi_\perp \ll \Psi_\parallel$$

$$B_{ax}\,,\,B_\perp\,,\,A_{ax\perp} \ll A_{\parallel\perp} = A\,,\,A_{ax\parallel} = 3\,A\,,\,-D = \sqrt{3}\,A \ll B_\parallel \tag{38}$$

Moreover, assuming either a Fisherian distribution of the c-axes, or a simpler power-law distribution, it is found:

$$\frac{\Psi_\perp}{\Psi_\parallel} = \frac{A}{2B_\parallel} = \frac{1 - \cos\theta_{1/2}}{\text{Log }2} \tag{39}$$

where $\theta_{1/2}$ denotes the median of the angles between the c-axes and the z-axis.

In the 2-D problem, $s_4 = s_6 = 0$, and $\dot\gamma_d + \dot\gamma_{ax}/\sqrt{3} = 0$. In the considered case of very anisotropic ice it is found:

$$\sigma_y = \sigma_z \left(\text{whereas for isotropic ice } \sigma_y = \frac{\sigma_x + \sigma_z}{2} \right)$$

$$\dot\varepsilon_{xx} = -\dot\varepsilon_{zz} = \left(\Psi_\perp + \frac{3\,A}{2}\,\tau_{xz}^2 \right) \frac{\sigma_x - \sigma_z}{2} \tag{40}$$

$$2\,\dot\varepsilon_{xz} = \left(\Psi_\parallel + B_\parallel\,\tau_{xz}^2 \right) \tau_{xz}$$

12. CONCLUSIONS

This review has been longer than expected by the organizers of the ASI, because actual modelling of ice-sheets is unsatisfactory. The problem is much more complex, and conceptually difficult, than many investigators imagine, even when external feedbacks, on the ice age time scale, are not considered.

In the fifties, glaciologists realized that ice rheology was not linear, and strongly temperature-dependent. In the seventies, the important role of subglacial water was recognized, and in the eighties the role of subglacial soft sediments. It appears that any progress is still hampered by our poor quantitative knowledge about the rheology of anisotropic polar ices.

Thus, starting from well-funded studies on climatic changes, glaciologists have to face problems in materials science. It was the topic that interested physicists since the beginning (cf. an experiment by Lord Kelvin in 1849). Unfortunately, deep polar ice cores are a scarce and precious material, that paleoclimatologists keep for their own studies, and that cannot be reproduced in the laboratory.

13. REFERENCES

Alley, R.B. (1988) Fabrics in polar ice sheets: development and prediction. Science 240, 493-495.

Alley, R.B. (1991) Deforming-bed origin for southern Laurentide till sheets? J. Glaciol. 37, 67-76.

Alley, R.B. and C.R. Bentley (1988) Long term climate changes from crystal growth. J.R. Petit et al., Reply. Nature 332, 593-594.

Alley, R.B., D.D. Blankenship, C.R. Bentley and S.T. Rooney (1986) Deformation of till beneath ice stream B, West Antarctica. Nature 322, 57-59.

Blankenship, D.D., C.R. Bentley, S.T. Rooney and R.B. Alley (1986) Seismic measurements reveal a saturated porous layer beneath an active Antarctic ice stream. Nature 322, 54-57.

Boulton, G.S. and A.S. Jones (1979) Stability of temperate ice caps and ice sheets resting on beds of deformable sediment. J. Glaciol. 24, 29-43.

Budd, W.F. and T.H. Jacka (1989) A review of ice rheology for ice-sheet modelling. Cold Regions Sci. Technol. 16, 107-144.

Budd, W.F., D. Jenssen and U. Radok (1971) Derived Physical Characteristics of the Antarctic Ice Sheet. Meteor. Dept., Univ. of Melbourne, Publ. 18.

Budd, W.F., D. Jenssen and N.W. Young (1973) Temperature and velocity interaction in the motion of ice sheets. 1st Australian Conf. on Heat and Mass Transfer, Monash Univ., Melbourne, 17-24.

CLIMAP Group (1976) The surface of the ice age earth. Science 191, 1131-1136.

Dowdeswell, J.A., D.J. Drewry, O. Liestøl and O. Orheim (1984) Radio echo-sounding of Spitsbergen glaciers. J. Glaciol. 30, 16-21.

Duval, P. (1981) Creep and fabrics of polycrystalline ice under shear and compression. J. Glaciol. 27, 129-140.

Gow, A.J. (1970) Preliminary results of studies of ice cores from the 2164 m deep drill hole, Byrd station, Antarctica. Int. Symp. on Antarctic Glaciological Exploration, IAHS Publ. 86, 78-90.

Gundestrup, N.S. and B. Lyle Hansen (1984) Bore-hole survey at Dye 3, south Greenland. J. Glaciol. 30, 282-288.

Higashi, A. (ed.) (1988) Lattice Defects in Ice Crystals. Hokkaido Univ. Press, Sapporo. 156 p.

Lipenkov, V.Ya., N.I. Barkov, P. Duval and P. Pimienta (1989) Crystalline texture of the 2083 m ice core at Vostok station, Antarctica. J. Glaciol. 35, 392-398.

Lliboutry, L. (1979) A critical review of analytical approximate solutions for steady state velocities and temperatures in cold ice sheets. Z. Gletscherkde. Glazialgeol. 15, 135-148.

Lliboutry, L. (1987a) Very Slow Flows of Solids, Basics of Modeling in Geodynamics and Glaciology. Martinus Nijhoff/Kluwer Acad. Publ., Dordrecht. 510 p.

Lliboutry, L. (1987b) Sliding of cold ice sheets. In: The Physical Basis of Ice Sheet Modelling. IAHS Publ. 170, 131-143.

Lliboutry, L. and P. Duval (1985) Various isotropic and anisotropic ices found in glaciers and polar ice caps and their corresponding rheologies. Annales Geophysicae 3, 207-224.

Nye, J.F (1969) A calculation on the sliding of ice cover over a wavy surface using a Newtonian viscous approximation. Proc. Roy. Soc. London A 311, 445-467.

Pettre, P., J.F. Pinglot, M. Pourchet and L. Reynaud (1986) Accumulation distribution in Terre Adélie, Antarctica: effect of meteorological parameters. J. Glaciol. 32, 486-500.

Philberth, K. and B. Federer (1971) On the temperature profile and the age profile in the central part of cold ice sheets. J. Glaciol. 10, 3-14.

Radok, U., D. Jenssen and B. McInnes (1987) On the Surging Potential of Polar Ice Streams. DOE/ER/60197-H1 (Distributed by NTIS, Springfield VA 22161).

416

Ritz, C. (1987) Time dependent boundary conditions for calculation of temperature fields in ice sheets. In: The Physical Basis of Ice Sheet Modelling. IAHS Publ. 170, 207-216.

Ritz, C., L. Lliboutry and C. Rado (1982) Analysis of a 870 m temperature profile at Dome C. Annals Glaciol. 3, 284-289.

Robin, G. deQ. (1955) Ice movement and temperature distribution in glaciers and ice sheets. J. Glaciol. 2, 523-532.

Robin, G. deQ., D.J. Drewry and D.T. Meldrum (1977) International studies of ice sheet and bedrock. Phil. Trans. Roy. Soc. London B 279, 185-196.

Russell-Head, D.S. and W.F. Budd (1979) Ice-sheet flow properties derived from bore-hole shear measurements combined with ice-core studies. J. Glaciol. 24, 117-130.

Shabtaie, S., C.R. Bentley, R.A. Bindschadler and D.R. McAyeal (1988) Mass-balance studies of ice streams A, B and C, West Antarctica, and possible surging behaviour of ice stream B. Annals Glaciol. 11, 137-149.

Weertman, J. (1957) On the sliding of glaciers. J. Glaciol. 3, 33-38.

Weertman, J. (1972) General theory of water flow at the base of a glacier or ice sheet. Rev. Geophys. Space Phys. 10, 287-333.

GLACIER FLOW MODELING

BARCLAY KAMB
Division of Geological and Planetary Sciences
California Institute of Technology
Pasadena, CA 91125

Contents

D. B. Stone and S. K. Runcorn (eds.),
Flow and Creep in the Solar System: Observations, Modeling and Theory, 417–506.
© 1993 *Kluwer Academic Publishers. Printed in the Netherlands.*

418

1. Introduction

As a geophysical fluid dynamics phenomenon, glacier motion has features in common with some of the other terrestrial and planetary flows considered in this book, but it also has a number of special features that give it a character of its own. It can be compared to mantle flow and convection: both involve slow flow of essentially incompressible fluids of very high viscosity, at very low Reynolds number, with negligible inertial, magnetic, and Coriolis forces. The rheology is similar in terms of temperature dependence and nonlinear stress dependence. Also, coupling of fluid flow and heat flow is usually important. A notable difference between mantle flow and glacier flow problems is in the nature of the flow boundary conditions: in convection one usually assumes either a no-slip condition or a no-shear-stress condition at the base of the mass, whereas in glacier flow a stress-controlled slip-rate condition is required in many cases; also in convection it is common to place a no-slip or specified-slip condition at the surface, whereas in glacier flow the surface boundary condition is always stress free. Another difference is that, while both types of flow are in response to gravity as the driving force, in mantle flow it is usually expressed either as a buoyancy force or as a surface mass loading, whereas in glacier flow it usually acts as a downslope body force. Glaciers are often, and ice sheets always, thin compared to their horizontal dimensions, and in this respect their flow has geometrical similarity with "thin-skinned tectonics," or, on an even larger scale, the tectonics of the earth's crust as distinct from that of the mantle. Among tectonic phenomena glacier flow is most closely related to detachment thrusting and other forms of gravitational tectonics (Bucher, 1956; Hills, 1963, p. 79, 336; Suppe, 1985, p. 286). And of course glacier mechanics and mantle mechanics share common ground in dealing with isostatic adjustment to ice-sheet loading.

The Rayleigh number in large polar ice masses may be of order 10^4-10^5, which led Hughes (1972, 1976) to propose that convection occurs in the ice sheets, but no actual evidence of convection has been seen (Paterson, 1981, p. 170; for a contrary view see Hughes, 1985).

Glacier flow modeling is done principally for two general reasons: (1) verification: to find out whether concepts of glacier mechanics, such as those in Section 2, can account quantitatively for observed features of actual glacier flow; and (2) prediction: to predict the response of glaciers and ice sheets to external or internal influences such as climatic change or surge initiation. Modeling of the great ice sheets is a means of extending the range of applicability of sparse observations and of reconstructing the vanished Pleistocene ice sheets (Hughes, 1981; Budd and Smith, 1987). Flow modeling is specifically needed in interpreting ice core samples and borehole temperature profiles in terms of the climatic and topographic history of the ice sheets as well as the chemical history of the atmosphere (Robin, 1983; Oeschger and Langway, 1989). It was also undertaken to help resolve the question as to whether the ice sheets are currently in a state of mass balance or are growing or shrinking (Budd et al., 1971, pp. 4, 145, 163).

In a broader perspective of earth and planetary science, glacier modeling provides a natural test-bed with which to ascertain how nonlinear flow mechanics can handle quantitatively the flow behavior of large solid masses on geophysically interesting scales of space and time, in which actual measurements of flow can be made and flow processes can be directly observed. Glacier-type modeling has been applied to the Martian polar caps (Budd et al., 1986b; Clifford, 1987; Hofstadter and Murray, 1990) and "glaciotectonic" flow phenomena in the icy satellites of the major planets have been modeled, including the role of high pressure ice phases and ammonia-containing ices (Poirier, 1982; Echelmeyer and Kamb, 1986; Stevenson and Lunine, 1986; Kirk and Stevenson, 1987; Durham et al., 1988; Herrick and Stevenson, 1990).

This article reviews salient principles and physical foundations of glacier flow modeling with some comparison of modeling results and observations. Details of the numerical methods used in modeling are not treated; for them, refer to the original papers cited. Another current review of the subject is by van der Veen (1991). "The Physical Basis of Ice Sheet Modelling" was the subject of a recent international symposium (Waddington and Walder, 1987), and a textbook of ther-momechanical flow modeling in both glaciology and geodynamics has been written by Lliboutry (1987). Van der Veen (1992) gives a condensed overview of glacier and ice-sheet modeling as a component of coupled atmosphere-cryosphere modeling of climatic change. Atmospheric modeling is not covered in the present review.

2. Basis of ice flow modeling

On the basis of laboratory experiments and analogy with high-temperature creep of metals, ice is flow modeled as an isotropic, nonlinearly viscous fluid of Reiner-Rivlin type

$$\dot{e}_{ij} = \tau'_{ij}/2\eta \tag{1}$$

with effective viscosity η that varies with strain-rate $\dot{e}_{ij}$ or deviatoric stress τ'_{ij} as

$$2\eta = B\dot{\varepsilon}^{-(1-1/n)} = B^n \tau^{-(n-1)}. \tag{2}$$

Here $\dot{\varepsilon}$ is the second invariant of the strain-rate tensor,

$$\dot{\varepsilon}^2 = \tfrac{1}{2}\,\dot{e}_{ij}\,\dot{e}_{ij} \tag{3}$$

(with summation convention for repeated indices), and τ is the second invariant of the deviatoric-stress tensor

$$\tau^2 = \tfrac{1}{2}\,\tau'_{ij}\,\tau'_{ij}\,, \tag{4}$$

the stress deviators τ'_{ij} being the stress components τ_{ij} reduced by the mean stress,

$$\tau'_{ij} = \tau_{ij} - \tfrac{1}{3}\,\tau_{kk}\,\delta_{ij} \tag{5}$$

where δ_{ij} is the Kronecker delta. $\dot{\varepsilon}$ and τ are measures of the overall deformation rate and deforming stress state, respectively, and, according to (1)-(4), are related by the flow law

$$\dot{\varepsilon} = f(\tau) = \tau/2\eta = (\tau/B)^n = A\,\tau^n \tag{6}$$

(see Paterson, 1981, p. 30). A is often called the softness parameter, and $B = A^{-1/n}$ the hardness parameter. The flow-law exponent n is commonly assumed to have the value 3.

Rarely, in place of the power-law relation (6) a hyperbolic sine function is advocated, and sometimes empirical tables representing the flow law are used (Budd and Jacka, 1989). The latitude in what has been used for the flow law in modeling is rather large (see e.g. Radok et al., 1982, p. 168).

The parameter A in the flow law (6) depends on temperature T and other factors approximately via

$$A = A_o \exp\left(-T_A/T\right) \tag{7a}$$

$$\cong A'_o \ni \exp \frac{T_A(T-T_o)}{T_o^2} \tag{7b}$$

in which the constant T_A is proportional to the flow activation energy ($T_A = E_A/R$), and the constants A_o and A'_o are related by $A'_o = A_o \exp(-T_A/T_o)$, where T_o is a reference temperature usually taken to be the melting point ($-5°C$ is taken by Lliboutry [1968] and Hooke et al. [1979, p. 134]). Although T_A is usually taken constant (approximately 10^4 K) in modeling, the experimentally observed temperature dependence of A actually requires T_A to increase markedly with T above $-10°C$; sometimes an empirical table of the temperature dependence is used (Budd and Jenssen, 1989, p. 21), and a variant from the Arrhenius formulation (7a) has been suggested by Hooke (1981, eqn. (6)). The pressure dependence is small but can be taken into account by using (7b) with the melting temperature T_o dependent on pressure, so that A depends only on the difference between the temperature and the pressure melting point (for this purpose the T_o^2 in the denominator in (7b) is treated as a constant). A can also vary because of differences in the internal structural and/or chemical state of the ice, this variation being incorporated into the "enhancement factor" $\ni$ in (7). A strong influence of this type is crystal orientation anisotropy ("fabric") in the polycrystalline material (Lile, 1978), which can enhance shear flow by as much as a factor $\cong$ 8-12 (Budd and Jacka, 1989, p. 126; Shoji and Langway, 1985, p. 46). Strictly, anisotropy invalidates the foregoing flow formulation based on isotropy, and requires a more elaborate treatment (Lliboutry and Duval, 1985; van der Veen and Whillans, 1990, p. 334; Lliboutry, this volume), but glacier flow modeling to date has used (1)-(7), with allowance for a spatially varying $\ni$ in some cases. In principle, allowance should also be made for the possibility of different n values under different flow regimes or under differences in ice composition or structural state (Alley, 1992).

A glacier flow model must obey: I. stress equilibrium (momentum conservation) under the gravity body force; II. mass continuity (conservation of ice and meltwater); and III. energy conservation (heat flow linked to heat sources and sinks). These physical conditions are written as differential equations (all with summation convention):

$$\text{I. Stress equilibrium: } \partial\tau_{ij}/\partial x_j + \rho g_i = 0 \tag{8}$$

where the ρg_i are the components of the body force ρg, and ρ is ice density.

$$\text{II. Ice conservation: } \partial u_i/\partial x_i = 0 \tag{9}$$

where the u_i are the components of the ice flow velocity vector u, and where density changes (such as compaction) are neglected. (9) contains implicitly (by integration) the overall mass balance condition (growth or shrinkage) of the ice mass and its component parts. If ice melting is involved, (9) must be supplemented to include meltwater generation and flow (Section 10).

$$\text{III. Heat flow: } \rho c\left(\frac{\partial T}{\partial t} + u_i\frac{\partial T}{\partial x_i}\right) = k\frac{\partial^2 T}{\partial x_i \partial x_i} + \tau_{ij}\dot{e}_{ij} \tag{10}$$

where $\tau_{ij}\dot{e}_{ij}$ is the strain heating rate, and the parameters ρc (density times heat capacity) and k (thermal conductivity) are assumed constant. (Dependence of k on T must be taken into consideration in refined modeling. Beware the sign of the advection terms $u_i\partial T/\partial x_i$ given in some of the ice-sheet modeling references cited later.) The coupling of heat flow and ice flow are evident through the appearance of u_i and $\dot{e}_{ij} = \frac{1}{2}(\partial u_i/\partial x_j + \partial u_j/\partial x_i)$ in (10) and of T in (7). If ice melting or refreezing occurs, a corresponding heat sink/source term must be added to (10) and tied to the volume source/sink term(s) in the supplementation to (9) mentioned above. As boundary conditions for I-III a glacier flow model must have (i) a stress-free boundary at the top surface (atmospheric pressure can be disregarded), (ii) appropriate slip/no-slip boundary conditions at the bottom of the ice, and (iii) temperature or heat-flow boundary conditions at top and bottom. In some cases, flow conditions on the top or side boundaries are also applied in conjunction with (i) and (ii). On solid boundaries (bottom or sides) the no-slip condition is applied if the temperature there is below freezing; if at the melting point the ice can slide along the boundary and a slip-rate condition is required, as discussed in Sections 8 and 12.

The way in which the physical conditions I-III and associated boundary conditions (i)-(iii) are formulated and/or approximated, and the way in which they are combined or manipulated in obtaining a temperature and flow solution, is particular to each type of glacier flow model. A number of illustrative examples are described in the following sections.

3. Flow-plane modeling

Hooke et al. (1979) used nearly the full physical foundation of Section 2 in two-dimensional modeling of a section of the Barnes Ice Cap, Baffin Island, along a flow plane (vertical section parallel to flow) from center to margin (Fig. 1). Such two-dimensional models are appropriate for

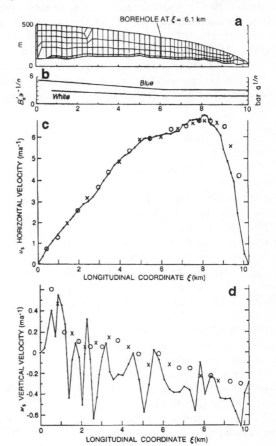

Figure 1. Flow-plane modeling of the Barnes Ice Cap (Baffin Island) by Hooke et al. (1979, Figs. 2, 3, 4). (a) Finite-element grid, from ice surface to bed; the dotted quadrilaterals at the base were assigned the flow properties of "soft," "white" Pleistocene ice. (b) model flow "hardness" parameter $B_o\text{э}^{-1/n}$ (corresponding to $A'_o\text{э}$ in (7b), with $T_o = -5°C$), as a function of longitudinal coordinate ξ measured from the summit of the ice cap; the two separate curves are for "white" Pleistocene ice and for "blue" Holocene ice. (c) Horizontal surface velocity u_S as a function of ξ, the model-calculated values are shown by dots along the continuous curve, while observed velocities are shown with open circles and crosses. (d) Vertical surface velocity component, calculated and observed as in (c) above.

glaciers that are wide compared to their thickness – for example, the Greenland Ice Sheet in mid latitudes. The finite-element grid used in the numerical modeling is shown in Figure 1a, and the model results are shown in Figure 1b,c,d. The finite-element method used is detailed by Raymond (unpublished).

The shapes of the surface and basal boundaries (the latter from radio echo sounding) are traced by the boundary mesh points in Figure 1a. At the ice divide, at longitudinal (horizontal) coordinate $\xi = 0$ in Figure 1a, the horizontal velocity u_ξ and shear stress $\tau_{\xi z}$ (where z is the vertical coordinate) are set to 0 on grounds of flow symmetry. On the base the no-slip condition is applied. The flow calculation was done by a full finite-element formulation of I and II (Section 2), while the temperature distribution was obtained in a separate calculation based on III. In a later version of the model (Hanson, 1990), the ice flow and temperature distributions are obtained in a single, coordinated finite-element calculation based on I-III, and the restriction of two-dimensional plane flow is lifted to the extent of allowing non-zero extension rates perpendicular to the flow plane, which allows the model to treat radial flow in an ice sheet of circular plan, more appropriate to the Barnes Ice Cap.

The results in Figure 1c, based on $n = 2.6$ in the flow law, show that the observed forward (horizontal) velocity component at the ice surface can be well accounted for by the model, except within about 2 km of the terminus ($\xi > 8$ km). To accomplish this it was necessary to introduce a 3.5-fold spatial variation in enhancement factor ɘ along the flow-plane from $\xi = 0$ to $\xi = 6$ km, as shown in Figure 1b; with constant ɘ, no homogeneous, uniform flow law of the type (6)+(7) could account for the observed flow. It was also necessary to assign to the layer of white, Pleistocene (ice-age) ice at the bottom of the ice cap an enhancement factor of 3.9 to 4.4 relative to the overlying blue ice (Fig. 1a). The significance of the "soft" Pleistocene ice, which is also observed at depth in the Antarctic and Greenland Ice Sheets, is discussed by Paterson (1991).

The model is less successful in accounting for the vertical component of ice flow (Fig. 1d) and for the pattern of flow at depth (Fig. 2). Although vertical profiles of calculated horizontal flow velocity at depth (Fig. 2, dashed curves) show the qualitatively expected, typical pattern – a higher-order parabola with shear concentrated near the bottom – the velocity profile measured in a borehole (Fig. 2, solid curve) does not agree closely with the calculated profile there (Fig. 2, dashed curve labeled "$\xi \approx 6$ km"). These problems are ameliorated in the later version of the model (Hanson, 1990), which used

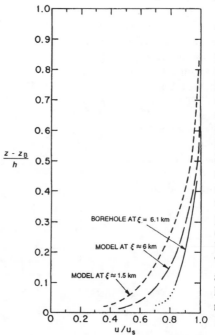

Figure 2. Vertical profiles of horizontal velocity in the Barnes Ice Cap: comparison of modeled and observed profiles, from Hooke et al. (1979, Fig. 5). Horizontal velocity $u(z)$ is normalized to the surface velocity, u/u_S. Vertical elevation above the bed is normalized to the ice thickness $(z-z_B)/h$. The profile $u(z)$ observed in a borehole at $\xi = 6.1$ km (see Fig. 1a) is plotted as a solid line, extrapolated below the bottom of the borehole as a dotted line. Model velocities $u(\xi,z)$ at longitudinal coordinate $\xi = 1$ and 2 km cluster around the short-dashed curve (individual points not shown); for $\xi = 3$ to 9 km the model values cluster around the long-dashed line.

BOREHOLE AT $\xi = 6.1$ km

MODEL AT $\xi \approx 6$ km

MODEL AT $\xi \approx 1.5$ km

$\dfrac{z - z_B}{h}$

u / u_s

$n = 3$. To successfully model the known velocity field at the surface and in boreholes it was found necessary to increase з for the basal white ice to 5.8 and to assign з = 0.6 to the overlying blue ice, an actual hardening relative to the assumed "standard" flow law. In fact, the "hardness parameter" value $B = 3.2\,\mathrm{b}\,\mathrm{a}^{1/n}$ (corrected to $T = -5°C$), indicated for blue ice on the left in Figure 1b, corresponds to з = 0.91 in relation to the standard flow law of Paterson (1981, p. 39) or з = 0.76 in relation to that of Hooke (1981, Fig. 2).

The way in which the calculated velocity profile pulls to the left and becomes less concave near the flow divide in the calculated profile at $\xi \approx 1.5$ km in Figure 2 proves to be a general feature associated with flow divides, revealed by modeling (Raymond, 1983).

A flow-plane model of the above type (Schøtt, Waddington, and Raymond, 1991) is being used to predict the flow at depth under the crest of the Greenland Ice Sheet near Summit, where two deep core holes (Fig. 3a) are currently being drilled to obtain a time history of the atmosphere from the chemical record in the ice. Knowledge of the ice flow at depth is necessary in order to assign ages to the deep ice core samples, which are beyond the range of existing absolute dating methods. The modeled streamline pattern in the flow plane containing the boreholes is shown in Figure 3b, revealing the geometry of the flow divide, which lies at the position of the easterly borehole. Figure 4 shows the model-calculated age of the ice as a function of depth in the two holes, for particular

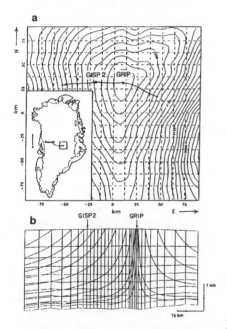

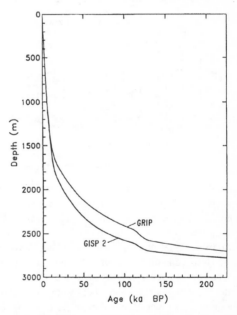

Figure 3. Flow-plane model of the crestal region of the Greenland Ice Sheet, from Schøtt et al. (1991, Figs. 1 and 3b). (a) shows the location of the flow plane, passing through the borehole sites GISP2 and GRIP. (b) shows the model-calculated flowlines, fanning downward and outward from the crest at GRIP; the finite-element mesh is also shown.

Figure 4. Model age-vs.-depth relations for ice of the GISP2 and GRIP boreholes in Greenland, calculated by the flow model of Schøtt et al. (1991, Fig. 2, curves 3). The assumed depth of bottom is 2975 m at the GRIP site and 2920 m at the GISP2 site, from sounding data. The calculation assumes a space- and time-varying accumulation rate a: increasing gently westward from GRIP, and reduced to a third of its current value during the glacial time from 18 to 110 k a ago, with no flow enhancement of Pleistocene ice.

424

assumptions as to changes in A_oэ with depth and as to the history of snow accumulation rate as a function of time (see figure caption). Other modeling efforts to predict ice sheet characteristics important in the selection of and planning for deep core drilling sites, such as basal temperature and the possibility of basal melting, are those of Dahl-Jensen (1989), Jenssen et al. (1989), and Firestone et al. (1990). Such predictions are further considered in Section 7.

Flow-plane modeling to interpret seemingly anomalous glacier flow behavior is illustrated by the work of Budd and Rowden-Rich (Budd and Jacka, 1989, p. 137) on a flowline of the Antarctic Ice Sheet from the summit of Law Dome to Cape Folger, a portion of which is shown in Figure 5a.

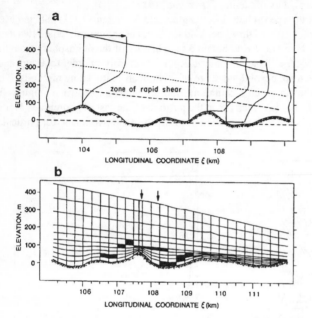

Figure 5. Ice flow and stress along part of the Law-Dome-to-Cape-Folger flow plane, East Antarctica, from Budd and Jacka (1989, Fig. 13). (a) shows three borehole-observed vertical profiles of horizontal flow velocity $u(z)$, with indication of a zone of rapid shear at intermediate depths. (b) shows the finite-element grid of a flow model that gave maxima (as a function of z) in the base-parallel shear stress at the depths represented by the black quadrilaterals. Calculated maxima were in the range 0.6-1.5 bar.

Vertical profiles of horizontal flow velocity observed in three boreholes along the section in Figure 5a show maximum bed-parallel shear in a zone about 100-200 m above the bed, which is quite anomalous by comparison with the velocity profiles in Figure 2 showing the typical concentration of shear near the base. The zone of rapid shear well above the bottom can be explained to some extent by the occurrence of a zone of strong c axis fabric and high enhancement factor (э ≈ 4) at this level. But Budd and Jacka believe that the bed-parallel shear stress must also reach a maximum in this zone and decrease below it, to explain the concentration of fabric development in the zone and the decrease in shear strain rate below it, which is larger than can be explained by э alone if the shear stress increases toward the bed in the normal way (see (13)). A finite element flow model using the grid in Figure 5b, under the assumption of isotropic (and presumably homogeneous) ice, gave shear-stress maxima at the depths represented by the blacked-in quadrilaterals, most of which are above the base (although not as high as the zone of maximum shear strain). This is surprising, especially over the crests of the bedrock highs, where maximum shear stress at the base would be expected. The modeling has been extended to the more realistic case of anisotropic ice (perhaps with inhomogeneous э), but results have not been reported as yet (Budd and Jacka, 1989, p. 135).

As the foregoing and later examples indicate, numerical modeling of glaciers has progressed to a resolution corresponding to 5-50 mesh nodes in vertical profiles through the ice, which is comparable to the resolution in some global atmosphere circulation models and mantle-flow models.

4. Flow cross-section models

For modeling the flow of valley glaciers it is necessary to deal not only with a longitudinal flow plane but with the effects of the valley walls; these are best examined in a cross-section perpendicular to flow. The idealized case in which these effects are seen in "pure" form, unmodified by longitudinal influences, is the case of flow in a channel of constant cross-section and constant slope, filled with ice to a constant depth. Flow in straight channels of idealized shape (rectangular, elliptical, and parabolic cross-section) was modeled by Nye (1965), and flow in straight and curving channels of arbitrary cross-section was modeled by Echelmeyer (1983) and Echelmeyer and Kamb (1987). Harbor (1992) extended the modeling of arbitrary straight channels with the admission of basal sliding (Section 8), which was not included in the earlier treatments.

As an example of this type of modeling we here consider the effect of channel curvature on flow. The flow geometry is idealized as helical, with constant channel shape as seen in cross sections containing the helical axis (vertical axis through center of curvature in map view). The flow and the flow stresses are described in a cylindrical coordinate system (r, ξ, z) with z along the helical axis, r radially outward from the axis, and ξ in the flow direction (horizontal component), taking the place of the polar angle $\theta = \xi/r$. If the slope of the channel is low, so that the pitch of the helix is small and the flow follows helical flow lines of this same small pitch, the main flow stresses (shear stress components) are $\tau_{\xi r}$ and $\tau_{\xi z}$. The other shear stress components can be neglected, and $\tau_{\xi r}$ and $\tau_{\xi z}$ satisfy the following simplification of the ξ equilibrium equation from (8):

$$\frac{\partial \tau_{\xi r}}{\partial r} + \frac{\partial \tau_{\xi z}}{\partial z} + 2\frac{\tau_{\xi r}}{r} + f_\xi = 0 \tag{11}$$

where f_ξ is an effective body force component that results from the pressure distribution in the ice mass; in the approximation of small slope α of the ice surface, this body force can be written $f_\xi = \alpha_0 r_0/r$ where α_0 is the surface slope at a particular radial coordinate r_0. The main strain-rate components are

$$2\,\dot{e}_{\xi r} = \frac{\partial u}{\partial r} - \frac{u}{r}, \quad 2\,\dot{e}_{\xi z} = \frac{\partial u}{\partial z} \tag{12}$$

where u is the (ξ-directed) flow velocity. The terms involving $1/r$ in (11) and (12) contain the effects of channel curvature, and drop out in the case of a straight channel ($r \to \infty$).

An example of the results of a cross-sectional flow model based on (11), (12), and (1)-(6) with constant A (isothermal) is given in Figure 6, showing observations and model calculations for a cross-section across Blue Glacier, Olympic Mountains, Washington (Echelmeyer and Kamb, 1987, p. 10). Figure 6a shows the measured profile of surface velocity (downglacier component) across this section, and Figure 6c shows the channel cross-section and finite-element mesh. For a straight channel with no basal slip the calculated surface velocity profile is shown in Figure 6b with a dashed curve. In actuality the channel curves strongly to the left, with a center of curvature lying about 1000 m west of the west margin of the glacier. The effect of this curvature changes the modeled surface velocity profile to the solid curve in Figure 6b. It agrees with the observed profile (Fig. 6a) definitely better than the dashed curve. In this example, the effects of the actual channel shape, by comparison with a parabolic or semi-elliptical channel that approximates it, are comparable to the further effects of channel curvature in modifying the velocity profile. Figures 6d and e give the distribution of shear stress $\tau = (\tau_{\xi r}^2 + \tau_{\xi z}^2)^{1/2}$ over the glacier cross-section, comparing a straight and curving channel; they show that the "stress centerline," where $\tau = 0$ (center of the contour bulls eye), shifts considerably toward the inside of the bend as a result of the channel curvature. This

426

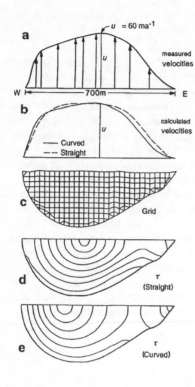

Figure 6. Flow cross-section model showing effects of curving flow in Blue Glacier, Washington, from Echelmeyer and Kamb (1987, Fig. 13). (a) shows the measured longitudinal velocity component u in a transverse profile across the glacier, at the surface, where the glacier width is 700 m; the velocity scale is indicated by the maximum velocity of 60 m a-1. (b) shows model-calculated velocity profiles for a straight channel (dashed line) and for curving channel (continuous line) with the center of curvature lying 1000 m west of the west margin, as determined from plan views of the glacier. (c) shows the shape of the channel cross-section (from sounding) and the grid used in the finite element model. (d) and (e) give the modeled distribution of the total shear stress $\tau = (\tau_{\xi z}^2 + \tau_{\zeta z}^2)$, for straight and curved channel.

causes the marginal crevasse zone on the inside of the band to be narrow and on the outside to be wide, an effect that is clearly seen in aerial views of the glacier (Echelmeyer and Kamb, 1987, Fig. 1).

5. Steady-state simple-shear models with vertical integration

In modeling the flow of large ice sheets such as those of Antarctica and Greenland, it is standard practice to sidestep a solution of the complete flow problem (Section 2) by using an approximation in which the vertical profile of flow at each point in map view is the same as it would be if the ice sheet were a planar slab of infinite lateral extent with thickness and surface slope equal to the local values. In this approximation, the flow in each vertical column does not have significant mechanical interaction laterally or longitudinally with adjacent columns, beyond that ascribable to hydrostatic pressure in the columns; gravitational equilibrium is provided by the surface-parallel shear stress, which is determined at all heights z by a much-simplified "local" form of the equilibrium equations (8) applicable in this case:

$$\tau_{\xi z} = \rho g \alpha (z_S - z) \qquad (z_B \leq z \leq z_S) \tag{13}$$

where z_S is the vertical (elevation) coordinate of the ice surface and α is the surface slope $\alpha = -\partial z_S / \partial \xi$, the coordinate ξ being in the horizontal flow direction, which is the direction of the surface slope $\nabla_2 z_S$. (13) involves the approximation that α is small, so that sin α can be replaced by α and the stress across planes parallel to the surface can be replaced by the horizontal shear stress

$\tau_{\xi z}$. (13) holds independent of the slope β of the bed as long as β is small. At the bed (at elevation z_B) the basal shear stress τ_B is

$$\tau_B = \tau_{\xi z}(z_B) = \rho g \alpha (z_S - z_B) = \rho g \alpha h \tag{14}$$

where h is the ice thickness. The nature of the mechanical approximation involved in using the "local equilibrium" formulas (13) or (14) in an ice sheet of actual, non-planar-slab shape has been much discussed, particularly by Hutter (1983, Ch. 5), who uses the name "shallow-ice approximation" for this approach and associated refinements such as (19) in Section 6. These approximations are good when the lateral variations in ice thickness and in surface and bed slope are small over distances of the order of the ice thickness.

In models based on (13) and (14), the vertical gradient of the ξ-directed ice flow velocity u_I implied by (13) is taken to be

$$\frac{\partial u_I}{\partial z} = 2\,\dot{e}_{\xi z} = 2\,A\,\tau_{\xi z}^n = 2\,A\,[\rho g \alpha (z_S - z)]^n \tag{15}$$

which follows from (1)-(6) when $\partial u_z / \partial \xi = 0$ and all strain rates except $\dot{e}_{\xi z}$ are neglected. This can be called simple shear flow. (It is often called "laminar flow," but this is a malapropism since all glacier flow is technically laminar, at very low Reynolds number.) (15) is integrated once to get the flow-velocity profile $u_I(z)$ (with $u_I(z_B) = 0$) and twice to get its vertical average $\bar{u}_I$; if A is constant,

$$\bar{u}_I = \frac{2}{n+2} A\,\tau_B^n\,h \tag{16}$$

As noted above, the direction of the velocity $\bar{u}_I$ and the shear stress τ_B is the direction of the surface slope, normal to elevation contours on the ice sheet. Subscript I on u_I identifies this quantity as the contribution to flow from internal deformation of the ice mass as distinct from a contribution from basal sliding u_B (Section 8). The effect of temperature on $\bar{u}$ is discussed in Section 7.

The continuity condition (item II of Section 2) is introduced into vertically integrated, steady-state glacier flow models in two rather different ways: (a) by calculating "balance velocities" from continuity, which can be compared with modeled flow velocities ("dynamics velocities") from (16) to obtain a test of the model including the assumption of a steady state; (b) by solving the vertically integrated continuity equation in combination with the "dynamics" equation (16) to obtain model values of surface elevation z_S, which can be compared with the actual surface, again as a test of the model and the steady-state assumption. Going beyond (b), by introducing the non-steady-state continuity condition, time-dependent models can be calculated, which predict changes in glaciers and ice sheets taking place through time; this extensive application of modeling, which can also be used to obtain steady-state models, is discussed in Section 9.

The balance velocity $\bar{u}_a$ at any point (x,y) is the vertically averaged flow velocity $\bar{u}$ that is just sufficient to flux away the volume of ice being currently deposited upstream from that point by snowfall with a given areal distribution of accumulation rate $a(x,y)$, the ice flow being along the flowlines governed by the surface topography $z_S(x,y)$ of the existing ice mass (flowlines normal to the surface contours). With flow at the balance velocity throughout, the ice mass is in a steady state, input (accumulation) being balanced by outflow. Balance velocities serve to provide a measure of the glacier flow when direct measurements of the actual flow are lacking, as they are in most of Antarctica and Greenland. Calculation of balance velocities, based implicitly on a vertically integral of the continuity equation (9), involves the following steps. First the two-dimensional (map-view) pattern of flowlines is constructed from the ice surface topography (Fig. 7a). Then the accumulation

428

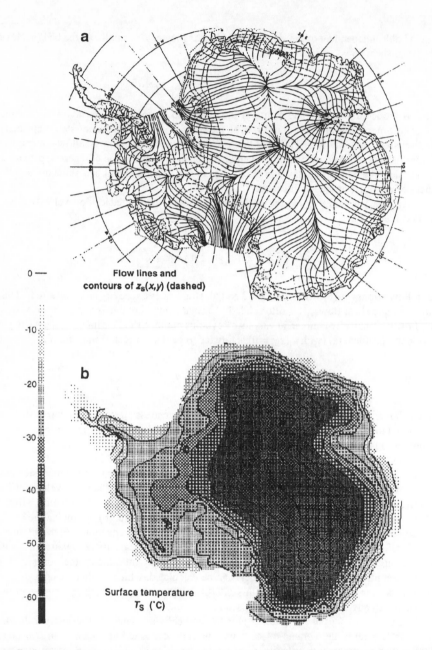

0 —

**Flow lines and
contours of $z_S(x,y)$ (dashed)**

-10

-20

-30

-40

-50

-60

**Surface temperature
T_S (˚C)**

Figure 7. Input to flow models of the Antarctic Ice Sheet. (a) Surface elevation $z_S(x,y)$ contours (dashed curves) and flowlines (continuous curves) derived from the z_S contours by the perpendicularity condition, from Budd et al. (1971, Map 2/1); the z_S contour interval is 500 m. (b) surface temperature $T_S(x,y)$, contoured and shaded as indicated by the bar at left; courtesy of Dick Jenssen (pers. com.). Coordinate lines in (a) (latitude and longitude lines, the 70°S circle being the one bearing the longitude labels) can be transferred to (b), which is at the same scale and in the same orientation.

rate $a(x,y)$ (expressible in meters of ice per year) is integrated over the area between two adjacent streamlines (called a "flowband"), starting at any point (x_1,y_1) in the flowband and integrating upward to the head of the flowband at an ice divide. This gives an ice-volume input per year, which is then divided by the local width $b(x_1,y_1)$ of the flowband at (x_1,y_1) and by the local ice thickness $h(x_1,y_1)$ to get the balance velocity $\bar{u}_a(x_1,y_1)$. This is repeated for chosen points (x_1,y_1) throughout the area of the ice sheet.

Calculation of balance velocities and comparison with modeled flow velocities ("dynamics velocities") has been carried out for the Antarctic Ice Sheet by Budd, Jenssen, and Radok (1971), Budd, Jenssen, and Smith (1984), Radok, Jenssen, and McInnes (1987, p. 12-35), and Budd and Jenssen (1989), and for the Greenland Ice Sheet by Budd et al. (1982) and Radok et al. (1982). Figure 8 shows a comparison for Antarctica. The overall features of the balance-velocity pattern (Fig. 8a) are reproduced in the dynamics velocity pattern (Fig. 8b), particularly the general increase from low velocities (<10 m a^{-1}) in the central regions of the East Antarctic and West Antarctic Ice Sheets to high velocities (≥100 m a^{-1}) near the coast, with very high velocities ~1000 m a^{-1}) in the Ross, Ronne-Filchner, and Amery ice shelves. The overall agreement has been achieved by tuning the flow law parameters (including ɘ in (16) via (7), and others as discussed in Section 8). Because of this tuning, the overall agreement does not provide an indication that the ice sheet is in or near a steady state. Moreover, there are numerous interesting discrepancies in detail between the model flow velocities and balance velocities, and there are extensive areas over which the two differ by more than a factor of 2 (Radok et al., 1987, p. 35). In general, the modeled velocities show less systematic fine-grained lateral variations than the balance velocities do. These discrepancies indicate that something is amiss in the

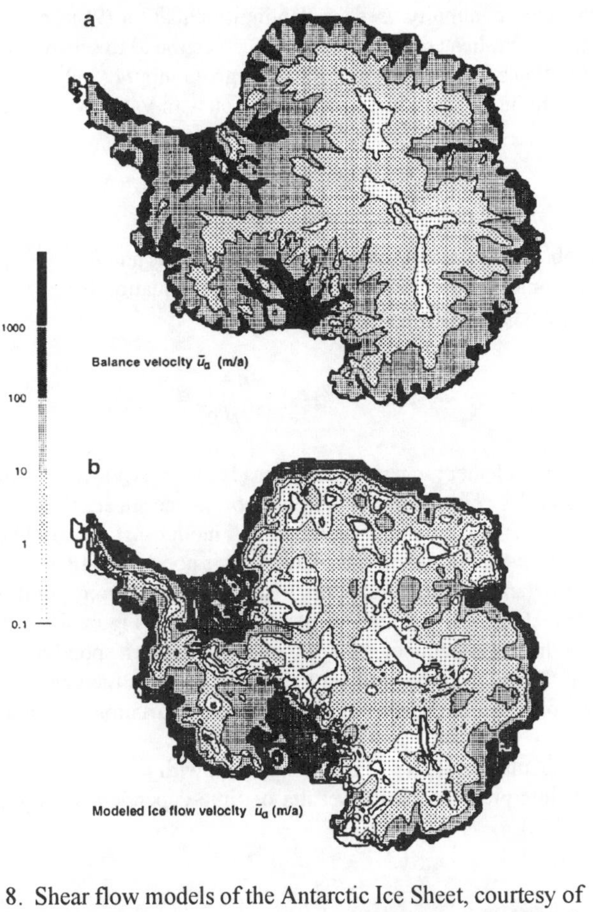

Figure 8. Shear flow models of the Antarctic Ice Sheet, courtesy of Dick Jenssen. (a) Balance velocity $\bar{u}_a$. (b) "Dynamics" velocity $\bar{u}$, calculated essentially by (16), or rather by integration of (15), with the refinements considered in Sections 7 and 8 (which include addition of the sliding velocity component u_B from Fig. 17 to the u_I from Fig. 16). The logarithmic velocity scale (bar at left applies to both (a) and (b). Map scale and coordinates as in Fig. 7a.

modeling, but one cannot tell immediately whether the trouble is more in the flow-mechanics model or the balance velocity model. The balance velocities and the actual flow velocities would differ if the steady-state assumption were invalid (i.e., if local growth or shrinkage of the ice sheet were occurring so that $\dot{h}(x,y) \neq 0$), or if there were errors in the assumed accumulation distribution $a(x,y)$. However, it seems unlikely that undetected fine-grained local variations in $\dot{h}(x,y)$ or $a(x,y)$, departing from the assumed $\dot{h} = 0$ or from the assumed smooth, broad variation in $a(x,y)$, could be large enough and numerous enough to explain the discrepancies. Hence there is probably trouble in the flow-mechanics model. Although the simplifying assumptions of the vertically integrated model as discussed above (see also Section 6) make (14) and (16) an approximation, most glacier modelers appear to think that the trouble is not in this flow-mechanical foundation. It is often attributed formally to lateral variations in the enhancement factor ǝ in (7), such as those assigned in Figure 1 to the Barnes Ice Cap model.

The second method of introducing continuity uses the continuity equation (9) in combination with the flow relation (16) (or more complicated flow relations – see Section 8) to set up an equation for the ice-sheet surface configuration $z_S(x,y)$ in steady state under given input $a(x,y)$. In vertically integrated form, the steady-state continuity equation (in vector notation with ∇_2 the two-dimensional gradient operator) is

$$\nabla_2 \cdot (h\,\overline{u}) = a \tag{17}$$

When the (vector) velocity $\overline{u}$ is obtained from (14) and (16), with $\alpha = -\nabla_2 z_S$ and $h = z_S - z_B$, and is introduced into (17) the latter becomes a non-linear partial differential equation for $z_S(x,y)$, given the functions $z_B(x,y)$ (bedrock topography) and $a(x,y)$ (accumulation):

$$-\nabla_2 \cdot (A(z_S - z_B)^{n+2} \,|\, \nabla_2 z_S |^{n-1} \, \nabla_2 z_S) = \frac{n+2}{2\rho^n g^n} \, a \tag{18}$$

Fastook (1990) has developed a finite-element scheme that solves (18) for $z_S(x,y)$ with given $a(x,y)$, $z_B(x,y)$, and boundary conditions on h or $h\overline{u}$ around the periphery of the ice sheet. For Antarctica, results are shown in Figure 9a in terms of contours of the "best fit" model surface $z_S(x,y)$ and, for comparison, contours of the actual ice-sheet surface. The agreement is good. The fit was achieved by adjusting the flow parameter A (and related parameters – see Section 8) from point to point throughout the model, in effect making A a function of position, $A(x,y)$. Figure 9b shows this function expressed in terms of the hardness parameter $B = A^{-1/n}$ in (6). The corresponding variation of A is by a factor of up to about 80, but over much of central East Antarctica the variation is less than a factor of 4. The variation of A can be thought of as representing variation of ǝ and/or T in (7).

This second method – model calculation of $z_S(x,y)$ and comparison with the observed surface – is also exploited extensively in interpretation of the results of time-dependent flow modeling (Section 9; Huybrechts, 1992, p. 140).

6. Longitudinal and lateral coupling

When valley glaciers are modeled by the methods of the last section (with correction for the drag of the valley walls as mentioned in Section 4), the models typically show much larger longitudinal fluctuations in flow velocity than are actually observed (Fig. 10). The fluctuations are damped by longitudinal stress coupling within the ice mass, which is strong enough that the approximation of "local equilibrium" in (14) and (16) breaks down significantly. The longitudinal coupling can be

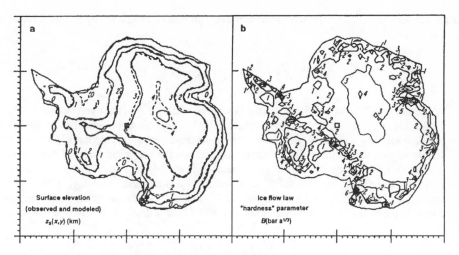

Figure 9. Steady-state model of the Antarctic Ice Sheet surface $z_S(x,y)$, from Fastook (1990, Figs. 7 and 9), based on numerical solution of (18). (a) Model-calculated contours of surface elevation $z_S(x,y)$, in km, shown with dashed lines, are compared with the actual surface contours, shown with continuous lines. (b) Areal distribution of flow-law parameter values used in the model: this is represented by contours of the "hardness" parameter B, in bar $a^{1/3}$; the actual numerical value corresponding to each contour is the labeled integer plus 0.5 bar $a^{1/3}$.

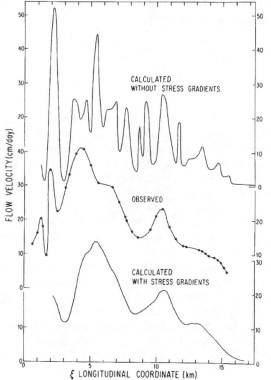

Figure 10. Effect of longitudinal stress gradients on the flow of Variegated Glacier, Alaska, from Kamb and Echelmeyer (1986, Fig. 9). The flow calculated from (16), without taking into account longitudinal stress gradients (upper curve), is compared with the observed center-line velocity (middle curve) and with the flow calculated from (20), based on a model that includes the effect of longitudinal stress gradients. In the calculation of flow from (16) the longitudinal coupling length l is taken to be 0.5 km, which is 1.2 to 2.5 times the ice thickness h. The local slope values from which the upper curve is calculated by (16) are averages over 0.25 km longitudinal intervals.

taken into account to a first approximation in the equilibrium equations (8) by adding a correction term to (14):

$$\tau_B = \rho g h \alpha + 2\frac{\partial}{\partial \xi}(h\overline{\tau}'_{\xi\xi}) = \rho g h \alpha + 4\frac{\partial}{\partial \xi}(\overline{\eta h u}) \tag{19}$$

where ξ is the longitudinal coordinate (in the flow direction), η is the effective viscosity in (1) and (2), and a bar over denotes a vertical average. The last term in both equalities in (19) is the longitudinal coupling term and represents the longitudinal gradient of the longitudinal normal stress $\tau_{\xi\xi}$ integrated over the ice thickness, or, rather, it represents the integrated effect of that part of $\tau_{\xi\xi}$ that results from ice deformation and is not contained in the "driving stress" $\rho g h \alpha$; in (19), $\overline{\tau}'_{\xi\xi}$ is the vertical average of the longitudinal deviatoric stress $\tau'_{\xi\xi}$. Kamb and Echelmeyer (1986) showed that the effect of the longitudinal coupling term in (19) on the flow can be rather simply expressed as follows in the case of a small flow perturbation $\Delta\overline{u}(\xi)$ from a datum state with velocity $\overline{u}_0$:

$$\Delta \overline{u} = \overline{u}(\xi) - \overline{u}_0 = \frac{\overline{u}_o}{2l}\int_0^L \Delta\ln(\alpha^n h^{n+1})\exp\left(-\frac{|\xi'-\xi|}{l}\right)d\xi' \tag{20}$$

where $\overline{u}_0$ is related to the datum-state α_0 and h_0 by (14)+(16), and the integration is over the glacier length L. α_0 and h_0 may be taken to be longitudinally averaged values for a particular glacier. The quantity $\Delta\ln(\alpha^n h^{n+1})$ in (20), a function of the integration variable ξ', is the local deviation from the datum value:

$$\Delta\ln(\alpha^n h^{n+1}) = \ln[(\alpha(\xi'))^n (h(\xi'))^{n+1}] - \ln(\alpha_0^m h_0^{n+1}) \tag{21}$$

and l in (20) is the *longitudinal coupling length*, given by

$$l = \sqrt{4n\overline{\eta h u}_0/\tau_0} \tag{22}$$

where τ_0 is the datum-state value of τ_B and $\overline{\eta}$ is a weighted vertical average of η (Kamb and Echelmeyer, 1986, eqn. (6)). Thus the consequence of longitudinal coupling is that the local ice thickness h and surface slope α influence the flow over a longitudinal distance range of order $2l$, the influence dying off exponentially, with scale length l. The longitudinal coupling length l is usually a few times the ice thickness. The result of applying (20) is shown by the curve labeled "with stress gradients" in Figure 10; it is a strong improvement over (14)+(16) ("without stress gradients") in reproducing the observed flow.

Longitudinal-coupling effects are included implicitly in the two-dimensional models discussed in Section 3, and explicitly in treatments of the ice-sheet/ice-shelf transition zone in Section 9.4.3 and of glacier surging in Section 11.2.

If there are significant transverse-to-flow variations in the flow of a glacier or ice sheet, a lateral flow coupling effect arises that is the counterpart to the longitudinal coupling considered above. One clear manifestation of lateral coupling is the effect of drag of the valley walls in valley glaciers, discussed for the idealized models of cylindrical or helical channels in Section 4. A more general model that includes the effects of both lateral and longitudinal coupling uses the two-dimensional generalization of (19), which is discussed in Section 12. Another general approach to modeling these coupling effects is the "force budget" theory of van der Veen and Whillans (1989, 1990).

7. Thermal models and temperature/ice-flow interactions

In order to take into account the effects of temperature in the vertically integrated flow models of large ice sheets discussed in Section 5, as well as in fully three-dimensional flow models, it seems to be necessary to have a three-dimensional temperature model, that is, one in which the temperature distribution is resolved vertically (in z) as well as horizontally (x,y). At least, no vertically-integrated thermal model has been developed for the purpose, although a simple globally integrated thermal model exists (Oerlemans and van der Veen, 1984, p. 84). Principles and applications of thermal modeling in ice sheets were summarized by Budd and Young (1983a,b). By "temperature" is here meant "mean annual temperature"; the seasonal and short-term temperature fluctuations, which affect only the upper 10 m or so of a glacier or ice sheet, are here disregarded.

In most ice sheet models up to now the temperature distribution has been determined as a vertical profile $T(z)$ at particular map points (x,y) by solving the heat flow equation (10) in the approximate form

$$\kappa \frac{\partial^2 T}{\partial z^2} - w \frac{\partial T}{\partial z} + \frac{\tau_{\xi z}}{\rho c} \frac{\partial u}{\partial z} = \bar{u} \frac{\partial T}{\partial \xi} \tag{23}$$

which assumes a steady state $(\partial T/\partial t = 0)$, neglects horizontal heat conduction $(\kappa \nabla_2^2 T)$, disregards shear heating from strain components other than $\partial u/\partial z$, and treats the flow as "block flow" $(u = \bar{u}$, independent of $z)$ as far as the ξ-advection term (right-hand term in (23)) is concerned. The ξ axis is the local flow direction; thus there is no horizontal advection perpendicular to it. In (23), κ is the thermal diffusivity $(k/\rho c)$ and w is the vertical component of ice flow velocity. The boundary conditions for the solution of (23) are, at the ice surface, the current temperature T_S and ice accumulation rate a (that is, $w(z_S) = -a$), and, at the base, the geothermal heat flux (except as noted below). Since there are extremely few measured values of the geothermal heat flux under ice sheets, it is usually assumed constant throughout the model area or a few subareas, as West vs. East Antarctica (Budd et al., 1971, p. 70, 118), and separate temperature models are often calculated for a few different values of the heat flux, corresponding to basal temperature gradients $-(\partial T/\partial z)_B$ usually in the range 15-35°C km^{-1}. The larger is T_S, or $-(\partial T/\partial z)_B$, or h, the higher will be the basal temperature T_B given by the solution of (23). If T_B reaches the melting point, then the flux condition at the base has to be replaced by a temperature condition there, which specifies T_B to be the melting temperature under the ambient basal pressure. In a few cases, a thermal model of the bedrock below the ice is adjoined to the ice thermal model (e.g. Hanson, 1990, p. 50; Huybrechts, 1992, p. 107).

At an ice divide, where $\bar{u} = 0$ and $\partial u/\partial z = 0$ so the ξ-advection and shear-heating terms vanish, (23) becomes an ordinary differential equation in z and is readily solved if $w(z)$ is specified. This $w(z)$ should come from the flow model, as it does in the fully resolved models of Section 3. In the simple-shear models of Section 5, however, the practice has generally been to assume that $w(z)$ is linear in z: $w(z) = -a(z - z_B)/h$ (Robin, 1955; Paterson, 1981, p. 201; Oerlemans and van der Veen, 1984, p. 79); but the flow-plane modeling results of Raymond (1983, p. 365, 370) mentioned in Section 3 indicate that it is instead approximately quadratic:

$$w(z) = -a (z - z_B)^2/h^2 \tag{24}$$

For a general location, not at an ice divide, the shear-heating and ξ-advection terms in (23) need to be evaluated as functions of z, and $w(z)$ needs to be reconsidered. Often $w(z)$ is again assumed linear in z, but more appropriate is the assumption that $\partial w/\partial z$ is proportional to the flow velocity $u(z)$ (Lliboutry, 1979, p. 145; Budd et al., 1982, p. 28; Budd and Young, 1983a, p. 146), which follows from incompressibility (9) if in the dependence of u on ξ and z the variables are functionally separated, $u(\xi,z) = f(\xi)g(z)$; such separation of variables would hold for $u(\xi,z)$ derived from (15) if the ξ-dependence were mainly due to $\alpha(\xi)$, other sources of ξ-dependence being negligible. This assumption is confirmed to a good approximation by the model results of Raymond (1983, Fig. 4a and p. 366), although the isothermal nature of the model makes applicability of the results subject to some question (Raymond, 1983, p. 370). The proportionality constant between $\partial w/\partial z$ and $u(z)$ is fixed by the surface boundary condition $w(z_S) = -a$. Then by integration of $\partial w/\partial z$,

$$w(z) = -a \int_{z_B}^{z} u(z)\, dz \Big/ \int_{z_B}^{z_S} u(z)\, dz \tag{25}$$

Lliboutry (1979b, p. 145) proposed that the effect on $u(z)$ of the temperature dependence of A in (15) can be approximated by (15) with fixed A by taking an artificially large value of n, call it m, given by $m = n - 1 - hT_A(\partial T/\partial z)_B/T^2$. In this case the vertical velocity profile from (25) is

$$w(z) = -a\,\frac{m+2}{m+1}\left[\frac{z - z_B}{h} - \frac{h^{m+2} - (z_S - z)^{m+2}}{(m+2)\,h^{m+2}}\right] \tag{26}$$

Ritz (1987, 1989) has applied (26) with $m = 11$, which she considers appropriate for central Antarctica (Vostok). A more complicated formula for $w(z)$ can be derived from (25) by assuming that the temperature profile $T(z)$ is linear in z (Philberth and Federer, 1971), which is a good approximation for the purpose in the lower half of most ice sheets, where most of the shear deformation occurs; the formula has been little used in modeling, however. The formula given by Budd and Jenssen (1989, p. 21, eqns. (d) and (e)) appears to reduce to $w(z) = -\frac{1}{2}a[h^2 - (z_S - z)^2]/h^2$, which differs substantially from (24) and (26) and gives $w(z_S) = -\frac{1}{2}a$. The formulas given by Jenssen (1989, eqn. (12)), and by Huybrechts and Oerlemans (1988, eqn. (5)) for calculating $w(z)$ numerically from model values of $u(x,y,z)$ or $u(\xi,z)$ are complicated due to the use of the relative vertical coordinate ζ.

The formulation of $w(z)$ is needed not only for the thermal models but also to obtain the age of the ice at depth by integrating downward from the surface:

$$\text{age of ice at height } z = \int_{z_S}^{z} dz/w(z;t) \tag{27}$$

This is important in the attempt to assign ages to ice samples from deep core drilling, as noted in Section 3. For this purpose Dansgaard and Johnson (1969) assumed for the Greenland Camp Century drill site that $u(z)$ increases linearly with z to a height 400 m above the bed and is constant thereafter, which is an approximation to $u_1(z)$ from (15) and therefore from (25) gives a $w(z)$ similar to (26) with $m = n$. Reeh (1989) summarizes this and other efforts to do ice-core dating by flow modeling. The dependence of w in (27) on (past) time t enters if the ice sheet has not been in a steady state since ice now at depth $z_S - z$ was originally deposited; an example of the consequences of this is the wiggle in the curves in Figure 4 at a depth of about 2600 m (see figure caption).

The shear heating $\tau_{\xi z}\partial u/\partial z$ for (23) is readily obtained from (13) and (15). Often it is simplified by noting that because this distributed heat source is concentrated near the bottom it can to a good approximation be replaced by an integrally equivalent heat source $\tau_B\bar{u}$ at the base. The basal boundary condition on $(\partial T/\partial z)_B$ then becomes

$$-(\partial T/\partial z)_B = -(\partial T/\partial z)_G + \tau_B\bar{u}/k \tag{28}$$

where G refers to the geothermal gradient. (28) amounts to introducing some vertical integration into the thermal model, as does the "block flow" assumption in the ξ-advection term in (23), which treats the shear flow $\partial u/\partial z$ as if concentrated entirely at the bed.

The simplification (28) puts aside the possibility that strain heating in the region of rapid shear near the glacier base may raise the ice temperature to the melting point in a basal zone of appreciable thickness, which seems to have been observed in a few actual examples (basal temperate layer 20-40 m thick: Classen, 1977; Blatter, 1987; Bamber, 1987; Kotlyakov and Macheret, 1987; Blatter and Kappenburger, 1988). The modeling of such "polythermal" (partly cold, partly temperate) glaciers involves a number of additional complications in locating the cold-temperate boundary within the ice mass and in handling the water content of the temperate ice (Fowler and Larson, 1978; Hutter, 1982; Fowler, 1984; Hutter et al., 1988; Blatter and Hutter, 1991). This modeling, and indeed the modeling of fully temperate glaciers also, should probably include consideration of the effect of water content on ice rheology, as discussed by Lliboutry and Duval (1985, p. 215), but as far as I am aware this has not been done in such modeling.

The main problem in solving (23) at a general location is to obtain the along-flow temperature gradient $\partial T/\partial \xi$ so that the ξ-advection term (right side of (23)) can be formulated as a function of z. Extensive discussion of this problem is given by Budd, Jenssen, and Radok (1971, p. 47-120). At or very near the surface the ice temperature is controlled by the surface temperature T_S, which varies with altitude z_S in accordance with the local atmospheric lapse rate $\Lambda = -\partial T_{atm}/\partial z$. Thus if T_S depends mainly on z_S, $(\partial T/\partial \xi)_S \cong (\partial T_S/\partial \xi) = \alpha \Lambda$. In the "simple column model" of Budd, Jenssen, and Radok (1971, pp. 64, 76), $\partial T/\partial \xi$ at all depths was in effect assumed constant, equal to $\alpha \Lambda$, although there seems to be no clear physical basis for this assumption, and it does not give good results when the along-flow gradients of T_S, z_S, and a are large (Budd et al., 1971, p. 59; Radok et al., 1982, p. 155). It does, however, have the advantage of allowing an analytic solution of (23) and avoiding lengthy numerical computations (Radok et al., 1982, p. 69-72), and it is considered useful for quick preliminary estimation of the temperature distribution in an ice sheet (Budd and Young, 1983a, p. 148).

The assumption of constant $\partial T/\partial \xi$ is avoided in a different approach generally known as the "moving column model" (Budd et al., 1971, pp. 59, 63; Lingle and Brown, 1987, p. 257)[§]. In this method, attention is focused on a vertical column of ice moving with velocity $\bar{u}$ along a flowline, and the vertical temperature profile $T(z)$ in the column is followed as a function of time or, equivalently, of distance ξ along the flowline. That the column remains vertical is an expression of the "block flow" approximation in (23), which neglects the shear distortion of the column due to $\partial u/\partial z$. The moving column model starts with the column at the head of a chosen flowline, at an ice divide, with initial $T(z)$ that is the steady-state profile there, obtained as described above. The column then moves outward from the ice divide at velocity $\bar{u}(\xi)$ along the flowline, and its thermal evolution is followed by using (23) in a form obtained by transforming to a coordinate system moving with the column:

[§]There is some terminological confusion in the literature on this model. Budd et al. (1971, pp. 47, 63, 106) called it the "flowline model" or "flowline column model," and also the "transient state computation," a term also used by Radok et al. (1982, p. 73), along with "Lagrangian solution" and "flowline solution" (p. 156). What Paterson (1981, p. 206) calls the "moving column model" is in fact essentially the simple column model of the previous paragraph because it takes $u\partial T/\partial \xi$ (called "$\partial T/\partial t$") to be constant, independent of z. Other names that have been used for the simple column model are "fixed column model" (Budd et al., 1971, p. 47), "column model" (p. 64), "spot value column model" (p. 120), "single column"/"constant warming rate" model (Budd and Young, 1983a, p. 147, 148), and "Eulerian model" (Budd et al., 1982, p. 70).

$$\frac{DT}{Dt} = \bar{u}\frac{DT}{D\xi} = \kappa\frac{\partial^2 T}{\partial z^2} - w\frac{\partial T}{\partial z} \tag{29}$$

The D here means differentiation following the moving column along the flowline; because of the assumed steady state, the differentiation can be expressed with respect to ξ, independent of t, as given in the first equality in (29). The shear heating term does not appear because it has been incorporated into the basal boundary condition via (28). Starting from the initial $T(0,z)$ at the ice divide, (29) is integrated forward numerically in ξ (or equivalently in t) to get $T(\xi,z)$, the vertical temperature profile at any position ξ along the flowline. In the integration the surface temperature boundary condition $T_S(\xi)$ and the surface and bed elevations z_S and z_B are progressively updated as the column moves. The changing column height $h = z_S - z_B$ causes complications in the numerical integration of (29), which are dealt with by transforming the z coordinate to a relative-depth coordinate ζ in the ice column, $\zeta = (z_S(\xi) - z)/h(\xi)$, a procedure that involves a number of technical details (Budd, Jenssen, and Radok, 1971, p. 91; Radok et al., 1982, Appendix 2; Huybrechts and Oerlemans, 1988, p. 53-54; Budd and Jenssen, 1989, p. 21; Huybrechts, 1992, p. 117).

Despite the ostensible derivation of (29) from (23), the validity of (29) is general, for time-dependent as well as steady-state temperature distributions. Thus the moving column model has often been used for the time-dependent glacier models considered in Section 9. When there is time dependence, different columns moving along a given flowline at different times will have different column heights h and will see different surface boundary conditions T_S (and a change in the basal boundary condition if a dry base starts to melt or a wet base freezes). Also, the initial temperature profile at the ice divide has to be calculated with the time dependence of h and T_S there taken into consideration. Finally, a in (24)-(26) has to be replaced by $a - \dot{h}$ in time-dependent models.

The ice flow velocity components $\bar{u}$ and w needed to obtain the temperature model from (23) by the simple column method or from (29) by the moving column method cannot be calculated from (7), (16) and (25) unless the temperature distribution is already known. This difficulty is usually sidestepped by taking for the column velocity $\bar{u}$ the balance velocity $\bar{u}_a$, which assumes a steady state system and in effect uncouples the heat flow model from the ice flow model (16). The resulting model temperatures are called "balance temperatures" (e.g. Radok et al., 1982, Map 4/9).

Results of the moving column method are illustrated in Figures 11 and 12. Figure 11 gives balance temperature profiles in the Greenland Ice Sheet along a flowline from the crest of the ice sheet to

Figure 11. Vertical temperature profiles through the Greenland Ice Sheet along the flowline terminating in Jakobshavns Glacier. (a) shows four modeled profiles, at distances $\xi = 0$, 200, 400, and 600 km from the crest, along the vertical lines shown in relation to the surface and bed profiles (drawn with vertical exaggeration ×125); the modeling was done by the moving column method (Budd et al., 1982, Fig. 7.5). Temperature scales are at the top. (b) shows the temperature profile measured by Iken et al. (1992), Fig. 4a) about 60 km from the terminus of Jakobshavns Glacier; this is to be compared with the modeled profile at $\xi = 600$ km in (a).

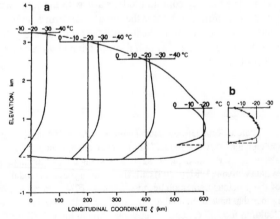

the Jakobshavns outlet glacier on the west coast. The calculated profiles in Figure 11a (Budd et al., 1982, Fig. 7.5) display the typical features that result from the surface and basal boundary conditions and from vertical advection (major effect) and horizontal advection (minor effect), as discussed by Paterson (1981, p. 197-211). Figure 11b shows a measured profile in Jakobshavns Ice Stream 50 km from the ice sheet margin (Iken et al., 1992, Fig. 5); it can be compared with the modeled profile

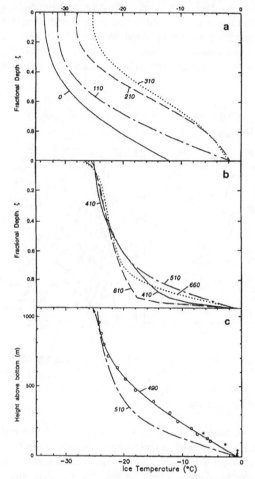

at $\xi = 600$ km in Figure 11a. Grossly there is agreement, but in detail there are substantial discrepancies, which will require further, more detailed modeling to understand. The effect of vertical advection on the temperature profiles can be seen by comparing Figure 11a (for high accumulation a and therefore large vertical advection) with Figure 23g, which is for the low-accumulation conditions of interior Antarctica.

Figure 12 contains results of applying the moving column method to an ice stream in the West Antarctic Ice Sheet (Lingle and Brown, 1987, Fig. 5). The (balance) temperature profiles are plotted on a relative-depth scale, $T(\zeta)$. The model T_B reaches the melting point (–2°C under the ambient pressure) only 100 km from the ice divide and remains at melting over the further 560 km of the profile. The reversed curvature near the bottom in the profiles at $\xi = 210$ and 310 km would be the expected effect of distributed shear heating, and the great change in $T(\zeta)$ from $\xi = 310$ to 410 km (with elimination of the reversed curvature) is what would be expected if there were a switch from internal deformation to basal sliding (Section 8), but since the simplification (28) was used, a different explanation is needed. In Figure 12c the model profile at $\xi = 510$ km is compared with a measured profile at $\xi = 490$ km (Engelhardt et al., 1990, Fig. 2). The agreement is fair. Other comparisons of modeled and observed temperature profiles are given by Budd and Young (1983b), Radok et al. (1982, p. 170), and Ritz (1989).

Figure 12. Vertical temperature profiles through Ice Stream B, West Antarctica, as modeled by Lingle and Brown (1987, Fig. 6) by the moving column method. (a) gives modeled profiles at locations $\xi = 0$ to 310 km from the flow divide, and (b) gives profiles at locations from $\xi = 410$ bar to the (un)grounding line at $\xi = 660$ km; the ξ coordinate of each profile is marked alongside the corresponding $T(z;\xi)$ curve. The vertical coordinate in (a) and (b) is fractional depth $\zeta = (z_S - z)/h$. (c) compares the modeled profile at $\xi = 510$ km, from (b), with the measured profile at $\xi = 490$ km, from Engelhardt et al. (1990), Fig. 2).

438

Thermal models of the entire Greenland and Antarctic Ice Sheets were generated in the '70's and early '80's from simple-column balance temperature profiles on an areal grid of ~300 points and from moving-column balance temperature profiles on an array of flowlines spanning the ice sheets (27 flowlines in Greenland, ca. 52 in Antarctica) (Budd, Jenssen, and Radok, 1971; Budd et al., 1982; Radok et al., 1982). The results were presented in profiles along selected flowlines (e.g. Fig. 11a) and in the form of maps showing input quantities such as the surface temperature distribution $T_S(x,y)$ and output (model-calculated) quantities such as basal temperature distribution $T_B(x,y)$, column-mean temperature, column-minimum temperature, and depth location of minimum. Typical maps of $T_S(x,y)$ and $T_B(x,y)$ are given for Greenalnd in Figure 13 and for Antarctica in Figures 7b and 14, the latter from a more recent generation of models as explained below. The main overall

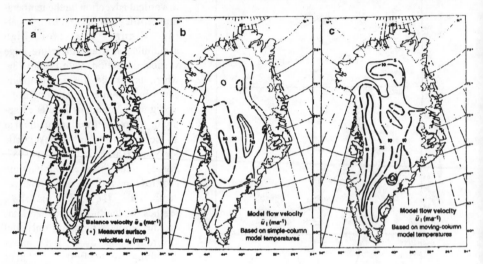

Figure 13. Comparison of observed surface temperature T_S of the Greenland Ice Sheet with the basal temperature T_B from modeling, by Budd et al. (1982, Maps 3/6, 6/1a, 6/2). (a) is $T_S(x,y)$, (b) is $T_B(x,y)$ calculated by the simple column model, and (c) is $T_B(x,y)$ from the moving column model. Areas of calculated basal melting are cross hatched and labeled "M."

feature seen in the models is that T_B does not decrease strongly toward the center of the ice sheets the way T_S does This is because the decrease in T_S, which is tied to the surface elevation z_S via the lapse rate Λ, is compensated by the effect of larger h in raising T_B (h ~3 km in the central part of the ice sheets). Increased h raises T_B by a simple "blanketing" effect: in a general way the temperature gradient $\sim(\partial T/\partial z)_B$ needed to conduct upward the geothermal and shear heat extends over a greater vertical interval and results in a correspondingly larger temperature rise in a thick ice mass than in a thin one. For the same reason, local cold vs. warm T_B areas generally correspond to relatively high vs. low areas of the bed, respectively (Budd et al., 1971, p. 119). This is particularly notable in the Antarctic models (Fig. 14), in which the basal topography is rather uneven. In the simple-column model of Figure 13b there is a band of relatively low T_B around the periphery of the Greenland Ice Sheet, whereas the moving-column model of Figure 13c has a peripheral warm zone. The difference probably arises as follows. The simple-column models' assumed z-independent horizontal advection term $\bar{u}\partial T/\partial\xi = \bar{u}\alpha\Lambda$ in (23) increases greatly and dominates the calculation of T_B toward the periphery, where $\bar{u}$ and α become relatively large. The effect of this advection term on T_B is opposed by the basal shear heating term $\tau_B\bar{u}$ in (28) or $\tau_{\xi z}\partial u/\partial z$ in (23), which also increases

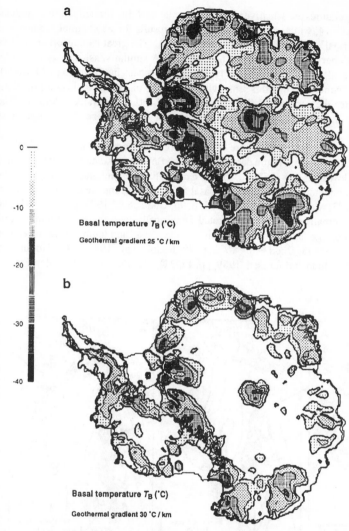

greatly toward the periphery. In moving-column models, the advection term is not forced to be large at depth, and the shear heating can dominate and produce a peripheral warm zone, in which the melting point is often reached. This is particularly prominent along the west and southeast coasts of Greenland as shown in Figure 13c (also in Fig. 11a from $\xi = 400$ to 600 km), where it is believed to correspond better to reality (e.g. Fig. 11b) than does the cold periphery given by simple-column models. The interpretation is however clouded by model results for the Antarctic Ice Sheet: the simple-column model of Budd et al. (1971, Map 4/1a) showed a distinct peripheral warm zone while the corresponding moving-column model (Map 4/1b) showed somewhat cooler peripheral temperatures; more recent models show a broad but somewhat discontinuous peripheral cool zone rimmed by a generally narrow warm band (Fig. 14) (see also Radok et al., 1987,

a Basal temperature T_B (°C)

Geothermal gradient 25 °C / km

b Basal temperature T_B (°C)

Geothermal gradient 30 °C / km

Figure 14. Results of three-dimensional thermal modeling of the Antarctic Ice Sheet, courtesy of Dick Jenssen (pers. com.). The maps show the modeled basal temperature, $T_B(x,y)$. (a) is for an assumed geothermal gradient of 25°C/km, and (b) for 30°C/km. The temperature contours and shading patterns are identified by the scale bar on the left. These are balance temperatures (see text). The maps of $T_B(x,y)$ here can be compared with the map of surface temperature $T_S(x,y)$ in Fig. 7b.

Maps 13 and 15). These differences make it appear that the model results are sensitive to the details of the modeling calculation. However, according to Budd et al. (1982, p. 59), "the discrepancies introduced by different [temperature] calculation schemes [(simple column vs. moving column)] are of less magnitude than those created by the uncertainty in the values of some [input] parameters," especially the geothermal gradient as discussed below.

440

The modeled basal temperatures are sensitive to the geothermal heat flux (or temperature gradient in (28)), as shown in Figure 14, which gives Antarctic model results for geothermal gradients of 25°C/km (considered "normal") and 30°C/km (considered high). The great expansion in the area of basal melting for the higher geothermal flux is noteworthy. A similar sensitivity is shown by Greenland models (Budd et al., 1982, Maps 6/1a-6/1f, for $-(\partial T/\partial z)_G$ varying from 10 to 30°C/km). Radok et al. (1982, p. 156) concluded that this sensitivity constitutes one of the main complications of ice sheet modeling, since the geothermal flux under the ice sheets is generally unknown. The sensitivity is further enhanced by the feedback mechanism in heat-flow/ice-flow coupling as discussed later.

From the thermal models, ice flow models are calculated for the ice sheets by the methods of Section 5, taking into account the temperature dependence of the flow-law parameter A in (7). When the vertically integrated flow equation (16) is used, the temperature dependence is introduced by taking an average of the model temperature over the basal 5% to 50% of the ice column (Radok et al., 1987, p. 32; 1989, p. 133; McInnes and Budd, 1984, p. 96; Budd et al., 1984, p. 32; Lingle and Brown, 1987, p. 260). Alternatively, (15) is integrated numerically with the model temperature profile introduced into (7) or into an empirical tabulation of the stress- and temperature-dependent flow law (Budd et al., 1971, p. 143; Budd et al., 1982, p. 70; Radok et al., 1982, p. 67; Huybrechts and Oerlemans, 1988, p. 53; Budd and Jenssen, 1989, pp. 17, 21). Results are illustrated in Figures 15 and 16.

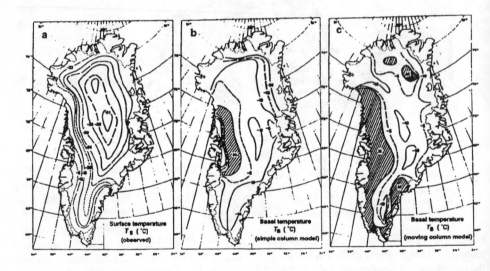

Figure 15. Flow models of the Greenland Ice Sheet, from Budd et al. (1982, Maps 4/2, 6/9, 6/10) (a) Balance velocity $\bar{u}_a$, and point measurements of surface velocity u_S along the EGIG profile. (b) Modeled mean flow velocity $\bar{u}_I$ based on the simple column thermal model of Fig. 13b. (c) Model $\bar{u}_I$ based on moving column model of Fig. 13c.

Figures 15b and c compare the flow velocities $\bar{u}_I(x,y)$ calculated for the Greenland Ice Sheet based on the simple-column vs. moving-column models (Budd et al., 1982, Maps 6/9 and 6/10). There are substantial differences, especially in the western half where the considerable differences in modeled T_B occur (Figs. 13b,c). It is very puzzling that the model flow-velocities along the west-central coast appear to be larger in the simple-column model (>20 m a^{-1}, Fig. 15b), in which T_B is well below melting ($\lesssim -10$°C, Fig. 13b), than in the moving-column model (<10 m a^{-1}, Fig

Figure 16. Modeled flow velocity $\bar{u}_I$ of the Antarctic Ice Sheet based on the thermal model of Fig. 14a, courtesy of Dick Jenssen (pers. com.). The flow is calculated from $T(x,y,z)$ at each point (x,y) by integration of a version of (15) in which the flow law dependence on stress $\tau_{\xi z}$ and on temperature T is given by empirical tables rather than by (2) and (7a). The velocity scale, given by the bar on the left, is logarithmic.

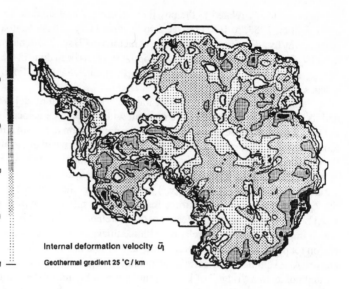

Internal deformation velocity $\bar{u}_I$

Geothermal gradient 25 °C / km

15c), where T_B is at melting (Fig. 13c). Comparison of model velocities with balance velocities (Fig. 15a) also reveals considerable discrepancies, especially along the west coast. This might be thought to indicate that the ice sheet is not in a steady state, but such a conclusion is not warranted because the two sets of model velocities differ between themselves as much as they differ from the balance velocities (Budd et al., 1982, p. 73). The latter, on the other hand, agree fairly well with measured surface velocities along the "EGIG" traverse (points with numbers in Fig. 15a), which implies that a steady state is a reasonable approximation. (Surface velocities are expected to differ only slightly from vertically averaged flow velocities.)

Coupling between thermal model and ice-flow model, which is required of a complete model as indicated in Section 2 but is overlooked in the balance temperature models as noted above, can be introduced by an iterative method in which, starting with the balance temperature distribution $T(\xi,z)$ based on $\bar{u} = u_a$, the ice flow $\bar{u}$ is recalculated from (15) or (16) with the current $T(\xi,z)$, and then $T(\xi,z)$ is recalculated from (23) or (29) with the new $\bar{u}$, this process being iterated until it achieves mutually consistent ice flow and heat flow models with common $T(\xi,z)$ and with $\bar{u} = \bar{u}_I$. (However, since the calculated $\bar{u}_I$ generally differs significantly from $\bar{u}_a$, as noted above and in Section 5, the result would not in general be a steady-state system, so the usual assumption of a steady state in calculating the thermal model would not seem justified.) Radok et al. (1982, p. 171) experimented with application of the iterative procedure to the thermal and ice-flow models of the Jakobshavns flowline (Fig. 11). In relation to the balance temperature model (Fig. 11a) the coupling resulted in a general slight warming in the upper half of the ice sheet and a cooling near the base. These changes occurred because the velocities $\bar{u}_I$ from the balance temperature model were generally somewhat smaller than $\bar{u}_a$: decreased basal shear heating in (28) caused the basal cooling, while the warming above was caused by reduced horizontal advection of colder temperatures from upstream.

There is a positive feedback process in this element of the model: if $\bar{u}_I < \bar{u}_a$ so that T_B is reduced in the first iteration, then in the next stage of iteration the ice flow velocities $\bar{u}$ are decreased further because of the temperature dependence of A in (7), which leads to further basal cooling (decrease in basal shear heating), and so on. A similar feedback process enhances the sensitivity of coupled heat-flow/ice-flow models to changes in the geothermal heat flux, noted above. (In the approximation of (28) the feedback process is the same in both cases.) This type of thermal-mechanical feedback is closely related to the possible phenomenon of thermal creep instability (Clarke et al., 1977; Hutter, 1983, p. 168-178; see also Section 11.1). These feedback effects can be offset in a general way by adjusting the enhancement factor $\ni$ in (7) so as to achieve general agreement between $\bar{u}_I$ and $\bar{u}_a$ (assuming a steady state is appropriate), as has been done in the Antarctic models of Figures 14 and 16. Further aspects of feedback in thermomechanically coupled glacier flow models are discussed in Section 9.5.

It would be of interest to know how important the thermal model is for the ice-sheet flow model and its extent of agreement or disagreement with the balance velocities. The published models do not provide a direct assessment of this, but the comparison in Figure 15b,c indicates that the thermal model can have a substantial effect. A related question is whether the effect of τ_B on flow dominates the effect of temperature (mostly T_B) in these flow models. Study of Figures 14, 16, and a map of $\tau_B(x,y)$ not reproduced here suggests that variations in $\bar{u}_I$ are in general correlated more strongly with variations in $\tau_B(x,y)$ than with $T_B(x,y)$, suggesting that the control by τ_B dominates, but features of the $T_B(x,y)$ pattern tend to reinforce features in the $\tau_B(x,y)$ pattern in generating peaks or lows in $\bar{u}_I(x,y)$. This is to be expected because of the correlation of T_B with bedrock topography noted earlier, which implies a positive correlation with ice thickness h and thus with τ_B via (14). An additional reason for positive correlation between τ_B and T_B and thence with $\bar{u}_I$ is the feedback effect discussed above. An example of the correlation among τ_B, T_B, and $\bar{u}_I$ is indicated by a small arrow in Figure 14 that points to a "ridge" of high T_B running roughly northwestward near lat. 80°S, long. 50°W, and a small arrow in Figure 16 that points to a corresponding "ridge" of high $\bar{u}_I$. The $\tau_B(x,y)$ map shows a weaker, more irregular high in τ_B in the same area. In this case, the temperature effect seems to dominate in influencing $\bar{u}_I$, but the τ_B effect reinforces it. The "ridge" in T_B and $\bar{u}_I$ apparently corresponds to a bedrock trough that runs roughly southeastward from a gap in the Transantarctic Mountains between the Neptune Range and the Patuxent Range. The "ridge" is not seen in the $\bar{u}_a(x,y)$ plot (Fig. 8a) and is thus an example of the "fine-grained" flow features generated by models but seemingly absent in reality, as noted in Section 5.

There is one type of observed flow feature for which the temperature model has the potential for making a strong improvement in the flow model: the small model $\bar{u}_I$ values in and around the large ice shelves (especially the Ross and Ronne Ice Shelves), where the highest $\bar{u}_a$ values occur (Fig. 8a). In these areas the modeled T_B is high, but the $\bar{u}_I$ response to the high T_B is evidently inadequate (Fig. 16). These areas are among those in which the thermal model indicates that T_B reaches the melting point, as shown in Fig. 14. A similar situation prevails along the central west margin of the Greenland Ice Sheet (Figs. 13c, 15). Effects of basal melting on flow models are considered in Sections 8, 10, 11, and 12.

The recent history of glacier and ice sheet modeling, going beyond the first-generation models discussed above, is one of great increase in computer power and capacity for modeling, and at the same time significant additions to the physical processes included as model components. The increase of capacity shows itself in greater spatial resolution, improved handling and display of input data and output results, and relaxation of restrictive assumptions, approximations, and other computational constraints. What we may call "second-generation" models, which appeared in the mid and late '80's, generally include one or more of the following new elements: basal sliding (see Section 8), time dependence (Sections 9 and 11), basal melting and water flow (Section 10), solution of the heat-flow problem without the approximations of "column" models, full coupling of the heat-flow and ice-flow models without dependence on "balance temperatures," gridded calculation of flowlines, flowline-independent solutions, longitudinal and lateral stress coupling (Sections 6, 9.4.3, 11.2, and 12), inclusion of mechanically coupled floating ice shelves (Section 9.4), and adjustment of flow-law and basal sliding parameters to optimize agreement between model ice-flow velocities and observed or balance velocities (Section 8). Examples include the Antarctic models of Budd et al. (1984), McInnes and Budd (1984), Jenssen et al. (1985), McInnes et al. (1986), Radok et al. (1987), Lingle and Brown (1987), Budd and Jenssen (1987), Budd et al. (1987), Huybrechts and Oerlemans (1988), Herterich (1988), MacAyeal (1989), Fastook (1989), Budd and Jenssen (1989), Böhmer and Herterich (1990), and Huybrechts (1990, 1992), and the Greenland models of Letréguilly et al. (1991). In their current state of development the models are approaching full three-dimensionality in their treatment of temperature distribution on the basis of (10), in their applicability to ice masses of arbitrary three-dimensional shape free of restriction to flowplanes or flowbands, and in their vertical resolution instead of vertical integration or parameterization (Section 5). The only non-three-dimensionality of possible significance retained in most of these models is the use of (13) (or in some cases (19) or its two-dimensional generalization (73)+(74)) for mechanical equilibrium instead of the full set of equations (8); the latter are, however, included in an unpublished three-dimensional valley-glacier model of Brian Hanson (pers. com., 1992) which has been applied to Störglaciaren in Sweden. The difficulties confronted by three-dimensional, time-dependent models have been discussed for some time (e.g. Jenssen, 1977, p. 388; Oerlemans and van der Veen, 1984, p. 87), but these models are now feasible. In Figures 8, 14, 16, 17, and 25 Dick Jenssen has very kindly allowed me to include here some not-yet-published results of his current Antarctic model, which is calculated on an x,y,z grid of $281 \times 281 \times 30$ points. In terms of detail and quantitative reliability this model is clearly much preferable to the first-generation models discussed above. The results appear to differ only modestly from those of the earlier models, and it appears that the earlier models still have educational value in building physical understanding of the inner workings of the ice sheets.

An exemplary collection of 24 maps, in color, displaying results of Antarctic modeling at a stage precursory to Dick Jenssen's current model is given by Radok et al. (1987, p. 12-35) and conveys a much more nearly complete picture of the content of the models than is possible to do in this review.

8. Basal sliding

When T_B reaches the melting point, sliding of the ice mass over its bed becomes possible. The no-slip condition at the bed, on which the first-generation models in Sections 5 and 7 were based, has to be replaced by a slip condition, called the *basal sliding law*, which specifies the basal sliding velocity u_B as a function of controlling variables. The mean flow velocity $\bar{u}$ is increased by the amount u_B over the contribution $\bar{u}_I$ from internal deformation of the ice mass in (16):

$$\bar{u} = \bar{u}_I + u_B \tag{30}$$

Because the sliding response to τ_B could in principle be anisotropic, $\bar{u}_I$ and u_B could have somewhat different directions, but in practice they are assumed to be the same and (30) is written on this basis.

It is often said that more rapid basal sliding results from increased "lubrication" of the bed by basal water. Four distinct types of the "lubricating" action by basal water can be recognized: (i) effect of basal water *pressure* P_w in promoting sliding over bedrock; (ii) effect of the *amount* of basal water on sliding; (iii) effect of the basal melting *rate* on sliding; (iv) effect of water on the role of basal rock debris in sliding. These effects are to some extent interrelated physically, and yet are distinct enough to be usefully distinguished. In this section and in Section 10 we consider effects (i) and (ii); effects (iii) and (iv) are discussed in Sections 11 and 12.

8.1 FORMULATION OF BASAL SLIDING LAWS

The first and simplest sliding law was derived theoretically by Weertman (1957) for clean ice sliding over rough bedrock:

$$u_B = S \, \tau_B^{\,p} \tag{31}$$

The exponent p and "sliding coefficient" S depend on the bedrock roughness, or, more precisely, on the distribution of roughness amplitude R over the roughness wavelength spectrum λ, the bedrock topography being thought of as analyzed into its Fourier components $R(\lambda)$. If the bed roughness is mostly at short wavelengths ($\lambda \lesssim 0.1$ m), then theoretically $p = 1$, and if it is mostly at long wavelengths ($\lambda \gtrsim 10$ m), then $p = n$, the exponent in (6); if R/λ is essentially constant, independent of λ, over a broad wavelength range 0.1 m $\leq \lambda \leq 10$ m, then $p = \frac{1}{2}(n + 1)$. The physical processes that contribute to the sliding and govern the sliding law (31) – ice regelation and ice flow around bedrock obstacles – were observed at the base of a glacier by Kamb and LaChapelle (1964) and the observations were interpreted quantitatively by Kamb (1970, Sects. 15-18) in terms of the theory that underlies (31).

The sliding law (31) applies for ice in intimate contact with bedrock. At high enough basal water pressure P_w the sliding ice mass partially decouples from the irregular bed by ice-bedrock separation, often called "cavitation" (but not to be confused with hydrodynamic cavitation). The partial decoupling enhances u_B over the velocity given by (31) for a given τ_B. The analysis of basal sliding with ice-bedrock separation, and the development of a sliding law for this process, proceeds along three rather different lines:

1. The formation of discrete water-filled cavities by ice-bedrock separation in the lee of bedrock protuberances is the mechanism treated extensively by Lliboutry (1959, 1968, 1979a, 1987b) in formulating theoretically a number of sliding laws. The mechanics of sliding with cavitation has been considered also by Weertman (1964), Fowler (1986, 1987a,b), Kamb (1987, Sect. 6), and Humphrey (1987). Iken (1981) has carried out numerical flow modeling calculations of the basal cavitation process.

2. The formation of a basal "water film" or "water sheet" separating ice and bedrock and "drowning" the smaller-scale roughness elements of the bed is a mechanism for enhanced sliding proposed and developed into a sliding law by Weertman (1962, 1964, 1969). The water film is a form of ice-bedrock separation that, as treated by Weertman (1964, p. 290, 300), is quite distinct from the cavitation-type ice-bedrock separation in item 1.

3. Laboratory experiments on the sliding of ice blocks at the melting point over artificial and natural rock surfaces (cement, pebble concrete, shale, granite) by Budd et al. (1979) provided the basis for proposing an empirical sliding law of the form

$$u_B = \hat{S} \, \tau_B^{\,p} \, \hat{P}^{-q} \tag{32}$$

where $\hat{P}$ is the effective normal stress (or effective pressure) across the sliding interface:

$$\hat{P} = P_I - P_w \tag{33}$$

Here P_I is the normal stress applied across the interface by the ice, and P_w is the pressure of water having access to the interface. The factor $\hat{P}^{-q}$ (with $q > 0$) in (32) expresses the effect of the ice overburden pressure in suppressing cavitation at the sliding interface and the opposing effect of water pressure in the cavities. Cavitation associated with the sliding apparently was not studied in the experiments, however, and P_w was 0; its role in (32), via the effective pressure (33), was assumed on general grounds (Budd et al., 1979, p. 167). The "sliding coefficient" $\hat{S}$ in (32) is not the same as S in (31); both must decrease with increasing bed roughness, but probably in somewhat different ways. For the exponents p and q in (32), Budd et al. (1979, p. 163) reported $p = 1$, $q = 1$ for the experimental results at low shear stresses and $p = 3$, $q = 1$ at high stresses ($\hat{P} > 5$ bar)[§]; Budd, Jenssen, and Smith (1984, p. 33) stated the low-stress results as $p = 1$, $q = 2$. The exponents evidently cannot be considered to be constants, which tends to call into question the use of the mathematical form of (32).

From a lengthy and powerful theoretical treatment of sliding with cavitation, cited in item 1 above, Fowler (1987a, p. 260, eqn. (3.37)) obtained a sliding law of the form (32) for beds with power-law roughness spectra $R(\lambda) \propto \lambda^s$, the theory giving $p = (2 - s)/(s - 1)$ and $q = p - n$; here s is restricted to the range $1 < s < (n + 2)/(n + 1)$, hence $p > n$ and $q > 0$. The theory gives a (complicated) prescription for $\hat{S}$, but Fowler (1987a, p. 261) concludes that $\hat{S}$ "can only be estimated in practice by fitting experimental data; in fact, the best one could do would be to fit the constants [$\hat{S}$, p and q] in a law of the form [(32)] to the data." This has been done extensively as noted above and in what follows. The reason for the needed empiricism is that in general too little is known about the roughness spectra $R(\lambda)$ of glacier beds.

The sliding laws derived theoretically by Lliboutry, cited in item 1 above, generally are expressed as $\tau_B = $ a sum of terms each of which is like (32) solved for τ_B, with various values of the exponents p and q; in addition they often contain a Coulomb-like term for which τ_B is proportional to $\hat{P}$ without any dependence on u_B. For example, Lliboutry (1979a, eqn. (110)) gives the following sliding law for a bed with roughness components of three widely different wavelengths:

$$\tau_B = F\hat{P} + S_1^{-1} u_B^{1/n} + S_2^{-1} u_B^{2/(n+1)} \tag{34}$$

where the coefficients F (called the "friction function"), S_1, and S_2 depend on the roughness amplitudes and wavelengths, and F and S_2 depend also on the sliding velocity; for some types of roughness wave forms (e.g. "Gaussian modulation"), F is essentially constant, independent of u_B, and thus F is effectively a coeffient of friction (Lliboutry, 1979a, Fig. 7). Although (34) and similar laws with several terms on the right-hand side seem rather different from the law (32) derived by Fowler (1987), especially when they contain the "solid friction" term $F\hat{P}$, Fowler (1987, pp. 256, 260, 264) has pointed out much similarity between his results and Lliboutry's, including the possibility of a "solid friction" type of behavior under some circumstances. A solid-friction-like law is also given rigorously by Iken's (1981, p. 409) model of sliding over a staircase-type bed (see (71), Section 12). Perhaps because of their complexity, Lliboutry's sliding laws do not seem to have been used in modeling, except by Renaud (1973) in applying the "solid friction law" to flow of a valley glacier.

In the water-film model of Weertman (item 2 above) a sliding law is formulated in terms of the effect not of water pressure P_w or effective pressure $\hat{P}$ but of water-film thickness δ (treated as uniform); this introduces the concept of control of sliding by the *amount* of water stored at the bed (storage volume δ per unit area). The effect is calculated on the basis of the "drowning" of

---

[§]The sliding law (32) with $p = 3$, $q = 1$ has been called "the classical Weertman-type sliding relation corrected for the effect of subglacial water pressure" (van der Veen, 1987, pp. 223, 237; Huybrechts, 1990, p. 82; 1992, p. 95), while (32) with $p = 1$, $q = 2$ has been called the "Budd-type sliding relation" (van der Veen, 1987, p. 238). From the discussion here it is seen that these designations make an inappropriate distinction between the two sliding relations, especially since Weertman's own "correction" (35) differs so greatly from (32).

small-scale roughness elements (of amplitude $< \delta$) by the water film, and leads to the following "first-approximation" modification of (32), as quoted by Weertman and Birchfield (1982, p. 137):

$$u_B = S \, \tau_B^2 (1 + 10 \, \delta/\delta^*) \tag{35}$$

Here $sS \, \tau_B^2$ is the ordinary sliding velocity when $\delta \to 0$, from (32) with $p = \frac{1}{2}(n + 1) = 2$, and δ^* is a "critical obstacle size" provided by the theory and estimated to be of the order of 2 to 5 mm.

The water-film model was modified by Weertman (1972, p. 312) in recognition of the fact that a continuous water layer ~1-10 mm thick between the ice and the bed would be impossible because it would prevent the basal shear stress τ_B from being supported across the ice-bed gap. In the modification, the water film is allowed to be very non-uniform in thickness and to be "squeezed out" in areas of high normal stress across the ice-bed interface, forming a "punctured water sheet" in which "modified sheet flow" takes place. However, the mechanical consequences of the non-uniformity, squeezing out, and sheet flow modification have not been incorporated into (35). Alley et al. (1989, p. 132) took the approach of retaining (35) while allowing the areal fraction of the partially "squeezed out" water film to be controlled by $\hat{P}$ and τ_B as discussed in Section 12.

A relationship of sliding to basal water storage, formulated for the water-film model in (35), has a counterpart in the cavitation model (item 1 above) because high basal water pressure causes both increased basal sliding and cavitation, so that it is perhaps somewhat moot whether the enhanced sliding should be thought linked more directly with the water pressure or with the volume of water stored at the glacier bed in water-filled cavities, within which "drowning" of small-scale roughness features occurs as it does in the water-film model. The same mootness perhaps applies also to the water-film model, because the above-noted approach of Alley et al. (1989) for the partially "squeezed out" film defines a relationship between δ and $\hat{P}$ (see Section 12). However, Weertman (1972, p. 313) stated that "the thickness [of the punctured water sheet] does not depend on the value of the water pressure." Observationally in some cases there seems to be a closer correlation between stored water and sliding velocity than between water pressure and sliding (Fahnestock, 1991, p. 72; Kamb et al., 1992, p. 5), whereas in other cases pressure clearly plays a strong role, although accompanied by storage (Kamb and Engelhardt, 1985, p. 40-42; Kamb et al., 1985, p. 474-475; Section 11.3 below). The possible distinction between water-pressure control and stored-water control on sliding is an active area of investigation.

8.2 APPLICATION OF SLIDING LAWS IN ICE-SHEET AND GLACIER MODELING

Sliding laws of the form (31) with $p = 2$ or 3 were used by Thomas and Bentley (1978), Thomas et al. (1979), and Hughes (1981) in models of the West Antarctic Ice Sheet. A sliding law of type (32), or elaborations thereon, has been extensively used in ice-sheet flow modeling. Usually it is rewritten

$$u_B = \hat{S}' \, \tau_B^p \, \hat{h}^{-q} \tag{36}$$

where $\hat{S}' = \hat{S}(\rho g)^{-q}$ and where $\hat{h}(= \hat{P}/\rho g)$ is the "ice thickness in excess of grounding," that is, the total ice thickness less the thickness that would just be afloat, given the elevation of the base of the ice z_B in relation to sea level at $z = 0$:

$$\hat{h} = h - (-z_B)\rho_w/\rho \tag{37}$$

where ρ and ρ_w are the densities of ice and (sea) water. For a bed above sea level ($z_B > 0$), (37) does not apply, and $\hat{h} = h$.

For historical perspective it may be noted that the earliest computer model of a glacier (Campbell and Rasmussen, 1969, 1970) used in effect a sliding law (36) with $p = 1$, $q = 1$, and $\hat{h} = h$; it combined this with linear ice rheology corresponding to taking $\overline{\eta}$ to be a constant in (19) but omitting the flow contribution $\overline{u}_I$ from (16) or in effect subsuming it into the u_B contribution.

According to Budd et al. (1984, p. 35), the parameter choice $p = 3$, $q = 1$ was used in early modeling of the Antarctic Ice Sheet with (36), but was not very effective in reproducing the low-τ_B, high-$\overline{u}_a$ zone of West Antarctica; the choice $p = 1$, $q = 2$ gave a much improved model. From modeling of the ice streams McInnes et al. (1986, p. 15) suggested choices $(p,q) = (1,1)$, $(2,1.5)$, and $(3,2)$ as equally satisfactory. McInnes and Budd (1984, p. 98) reported that a still better model was obtained with the sliding law

$$u_B = \hat{S} \, \tau_B \, (\hat{h}^2 + c_1 \hat{h})^{-1} \tag{38}$$

with c_1 a constant, and Budd et al. (1987, p. 334) reported a "reasonable match" with this model. $\hat{S}$ does not of course have in general the same numerical value for laws with different p or q or with elaborations as in (38) and (39).

In models of the Antarctic Ice Sheet and ice streams Radok et al. (1987, p. 33; 1989, p. 133) and Budd and Jenssen (1989, p. 22) went to a sliding law

$$u_B = \hat{S} \, \tau_B \, e^{\nu(T_B - T_o)} \, [(\hat{h} + c_4 \hat{h}^2)^2 + c_0]^{-1} \tag{39}$$

while Jenssen et al. (1985, p. 30) used the same law but with $c_0 = 0$. In (39), $(T_B - T_o)$ is the basal temperature T_B relative to the pressure melting point T_o (cf. (7b)), and $\nu = 0.1 \mathrm{K}^{-1}$. Radok et al. (1987, p. 33) note that the complicated functional form in (39) lacks physical justification.

The inclusion of the temperature-dependent exponential factor in the sliding law (39) is noteworthy in relation to the standard concept that significant amounts of basal sliding occur only when T_B reaches the melting point. This factor seems to have been first introduced by Budd et al. (1985, p. 134), with the following explanation: "because of the relatively low basal water film thickness the major control on the sliding rate, besides the basal shear and effective normal stresses, was supposed here to be the basal temperature." Since the sub-freezing sliding-friction experiments of Barnes et al. (1971) were also cited as a basis, it seems evident that the exponential factor was intended also to bring into the model the type of "anomalous" sliding at sub-freezing temperatures that has been found in field observations (Echelmeyer and Zhongxiang, 1987), laboratory experiments (Barnes et al., 1971; Kamb, 1972, p. 218), and theoretical analyses (Shreve, 1984; Fowler, 1986b). However, it is not known whether the form of the temperature dependence assumed in the exponential factor, with the assumed $\nu = 0.1 \mathrm{K}^{-1}$, actually describes correctly the magnitude of the "anomalous" sliding at sub-freezing temperatures relative to normal sliding at the melting point. For further information bearing on this question see the review by McInnes et al. (1985, pp. 15-24).

In the "second-generation" ice-sheet models the sliding-law parameters $\hat{S}$, c_0, c_1, c_4 have been adjusted to try to maximize the agreement between the balance velocity $\overline{u}_a$ and the modeled "dynamics" velocity $\overline{u}$ from (30); simultaneous adjustment of the constant in the flow law (16) for $\overline{u}_I$ was also in general involved, as discussed in Section 5. Sliding laws of types (31), (32), (38), and (39), with parameters adjusted as just stated, were discussed and evaluated by Bentley (1987, p. 8855) (see Section 10.1).

A related alternative to the foregoing empirical optimization of the sliding law by formula manipulation and parameter adjustment was pursued by Morland et al. (1984) using data from the EGIG profile in Greenland (Fig. 15a) and from a profile across part of the Barnes Ice Cap. From the observables $z_S(\xi)$, $h(\xi)$, and $u_S(\xi)$, and with assumed flow-law constants for ice, they calculated $u_B(\xi) = u_S(\xi) - u_I(\xi)$ where $u_I = \overline{u}_I(n+2)/(n+1)$ from (16). (Actually they used a more sophisti-

448

cated method to caluclate $u_1(\xi)$ from the data.) They assumed a sliding law of the form $u_B = \tau_B^p/f(\hat{h})$ which is like (36), (38), or (39) except that an initially unknown function $f(\hat{h})$ is introduced in place of the specific functions $\hat{h}^q$, $(\hat{h}^2 + c_1\hat{h})$, or $[(\hat{h} + c_4\hat{h}^2)^2 + c_0]$ in (36), (38), and (39). (Actually they used h instead of $\hat{h}$, disregarding P_w, but I presume that if they had thought P_w significant in the profiles they would have used $\hat{h}$.) Getting $\tau_B(\xi)$ from (14) they then plotted $\tau_B^p(\xi)/u_B(\xi)$ against $h(\xi)$ to reveal, for different choices of p (=1,2,3,4), the form of the function $f(h;p)$. (This function is not to be confused with the flow-law function $f(\tau)$ in (6).) However, the functions $f(h;p)$ so generated do not seem particularly reasonable, and none of the p values gives a result that stands out as particularly appropriate; moreover, the functions are quite different for the two profiles. Probably for these reasons the results of this approach do not seem to have been used as a sliding law in ice sheet modeling. Possibly some of the difficulty in this approach may be due to error in the assumed ice flow law or to neglect of the effect of P_w. Morland et al. (1984, p. 138) conclude that "there is no universal sliding law, and choice of a basal sliding condition will depend on the particular application."

The basal sliding contribution $u_B(x,y)$ obtained from (39) in the current Antarctic model of Jenssen (see Section 7), with the basal temperature distribution in Figure 14b, is shown in Figure 17. It is generally similar to that of earlier models (e.g. Budd and Jenssen, 1989, Fig. 4; Radok et al., 1987, Map 22). The role of the $\exp \nu(T_B - T_o)$ factor in (39) is seen in the lack of complete correlation between the areas of basal slding in Figure 17 and the areas of basal melting in Figure 14a; the model gives appreciable basal sliding in many areas where the modeled basal temperature is below freezing. When $u_B(x,y)$ in Figure 17 is added to $\bar{u}_1(x,y)$ in Figure 16, the complete "dynamics" velocity model $\bar{u}$ in Figure 8b is obtained. The contribution from u_B is responsi-

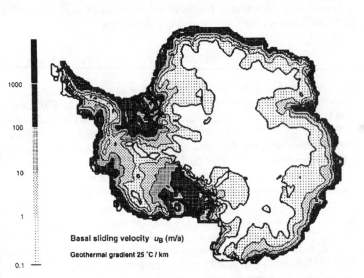

Figure 17. Basal sliding velocity u_B of the Antarctic Ice Sheet as modeled on the basis of (39) for the thermal model of Fig. 14a (courtesy of Dick Jenssen). The (logarithmic velocity scale is on the left.

ble for the high model velocities in the neighborhood of the large ice shelves and greatly improves the agreement between $\bar{u}$ and $\bar{u}_a$ in these areas, by comparison with the velocity due to $\bar{u}_1$ alone (Fig. 16). The overall agreement/disagreement between $\bar{u}_a$ and $\bar{u}$ for this model was discussed in Section 5. Budd and Jenssen (1989, p. 19) made a detailed test of their model in the Wilkes Land region of East Antarctica, where actual surface velocity measurements are available: over a range of two orders of magnitude the model velocities $\bar{u}$ tracked the observed velocities to within a factor of about ½ to 2; a similar range of agreement/disagreement was found between the balance velocities u_a and the observed velocities.

In areas of basal melting in the interior of the East Antarctic Ice Sheet (Fig. 14b) the modeled flow contribution from u_B (Fig. 17) is mostly small in relation to $\bar{u}_I$ (Fig. 16). This reflects the effect of the $\hat{h}$ factor in the sliding law (39). In models of this type, significant basal sliding occurs only where the ice is thin enough and its base is far enough below sea level that $\hat{h}$ from (37) is small.

Figure 18a, from Budd and Jenssen (1987, Fig. 13), shows results of a detailed model of the "Ross sector" of the West Antarctic Ice Sheet, modeled with the sliding law (38) multiplied by the temperature-dependent exponential factor used in (39), and with $c_1 = 0$. In this part of the ice sheet, even though τ_B is small (~0.1 bar), the balance velocities (Fig. 18b) reach high values (>100 m y^{-1}), which the model explains as due to extensive basal melting and rapid basal sliding. In detail, however, the high balance velocities extend farther inland and are more prominently concentrated into localized "currents" than the model is able to explain, even with inclusion of the temperature-dependent factor for this purpose (Budd and Jenssen, 1987, p. 315). These rapidly flowing

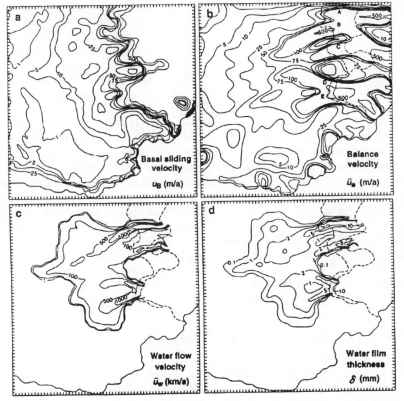

Figure 18. Models of ice flow, basal melting, and basal water transport in the "Ross sector" of the West Antarctic Ice Sheet, by Budd and Jenssen (1987, Figs. 13, 3, 11, 10). The modeling area is a 1200 km × 1200 km square centered on lat. 80°S, 135°W meridian running up and down on the page, south at the top. Ross Island is at right center near the edge, and western Marie Byrd Land occupies the bottom half. The area is contained in the central and lower left part of Fig. 19, with the square rotated 45° relative to the borders of Fig. 19. (a) Basal sliding velocity u_B, from (38) modified as noted in the text. (b) Balance velocity $\bar{u}_a$. The letters A, B,..., E identify Ice Streams A, B,..., E. (c) Water flow velocity $\bar{u}_w$ in a basal ice-bed gap of thickness δ of the gap or "water film," through which basal meltwater is conducted at velocity $\bar{u}_w$.

450

"currents" are the West Antarctic *ice streams* (Fig. 19; Bentley, 1987), whose existence points to the need for consideration of additional elements in the ice sheet model, as discussed in Sections 10 and 12.

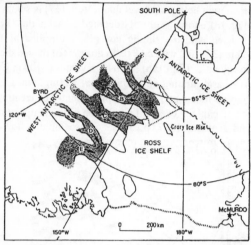

Figure 19. Location map showing the West Antarctic ice streams that flow into Ross Ice Shelf. The ice streams (stippled) are labeled with their letter designations (A, B,..., E), and their direction of motion is shown with arrows. The (un)grounding line marking the inner edge of the Ross Ice Shelf is drawn with a fine continuous line that terminates the stippling; the outer, calving edge is shown with a hachured line. The polyhedron bounded by the dotted and heavy dashed lines around parts of Ice Streams A and B is the area of the two-dimensional flow model of MacAyeal (1989a).

In modeling Antarctic ice streams, Lingle and Brown (1987) used (36) with $p = 1$, $q = 1$, while Lingle (1984) in effect used (36) with $p = 3$, $q = 1$ but with $\hat{h}$ replaced by h as though $P_w = 0$. They took an additional step in allowing $\hat{S}$ to be a function $\hat{S}(\xi)$ of position ξ along the modeled flowline, which allows for longitudinal variation in bed roughness. The variation of $\hat{S}(\xi)$ was rather extreme (four orders of magnitude) in the Lingle (1984) model, which is doubtless due to assuming (in effect) $P_w = 0$; near the grounding line P_w doubtless has a large effect on sliding. In the Lingle and Brown (1987) model the variation of $\hat{S}$ was relatively modest (one order of magnitude) and could have been practically eliminated if $q = 2$ had been used instead of $q = 1$. Spatial variation of $\hat{S}$ in (38) and of A in (6) was used by Budd et al. (1987, p. 340) in tuning a steady state flowband model of the West Antarctic Ice Sheet to current values of $h(\xi)$, $\alpha(\xi)$, and $\bar{u}_a(\xi)$.

Spatial variation of $\hat{S}$ and A was also introduced in the vertically integrated two-dimensional Antarctic model of Fastook (1987, p. 8942; 1990, p. 51) and Hughes (1992, p. 220), discussed in Section 5; it includes provision for basal sliding via a combined flow-plus-sliding law

$$\bar{u} = (1 - f)\frac{2A}{n+2}\,\tau_B^3\,h + f\hat{S}\,\tau_B/\hat{h}^{\,2} \tag{40}$$

in which f (with $0 \le f \le 1$) is an additional parameter, which sets the percentages of u_B from (36) and $\bar{u}_l$ from (16) that are included in the combination (30). The need for such a parameter presumably enters as a substitute for the effects of T_B, since temperature and basal melting are not modeled in this treatment. Hughes (1992, p. 221) related f to the areal fraction of the bed that is thawed (wet) when a patchy distribution of thawed and frozen areas exists, but he does not appear to specify fully how the thawed fraction is determined. By choosing f to be near 1 in areas where T_B is thought to reach the melting point, a full or nearly full contribution from basal sliding can be provided there. However, it seems inappropriate to multiply the internal-deformation contribution by $(1 - f)$ in (40), because this contribution should be greatest when T_B is at the melting point, according to (7)+(16). The model in Figure 9a results from substantial point-by-point adjustments not only in A as indicated in Figure 9b but also in the parameters $\hat{S}$ and f in (40).

The sliding law (36) has been used by Harbor (1992) to model the effects of basal sliding on the flow pattern in cross sections of valley glaciers, extending the treatment discussed in Section 4. For the effective pressure $\hat{P}$ at the bed, or equivalent $\hat{h}$, he used (37), with the datum for z_B being not sea level but the local phreatic surface (water table) in the water-conduit system at the base of the glacier. In this application (37) is clearer if rewritten without reference to the datum, as follows:

$$\hat{h} = h(y)\,(1 - \frac{\rho_w}{\rho}) + d_w\frac{\rho_w}{\rho} \qquad (h(y) \geq d_w) \tag{41}$$
$$\hat{h} = h(y) \qquad (h(y) \leq d_w)$$

where d_w is the depth of the phreatic surface below the ice surface and $h(y)$ is the laterally variable ice thickness, y being a transverse coordinate; ρ_w is the density of (fresh) water. Values of p and q from 1 to 5 were explored, as well as values of d_w/h_c from 0.1 to 0.4 (where h_c is the centerline ice thickness). The best agreement of the modeled flow pattern with the observations by Raymond (1971, pp. 69, 76) in Athabasca Glacier (Canada), where $d_w/h_c = 0.13$, was for $p = q = 2$. It was, however, difficult in the modeling to obtain sufficiently low marginal sliding velocities, and the agreement with observation was obtained by the rather non-intuitive step of changing the sliding law to (31) above the phreatic surface, which requires the bed roughness to increase there and not be counteracted by increasing cavitation near the surface.

Although the water-film model of basal sliding, expressed in (35), has not been extensively used in glacier or ice-sheet modeling, three applications in the West Antarctic Ice Stream region can be noted. Weertman and Birchfield (1982) argued that (35) could account for the high ice-streaming velocities, and they concluded that under certain conditions the flow of Ice Stream B could be unstable if controlled by basal sliding lubricated by the water-film mechanism. Alley (1989b, p. 120; also Alley et al., 1989, p. 132) used (35) to argue that basal sliding of Ice Stream B is small (~3 m a^{-1}). Budd and Jenssen (1987, p. 312) considered application of (35) to the ice-stream region in the light of model calculations of δ (see Section 10.1), but they did not actually calculate an ice-flow model based on $\delta(x,y)$, which leaves uncertain the ability of (35) to account for the ice flow in the region.

9. Time-dependent, predictive models

One of the main objectives in the development of ice modeling is to predict the course of growth or shrinkage of glaciers and ice sheets under assumed variations in climatic conditions through time. Most such time-dependent models are generated by applying the non-steady-state ice-flow continuity equation

$$\frac{\partial h}{\partial t} = a - \nabla_2 \cdot (h\overline{u}) \tag{42}$$

to ice-flow models of the types discussed in Sections 5, 7, and 8. (42), like (17), is obtained by vertical integration of (9), but without the steady-state assumption $\dot{h} = 0$. For flowband models, representing the flow between closely-adjacent flowlines in an ice sheet (or alternatively the flow of a valley glacier), (42) becomes

$$\frac{\partial h}{\partial t} = a - \frac{1}{b}\frac{\partial}{\partial \xi}(bh\overline{u}) \tag{43}$$

452

where ξ is as before the (curvilinear) longitudinal coordinate along the flowband and $b(\xi)$ is the flowband width (or effective valley-glacier width).[§]

By (42) or (43), $\partial h/\partial t$ can be calculated numerically from the model values of velocity $\bar{u}(x,y)$, ice thickness $h(x,y)$, and accumulation/ablation $a(x,y)$ (or $\bar{u}(\xi)$, $h(\xi)$, $a(\xi)$); the development of an ice mass through time can thus be followed. In the simplest procedure, (42) or (43) can be time integrated forward in steps Δt by incrementing h by $\Delta h = (\partial h/\partial t)\Delta t$ at each point of the model; at each step, of course, u is recalculated for the new h from (16) or equivalent, and $\partial h/\partial t$ is recalculated from (42) or (43). At each step also the coupled thermal model (Section 7) has to be recalculated, by time integration of (10) or (29). In this way $h(x,y;t)$ and $T(x,y,z;t)$ are stepped together through time. This "explicit" type of finite difference method as used for numerical integration of (42) or (43) in the isothermal case (constant A) is described by Oerlemans and van der Veen (1984, p. 68) and van der Veen (1991, eqns. (16)-(18)); a more complicated "implicit" type of finite difference method involving also the coupled integration of (10) (in the non-isothermal case) is described by Huybrechts (1992, p. 113-128). Use of the finite element method for this purpose is described by Fastook (1987b; 1990, p. 113-128), MacAyeal and Lange (1988), and Fastook and Chapman (1989).

An example of results from (43) is given in Figure 20, showing predicted changes in a profile of the West Antarctic Ice Sheet in response to a climatic warming (Budd et al., 1987, Figs. 9-11).

Time-dependent model calculations are frequently used to generate steady-state models, an alternative route to models of the types discussed in Sections 5 and 7. To generate a steady-state (or "equilibrium") model for a set of climatic conditions and other boundary conditions of interest, an arbitrary starting model is integrated forward in time until there is no further change. Starting with no ice, an integration over ~50,000-300,000 years of model time is needed to reach equilibrium for continental ice sheets (Oerlemans, 1988, p. 54; Budd and Jenssen, 1989, p. 16), whereas for a shift of a few °C in temperature from a starting model of an equilibrium ice sheet ~30,000 years is said to be sufficient (Letréguilly et al., 1991, p. 155).

In this way Letréguilly et al. (1991) predicted changes that would occur in the steady-state configuration of the Greenland Ice Sheet as a result of climatic warming (Fig. 21). They found a surprising "hysteresis": the equilibrium models obtained by starting with a model of the present ice sheet and increasing the temperature by a few degrees (2 to 5°C) were considerably larger in ice area and volume than equilibrium models obtained at the same temperature but starting with no ice present (compare Figs. 21a-c and d-f). This hysteresis could have been caused by a failure of the latter models to reach equilibrium in the integration time used (50,000 years), and Letréguilly et al. (1991, p. 154) noted that the model in Figure 21g did not in fact reach equilibrium. They believe, however, that the hysteresis is a real effect of the type found by Oerlemans and van der Veen (1984, p. 102) in an idealized Antarctic ice sheet model, which for certain parameter ranges could exist in two quite different steady states under the same climatic conditions, one an internally "cold" state of frozen bed and slow ice flow, the other a "warm" state of basal melting and rapid flow (see Section 10.1). Letréguilly et al. (1991) did not however show that their "hysteresis" involves cold vs. warm states of this type.

[§]In modeling valley glaciers one-dimensionally with (43) it is necessary to correct (13)-(16) for the drag of the valley walls, which is normally done in an approximate way by multiplying the right-hand side of (13) and (14) by a "shape factor" f based on the models of Section 4 (Raymond, 1980, p. 101; Paterson, 1981, p. 103; Echelmeyer, 1983, p. 263); the right-hand sides of (15) and (16) are thus multiplied by f^n. Typically $f \sim 0.6$ for valley glaciers. This f is not to be confused with the flow-law function $f(\tau)$ in (6).

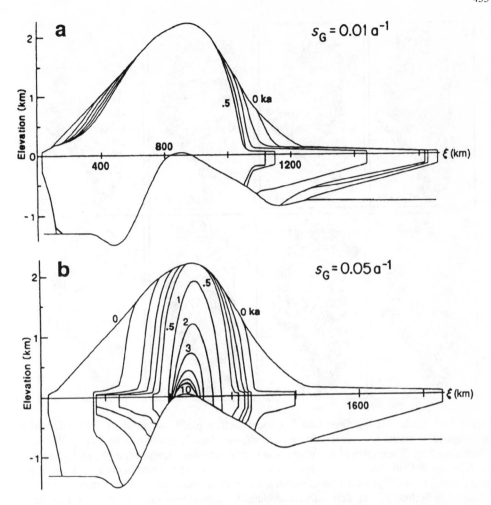

Figure 20. Time-dependent flowband model of the West Antarctic Ice Sheet in vertical cross section, showing the effect of the ice-shelf settling/spreading rate s_G at the grounding line, from Budd et al. (1987, Figs. 9b, 10a, 11). The figure shows the change with time in the surface of the ice sheet (z_S), starting from the present surface (labeled 0 ka) and evolving in time steps of 0.1 ka (or 1 ka in (b) from 1 ka onward). The longitudinal coordinate scale ξ (in km) starts from 0 near the initial position of the eastern edge of the ice sheet, in Pine Island Bay. The surface and bed profiles, $z_S(\xi)$ and $z_B(\xi)$, are plotted over the length of the flowband, with vertical exaggeration 350×. The flow and sliding law parameters were adjusted to give a steady-state fit to the 0 ka profile as explained in the text, with an assumed $s_G = 0.0042$ a^{-1} at the Pine Island Bay grounding line and 0.0016 a^{-1} at the Ross Ice Shelf grounding line. In (a), the evolution of the ice sheet is followed with s_G changed at $t = 0$ a to 0.01 a^{-1} at both grounding lines, and (b) gives the evolution for $s_G = 0.05$ a^{-1} at both grounding lines. Details of how the bounding ice shelves were modeled to not seem to be given by Budd et al. (1987) but are probably similar to those in other comparable examples (e.g. McInnes and Budd, 1984, p. 98).

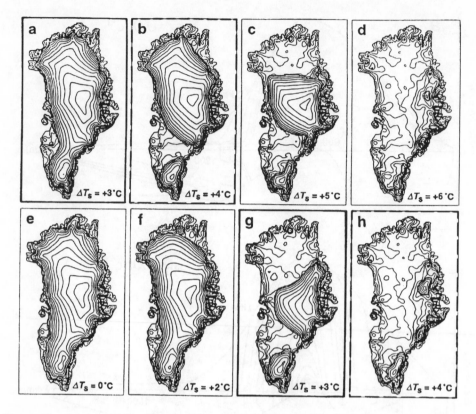

Figure 21. Models of the Greenland Ice Sheet surface profile obtained by integration of a time-dependent model to a steady state under different climatic conditions, by Letréguilly et al. (1991, Figs. 5, 6). The models of $z_S(x,y)$ were run at different surface temperatures T_S, shifted upward by $\Delta T_S = 0$ to 6°C relative to present-day $T_S(x,y)$, with corresponding shifts in the accumulation/ablation distribution $a(x,y)$ based on climatological relations between T_S and a, similar to those discussed in Section 9.3. Models a-d were obtained by starting with the present $z_S(x,y)$ (similar to model e) and changing T_S by the ΔT_S amounts indicated in the upper half of the figure. Models e-h were generated by starting with an ice-free Greenland with isostatically adjusted (rebounded) bedrock surface, and allowing the model to start up and grow under the different T_S regimes indicated in the lower half of the figure. The difference between the resulting $z_S(x,y)$ obtained for a given ΔT_S, such as between models a and g or between models b and h, is called "hysteresis" by Letréguilly et al. (1991).

9.1 ISOSTATIC ADJUSTMENT; FLOW MODEL OF THE ICE AGES

Time-dependent models of large ice masses have to include modeling of the isostatic adjustment of the bed to the weight of the overlying ice, because the resulting changes in elevation of the ice surface z_S and in submergence of the bed below sea level ($-z_B$) can have quite appreciable effects on accumulation/ablation a, on flow of the ice $\bar{u}$, and on ice flotation. Isostatic adjustment has three interrelated aspects: (i) the magnitude of the equilibrium bed depression per meter of overlying ice (normally taken to be about 0.27 m per m of ice thickness in excess of grounding ($\hat{h}$)); (ii) the localization vs. delocalization of the equilibrium depression in relation to the local distribution of ice load over the bed; and (iii) the time dependency and time scale or time delay of the isostatic adjustment. In simpler modeling treatments the adjustment is taken to be local and either an exponential relaxation with e-folding time constant ~5 ka (e.g. Oerlemans, 1982, eqn. (3)) or a delayed response with a delay time ~5 ka (e.g. Budd and Smith, 1982, eqn. (6)). More elaborate treatments take into account the effect of the elastic flexural rigidity of the lithosphere, which leads to a delocalized response that evolves in time via a diffusion equation (Oerlemans and van der Veen, 1984, Ch. 7; Huybrechts, 1992, p. 109-113) or via a Heaviside space-time Green's function (Lingle and Clark, 1985, p. 1103-1106); for ice loads of horizontal dimension ~1000 km the resulting isostatic adjustment time scale is ~20 ka. Many models do not include isostatic adjustment at all (e.g. Fig. 20); in the case of relatively small ice masses well above sea level, like most valley glaciers, the adjustment is too small to matter.

The coupling of time-dependent glacier dynamics, represented by (42), with a time dependent isostatic adjustment equation of Green's-function type constitutes the basis for a mathematical model of the response of an idealized north-circumpolar ice sheet (Greenland+Scandinavian+North American ice sheets) to time variation in climate through the ice ages, expressed in terms of the space and time variation of a in (42). Such a model (Peltier and Hyde, 1987; Hyde and Peltier, 1987; Deblonde and Peltier, 1990) has proved capable of producing an ice volume response to Milankovitch-type forcing (variations in received solar radiation due to celestial-mechanical variations in the earth's orbit and rotation-axis tilt) that remarkably resembles the geological record of ice volume through time during the ice ages. The model behaves in effect like a relaxation oscillator of frequency 100 ka, which is excited by the orbital eccentricity variation at this frequency and thus gives the predominantly 100-ka glacial-interglacial cycle observed in the ice-volume-vs.-time record. If this result holds up under scrutiny it will be a most important accomplishment of combined glacier and deep-earth flow modeling.

9.2 BOUNDARY CONDITIONS FOR GLACIERS TERMINATING ON LAND

The boundary conditions that need to be applied to (42) or (43) in time-dependent modeling are somewhat problematical and pose different problems for glaciers or ice sheets that terminate on land vs. in water. These two situations are considered separately below.

On land, the boundary condition is obviously $h = 0$ at the terminus or ice margin, but in principle this is incompatible with flow or sliding laws (16), (32), or (36), because with $h = 0$ they give $\bar{u} = 0$ at the margin and thus prevent the ice from advancing or even maintaining its marginal position if $a < 0$ (net ablation) there, as is normally the case. Morland et al. (1984, p. 132), Boulton et al. (1984, p. 141), and Hindmarsh et al. (1988, p. 61) take this as a basis to require that the sliding law (Section 8) have a form like (36) with $p = 1$ and $q = 1$ as $\hat{h} \to 0$, so that u_B will tend to a finite non-zero value as $\hat{h} = h \to 0$. (The more general condition $p = q$ would seem sufficient for the purpose.) But in actuality, longitudinal coupling (Section 6) with the ice behind maintains ice motion at the terminus even though the "driving stress" τ_B from (14) goes to zero there. Mahaffy (1976, p. 1062) mimicked this in his finite-difference model of the Barnes Ice Cap (Baffin Island) by assigning to each outermost grid point within the ice cap the same ice velocity as given by the model at the adjacent next-to-outermost point, and assuming a particular (but unspecified) wedge-shaped configuration for the ice edge. The related problem of how to determine in a gridded model when the margin should jump from one grid point to the next (or from a position between one pair of grid points to between the next pair) is discussed by Kruss (1984) and Huybrechts et al. (1989, p. 378). Oerlemans and van der Veen (1984, p. 69-71) conclude that in spite of the problems just discussed, gridded models in which the flow is controlled by τ_B as in (16), (31), and/or (32) without any modification near the margin appear to work and to produce a well-defined margin, for a reason that seems rather extraneous – truncation error in the numerical methods. Perhaps this is why the problems are rarely mentioned in descriptions of glacier and ice sheet modeling. Such problems.

or indeed any aspects of ice-margin migration, do not seem to be mentioned by Fastook (1987b, 1990) in discussing time-dependent finite element models, and there appears to be no indication that the margin migrates spontaneously in these models.

9.3 RESPONSE OF VALLEY GLACIERS TO CLIMATIC VARIATIONS

Time dependent models of the climatic response of three temperate valley glaciers in the Alps – Rhone Glacier (Stroeven et al., 1989), Argentière Glacier (Huybrechts et al., 1989) and the Hintereisferner (Gruell, 1992) – and one in Scandinavia – Nigardsbreen (Oerlemans, 1986) – provide tests of the predictive power of glacier modeling based on (43) and the associated foundation of Section 5, without the complication of the coupled thermal modeling of Section 7. A comparison is made between historic records of the glaciers' terminus position through time $\xi_T(t)$ and the results of time-dependent modeling based on meteorological records. For the modeling of temperate glaciers, climate manifests itself only in terms of the distribution of ice accumulation/ablation $a(\xi;t)$. (A separate effect of water input, which might affect basal sliding as discussed in Section 10.2, has not been considered.) If a historic record of $a(\xi;t)$ were available, it would be used directly as input to the time-dependent model, but this is rarely the case. Assumptions therefore have to be made to relate one or more recorded meteorological or climatological variables to $a(\xi;t)$. This is usually done in terms of an assumed or observed functional relation between accumulation/ablation and altitude, $a(z_S)$. (Dependence of a on ξ enters via $z_S(\xi)$.) One of two simple assumptions is usually made as to the effect of a climatic shift on $a(z_S)$: (i) the function is shifted parallel to the a axis by a time-dependent shift $\Delta a(t)$ that is independent of altitude: $a(z_S;t) = a(z_S;0) + \Delta a(t)$ (ii) the function is shifted parallel to the z_S axis by a time-dependent, elevation-independent shift $\Delta z_E(t)$ which is equated to a shift in the "equilibrium line altitude" (firn line; snow line at the end of the ablation season): $a(z_S;t) = a(z_S + \Delta z_E(t))$. The shift function $\Delta a(t)$ or $\Delta z_E(t)$ then has to be related to a climatic variable, for example temperature $T(t)$ at an appropriate recording station. In the tests discussed here a simple linear relation was used: (i): $\Delta a(t) = -c_a \Delta T(t)$; (ii): $\Delta z_E = -c_E \Delta T(t)$, where c_a and c_E are positive constants. Method (i) was used for the Argentière Glacier and method (ii) for the Rhone Glacier, but since the function $a(z_S)$ was assumed linear in z_S except near the terminus in both cases there is a difference between the two methods only near the terminus. For the Hintereisferner, (i) was used with an additional linear term proportional to total yearly snowfall at a village 10 km distant. For Argentière Glacier, flow law (16) was used, and for the Rhone Glacier and Hintereisferner sliding law (32) in addition (with $p = 3$, $q = 1$); the parameters ə and/or $\hat{S}$ were adjusted to give a reasonable match between the observed and model calculated profiles $h(\xi)$ and current terminus position ξ_T, and the value of c_a or c_E was chosen such that the range of terminus-position fluctuations produced by the models is close to the historically observed range. Using historical records of temperature $T(t)$, the time-dependent models were run starting from no ice at $t = -1000$ a (b.p.). The time scale for terminus response to modest changes in $a(\Delta a \sim 0.5$ m a^{-1}) was 35-200 years, so transients from the starting condition had died out before the time frame of interest, -400 a (b.p.) to the present. The model results in terms of the terminus position $\xi_T(t)$ versus time are shown in Figure 22. They are disappointing in the lack of clear correlation between predicted and observed records. In particular, the pronounced glacier retreat observed from 1850 onward is poorly reproduced by the models, or is reproduced in the Alps only with a delay of 75-100 years; in Scandinavia the onset of the retreat seems to be correctly predicted but its subsequent course, particularly the predicted readvance starting in 1900, does not match the observations. Other climatological time-series records were tried in place of $T(t)$ as forcing functions for $a(z_S;t)$ for the Rhone Glacier and Hintereisferner, without any particular improvement in the predictions. However, by running the Hintereisferner model only from 1894 onward, starting from a steady state, a reasonable agreement with observation was obtained (Gruell, 1992, Fig. 15). The authors of these

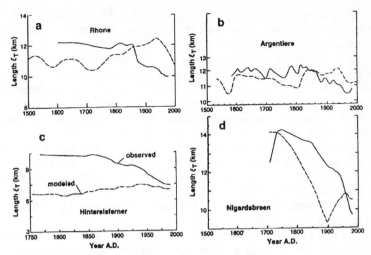

Figure 22. Comparison of observed and modeled time variations of the terminus positions of four valley glaciers. ξ_T is the longitudinal coordinate of the glacier terminus measured from the head of the glacier. The observed $\xi_T(t)$ is shown with a solid line, and the calculated position, obtained by flow modeling as described in the text, is shown with a dashed line. (a) Rhone Glacier, Switzerland, from Stroeven et al. (1989, Fig. 7). (b) Argentière Glacier, France, from Huybrechts et al. (1989, Fig. 8). (c) Hintereisferner, Austria, from Gruell (1992, Fig. 14). (d) Nigardsbreen, Norway, from Oerlemans (1986) as reproduced by van der Veen (1991, Fig. 2).

predictions believe that the lack of success is due not to faults in the glacier flow modeling but rather in the relation between climatological variables and mass-balance history $a(z_S;t)$ (Huybrechts et al., p. 387; Stroeven et al., p. 401). Van der Veen (1992, p. 6) and Oerlemans (1992) describe recent efforts to improve the micrometeorological basis of climate/mass-balance relations for glacier modeling, and Oerlemans (1989) calls attention to related complexities in the proper modeling of glacier response to climatic change.

The foregoing examples represent the numerical-modeling counterpart of the mathematically more sophisticated theory developed by Nye (1960, 1963, 1965a) for the response of glaciers to climatic change and applied by him (1965b) to two valley glaciers. (For a summary see Paterson, 1981, p. 242-267.) This theory has the limitation of being restricted to small perturbations from a datum state, a restriction not imposed on the numerical models discussed here.

9.4 GLACIERS TERMINATING IN WATER; ICE SHELVES

For glaciers terminating in sea water (or large lakes) the modeling complications are greater. Special attention to this modeling is warranted, because of its importance in predicting the future behavior of the West Antarctic Ice Sheet, for which a catastrophic collapse with worldwide consequences has been predicted and much discussed (Hughes, 1977, p. 44; Thomas, 1979, p. 174; Stuiver et al., 1981, p. 390-431; van der Veen, 1987, p. 8; Lingle and Brown, 1987, p. 279; Alley, 1990; MacAyeal, 1992).

If the termination is in water but the ice mass does not go afloat, as is the case for many large Alaskan tidewater glaciers, the terminus boundary condition (calving rate) is determined by the calving mechanics of grounded ice. This is poorly understood, but empirical "calving laws" have been inferred from observational data and used in models to predict the catastrophic retreat that can occur in tidewater glaciers and is currently underway in Columbia Glacier, Alaska (Post, 1975; Meier and Post, 1987; Meier et al., 1980; Brown et al., 1982; Krimmel, 1992, Fig. 13). A theory for calving of grounded ice has recently been given by Hughes (1992b).

If instead the ice goes afloat before terminating, as is generally the case in Antarctica (except in the dry valleys), its ultimate termination is by calving of ungrounded ice (probably with a different "calving law"), and an important additional element enters: between the (un-)grounding line and the calving ice edge is a floating *ice shelf* (or what is called an "ice tongue" if a glacier of limited width goes afloat and protrudes out from the coastline). In this case the boundary conditions on the ice sheet are applied by the ice shelf (or ice tongue) and express its mechanics. Ice shelves are outstanding objects for model analysis in their own right and have been the subject of a number of flow models, both steady-state and time-dependent (e.g. MacAyeal and Thomas, 1982, 1986; Oerlemans and van der Veen, 1984, p. 61-66; van der Veen, 1986; Lindstrom and MacAyeal, 1987; Lange and MacAyeal, 1988; MacAyeal, 1989b; Lingle et al., 1991). The emphasis here is on their role in the boundary conditions for ice sheets. Essentially four such conditions have been used, as discussed in the following subsections.

A well known problem in ice sheet modeling near the grounding line is that if basal sliding is modeled with sliding law (36) with $q > 0$, as appropriate for basal decoupling as the ice sheet approaches flotation, then as $\hat{h} \to 0$ at the grounding line, $u_B \to \infty$. If also $\tau_B \to 0$, as we expect, then $u_B \to \%$, which though in principle possibly finite may still give numerical problems in the modeling. If τ_B is calculated from (14) and if the model takes into account the non-zero surface slope of the ice shelf (because of seaward thinning) then again $u_B \to \infty$ at the grounding line. To avoid this problem a lower limit on $\hat{h}$ is sometimes arbitrarily specified (e.g. McInnes et al., 1986, p. 9); in (39) the lower limit is incorporated into the sliding law by means of the constant c_0. But a physical basis for this lower limit does not exist. McInnes and Budd (1984, p. 98) dealt with the problem by placing an upper limit on $\partial \bar{u}/\partial \xi$ equal to a prescribed ice-shelf spreading rate at the grounding line (see Section 9.6 below). The problem may also be avoided if no mesh point falls very close to the grounding line, so that $\hat{h}$ never gets arbitrarily small at a mesh point on the ice-sheet side of the grounding line. This may explain why the problem is not mentioned in many of the modeling studies cited below.

9.4.1. *Boundary condition I: grounding; marine ice sheet instability.* – Condition I is that the ice thickness at the grounding line h_G is given by (37) with $\hat{h} = 0$:

$$h_G = (-z_B)\rho_w/\rho \tag{44}$$

where $(-z_B)$ is the local depth of the bedrock surface below sea level.[§] (44) plays a curiously double role: it both prescribes the boundary value of h and is also the condition that determines the location of the boundary (the grounding line, at longitudinal coordinate ξ_G on a given flowline). It thus determines the rate of advance or retreat of the grounding line:

$$\dot{\xi}_G = c_{R,A} [\dot{h} + \dot{z}_B (\rho_w/\rho)] \tag{45}$$

[§]McInnes and Budd (1984, p. 98) uses the condition $z_S \geq 0.11h + 20$ m, or equivalently $h \geq 1.12(-z_B) + 22.5$ m, for going aground. McInnes et al. (1986, p. 8) introduce in addition the separate condition $z_S \leq 0.95(0.11h + 20$ m) for going afloat; this is done to avoid the model flip-flopping between the grounded and ungrounded states. The additional 20 m (or 22.5 m) is to allow for the lower density of the firn zone. For ice density 0.92 g cm^{-3} and seawater density 1.03 g cm^{-3}, (44) is $h = 1.12(-z_B)$.

where for $\dot{\xi}_G < 0$ (grounding line retreat)

$$c_R = [\alpha + \beta(\rho_w - \rho)/\rho]^{-1} \tag{46}$$

while for $\dot{\xi}_G > 0$ (grounding line advance)

$$c_A = c_R(\rho_w - \rho)/\rho = [\gamma + \beta(\rho_w/\rho)]^{-1} \tag{47}$$

Here $\gamma = -\partial h/\partial x$ is the longitudinal taper of the floating ice shelf. ($\gamma \neq \alpha - \beta$ because the base of the ice shelf does not generally rest on the "bed" [sea bottom] downstream from the grounding line.) In (45), $\dot{z}_B$ is the rate of rise of the sea bed relative to sea level either through isostatic adjustment or through a eustatic fall of worldwide sea level. (45) is derived from (44) by comoving differentiation D/Dt moving with the grounding line. Except for inclusion of the effect of $\dot{z}_B$, (45)+(46) for grounding line retreat is the same as equation (A6) of Thomas and Bentley (1979, p. 166), who did not give a separate result (47) for grounding-line advance.

If the bed slopes back upstream ($\beta < 0$) steeply enough, c_R in (46) can become large, so that a small thinning rate $-\dot{h}$ can be amplified via (45) to a rapid grounding-line retreat. This is part of the basis for the much discussed concept of "marine ice sheet instability" (Thomas, 1979), explained further in the next subsection.

9.4.2. Condition II: ice-shelf control of grounding line migration; "buttressing." – Condition II equates the vertically averaged vertical strain rate $\bar{\dot{e}}_{zz}$ at the grounding line to $\dot{e}_{zz}$ in the immediately adjacent ice shelf, which is determined by the settling and spreading of the ice shelf under its own weight; the $\bar{\dot{e}}_{zz}$ then controls, via (43) and (45), the migration rate of the grounding line $\dot{\xi}_G$. Although this condition, introduced by Thomas and Bentley (1979, p. 167), has been bypassed in some recent models as discussed later (Section 9.4.5), it has been extensively used in ice-sheet modeling and has guided much interpretation of ice-sheet/ice-shelf behavior. Condition II can be introduced into (43) by noting that the last two terms on the right side of (43) can, from incompressibility (9), be rewritten $\partial \bar{u}/\partial \xi + (\bar{u}/b)\partial b/\partial \xi = \bar{\dot{e}}_{\xi\xi} + \bar{\dot{e}}_{\eta\eta} = -\bar{\dot{e}}_{zz}$ (where η is a transverse coordinate). Call $s_G = -\bar{\dot{e}}_{zz}$ the "settling/spreading" rate at the grounding line (normally a positive quantity). Then from (43), at the grounding line

$$\dot{h} = a - \bar{u}\frac{\partial h}{\partial \xi} - h_G s_G \tag{48}$$

and hence, from (45),

$$\dot{\xi}_G = c_{R,A}\left(a - \bar{u}\frac{\partial h}{\partial \xi} - h_G s_G + \dot{z}_B(\rho_w/\rho)\right) \tag{49}$$

The effect of the ice shelf on migration of the grounding line is contained in the $h_G s_G$ term; (49) together with (50) below is the quantitative statement of condition II.

The settling/spreading rate s_G can be as large as that of an unconfined, freely floating ice shelf, but in general it is less than that because of impediments to the forward motion of the ice shelf, which result in a "backpressure" σ_b. For an ice shelf of thickness h spreading in the longitudinal (ξ) direction under its own weight, restricted by a uniform backpressure σ_b, the spreading rate s (assumed independent of z) is given by

$$s = -\dot{e}_{zz} = \bar{A}\left[\frac{1}{4}(1 - \frac{\rho}{\rho_w})\rho gh - \frac{1}{2}\sigma_b\right]^n \qquad (50)$$

(Weertman, 1957; Thomas, 1979b, p. 281). Thus the local settling/spreading rate s depends only on the local ice thickness h, and not on the extent of the ice shelf or on its longitudinal taper except to the extent that these affect σ_b. If there were no backpressure ($\sigma_b = 0$), an ice shelf of thickness $h \sim 800\text{-}1000$ m, typical of the grounding zone of the Ross Ice Shelf, at an ice temperature of $-15°C$ (vertical average, giving $\bar{A} = 0.01$ a^{-1} bar^{-3} from Paterson, 1981, p. 39), would by (50) have a settling/spreading rate $s_B \approx 0.1\text{-}0.2$ a^{-1}, or $h_G s_G \approx 100\text{-}250$ m a^{-1}. Such values are much larger than the other terms in (49) and would completely dominate the time-dependent behavior of the adjacent ice sheet, driving the grounding line rapidly inward according to (49). This effect is seen in the model of Figure 20b, calculated for an assumed $s_G = 0.05$ a^{-1}; the West Antarctic Ice Sheet rapidly disappears in this model.

Condition II was used by Thomas and Bentley (1979) in a time-stepped but spatially integrated (not discretized) model of the West Antarctic Ice Sheet and by Lingle (1984, p. 3527) in a spatially discretized model. In both of these models, $\bar{u}$ and $\partial h/\partial\xi$ at the grounding line for use in (49) were determined by a global ice-volume-conservation condition rather than from discrete $h(\xi)$ values adjacent to the grounding line, as could have been done in the spatially discretized model; grounding line motion $\xi_G(t)$ was calculated directly from (49), stepped through time, rather than by calculating $h(\xi,t)$ from (43) and applying (44) to find the grounding line $\xi_G(t)$. Exactly how in these respects condition II was applied in other ice-sheet/ice-shelf models with a grounding-line constraint based on s_G, such as the model of Figure 20, is not clear from the published descriptions.

(50) makes evident a second element in the "marine ice sheet instability" concept mentioned above (Section 9.4.1). When for negative β (inward sloping bed below sea level) the grounding line is driven inward by the effect of the spreading term $h_G s_G$ in (49), the ice thickness h increases as the grounding line retreats (because $\beta < 0$ and $\alpha > 0$). If the fringing ice shelves can spread freely so that s_G is given by (50) with $\sigma_b = 0$, then s_G will increase rapidly with h, which from (49) will result in further acceleration in the retreat of the grounding line. The instability concept is that essentially the entire ice sheet can go rapidly afloat and disintegrate in this way.

To protect a marine ice sheet like that modeled in Figure 20 from the effect of a rapid spreading rate s_G in (49) requires a substantial backpressure $\sigma_b \sim 4$ bar in (50), to reduce s_G to ~ 0.002 y^{-1}, the typically observed level. This is often called "buttressing" of the ice sheet by the ice shelf. The backpressure is thought to arise from several types of impediments to forward motion of the ice shelf: peripheral drag at the lateral margins of an ice shelf in a coastal embayment, where the ice shelf shears past bedrock or grounded ice on the sides of the embayment; basal drag from seafloor shoals where the ice shelf goes aground and overrides the shoals; and the head-on compressive resistance and side drag generated by islands that poke up through the ice shelf, forcing the ice to split and flow around them on two sides. The latter two types of impediments are often called "pinning points" and are regarded as crucial to the stabilization of large ice shelves. Several models have been constructed to calculate the effects of such impediments on ice shelf motion, which determines the resulting backpressure σ_b. Calculated σ_b values from these models are given explicitly by Thomas (1973, p. 67) and Lingle et al. (1991, Fig. 8a ff.), and implicitly by MacAyeal and Thomas (1982, Fig. 7; 1986, Fig. 12) and MacAyeal et al. (1986, Figs. 7, 8) in terms of modeled principal strain rates (or principal deviatoric stresses) that could be converted to σ_b via (50) (or via $\sigma_b = \frac{1}{2}(\rho_w - \rho)\rho gh/\rho_w - 2\tau'_{\xi\xi}$, which follows from (50)). Although comparisons between calculated and observed strain rates are made, no comparison seems to have been made between these results and the map of $h\sigma_b(x,y)$ in the Ross Ice Shelf produced by Thomas and MacAyeal (1982, Fig. 8) by using measured values of s and h to solve (50) for σ_b in the flow direction. (Actually, the three-dimensional equivalent of (50) has to be used in these calculations.) One such comparison can be made on the basis of the "extra back-pressure force" transmitted across the grounding line of Ice Streams A and B, which was calculated from strain rate data by MacAyeal et al. (1987, p. 225 and Table IV, columns 4-6, entry 24) (see also MacAyeal et al., 1989, Table II); the calculated force corresponds to an average backpressure $\sigma_b \approx 3.2$ bar across the grounding line, compared to $\sigma_b \approx 4$ bar from the map of Thomas and MacAyeal (1982, Fig. 8). (These figures assume an average ice thickness of 700 m at the grounding line.) MacAyeal et al. (1987, p. 225) have calculated that approximately half of the backpressure results from the impediment to ice shelf motion represented by the Crary Ice Rise that extends from about 50 to 150 km downstream from the mouth of Ice Stream B.

Examples of time-dependent ice-sheet models that include the imposition of condition II at a grounding line, with s_G determined by (50) and the backpressure σ_b controlled by ice-shelf mechanics as just described, are the already-mentioned models of Thomas and Bentley (1978), Lingle (1984), and Lingle and Clark (1985), which address the question of how the West Antarctic Ice Sheet responded to the rise in sea level at the end of the last glaciation. The models follow the gradual shrinkage of the ice sheet and evolution of the ice shelf as the grounding line gradually retreated to its present position from the edge of the continental shelf, where it was located at the glacial maximum when sea level was 120 m lower than now.

With the model shown in Figure 20, Budd et al. (1987) considered the related question of how the West Antarctic Ice Sheet would react to increased bottom melting under the Ross Ice Shelf due to global warming. Bottom melting (caused by the circulation under the ice shelf of somewhat warmer water from offshore) is thought to affect the grounding line by thinning the ice shelf and thereby decreasing its contact with pinning points and embayment margins, which results in reduction of backpressure and consequent retreat of the grounding line as discussed above. Budd and Jenssen did not model the processes leading to a reduction in backpressure (loss of buttressing); instead they simply chose an arbitrary series of values of the ice-shelf settling/spreading strain rate s_G, and calculated time-dependent models for each of these s_G values, held constant at the (movable) grounding line. The model was first "tuned" to the current profile $h(\xi)$: for $s_G = 0.002$ a^{-1}, approximately the current value, $\ni$ in (7) and $\hat{S}$ in (36) were adjusted point by point through the profile until the model, run to steady state, reproduced the observed profile (Figure 20, curves 0). Changing to $s_G = 0.01$ a^{-1} and $s_G = 0.05$ a^{-1} then resulted in the shrinking models in Figure 20.

The outer (calving) edge of the coupled ice shelf in such models should in principle be determined by a calving law for floating ice, for which some modeling has been done by Reeh (1968). In practice the calving margin is usually taken at the point where the ice has thinned by settling/spreading to a "critical minimum thickness" of about 200-250 m (Paterson, 1980, p. 50; Oerlemans, 1982, p. 551; McInnes et al., 1986, p. 12; Radok et al., 1987, p. 40; Oerlemans and van der Veen, 1984, p. 74; Böhner and Herterich, 1990, p. 17; Huybrechts, 1992, p. 99). The position of this calving margin has no effect on the ice sheet model except in so far as it affects σ_b in (50).

The large influence of the grounding line, illustrated in Figure 20, applies as already noted to marine ice sheets – those with base largely below sea level, and with $\beta < 0$ at the periphery (bed sloping down toward the interior). In the modeling of terrestrial ice sheets that go afloat and terminate in tidewater with $\beta > 0$ at the periphery, such as much of the East Antarctic Ice Sheet, the detailed conditions I and II are generally ignored, it appears, and the grounding line is taken simply to lie at or a short distance seaward of the bedrock coastline, with h set to a small value ($\lesssim 300$ m) similar to the thickness near current grounding lines or termini (e.g. Alley and Whillans, 1984, p. 6489). Detailed model calculations based on (44), (49), and (50) are compatible with this simplification. In some time-dependent models $h = 0$ has been taken as the condition at the margin (e.g. Huybrechts and Oerlemans, 1988, p. 57), but this is subject to the same objection raised earlier for termination on land. Except where there are large coastal embayments, the ice shelves fringing terrestrial ice sheets are generally narrow because pinning points on the continental shelf tend to be few; the shelves therefore thin rapidly seaward and calve off not far from the grounding line, as soon as the "critical minimum thickness" is reached.

9.4.3. *Boundary stress (condition III): the transition zone.* – Although boundary condition II involves the backpressure from the ice shelf, its role is not that of a stress boundary condition; thus the term "buttressing," which seemingly indicates mechanical support of the ice sheet by the ice shelf, is a misnomer as applied to condition II. The stress boundary condition itself, here called condition III, is that the vertical average of the horizontal deviatoric stress $\bar{\tau}'_{\xi\xi}$ at the boundary is equal to $\tau'_{\xi\xi}$ in the immediately adjacent ice shelf; for an ice shelf spreading in the flow direction ξ only, it is

$$\bar{\tau}'_{\xi\xi} = \frac{1}{4}\, \rho g h \,(\rho_w - \rho)/\rho - \frac{1}{2}\sigma_b \tag{51}$$

i.e., the quantity in the square brackets in (50). This is a physically rigorous condition, based on balance of horizontal forces between the grounding line and the calving face of the ice shelf (e.g. Paterson, 1981, p. 173). Condition III has to deal with the mechanical consequences of the changeover from a shear-stress-supported flow regime in the ice sheet (τ_B dominant; $\tau'_{\xi\xi} \cong 0$) to a longitudinal-stress-supported regime in the ice shelf ($\tau'_{\xi\xi}$ dominant; $\tau_B = 0$). Because of stress continuity this changeover cannot take place abruptly but must be distributed over a *transition zone* between the ice sheet and the ice shelf. This zone was originally thought to be up to 150 km in width parallel to flow (Oerlemans and van der Veen, 1984, p. 74; van der Veen, 1985, p. 262). From the upper edge of the transition zone, where $\tau'_{\xi\xi} \cong 0$ and the longitudinal compressive stress has its normal "glaciostatic" value $-\tau_{\xi\xi} = \rho g(z_S - z)$, to the ungrounding line at the lower end of the transition zone, the longitudinal compressive stress must progressively decrease (in vertical average) by the amount $2\bar{\tau}'_{\xi\xi}$ from (51). This reduction in $-\tau_{\xi\xi}$ at the grounding line acts upon the ice sheet as a removal of mechanical support ("unbuttressing") due to the ice going afloat to form the ice shelf. To satisfy equilibrium (19), the decrease in $-\tau_{\xi\xi}$ or increase in $\bar{\tau}'_{\xi\xi}$ through the transition zone must be accompanied by an increase in τ_B over the gravitationally driven value $\rho g h \alpha$ from (14). At the same time, τ_B must decrease progressively to zero as the ice goes progressively afloat (decoupling from the bed) and hence α must decrease to (nearly) 0 through the zone. An increase in τ_B, which would result via (19) from the positive stress gradient $\partial\bar{\tau}'_{\xi\xi}/\partial\xi$ in the transition zone, would represent locally increased outflow from the ice sheet, and in a steady state must be offset by other factors including the decrease in α through the zone. An increase in the backpressure σ_b decreases $\partial\bar{\tau}'_{\xi\xi}/\partial\xi$ and τ_B and thus reduces the outflow; this is how a true mechanical buttressing effect of σ_b on the ice sheet works. The net mechanical-support effect of unbuttressing by going afloat and true buttressing by ice-shelf back pressure σ_b, and the action of these effects over the transition zone, will be here designated condition III associated with the grounding line. These complexities were not considered in the models discussed above, which did not impose a condition III.

Models of the transition zone endeavoring to address the complexities just discussed have been developed by van der Veen (1985, 1987), Herterich (1987), Böhner and Herterich (1990), and Huybrechts (1992, p. 100). Only the Herterich (1987) model has a horizontal resolution (numerical grid spacing) fine enough to reveal how the complexities interact to give the detailed flow velocity and stress distribution through the zone. However, the flow and stress were calculated only for a simple, arbitrary ice surface profile $z_S(\xi)$ and ice thickness profile $h(\xi)$, and the time-dependent development of the profiles was not followed. The model gave very large longitudinal gradients in flow velocity in the transition zone, the ice flux increasing by a factor of ~8 from the input (ice-sheet) end to the output (ice-shelf) end. Consequently, very large adjustments in $h(\xi)$ and $z_S(\xi)$ would have been produced in a time-dependent model, and the steady state would probably look very different from the simple initial model. We cannot therefore draw from the model any very certain conclusions about the velocity and stress distribution in the transition zone of an actual ice-sheet/ice-shelf system near steady state. Nevertheless, it seems likely that we can accept Herterich's conclusion that the width of the transition zone (in the flow direction) is of the order of the ice thickness, and increases considerably ($\sim \times 2$) if basal sliding is admitted with an effective-pressure-dependent sliding law qualitatively similar to (36). This conclusion is compatible with the longitudinal coupling model of Section 6, which would predict a transition-zone width of the order of the longitudinal coupling length l, increasing with the amount of basal sliding (Kamb and Echelmeyer, 1986, p. 275). The conclusion is also compatible with the finding in van der Veen's (1987, p. 234) model that the results depended only on $\tau'_{\xi\xi}$ at the grounding line, not upstream (except perhaps in the model with basal sliding [ibid., Fig. 8]): since the grid spacing was $\sim$30 times the grounding-line ice thickness in this model, only one grid point could fall within a transition zone having a width of the order of the ice thickness. (The model results were in this case expressed as ice surface profiles $z_S(\xi)$ from time dependent models run to a steady state.) A further, somewhat contrary modeling conclusion on the width of the transition zone for ice streams is discussed in Section 12.

Both the Herterich (1987) and van der Veen (1987) models indicated a large effect of $\tau'_{\xi\xi}$ at the grounding line on the flow or the profile of the ice sheet. Herterich (1987, p. 194) credits this effect to the "gravitational pull from the ice shelf" and thus seems to suggest that it is caused by the mechanical unbuttressing discussed above (i.e. condition III). Van der Veen (1987, p. 241) concludes on the contrary that the gradient of the longitudinal deviatoric stress can be neglected. However, the evidence cited for this conclusion is flawed by the large grid spacing used in van der Veen's models, which effectively widens the transition zone to the grid spacing and reduces the longitudinal gradient $\partial \bar{\tau}'_{\xi\xi}/\partial \xi$ proportionally. If the transition-zone length is $\sim h$, it follows that $2\partial (h \bar{\tau}'_{\xi\xi})/\partial \xi$ in the transition zone is $\sim 2 \bar{\tau}'_{\xi\xi}$ at the grounding line, and from (19) there is a corresponding contribution $2 \bar{\tau}'_{\xi\xi} \sim 9$ bar to τ_B in the transition zone, which is far from negligible. It only appears negligible when reduced by a factor $\sim$30 (ratio of grid spacing to h) in the model.

9.4.4. Stress softening (condition IV). – It seems likely that the large effect of $\bar{\tau}'_{\xi\xi}$ at the grounding line on the model results obtained by van der Veen (1987) is, as pointed out by Huybrechts (1992, p. 103), due to "softening" of the ice by the contribution of $\bar{\tau}'_{\xi\xi}$ to τ^2 in the flow law (1)-(6), increasing the flow velocity and ice flux over what would be calculated from (15) or (16), which contain only the "softening" due to $\tau_{\xi z}$. Huybrechts (1992, p. 103) says that $\bar{\tau}'_{\xi\xi}$ may be up to 3 times $\tau_{\xi z}$, so that the extra "softening" effect on the flow, which from (1)-(4) goes as $\tau'_{\xi\xi}{}^2 + \tau_{\xi z}{}^2$ (for $n = 3$), is up to a factor of 10. This softening effect may be called "condition IV" associated with the grounding line, to keep it distinct from the other conditions discussed above. Its detailed formulation is rather complicated and is in fact very different in the approach of Alley and Whillans (1984, p. 6488) and van der Veen (1987, p. 239) on the one hand and of Herterich (1987, p. 186-188) and Huybrechts (1992, p. 102) on the other; we cannot delve into the complex details here.

If in low-resolution models the softening effect is attributed to a full grid interval ($\sim$50 km) around the grounding line, rather than to only a fraction of the width of the transition zone (allowing for the fact the $\bar{\tau}'_{\xi\xi}$ goes progressively to zero across the zone), then the softening effect will probably be much overestimated. This appears to be the case in the models of van der Veen (1987) and Huybrechts (1992).

The detailed effect of conditions III and IV on the ice-sheet/ice-shelf profile in the transition zone and the magnitude of their importance for ice sheet modeling are not yet particularly clear. The tendency of conditions III and IV to increase the flow in the transition zone should tend to cause a widening of the zone, as does basal sliding as noted in the last subsection (9.4.3), but such an effect has not been reported in the modeling. Transition-zone models that include sliding (van der Veen, 1987, p. 238) proved to be rather sensitive to the detailed sliding law: the addition of sliding by the relation (36) with $p = 3$, $q = 1$ had an effect on the ice-sheet/ice-shelf profile little different from an increase of A in (16), with no sliding, whereas addition of sliding by (36) with $p = 1$, $q = 2$ had a large and different effect in which the seaward part of the ice sheet profile became dominated by the effects of sliding. This suggests the possibility of a comparable sensitivity to the application of conditions III and/or IV.

9.4.5. Models without condition II. – Huybrechts (1990, 1992) has developed a time-dependent three-dimensional Antarctic model in which conditions I and IV are applied at the ice-sheet/ice-shelf boundary while conditions II and III are omitted, following the suggestions of van der Veen (1986, p. 45; 1987, p. 241). This approach is said (Huybrechts, 1990, p. 80) to be a "rigorous treatment of the flow across the grounding line," in presumable contrast to the earlier treatments that used condition II. However, omission of condition III "may be somewhat doubtful" (Huybrechts, 1990, p. 83; 1992, p. 101), as argued above. The application of condition IV is not completely rigorous because it assumes without proof that $\tau'_{\xi\xi}(z)$ can be replaced in the analysis by its vertical mean $\bar{\tau}'_{\xi\xi}$ (and similarly for $\tau'_{\eta\eta}$ and $\tau'_{\xi\eta}$) (Huybrechts, 1990, p. 83; 1992, p. 101); this is called "the major problem" and "the most dubious approximation" of the treatment (van der Veen, 1987, p. 224; 1985, p. 259). Another doubtful feature, noted above, is the apparent attribution of condition IV to the full width of the grid spacing instead of to an appropriate part of the width of the transition zone only.

An unusual and seemingly attractive feature of the Huybrechts (1992, p. 99) model, if the ice-sheet/ice-shelf interaction is correctly handled, is the way it surrounds the entire continent with an ice shelf and applies boundary conditions only at the seaward edge of the ice shelf, far from shore, allowing the ice shelf to apply grounding-line conditions on the ice sheets "automatically," without "operator" intervention (such as specification of σ_b in (50) or s_G in (49)). Of course, this can only be expected to work if the ice shelf interaction with all significant pinning points is properly handled mechanically, a seemingly formidable task that appears not to be mentioned in the otherwise quite thorough description of the model by Huybrechts (1992).

In relation to the earlier ice-sheet/ice-shelf models, with their emphasis on condition II, the main question is whether or not the Huybrechts and/or van der Veen models implicitly contain condition II or an approximation to it. A full answer would require going thoroughly into the technical details of the numerical procedures used, but I think the following can be said without going that far. For simplicity I will speak as though we are dealing with two-dimensional flowband models, although the Huybrechts model is three-dimensional. In this model the grounding line is replaced by a transition zone represented by a single ξ grid point, at which condition IV is applied (Huybrechts, 1992, p. 100). At all points seaward of the transition zone point the model uses an ice-shelf stress equilibrium condition (actually (73)+(74), Section 12) that guarantees that (50) will hold throughout the shelf (Huybrechts, 1992, p. 97); hence condition II is in fact applied to the ice sheet at a point one grid spacing beyond the transition-zone point. If the ice sheet and ice shelf are "fully coupled" (Huybrechts, 1990, p. 83; 1992, p. 103) this probably assures that condition II will hold to a good approximation at the grounding line, which must lie between the two grid points named.

In the van der Veen (1987) model there is a ξ grid point at the grounding line, and the transition zone extends upstream an unspecified distance from there. There are no grid points seaward of the grounding line, at which an "external" application of condition II could be made. At the grounding line the only condition imposed is that $\bar{\tau}'_{\xi\xi}$ is given by (51), with σ_b taken to be 0. In the model formulation I see no indication that the condition on $\bar{\tau}'_{\xi\xi}$ imposes condition (50) on $\bar{e}_{\xi\xi}$ at the grounding line. Although the flow law relating $\tau'_{\xi\xi}$ and $\dot{e}_{\xi\xi}$ is stated (van der Veen, 1987, eqn. (6)), it appears to me that the complicated stress condition based in part on it (van der Veen, 1987, eqn. (21) or (32)) could not be applied at the boundary, where it would be in conflict with the boundary condition on $\bar{\tau}'_{\xi\xi}$ just stated. It is presumably applied at the first grid point upstream from the grounding line (and all points beyond), but according to van der Veen (1987, p. 234), $\tau'_{\xi\xi}$ has already dropped to a negligible value at that first upstream point. Hence it appears that condition II or an approximation to it is not used in this model.

Huybrechts (1992, p. 100) states that "the thickness at the grounding line results from a subtle balance between a... number of competing mechanisms and feedback loops, ...so that by excluding one stress component almost any result can be obtained. Modelling the transition zone is thus a delicate matter and requires careful consideration of the various terms in the force balance." This aptly states the level of difficulty involved in treating the grounding line in ice-sheet/ice-shelf models. The foregoing discussion suggests that condition II should still have an important role in such modeling.

9.5 ROLE OF THERMOMECHANICAL COUPLING IN CLIMATIC RESPONSE

As an example of results obtained with the Huybrechts (1990, p. 88) model just discussed, Figures 23 and 24 show tests of the sensitivity of the modeled Antarctic ice-sheet surface to three separate

Figure 23. Antarctic Ice Sheet model results of Huybrechts (1992, Figs. 6.2, 5.7), obtained by integrating a time-dependent model to steady state (see Fig. 24) with change from interglacial to glacial climatic conditions. (a) Interglacial reference model (essentially the present). (b) Model obtained by lowering a to about 50 to 60% of the reference-state values. (The symbol "Δa" in the upper right is a reminder that only a was changed in this model.) (c) Model with T_S decreased by 10°C (note "ΔT_S" in upper right corner). (d) Sea level lowered by 130 m ("$\Delta s.l.$"). (e) Model with both Δa and ΔT_S applied ("$\Delta a \& T_S$"). (f) Model with all three changes applied ("Δ all"). (g) Model temperature profiles $T(z)$ at Vostok in East Antarctica (near the crest of the ice sheet): separate curves for the above models are indicated by the curve labels "(a)," "(b)," etc. (h) Model flow-velocity profiles $u(z)$ for the above models, labeled as in (g).

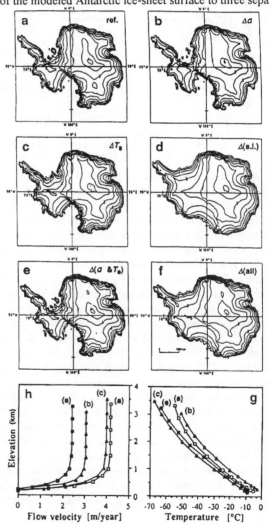

466

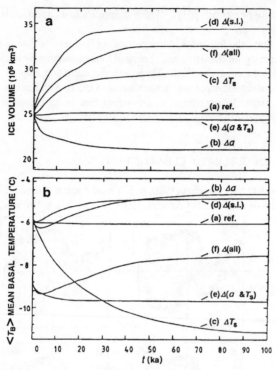

Figure 24. Time dependence of the models whose steady-state form is shown in Fig. 23, from Huybrechts (1992, Fig. 6.5). Starting with the reference model (Fig. 23a), a change in boundary conditions is introduced at $t = 0$ and the time evolution of the model is followed in terms of two global parameters: (a) total volume of the model ice sheet, and (b) mean basal temperature $\langle T_B \rangle$. The response curves are labeled "(a)," "(b)," etc., corresponding to the models in Fig. 24a, 24b, etc., and the reminders "Δa," "ΔT_S," etc., from the upper right corner of these figures are also given.

manifestations of the change from interglacial to glacial climate. Figure 24 gives transient, time-dependent responses and Figure 23 shows the ultimate steady-state ice profiles to which the time-depended responses lead. A steady-state "reference interglacial" model $z_S(x,y)$ (essentially the present) in Figure 23a is compared with steady-state models calculated for (b) ice-age accumulation $a(x,y)$, (c) ice-age surface temperature $T_S(x,y)$, (d) ice-age sea level, (e) combination of b and c, and (f) combination of b, c, and d. Figure 23g gives vertical temperature profiles $T(z)$ near Vostok for a, b, c, and e. Figure 23h shows how the ice flow velocity $u(z)$ responds to the climatic changes b, c, and e. The comparisons in Figure 23 show that sea level dominates the profile of the West Antarctic Ice Sheet, whereas the East Antarctic Ice Sheet is much less responsive to sea level change. The effect of sea level on the West Antarctic Ice Sheet and Ross Ice Shelf is similar to that studied by Lingle (1984) and Lingle and Craig (1985) in the flowband models discussed in Section 9.4.2 above, and its effect in East Antarctica is similar to that found in a flowband model by Alley and Whillans (1984). The ice-age $z_S(x,y)$ in Figure 23f is roughly similar to the reconstruction by Hughes et al. (1981, p. 267), except for some details along the Transantarctic Mountains and in the Ross and Weddell Embayments. Figures 23e, g, and h show that in the Huybrechts model the effects of decreased ice-age accumulation and temperature almost compensate one another in the model response of the ice surface $z_B(x,y)$ and to a lesser extent the temperature $T(z)$, but not, of course, in the flow velocity. The decrease in ice-age accumulation $a(x,y)$ to about 50-60% of Holocene values was based on an ice-age temperature decrease of 10°C (at sea level) and an assumed relation between accumulation and saturation vapor pressure of water as a function of temperature T_S (Huybrechts, 1992, p. 133, 168). (An analogous assumption as to the climatic shift of $a(\xi,t)$ for temperate glaciers was discussed in Section 9.3, except that the sign of the shift is opposite, decreased temperatures here causing decreased a in contrast to a net increase in a for temperate glaciers.)

The models in Figure 23 were obtained by time integration starting with the interglacial reference model and making an abrupt change at $t = 0$ to glacial boundary conditions, singly or in combination. Figure 24 gives the time development of the models as represented in integrated form by total ice volume (Fig. 24a) and by mean basal temperture $<T_B>$ (Fig. 24b). It shows that the responses to the different climatic changes in boundary conditions have different response times or combinations of response times, generally in the range 20-90 ka; the temperature can even show an initial response (for the first ~5 ka) that is opposite to the ultimate steady-state response (Fig. 24b, curves (b) and (f)).

The climatic model responses depicted in Figures 23 and 24 exhibit the results of thermomechanical coupling in ice-sheet flow and prompt an enlargement of the brief discussion of this subject near the end of Section 7.

In an effort to sort out the feedback effects due to thermomechanical coupling under external forcing by climatic changes it is helpful to consider the results of the Antarctic flowline model of Huybrechts and Oerlemans (1988, Fig. 1), which was a precursor to the three-dimensional Huybrechts (1990, 1992) model: a 10°C decrease in T_S without change in $a(\xi)$ led to an 8°C decrease in steady-state $<T_B>$. (Cf. the 5°C decrease in Fig. 24B, curve (c), and the 4°C drop between curve (a) and curve (c) at the ice-sheet base in Fig. 23g.) Thus the net result of the modeled thermomechanical coupling was a negative feedback, contrary to the positive feedback pointed out near the end of Section 7. This negative feedback is probably caused by the thermal blanketing effect associated with the increase in steady-state ice thickness that results from decreased ice temperatures and consequently decreased A in (7). The blanketing effect was not included in the model response discussed in Section 7, which was a flow response obtained at fixed ice configuration $z_S(x,y)$, not constrained to remain at steady state. In the non-steady-state flow response, shear heating changed in response to changed ice flow, and the resulting change in temperature caused the ice flow to change further in the same direction, whereas in the steady-state response to a change in temperature there is no change in ice flux and therefore little change in shear heating, while the thermal blanketing effect arises as a result of change in ice thickness.

A further stabilizing influence implied by the Huybrechts and Oerlemans (1988) model is due to the decrease in a that is estimated to accompany the climatic decrease in T_S as noted above; it cuts the decrease in modeled steady-state T_B to 2°C (ibid., Fig. 1). A similar effect can also be seen in curve (e) of Figures 23g and 24b. The cause of warming here would seem to be a decrease in downward and/or forward advection of colder temperatures, which results from the decrease in flow u and w in (23) linked to the climatic decrease in a. The same type of response appears to have arisen in an idealized ice-sheet model of Oerlemans and van der Veen (1984, p. 105): the response of ice-sheet thickness h to accumulation a was about twice as great when the temperature dependence of A in (7) was included as when it was not. It is not clear why the advection effect should in these cases appear to dominate the shear-heating effect (which has the opposite influence on T_S), contrary to the feedback situation discussed in Section 7.

Huybrechts (1992, p. 146) gives the following perspective on these complicated thermomechanical coupling effects in the response of an ice sheet to climatic change:

"Changes in the surface climate not only alter the ice sheet's mass balance, but also its thermal regime, which affects the flow properties of the ice. A whole number of feedbacks are involved, and different modes of heat transfer operate on different time scales in different parts of the ice sheet. The ultimate outcome is generally a complicated matter and simple reasoning is inadequate to predict a priori what will happen to the ice sheet in a different climate. However, in order to aid the interpretation of the modelling results, it is useful to give a brief qualitative description of basic mechanisms involved. Consider a situation in which a temperature lowering of typical glacial-interglacial magnitude (say 10°C) is applied at the ice sheet's surface. A first consequence is that a cold wave will be slowly conducted downwards. This wave will at some stage reach the basal layers, where ice deformation is concentrated."

Cooler ice deforms less readily, leading to decreased strain rates and a general thickening of the ice sheet. However, colder air carries less precipitable moisture, so a cooling of the surface climate in the Antarctic is also accompanied by lower accumulation rates. This has a number of additional consequences. First, the mass balance of the ice sheet decreases. This is a thinning effect and will counteract the thickening caused by the colder base. Second, less ice needs to be transported towards the ocean, so lower accumulation rates also imply lower velocities. A direct effect is that less heat will be generated by dissipation, especially at the margin, where friction is the main heat source. Additionally, advection of cold ice towards the basal layers becomes less effective, which should result in a warming. All of these mechanisms ultimately cause a change in the thickness of the ice sheet and this will in turn feed back on the surface boundary conditions. A thickening of the ice sheet implies higher elevations and, consequently, lower surface temperatures. This is a positive feedback. A lowering of the surface because of lower deposition rates, on the other hand, leads to higher surface temperatures (positive feedback), but it also implies higher accumulation rates, which is a negative feedback. Changes in ice thickness are also important for the vertical heat diffusion process. Thicker ice, for instance, enhances the insulating effect of the ice sheet and makes the evacuation of geothermal heat towards the surface more difficult. In view of all these (in part) counteracting effects, it is clear that numerical calculations are needed in order to find out what the ultimate response of the modelled ice sheet will be."

The Huybrechts and Oerlemans (1988) thermomechanically coupled model was designed to test the possibility of a runaway creep instability due to shear heating (Clarke et al., 1977). It led to the conclusion that because of the predominance of negative feedbacks the Antarctic ice sheet is not subject to this type of instability.

Ice sheet modeling with special focus on thermomechanical coupling has also been done by Hindmarsh et al. (1988).

10. Basal melting, water conduit systems, and "active" basal water pressure

In order to model glacier flow with a P_w-dependent or δ-dependent sliding law (Section 8) it is necessary to model the process(es) by which basal water pressure P_w is generated or δ is controlled. In most of the models in Sections 8 and 9, P_w is taken either to be "passive hydrostatic," tied to sea level by (37), or else equal to P_1 as required for the uniform water film on which (35) is based. However, local sources of basal water can actively raise P_w above the "passive hydrostatic" level, though not in general all the way up to the flotation pressure P_1. The introduction of "active" basal water pressure greatly affects glacier flow modeling.

Direct evidence that P_w is substantially above "passive hydrostatic" has been obtained from the base of the Antarctic Ice Sheet at lat. 83.5°S, long. 138.2°W in Ice Stream B (station "Up B," Fig. 19), where P_w was found to be about 25 bar above "hydrostatic" and within about 1 bar of flotation (Engelhardt et al., 1990). In places where basal "lakes" have been detected under the ice sheet (Oswald and Robin, 1973; Robin et al., 1977) P_w must equal P_1, even though the base of the ice may be above sea level.

Bentley (1987, p. 8855) showed that when the value $\hat{P} = 0.5$ bar ($\hat{h} = 5$ m), which is realistic for Ice Stream B, is inserted in the sliding laws (36)-(39) that had been used in modeling the Antarctic Ice Sheet they predict large to absurdly large flow velocities, in the range 1.4-870 km a^{-1}, whereas the observed ice streaming velocities are 0.4-0.7 km a^{-1} (Whillans, Bolzan, and Shabtaie, 1987). The huge discrepancy comes about because the flow-law constants $\hat{S}$, c_0, c_1 and c_4 had been adjusted to reproduce the balance velocities under values of effective pressure that were based on the "passive hydrostatic" model (37) and were much larger than the observed values (~25 bar vs. ~0.5 bar). The discrepancy indicates that the sliding laws (36)-(39) that have been used in ice sheet modeling do not have general validity or at least do not apply without modification to the ice streams.

Active P_w is determined by the magnitudes of the basal water sources in relation to the continuity condition on water flow and a water transport law based on the details of a conduit system through which water is conducted from the sources to the sinks – the glacier terminus, the (un-)grounding line, or areas of basal freeze-on. Important is the source of the basal water, whether by basal melting only, as in Antarctica and Greenland for the most part, or by input of water from the glacier surface in addition, as in temperate glaciers (those at the melting point throughout) and along the western fringe of the Greenland Ice Sheet. Where input of surface meltwater or rainwater to the glacier bed is involved, it usually completely dominates as the water source, because surface ablation is usually several meters per year whereas basal melting is usually at most a few centimeters per year. In the depths of winter, however, when the supply of surface water is shut off, temperate glaciers tend to approach the "subpolar" condition (water source at the base only). Very different considerations are involved in modeling the water-conduit systems of temperate and subpolar glaciers, and the discussion that follows is bifurcated along these two lines.

10.1. SUBPOLAR GLACIERS

Basal sliding produces shear heating at the sliding interface, the rate of heat generation being $\tau_B u_B$ per unit area of the bed. This heat generally causes melting of ice from the bottom of the glacier, because the basal temperature T_B is already at the melting point for significant sliding to occur (Section 8). The rate of bottom melting m is determined by the heat balance condition that is an enlargement of (28):

$$\rho H m = k\left[-(\partial T/\partial z)_G + (\partial T/\partial z)_B\right] + \tau_B u_B \tag{52}$$

where H is the latent heat of melting; the subscript G refers to the geothermal gradient as in (28), and $(\partial T/\partial z)_B$ is the (always negative) temperature gradient in the ice just above the bottom. In modeling, $(\partial T/\partial z)_G$ is input and $(\partial T/\partial z)_B$ is output from the thermal model (Section 7). If $-(\partial T/\partial z)_B$ is large enough, m from (41) will be negative; in this case, water at the bed freezes onto the base of the ice at the indicated rate.

Figure 25 shows the distribution of basal melting rate $m(x,y)$ given by the thermal models of the Antarctic Ice Sheet in Figure 14. Bottom melting at a rate of up to 10 mm a^{-1}, locally larger, occurs in a number of areas, particularly in areas adjacent to large outlet glaciers or ice shelves, such as the West Antarctic ice-stream area mentioned in Section 8 (Fig. 19). Raising the geothermal gradient from 25° to 30°C km^{-1} in the model causes a large expansion in the area and rate of basal melting, which now involves most of the interior of the ice sheet (Fig. 25b). Areas of negative m (basal freeze-on) are not plotted in these diagrams.

In the early models of basal meltwater transport under the ice sheets, water pressure was not explicitly involved, and the water was assumed simply to be carried along with the motion $\bar{u}$ of the ice "column" (Budd et al., 1982, p. 59; Oerlemans, 1983, p. 83). But because of the great difference (15 orders of magnitude) in the viscosities of water and ice this model is highly improbable. Instead, water will move through available passageways by viscous or turbulent flow under the propulsion of a P_w gradient and gravity. The reason that basal melting causes P_w to increase over "passive hydrostatic" is presumably that a hydraulic gradient has to build up to a sufficiently high level to conduct the meltwater away so that it does not accumulate except to a limited extent at the base of the ice (Bentley, 1987, p. 8854). Evidently this conduction presents hydraulic resistance, whose magnitude (and hence the magnitude of P_w) depends on the nature of the water-conduction pathways. The likely pathways are of two general types: conduits at the base of the ice, and subglacial aquifers.

The generation of P_w in a subglacial-aquifer-type conduction system is illustrated by Lingle and Brown's (1987) model for Ice Stream B, which endeavors to explain with the help of active water

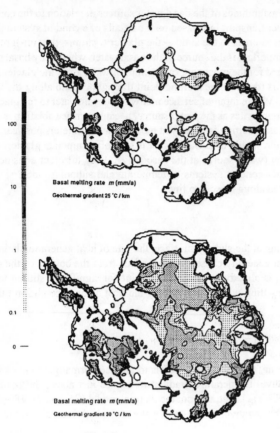

Figure 25. Basal melting rate $m(x,y)$ in mm a^{-1} under the Antarctic Ice Sheet, calculated from the thermal models of Fig. 14 for two values of the geothermal gradient: (a) 25°C/km, and (b) 30°C/km. Courtesy of Dick Jenssen. The contour values of m, spaced on a logarithmic scale, are given by the bar on the left.

Basal melting rate m (mm/a)
Geothermal gradient 25 °C / km

Basal melting rate m (mm/a)
Geothermal gradient 30 °C / km

pressure the glaciologically remarkable feature of ice stream, namely, that while the basal shear stress decreases downstream, the flow velocity increases substantially (Figure 26a,b). The model

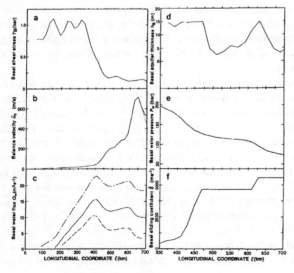

Figure 26. Results of an ice-stream flowband model with basal sliding enhanced by basal water pressure generated by basal melting and water flow in a subglacial aquifer, from Lingle and Brown (1987, Figs. 7, 4, 9, 12d, 12a, 12c). (a) Basal shear stress $\tau_B(\xi)$. (b) Balance velocity $\bar{u}_a$. (c) Basal water flux $Q_w(\xi)$ calculated from the basal melt rate $m(\xi)$ given by (52) for the temperature profiles in Fig. 12; the dashed curves show the estimated range of possible Q_w values. (d) Aquifer thickness $h_B(\xi)$. (e) Basal water pressure $P_w(\xi)$. (f) Sliding coefficient $\hat{S}(\xi)$ in (32).

follows a 610-km-long flowband from the ice divide at the center of the West Antarctic Ice Sheet
to the (un-)grounding line where the ice enters the Ross Ice Shelf. Temperature profiles (Fig. 12)
at a sequence of points along the flowband (longitudinal coordinate ξ), calculated by the moving
column method as discussed in Section 7, give values of $(\partial T/\partial z)_B$ and thence $m(\xi)$ in the zone of
basal melting, from (52), with the shear heating term calculated from τ_B and $\bar{u}_a$ in Figure 26a,b.
Water is assumed to flow in the direction of the ice flowlines but not at the ice flow velocity. The
water flux $Q_w(\xi)$ is thus obtained by areally integrating $m(\xi)$ (assumed laterally constant) along
the flowband, starting at its head; the result is in Figure 26c. The water conduction system is modeled
as a Darcy aquifer of hydraulic conductivity K and thickness h_B, lying immediately under the ice.
A sliding law of type (32) with $p = 1$, $q = 1$ is assumed, and the sliding-law parameter $\hat{S}$ and aquifer
thickness h_B are allowed to be functions of ξ as noted in Section 8. A special procedure (Lingle and
Brown, 1987, p. 268) is used to calculate $P_w(\xi)$ and simultaneously adjust $h_B(\xi)$ and $\hat{S}(\xi)$ so as to
get the required transport of water (flux Q_w) and ice (velocity $\bar{u}_a$) while keeping h_B less than an
arbitrary maximum thickness of 15 m (Fig. 26d). The $P_w(\xi)$ values obtained (Fig. 26e) range from
"passive hydrostatic" at the grounding line to about 22 bar above hydrostatic (or about 3 bar below
flotation) some 200 km upstream. The $\hat{S}(\xi)$ values (Fig. 26f) increase by a factor 14 from
$\xi = 260$ km, where basal sliding commences, to the (un-)grounding line at $\xi = 610$ km. The increase
in $\hat{S}(\xi)$ is interpreted as due to a longitudinal decrease in bed roughness. Successful models have
aquifer hydraulic conductivities K in the range 0.02-0.06 m s^{-1} (Lingle and Brown, 1987, p. 274).
The subglacial sediment actually sampled at Up B has $K\sim10^{-9}$ m s^{-1} (Engelhardt et al., 1990, p.
248). This is so much smaller than the K required by the model that Darcy flow in a subglacial
aquifer is ruled out as the water-conduction mechanism. But the model is nevertheless interesting
as a concrete example of the relation between a basal water transport system and the corresponding
distribution of basal water pressure.

A water transport model with conduction path in the basal water film visualized by Weertman
and Birchfield (1982) is exemplified by Budd and Jenssen's (1987) areal model of the "Ross sector"
of the West Antarctic Ice Sheet, discussed in Section 8.2 (Fig. 18). The melting rate distribution
$m(x,y)$ is again obtained by (52) from a moving-column thermal model using a number of flow lines
(Section 7). The azimuthal direction of water flow in the film is taken to be that of the two-dimensional hydraulic gradient vector

$$\Gamma = \alpha + \beta(\rho_w - \rho)/\rho \tag{53}$$

where $\alpha = -\nabla_2 z_S$ and $\beta = -\nabla_2 z_B$. Γ has approximately the direction of the surface slope α. A map
of Γ defines water flowlines and flowbands, along which $m(x,y)$ is areally integrated to get the water
flux density $\Phi_w(x,y)$, by a procedure analogous to the calculation in Section 5 of the ice balance
velocity $\bar{u}_a$. Laminar viscous flow in the basal water film of thickness δ is governed by the Poiseuille
relationship among Φ_w, δ, and the mean water-flow velocity $\bar{u}_w$:

$$\Phi_w = \frac{\rho_w g}{12\eta_w}\Gamma\delta^3 = \bar{u}_w\delta \tag{54}$$

where η_w is the viscosity of water and Γ is given by (53). Thus the model $\Phi_w(x,y)$ can be expressed
in terms of the water-flow velocity $\bar{u}_w$ and the water-film thickness δ, as given in Figures 18c and
d. In the vicinity of Up B the model δ is 3 or 4 mm and the model $\bar{u}_w$ about 3 cm s^{-1}; these values
are not greatly different from the values inferred from borehole observations there ($\delta \sim 1$ mm,
$\bar{u}_w \sim 7$ mm s^{-1}: Kamb and Engelhardt, 1991). Budd and Jenssen (1987) do not give model results
for $P_w(x,y)$ nor apply them in sliding law (32) to calculate a model u_B, as might be expected in the
context of the foregoing discussion. The reason is that the assumed hydraulic gradient Γ in (53)

corresponds exactly to P_w at the flotation pressure, $P_w = P_I$ or $\hat{P} = 0$, which is incompatible with sliding law (32) since it gives $u_B \rightarrow \infty$. Instead of (32), Budd and Jenssen use sliding law (37)+(38) (with $c_1 = 0$), although this law assumes "passive hydrostatic" P_w and thus disregards the condition $P_w = P_I$ on which the water-transport analysis is based. Budd and Jenssen (1987, p. 312) also consider the use of the very different flow law (35), which is compatible with $P_w = P_I$, but they do not actually calculate an ice-flow model from it.

The required modification of the water-film model because of the impossibility of a continuous water film with $P_w = P_I$, which was discussed in Section 8.1 in relation to the sliding law (35), must result in modifications in the modeling of water transport in such a basal water system, since the water must now move through an incomplete ice-bed gap ("punctured water sheet") at pressure $P_w < P_I$ (Weertman, 1972, p. 312). "Incomplete-gap-conduit" systems of this type have been used in one-dimensional ice-stream models by Alley (1989a, p. 114; 1989b, p. 119) and Kamb (1991, p. 16,589). They are only very primitively formulated in terms of the physical parameters and relationships that control the conduit geometrical elements (gap thickness; gap width and spacing or areal fraction; conduit branching and sinuosity), which need to be known to calculate $P_w(\xi)$ in the model. But they give some idea as to how basal sliding in response to basal water pressure may be modeled bringing into consideration an "active" P_w and avoiding the generally unwarranted assumption (37) that the basal water pressure is "passive hydrostatic," tied to sea level. They are further discussed in Section 12.

For use in evaluating the role of basal water in the dynamics of the Antarctic Ice Sheet Oerlemans and van der Veen (1984, p. 107) have introduced the following "diffusive equation" for water transport and storage in a basal water layer of thickness $\delta(x,y;t)$:

$$\frac{\partial \delta}{\partial t} = D_w \nabla_2^2 \delta + m \tag{55}$$

where D_w is a "diffusivity" and m is given by (52). (I have changed the sign of the first term on the right from what appears in eqn. (6.4.1) of Oerlemans and van der Veen [1984].) A physical basis for (55) or for D_w was not given, but (55) is said to be "the simplest approach to deal with the flow of basal water." A role of P_w in relation to (55) is not specified by Oerlemans and van der Veen, but if δ is assumed to be a linear function of P_w,

$$\delta = g_0 + g_1 P_w \tag{56}$$

where g_0 and g_1 are constants, and if $\beta = 0$ so that $\Gamma = (\rho_w g)^{-1}(\partial P_w/\partial \xi)$ in (54), then (55) can be interpreted as expressing approximately the continuity condition for water flow in an ice-bed gap, by (54), with $D_w = \delta^3 (12\eta g_1)^{-1}$. D_w is either taken constant, 10^9 m^2 a^{-1} (ibid, p. 108), or "linearly proportional to normal load" (ibid., p. 191). Finally, a sliding law is taken in the form (31), modified as in (35):

$$u_B = \tau_B^p(S + S_1 \delta) \tag{57}$$

but with an adjustable constant S_1 in place of the 10 in (35); apparently p is taken to be 2.5 (ibid., p. 103). The above formulation, used by Oerlemans and van der Veen (1984, p. 108) in a time-dependent Antarctic-ice-sheet flowline model of the type discussed in Section 9, leads for certain values of the parameters to the alternative existence of two stable steady states, one a slow-flow state with the ice sheet frozen to its bed and the other a fast-flow state with basal melting and sliding, as noted near the beginning of Section 9 in connection with Figure 21.

10.2. TEMPERATE GLACIERS

When large volumes of water derived from surface melting are transported by the basal water conduit system, rather different considerations apply. This subject, a sub-topic of glacial hydrology, has been reviewed by Röthlisberger and Lang (1987) and by Hooke (1989). Walder (1982) showed that a water-conducting layer or film of the type advocated by Weertman (1969, 1972) at the base of a temperate glacier is unstable and will tend to break up into localized conduits incised by melting into the ice above. Typically these conduits are ice tunnels of roughly semicircular or flat elliptical cross section a few meters in size. As shown by Röthlisberger (1972), the tunnels are formed and controlled by the competition of two processes: (1) tunnel enlargement by melting of the ice walls due to generation of heat by viscous dissipation in the through-flowing water; (2) tunnel closure by ice creep driven by the difference between ice overburden pressure P_I and water pressure P_w. A somewhat non-intuitive consequence of this competition is that the water flux transported by such a tunnel in steady state is a decreasing function of P_w, from which it follows further that an anastomosing system of many such tunnels will tend to condense to a single trunk tunnel, fed by arborescent tributary tunnels leading from surface water sources (Röthlisberger, 1972, p. 180; Shreve, 1972, p. 209). Using the quantitative formulation of tunnel response to changing P_w and water flux, together with flow continuity, the time-dependent performance of this type of water transport system has been modeled for various inputs, particularly those that result in the famous jökulhlaup-type outburst floods (Nye, 1976; Spring and Hutter, 1981).

Bindschadler (1983) modeled on the basis of Röthlisberger's (1972) formulation the steady-state distribution of water pressure $P_w(\xi)$ in single-tunnel systems along the length of two temperate glaciers in Alaska, finding that P_w is generally in the range 15 to 20 bar below P_I in one case (Fig. 27a) and 3 to 5 bar in the other (Fig. 27b), except near the terminus where P_w either goes to zero

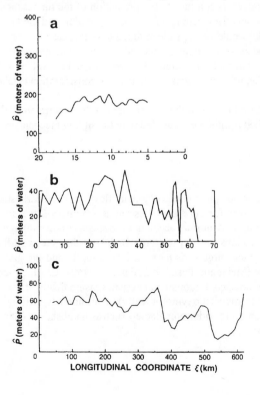

Figure 27. Basal effective pressure $\hat{P}$ (ice pressure less basal water pressure) calculated from the tunnel model of Röthlisberger (1972) by Bindschadler (1983, Figs. 1, 2, 4): (a) $\hat{P}(\xi)$ for Variegated Glacier, Alaska, under summertime surface-melting conditions; (b) $\hat{P}(\xi)$ for Columbia Glacier, Alaska, in summer; (c) $\hat{P}(\xi)$ for Ice Stream B, Antarctica, with water flux calculated from a basal melting rate m of 30 mm a^{-1}

(terminus on land) or to P_1 (tidewater terminus afloat). Observations of P_w in boreholes are in general agreement with these P_w models (especially the first case) for "normal" (nonsurging) glaciers (Kamb et al., 1985, p. 474; Kamb, 1987, p. 9096). Bindschadler used the modeled $P_w(\xi)$ in Variegated Glacier (Fig. 27a) to choose a sliding law of type (32) by fitting observed and calculated surface flow velocities on the assumption that most of the observed motion is due to basal sliding. He got an approximate fit to (32) with $p = 3$, $q = 1$. This sliding law, together with u_1 from (15) and $P_w(\xi)$ values from Bindschadler (1983), was used by Raymond and Harrison (1987) in a model of Variegated Glacier with the objective of determining whether the increase in surface velocity in the build-up toward a surge was primarily due to an increase in internal deformation or in basal sliding. The distinction proved to be difficult to make on the basis of the observed changes in ice thickness, surface slope, and surface velocity in relation to the above type of flow-plus-sliding law, because the model results were not very sensitive to the detailed features of the flow and sliding laws. Much greater sensitivity would be shown if wide ranges in $\hat{P}$ were involved, as would probably be the case in comparing continental and tidewater glaciers and as shows strongly in the Antarctic model of Figure 17.

Bindschadler (1983, p. 11) applied the same type of one-dimensional tunnel model to the subpolar situation in Antarctic Ice Stream B, obtaining a $P_w(\xi)$ with wild fluctuations and discontinuities, which could be damped by smoothing the basal topography (Fig. 27c). As so smoothed, the results ($\hat{P}$ range 2-7 bar) are not greatly different from those in Figure 27b. At Up B the predicted $\hat{P}$ is 4 bar (arrow in Fig. 27c), whereas the observed values are in the range 0-1.6 bar (Engelhardt et al., 1990, p. 58), which tends to rule out the tunnel model in this application. Another observation that rules it out is that upon reaching the bed almost all boreholes made immediate connection with the water conduit system (loc. cit.), which would be highly improbable if there were only one tunnel conduit in the 35-km width of the ice stream. The observation would be explainable if there were many conduits, closely spaced, which, however, is contrary to the prediction of the tunnel model mentioned above. In such a many-tunnel model the calculated $\hat{P}$ would be smaller, and the tunnel model and "incomplete-gap-conduit" model would perhaps blend together in this case.

A "linked-cavity" type of basal water system, compounded of lee-side ice-bed separation cavities of the kind considered in Section 8.1, constitutes a fourth alternative type of conduit system quite distinct from the systems discussed above (aquifer, basal gap, ice tunnel); its consideration is taken up in Section 11.1.

The discussion in this section and in Section 11.1 re-emphasizes the remark by Weertman (1979, p. 109) that "the water-flow process at the bed is intimately coupled with the sliding mechanisms."

11. Surging and unstable sliding

The modeling of glacier surges – the most dramatic phenomena of glacier flow – is in an early stage of development because rather little is known for sure about essential features of the surge mechanism(s). In a surge, a glacier looking initially normal accelerates to speeds of ~2-100 m d^{-1}, which continue with more or less wild fluctuations for ~1 y, during which time the glacier surface becomes extremely shattered by crevassing; the surge ends more or less abruptly and the glacier then resumes a seemingly normal flow state (Meier and Post, 1962; Paterson, 1981, Ch. 13; Kamb et al., 1985; Raymond, 1987). Glaciers that move continuously rather than spasmodically at speeds of ~1-10 m d^{-1} have sometimes been called (somewhat oxymoronically) "continuously surging" – for example, the Antarctic Ice Streams and some large tidewater glaciers such as Jakobshavns Glacier (Greenland) and Columbia Glacier (Alaska).

There are two distinct aspects of surge modeling: 1. modeling the mechanism(s) by which rapid surge motions are caused and controlled; 2. modeling the surge motions of idealized or actual glaciers on the basis of the mechanism(s) whose quantitative formulation is developed in 1.

11.1. SURGE MECHANISMS

It is generally thought that the rapid motion in surging is due to anomalously rapid basal sliding, although for Antarctic ice streaming Hughes (1977) proposed a model mechanism based on increase of ɔ in (7) due in part to ice crystal reorientation (fabric enhancement). Here we concentrate on models of basal sliding enhancement. Because it is necessary that the bed be at the melting point for substantial sliding to occur (Section 8), it has been argued that transition from frozen to unfrozen bed is the fundamental control mechanism for surging (Clarke, 1976). As a process that could promote such a transition, thermal creep instability has been suggested (Clarke et al., 1977) but also questioned (Fowler and Larson, 1980). Although models of thermal control of surging are directly related to the discussion in Sections 7 and 9.5, they will not be pursued here because they do not in themselves give the greatly enhanced basal sliding that, as distinct from normal sliding, is needed to explain surging. Also, surging is known to occur in fully temperate glaciers, to which a thermal control mechanism is not applicable because the ice temperature is fixed.

Most of the proposed mechanisms for surging by anomalous enhancement of basal sliding involve the role of ice-bedrock separation. A first example is the water-film model discussed in Section 8 (see equation (35)), which was developed as a mechanism of surging by Weertman (1962, 1969). In this model an increase in τ_B causes, first, an increase in water flux from (52) and thus in water-film thickness δ in (54), and, second, a decrease in the critical obstacle height δ^* as given by a theoretical relation between τ_B and δ^* (Weertman, 1979, eqn. (61)). These changes give an increase in u_B according to (35). If the increase in δ and decrease in δ^* are large enough, surge-type flow velocities can be obtained from this model. However, it would be quite difficult in an actual glacier to get an increase in τ_B as large and rapid as is needed to produce from (35) an increase in u_B as large and rapid as occurs in surge. The water-film model would seem to do better in accounting for a "permanent state of surging" (Weertman, 1969, p. 930) than for spasmodic surging.

The other concept of ice-bed separation – lee-side cavitation as distinct from a water film (see Section 8.1) – plays a central role in two models of the surge mechanism. The first is based on features of cavitation shown in Figure 28a,b. For an idealized bed with a single roughness wave-form (e.g. a sinusoid, sometimes called a "washboard" bed form) the relation between sliding u_B and shear stress τ_B that propels the sliding is as shown schematically in Figure 28a. At low τ_B, for which the extent of cavitation is nil, u_B rises with τ_B as τ_B^n for no cavitation (eqn. (31) with $p = n$) and somewhat more rapidly once cavitation has set in. As cavitation becomes extensive, τ_B reaches a maximum and then declines toward 0 with further increase of u_B, which represents unstable sliding. Under the name "subglacial supercavitation" Lliboutry (1964) proposed this instability as a mechanism for achieving very rapid sliding and hence surging. The decline of τ_B toward 0 at large u_B is said to result from "drowning" the roughness, although it is conceptually somewhat different from the "drowning" considered in the water-film model (Section 8.1). Subsequently Lliboutry (1968, p. 48; 1987b, p. 9106) added to the idealized bed model a second roughness wave at much longer (~30 times longer) wavelength, forming an "undulating washboard" bed, which leads to a u_B vs. τ_B relation shown schematically in Figure 28b; a second maximum in τ_B would in principle be encountered at very high u_B, beyond the range of Figure 28b, when cavitation of the long-wavelength roughness wave (which is nil in Figure 28b) becomes extensive. "Normal" sliding is represented by the segment OA of the u_B vs. τ_B curve. On the segment AB, the sliding is unstable, and on the segment BC the sliding is stable and the glacier is in surge. A surge is initiated when the ice thickness h and/or surface slope α increase to the extent that (τ_B, u_B) reaches the point A,

476

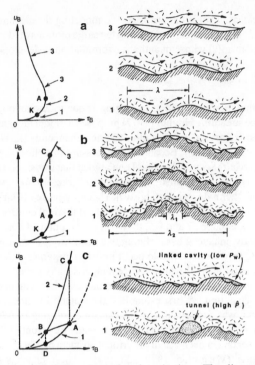

Figure 28. Models of enhanced sliding due to basal cavitation. The diagrams numbered from 1 to 3 on the right show schematically, in longitudinal vertical cross section through the glacier bed, the cavitation geometry at sliding velocities from low ("1") to high ("3") as suggested by the ice-motion arrows. On the left are corresponding relationships between basal shear stress τ_B and sliding velocity u_B for different bed geometries. The numbered arrows point to the parts of the $u_B(\tau_B)$ curves that correspond to the different cavitation geometries shown on the right. (a) Bedrock with a single roughness wavelength λ: diagrams 1, 2, and 3 on the right represent regimes of no cavitation, moderate cavitation, and extensive cavitation, respectively. The (water filled) cavities are the blank areas between bedrock (cross-hatched) and ice (patterned). Point A on the $u_B(\tau_B)$ curve on the left can be taken as the transition between regimes 2 and 3, and point K the transition between 1 and 2 (onset of cavitation). (b) "Undulating washboard" bed with two roughness wavelengths λ_1 and $\lambda_2 \gg \lambda_1$: diagrams 1, 2, and 3 on the right represent the same succession of regimes as in (a), with respect to roughness wave λ_1; onset of cavitation on roughness wave λ_2 would take place at a high u_B value, beyond the range in the $u_B(\tau_B)$ diagram on the left. (c) Bed of such a geometry that it can support alternatively two types of basal water conduit system as suggested schematically in diagrams 1 and 2 on right: 1, a tunnel system, resulting in high $\hat{P}$, little or no basal cavitation, and low u_B; and 2, a linked cavity system, resulting in low $\hat{P}$, abundant cavitation, and high u_B. The two systems correspond to the two separate $u_B(\tau_B)$ curves ODA... and OBC... in the u_B-vs.-τ_B diagram on the left which is drawn following Fowler (1987b, Fig. 4). The crosscutting curve AB represents a constant value of the "flow parameter" $u_B/\hat{P}^{\,n}$, below which the tunnel system is stable and above which the linked cavity system is stable, according to the Fowler (1987b) theory. The OB segment of curve 2 and the prolongation beyond point A of curve 1 (ODA...) are shown dashed, because these represent unstable states. The dotted lines AC and DB represent transitions from a tunnel system at A or D to a linked cavity system at C or B. Curve OA represents normal glacier flow and curve BC... represents surging flow.

whereupon u_B jumps up to C. Later, as the ice is drawn down in the surge, τ_B and u_B decrease along CB, and when B is reached u_B drops abruptly back to the "normal" segment OA. This cycle, based on a triple-valued $u_B(\tau_B)$ function, is a type of relaxation oscillation, as discussed by Fowler (1987b, p. 9112). A surge can also be initiated by an increase in P_w, which will shift the u_B vs. τ_B curve to lower τ_B, decreasing the value of τ_B at point A. Although Lliboutry (1987b, p. 9106) thinks that this undulating washboard model "might be possible" as an explanation of surging, he seems to consider that actual glacier beds are more likely to approximate an "undulating random bumpy profile" for which the $u_B(\tau_B)$ relation (Lliboutry, 1987b, p. 9107) is not triple valued and therefore a relaxation oscillation of the type discussed above cannot occur.

Observations of a surge in progress (Kamb et al., 1985) strongly suggest that the rapid motions were caused by abnormally high P_w. This led Kamb (1987) to introduce a cavitation-based surge model that explains the high P_w by a major change in the basal water conduit system in surge. When a certain stability criterion is met, a "linked cavity system" can form in place of the normal tunnel conduit system. In the linked cavity system, water-filled lee-side cavities are numerous over the bed, and basal water flows through them and from cavity to cavity through narrow connecting passageways (ice-bed gaps) also formed by cavitation. The conduction pathways are many and widely distributed over the glacier bed instead of being localized in one or a few large conduits as in tunnel systems. The model shows that the multiple-conduit linked-cavity system can persist stably because in such a system the water flux is an increasing function of P_w, in contrast to the inverse relation for tunnel systems noted in Section 10.2. The model also shows that under certain conditions of bed roughness the pressure P_w required to transmit a typical water flux under a given hydraulic gradient via a linked cavity system can be considerably greater than via a tunnel system, which can explain the high P_w and the resulting rapid basal sliding and extensive cavitation in surge. In the model, rapid sliding promotes stability of the linked cavity system vis à vis its reorganization into a tunnel system with drop of P_w, which represents termination of surge. The model has a "bootstrap" quality in explaining how surging conditions can maintain themselves once started but in giving only an incomplete explanation of how the glacier makes the transition from normal to surging state or back.

This incompleteness is avoided in the model of sliding and basal-water hydraulics in linked-cavity and tunnel systems by Fowler (1987a, p. 261 ff.; 1987b, p. 9113 ff.), which develops the proposition that on increasing u_B and/or P_w, transition from a tunnel system to a linked-cavity system occurs at a critical value of the "flow parameter" $u_B/\hat{P}^n$ and results in a change in $\hat{P}$ from a relatively high value for the tunnel system to a lower value for transport of the same water flux in the linked-cavity system. The $u_B(\tau_B)$ relation for this model is shown schematically in Figure 28c; details are explained in the figure caption. In a tunnel-to-linked-cavity transition (D to B or A to C in Fig. 28c) the tunnel system is said to collapse and be replaced by the linked-cavity system (Fowler, 1987a, p. 265; 1987b, pp. 9111, 9116). A detailed mechanism of collapse is not specified, but in the Kamb model two aspects are suggested: closure of the tunnels in winter when the surface sources of meltwater are shut off (Kamb, 1987, p. 9099), and destruction of tunnels at the surge front by the ice there being shoved forward against bedrock protuberances by the already rapidly surging ice behind (Kamb et al., 1985, p. 479). With inclusion of the connecting segment AB (which in the Fowler model represents constant $u_B/\hat{P}^n$), the $u_B(\tau_B)$) curve OABC... is triple valued, and can therefore generate a relaxation oscillation. In the Fowler model it is assumed that the transition from tunnel system to linked cavity system on surge initiation (at A) occurs at the same critical value of $u_B/\hat{P}^n$ as the back transition on surge termination (at B) (Fowler, 1987b, p. 9114), but from the reasoning in the Kamb model it seems unlikely that a "reversibility" of this sort could be involved (Kamb, 1987, p. 9098-9).

A very different approach to modeling the surge mechanism was taken by Budd and Radok (1971, p. 32) in introducing the concept of basal "lubrication" tied to the *rate* of meltwater production by frictional heating at the base of the ice. In their model, "lubrication" reduces the basal shear stress τ_B from the gravitationally imposed value $\rho g h \alpha$ given by (14) to

$$\tau_B = \tau^* \equiv \frac{\rho g h \alpha}{1 + \varphi \bar{u} \rho g h \alpha} \tag{58}$$

where $\bar{u}$ is the vertically averaged flow velocity. (Actually, Budd and Radok [1971, p. 32] and Budd and McInnes [1978, eqns. (3), (14)] used u_S instead of $\bar{u}$ in (58); Budd [1975, p. 7] and Radok et al. [1987, p. 37; 1989, p. 133] used $\bar{u}$, and McInnes et al. [1986, p. 37] used u_B.) In (58), φ is a "friction lubrication factor" that controls the magnitude of the lubricating effect that results from an areal heating rate $\bar{u} \rho g h \alpha$: "the lubrication factor φ...directly describes the effectiveness of the frictional production of basal water" (McInnes et al., 1985, p. 37).

Since the basal melting rate is tied by (52) to the basal heating rate $\bar{u} \tau_B$ (or rather $u_B \tau_B$), which differs from $\bar{u} \rho g h \alpha$ when $\tau_B < \rho g h \alpha$ on account of (58), the appearance of $\bar{u} \rho g h \alpha$ instead of $\bar{u} \tau_B$ or $u_B \tau_B$ in the denominator in (58) is subject to question (Lliboutry, 1979, p. 69). This question aside, the denominator of (58) can be seen as playing a role related to the role of P_w (via $\hat{P} = P_I - P_w$) in (32), since we can expect P_w to increase with an increase in the rate of basal meltwater production, according to the discussion of subpolar glaciers in Section 10.1. On the other hand, (58) differs fundamentally from (32) in the appearance, in the numerator, of $\rho g h \alpha$ instead of what would be expected from the discussion in Section 8.1, namely, a function of bed roughness and sliding velocity analogous to the various functions $\tau_B(u_B)$ in Figure 28, (34), and (31). Thus (58) is not a sliding law in the standard meaning of the term. In effect it takes an initial $\tau_B = \rho g h \alpha$ that is given by adjustment of the glacier configuration prior to onset of "lubrication," and, ignoring whatever ice-flow or sliding laws were involved in that adjustment, it then reduces τ_B from the initial value by the "lubrication denominator," with no provision for a compensating increase of τ_B by a resulting speedup of sliding. For any resulting flow adjustments it in effect assumes a velocity-independent basal drag at fixed basal meltwater generation rate, which is analogous to taking $p = \infty$ in (32). In fact the appearance of $\bar{u}$ in the denominator of (58) is analogous to $p < 0$ in (32).

Because the basal drag from (58) is a decreasing function of the velocity $\bar{u}$, as it is in Figure 28a at high u_B, the motion under control of (58) is rightly called "unstable sliding" by McInnes et al. (1986, p. 37) and distinguished from the "stable sliding" represented by sliding laws (32)-(39). As the basis for a surge mechanism (58) is thus analogous to the "subglacial supercavitation" instability of Lliboutry (1964), discussed above. Motion under (58) has also been called "self-induced sliding" (Radok et al., 1987, p. 37; 1989, p. 133) and "self-surging" (McInnes and Budd, 1984, p. 98). Strongly unstable sliding is obtained from (58) by choosing φ such that $\varphi \bar{u} \rho g h \alpha \sim 1$, which for a surge velocity $\bar{u} \sim 100$ m d^{-1} and $\rho g h \alpha \sim 1$ bar requires $\varphi \sim 3 \times 10^{-5}$ bar^{-1} m^{-1} a. The values of φ found to be optimum in surge models by Budd (1975, pp. 16, 18) and Budd and McInnes (1978, p. 247) are of this order.

The sliding experiments of Budd et al. (1979, p. 164) provided some basis for (58) in finding an upper limit to steady, stable sliding, which in different experiments occurred at about the same value of $u_B \tau_B$ (regardless, curiously, of the bed roughness); above this limit, sliding was found to accelerate unstably. According to Budd and McInnes (1978, Fig. 9a), for unstable sliding at speeds $\bar{u} > 500$ m a^{-1} the $\bar{u}$-dependence of τ_B in (58) with $\varphi = 3 \times 10^{-5}$ bar^{-1} m^{-1} a agrees with results of ice sliding experiments at $-0.5°C$ by Barnes et al. (1971); at $\bar{u} < 500$ m a^{-1}, for which the experiments gave stable sliding, there could not of course be agreement with (58), but reasonable agreement was found with the following modification of (58), which is said to be derived from theory (Budd and McInnes, 1978, eqn. (25)):

$$\tau_B = \frac{\kappa_1 u_B}{1 + \kappa_2 u_B} \cdot \frac{1}{1 + \varphi u_B \tau_B} \tag{59}$$

where κ_1 and κ_2 are constants. (59) is said to be under investigation as a basis for a surge mechanism, but I have not found any further mention of it. The occurrence of τ_B rather than $\rho g h \alpha$ in the second denominator of (59) is an improvement over (58) as noted above. The form of (59), which gives the function $u_B(\tau_B)$ implicitly, is a reasonable sliding law, with κ_1 presumably an increasing function of bed roughness. It makes $u_B(\tau_B)$ double valued, as in Figure 28a, and is thus more closely analogous to Lliboutry's (1964) "subglacial supercavitation" as a basis for surging than is (58). (It has not been shown that (58) [with or without (60) below] can be re-expressed as a double valued function $u_B(\tau_B)$, contrary to the apparent representations to this effect by Budd and McInnes [1979, pp. 97, 101, 102].)

In applying (58) in glacier surge models (Section 11.2), Budd and McInnes (1974, eqn. (8)) and Budd (1975, eqn. (14)) added a correction term:

$$\tau_B = \tau^* + (<\rho g h \alpha> - <\tau^*>) \tag{60}$$

where the brackets signify an average over the length of the glacier. The correction is needed to maintain global equilibrium, which requires $<\tau_B> = <\rho g h \alpha>$, but it furthers the estrangement of (58) from sliding laws, because it introduces a shift in the basal drag that is seemingly not tied to any feature of the local sliding process (sliding velocity, basal roughness, water-generation rate, water-film thickness,...).

11.2 SURGE MODELS

A model for the surging of a hypothetical or actual glacier by enhanced basal sliding should combine a model of the sliding enhancement mechanism, such as those discussed in Section 11.1, with a time-dependent glacier or ice-sheet model (Section 9), and possibly also with a basal-water-system hydraulic model (Section 10).

The glacier surge model most extensively developed to date is the model of Budd and McInnes (1974) and Budd (1975), based on the "frictional lubrication" model discussed above. In addition to the radical features of the "lubrication" model itself, noted already, the Budd and McInnes surge model takes a highly unorthodox approach to modeling longitudinal stress equilibrium in the surging glacier. When the "lubrication" effect is applied, the model drops the standard reliance on local gravitational equilibration via the basal shear stress τ_B as expressed in (13) and used throughout Sections 5 and 9. Instead, the model switches to equilibration via longitudinal stresses: the difference between the gravitational "driving stress" $\rho g h \alpha$ (computed for current values of h and α) and the lubrication-reduced τ_B is taken to be supported by a longitudinal (deviatoric) stress gradient according to (19), with $\bar{u}$ replaced by u_B and with the lubrication-reduced value of τ_B given by (60). From this form of (19) the model calculates $u_B(\xi)$ by two ξ integrations, with application of boundary conditions $\partial u_B/\partial \xi = 0$ at the terminus and $u_B = 0$ at the head of the glacier. To integrate (19), $\tau_B(\xi)$ needs to be known from (58)+(60), which means $\bar{u}(\xi)$ needs to be known; therefore, the integration is done iteratively starting with a trial $\bar{u}(\xi)$. Continuity (43) is then applied to get h, and the profile $h(\xi)$ is stepped forward in time as in the time-dependent models in Section 9. With suitable choice of the parameters φ and $\bar{\eta}$ in (58) and (19), this system has been able to produce periodic surging behavior in a hypothetical glacier model and two or three models based on actual surge-type valley glaciers (Budd, 1975, pp. 15, 17; Budd and McInnes, 1978), giving surge repetition periods and surge velocities comparable to what are actually observed. An example is shown in

Figure 29. The rapid retreat immediately preceding each forward surge in Figure 29 (with one exception) is puzzling and does not correspond to an observed feature of actual surging glaciers.

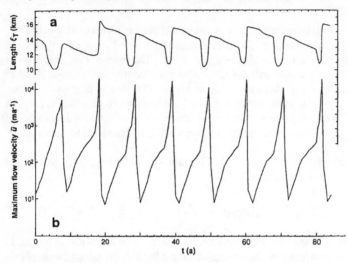

Beside the highly unorthodox calculation of $u_B(\xi)$ entirely from the longitudinal strain-rate gradient in (19), which represents the antithesis of the standard approach to glacier mechanics in Sections 5, 6, and 9, the Budd-McInnes surge model has a few other questionable features. (19) is only an approximate equation for longitudinal equilibrium, with a number of terms omitted (Kamb, 1986); it is satisfactory for a modest correction to (14), but may not be adequate in the case of a large departure from the "shallow ice approximation." The model's replacement of

Figure 29. Modeled surges of Medvezhy Glacier (F.S.U.) by Budd and Jenssen (1978, Fig. 3). (a) shows the position of the glacier terminus as a function of time, and (c) shows the maximum flow velocity (maximum along the length of the glacier at each given time). The time origin is arbitrary.

$\bar{u}$ by u_B in (19) also degrades its accuracy, as does the assumption that $\bar{\eta}$ in (19) is a constant. (Budd and McInnes [1978, eqns. (21), (22)] formulated a treatment with non-constant $\bar{\eta}$ but apparently did not put it into use.) The model inappropriately uses $\tau_B = \rho g h \alpha$ to calculate the internal deformation $\bar{u}_I$ from (16), rather than using the model's actual τ_B value, given by (58)+(60); elimination of this illogical feature was suggested by Budd and McInnes (1978, eqn. (16)).

However, the above defects probably have only relatively minor consequences compared to the dominant feature of the model, which is the way (58) forces τ_B to be substantially reduced from $\rho g h \alpha$ when $\bar{u}$ is large; by (19) this forces the mechanical support compensating for the loss of basal drag τ_B to be thrown onto longitudinal stress gradients and thence onto the terminal and head regions of the glacier. A natural and almost unavoidable consequence is the use of (19) to obtain $\bar{u}(\xi)$, by integration. The fundamental questions that must be raised in seeking the validity of this model are: (1) In surging glaciers, does a major redistribution of mechanical support take place, in which basal shear stresses are substantially reduced from $\rho g h \alpha$ and compensating mechanical support is provided by longitudinal stress gradients? (2) In surge is basal sliding governed by a relation that represents sliding instability, as (58) does? Question (1) is answered observationally for the 1982-3 surge of Variegated Glacier, Alaska, as follows: "[during the surge], over most of the surging length [the longitudinal stress gradient term in (19)] was quite small, and longitudinal [stress] gradients only slightly affected the basal shear stress. Therefore the ice was largely supported locally by [basal] shear stress..." (Raymond, 1987, p. 9129). This is contrary to the dominant feature of the model. An observational answer to question (2) does not seem to be available. The fact that no "blow ups" in $\bar{u}(\xi)$ were reported in the modeling by Budd and McInnes suggests that longitudinal stress gradients were able to stabilize model glacier systems with the unstable sliding relation (58), which suggests that there might be no direct observational manifestations of sliding instability. On th~

other hand, the observed wild velocity fluctuations of Variegated Glacier in surge (Kamb et al., 1985, Fig. 5) might be suspected as manifestations of some type of sliding instability.

The Budd-McInnes model was applied to several flowbands in the Antarctic Ice Sheet by Budd and McInnes (1979), McInnes et al. (1986, p. 37-52), and Radok et al. (1987, p. 36-57; 1989, p. 134) in an effort to assess the likelihood of ice-sheet surging. Numerous and varied types of velocity pulses and ice-thickness fluctuations were found in some of the models, namely for models with larger values of the parameter φ in (55) and smaller values of $\bar{\eta}$ in (19) (in fact for $\varphi/\bar{\eta} \gtrsim 3 \times 10^{-7}$ bar^{-2} m^{-1}) (McInnes et al., 1986, Figs. 6.2.10 and 6.2.11; Radok et al., 1987, Figs. 4.6 and 4.7). However, uncertainty or doubt as to the meaningfulness of these pulses and fluctuations has been expressed (McInnes et al., 1986, p. 51; Radok et al., 1987, p. 56), and a clear picture of ice-sheet surging or lack thereof does not seem to have been reached. The pulses are interpreted by Radok et al. (1989, p. 136) as "a controlled non-linear reaction to basal melt that cannot be dispersed by the model," although the model as formulated has no provision for accumulation of basal meltwater due to inadequate dispersal. For the extensive, intricate, and perplexing details of these models consult the three references cited above.

In the detailed model specifications given by McInnes et al. (1986, p. 37) a little confusion is created as to possible modification of the Budd-McInnes model for Antarctic application, because the specified procedural flow diagram (McInnes et al., 1986, Fig. 5.1) does not refer to eqn. (6.1) (a version of our (58) here) but instead apparently to the quite different equation (3.30) in McInnes et al. (1985, p. 43), whereas the description of results in McInnes et al. (1986) and Radok et al. (1987; 1989) makes it clear that equation (6.1) must have been used. Some modification of the model would seem appropriate to take into account the fact that much of the base of the ice sheet is probably below freezing and cannot therefore take part in the "lubrication" of the Budd-McInnes model, but there is no mention of that in the references. The use of u_B in place of $\bar{u}$ in (58), as specified in McInnes et al. (1986, eqn. (6.1)), is more appropriate physically than the use of $\bar{u}$, as noted earlier. It has the rather severe consequence of preventing in principle the development of the "lubrication" effect in any model for which initially $u_B = 0$, but if the iterative integration of (19) started with a trial $u_B > 0$, this difficulty would be sidestepped.

The surge model of McMeeking and Johnson (1986) is fundamentally similar to the Budd-McInnes model in specifying a reduction in τ_B from $\rho g h \alpha$ in such a way that the glacier has to compensate with longitudinal stress gradients, which require a changed velocity distribution $u(\xi)$. The reduction in τ_B is specified as a fixed quantity, rather than a variable determined by a velocity-dependent relationship such as (58). Also, the reduction is specified initially over a limited, arbitrary interval of the length of the glacier (so-called "surge nucleus"), rather than over its whole length; the glacier is allowed to slide over this interval (but with no sliding law specified), while outside of the interval it does not slide. A method is proposed for calculating a propagation of the sliding interval with time. In mathematical development the two models do not resemble one another hardly at all, the Budd-McInnes model being much clearer and less cumbersome. The McMeeking-Johnson model was applied to Medvezhi Glacier, a well known surge-type glacier in central Asia, but important details of how this was done were not given – for example, the sliding interval and the τ_B reduction were not specified. However, the greatest shortcoming of this model is its lack of a surge-mechanism model (Section 11.1) – the reduction in τ_B is just specified arbitrarily (and even inconsistently: first as "uniform" [McMeeking and Johnson, 1986, p. 125] and later [p. 128] as a fraction of $\rho g h \alpha$, which in general is non-uniform, being a function of ξ).

The early glacier-flow modeling scheme of Campbell and Rasmussen (1969, 1973) generated surges by arbitrarily making a glacier-wide 10 or 20 fold increase in the sliding-law parameter $\hat{S}$ in (36) (with $p = 1$ to 3, $q = 1$, $\hat{h} = h$) in models previously run to steady state. This of course caused the model glaciers to surge forward rapidly. This type of surge model is as arbitrary as the one discussed above, since no mechanism for occurrence of the large increase in $\hat{S}$ is provided; but it does not involve the radical change in glacier mechanics that is imposed in the Budd-McInnes and McMeeking-Johnson models by reducing τ_B below $\rho g h \alpha$ and throwing mechanical support of the glacier onto longitudinal stress gradients. In fact, in most of the surge models of Campbell and Rasmussen (1969, pp. 984, 985) the equivalent of $\overline{\eta}$ in (19) was set equal to 0, so there were no longitudinal stress gradients; also, models with $\overline{\eta} \neq 0$ differed little from those with $\overline{\eta} = 0$.

Oerlemans and van der Veen (1984, p. 191) were able to get repeating surge-like pulses from a vertically integrated, time-dependent Antarctic Ice Sheet model (Oerlemans, 1982) rather similar to those discussed in Section 9.5, to which they added their hydraulic model described in Section 10.1, consisting of basal melting, basal water transport by (55), and a dependence of sliding on δ as in (57), with S taken to be 0. Each surge is associated with a (rather abrupt) build-up of the water layer thickness, as visualized in the surge concept of Weertman (1969), and with a rather abrupt rise in basal temperature to the melting point, as visualized in the thermal-regulation surge concept of Clarke (1976) except for the abruptness, which cannot be due mainly to a heat conduction process. The model water layer thickness becomes huge, up to 1.5 m, but this could be scaled down by adjusting parameter values such as D_w in (55), whose value was chosen quite arbitrarily (Oerlemans and van der Veen, 1984, p. 108). The model gave surge pulse durations of about 1000 years and repetition times of 3000 to 12000 years, shortening with increasing strength of the melting-sliding feedback (S_1 in (57)). The extra ice volume discharged per surge was up to about half the total volume of the East Antarctic Ice Sheet. This model, contrary to the Budd-McInnes model discussed above, would indicate the possibility of major surges of the Antarctic Ice Sheet of the kind that were visualized by Wilson (1964; see also Flohn, 1974; Bowen, 1980) as the cause of ice-age initiation. Curiously, the West Antarctic Ice Sheet did not participate in the model surges, contrary to the prevailing idea that it is much more susceptible to "collapse" instability than is the East Antarctic Ice Sheet. In the model, the West Antarctic Ice Sheet remained in the steady fast-flow (melting and sliding) mode throughout (see end of Section 10.1).

The surge model of Fowler (1987b) is designed to describe the space-time development of a surge in a glacier system with a triple valued $u_B(\tau_B)$ sliding relation of the type shown in Figure 28c (see Section 11.1). Starting from the point of surge initiation, the model follows the enlargement and propagation of the surging sector in space and time. In these terms it is similar to the McMeeking-Johnson model discussed above, but the physical basis and mathematical development of the two models are very different. In its simplest form the (one-dimensional) Fowler model involves four physical relationships (Fowler, 1987, eqns. (24a)-(24d)): continuity (43) (with b constant), sliding law (32) (with p and q prescribed as noted in Section 8.1), longitudinal mechanical equilibrium (19), and a time-dependent relationship between u_B and $\hat{P}$ that governs transition of the basal water transport mechanism from that of a tunnel system to a linked-cavity system, with consequent change in u_B because the tunnel system is associated with high $\hat{P}$ and the linked-cavity system with low $\hat{P}$ (see Fig. 28c caption). Although a quantitative solution of the four equations is not given, Fowler (1987b, p. 9115-9118) sketches the behavior of the system qualitatively, in terms of $h(\xi)$ as a function of time (Fig. 30) and in terms of the behavior of system points in the (τ_B, u_B) diagram of Figure 28c. The model surge starts at the place where the critical point A in Figure 28c is first reached during the pre-surge build up (Fig. 30a). From there two "activation waves" spread rapidly up and down the glacier (Fig. 30b). At the passage of an activation wave the ice goes into surge and the corresponding system point (τ_B, u_B) in Figure 28c jumps up in u_B from curve DA to curve BC. This represents collapse of the tunnel system and formation of a linked-cavity system as the activation wave passes. The activation waves continue until the part of the glacier with (τ_B, u_B) lying initially along curve DA in Figure 28c becomes included in the surge. The foregoing steps are shown in terms of $h(\xi)$ at three closely successive times in Figure 30a,b,c. Thereafter the surge proceeds on a longer time scale to move ice downglacier as shown in Figure 30d, forming a steep surge front (actually a shock wave) that moves down into ice not previously activated and eventually beyond the initial terminus, producing a terminus advance (Fig. 30e). The activated ice is then progressively reduced in thickness until its (τ_B, u_B) points, descending along curve CB in Figure 28c, all reach critical point B, whereupon the surge terminates and (τ_B, u_B) jumps from B to D (deactivation) (Fig. 30f).

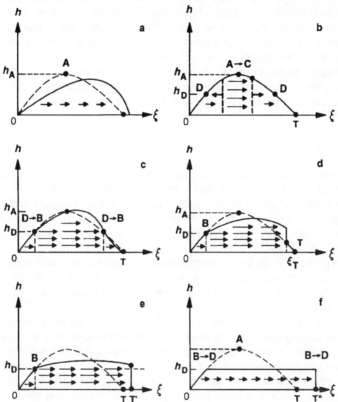

Figure 30. Surge propagation in the model of Fowler (1987b, Figs. 5-8). The sketches show schematically how a surge starts and runs its course, according to the model. The glacier is represented in terms of ice thickness $h(\xi)$ as a function of longitudinal coordinate ξ. Diagrams (a)-(f) represent snapshots at progressively increasing times. In (a) the continuous curve represents the glacier during build-up toward surge, and the dashed curve represents the glacier some time later when it reaches the point A (at critical thickness h_A, corresponding to the shear stress τ_B at point A in Fig. 28c) at which the transition to the surging state starts (A→C in Fig. 28c). In (b), the heavy vertical dashed lines represent the activation waves that spread rapidly (heavy arrows) up and down the glacier from the initiation point A. The fine arrows, long and short, within the ice mass are to suggest the rapid ice motion within the surging zone and the slow ice motion elsewhere. In (c) the activation waves have stopped at points D, with ice thickness h_D corresponding to point D in Fig. 28c, and longitudinal transport of ice in the glacier is now detectable by the change in the ice surface; the dashed curve shows for reference the glacier profile at the moment of surge initiation. In (d) a front (shock wave) has formed at the leading edge of the surge where it propagates forward into ice thinner than the critical thickness h_D. In (e) the front has overrun the initial terminus and the glacier has advanced to T'. In (f) the ice has thinned to the critical thickness h_D along the length of the surging zone and the surge stops (B→D in Fig. 28c), with the terminus at T"

The sketch of the solution given above and its representation in terms of $h(\xi,t)$ in Figure 30 is aided by approximating the four-equation system with a two-equation simplification (Fowler, 1987b, eqn. (28)) that has the form of the kinematic wave equation well known in glacier modeling (Paterson, 1981, p. 244, eqns. (1') and (6)), but with the unusual feature that the ice flux-vs.-thickness relation $Q(h)$ is triple valued, like $u_B(\tau_B)$ in Figure 28c. The approximation involves neglecting the dependence of u_B on surface slope α, as in standard kinematic wave theory (loc. cit.); therefore in sketching the solution in Figure 30 and relating it to Figure 28c the tacit assumption is made that h in Figure 30 can be directly identified with τ_B in Figure 28c, that is, $\tau_B \propto h$.

The surge model represented in Figure 30 is of course highly idealized, and it is not yet clear how many of its interesting features and predictions, such as the "activation waves," correspond to the reality of glacier surges. However, the predicted surge front has been clearly observed (Kamb et al., 1985, Fig. 4).

11.3 MINI-SURGES

In a few temperate glaciers, repeated "mini-surges" have been observed: in these events a rather sudden, short-lived (~1 d) speed up in glacier motion is caused by an abrupt rise in basal water pressure, which propagates somewhat as a shock wave downglacier at a speed of about 0.3 km h^{-1} (Kamb and Engelhardt, 1987). Because of the observed involvement of basal water pressure this phenomenon has the potential of providing further clues to the modeling of the basal water system in relation to basal sliding and possibly surging. One, or possibly two, of the mini-surging glaciers are of surging type, but it is not known whether there is a necessary physical connection between the surge and mini-surge phenomena. Nor is there necessarily any relation between the observed mini-surges and the small pulses in ice sheet models that have been called "minisurges" by McInnes et al. (1986, pp. 43,50).

Fahnestock (1991, Ch. 3) has developed a one-dimensional model of propagating mini-surge pressure waves by including a water-storage component in the hydraulic system (local water storage proportional to P_w) and by postulating that the effective hydraulic conductivity K of the system (ratio of water flux transported to pressure/potential gradient driving it) is a nonlinearly increasing function of P_w. The storage assumption is equivalent to (56), but in a temperate glacier the intraglacial water storage in moulins and other intraglacial porosity probably dominates over basal storage, and the storage assumption is thus equivalent to assuming that the intraglacial porosity in communication with the basal water system is uniform, independent of z or ξ. Figure 31a shows the propagating wave generated by a model with hydraulic conductivity inversely proportional to $\hat{P}$; it depicts, as a function of longitudinal coordinate ξ, the water level h_w of the basal-intraglacial water system (as it would be observed in a borehole) at successive times 4.8 hours apart. (The water level is plotted as height above the bed, thus $h_w = z_w - z_B$ where $z_w(\xi)$ is the hydraulic grade line; the ice thickness $h = z_S - z_B$ is taken constant at 350 m.) The flotation level, where $\hat{P} = 0$, is taken to be at height $h_w = 323$ m. The wave front propagates downglacier with a speed of about 0.35 km h^{-1} (toward the end of the run) and with approximately the observed shape; the observations are made in boreholes and therefore correspond to the model results $h_w(t)$ at fixed ξ, plotted in Figure 31b for $\xi = 13, 14$, ..., 22 km from the model origin.

It is of interest to compare this hydraulic model with one based on water transport by an ice tunnel system of the type discussed in Section 10.2. The tunnel model can generate a wave even remotely similar to the observed waves only if the intraglacial water-storage component, which is not included in the standard tunnel model, is added. Even so, the best wave that can be generated (Fig. 31c,d) does not closely resemble the observed waves, and it moves much more sluggishly (Fahnestock, 1991, p. 48). These results suggest that the basal water conduit system of a glacier that propagates mini-surge waves has features distinctly different from those of a standard tunnel system. The needed dependence of hydraulic conductivity on P_w in the model is perhaps an indication of the presence in the conduit system of some elements of the linked-cavity model described in Section 11.1.

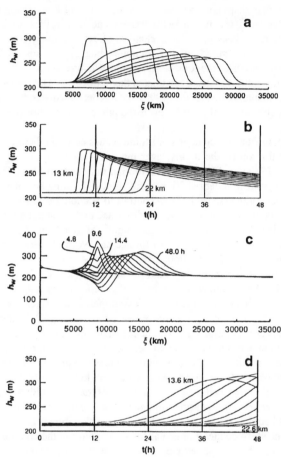

Figure 31. Mini-surge models of basal water pressure propagation, from Fahnestock (1991, Figs. 3.2-3.4). The basal water pressure as a function of ξ is represented by the height of the intraglacial water column or hydraulic grade line above the bed, $h_w(\xi)$. The models follow the evolution of $h_w(\xi)$ with time as a pressure wave propagates through the system, starting from an idealized source that initially injects a spatially Gaussian pulse of half width 120 m into the basal water system over the course of 10 hours. (a) shows a model in which the effective basal hydraulic conductivity is assumed to vary as $(h_\infty - h_w)^{-1}$ where $h_\infty (= 323$ m) can be regarded as corresponding to the basal water pressure at which the sliding is unstable and cavitation increases indefinitely. (b) shows the same model, replotted in terms of curves of $h_w(t)$ at fixed positions $\xi = 13, 14, ..., 22$ km, which corresponds to the mini-surge pressure observations in boreholes (Kamb and Engelhardt, 1987, Figs. 8b, 8f, 10b, 10f, 17). (c) and (d) show results corresponding to (a) and (b) for propagation of a water pressure wave in an optimized tunnel model as described in the text.

Short-term unstable glacier sliding of a kind far more drastic than anything seen in surges or mini-surges occurs in ice avalanches such as the catastrophic fall of the terminus of the Allalin Glacier, Switzerland, in 1965 (Vivian, 1966; Röthlisberger, 1980, p. 144). Modeling of the highly dynamic motions involved in such events would have to go well beyond the capabilities of the types of glacier models discussed here.

12. Role of rock debris; deforming bed models

Despite the beauty of the models of clean ice sliding over bedrock discussed in Section 8, actual glacier beds are usually littered or choked with rock debris (e.g. Engelhardt et al., 1978). From the comminuted sediment content of glacial outflow water it was shown by Kamb et al. (1976) and confirmed by Metcalf (1979) that rock frictional drag plays a significant role in glacier sliding mechanics. This frictional drag is not to be confused with the "solid friction" or "Coulomb friction" often referred to in the sliding theories of Lliboutry cited in Section 8.1; the latter is a sliding friction of clean, melting ice over rock and, unlike rock-rock friction, is not for the most part velocity independent and proportional to $\hat{P}$, except perhaps approximately in some cases as Lliboutry argues (see Section 8.1). If the frictional drag (call it τ_f^*) is between rock clasts embedded in the ice ("tools") and underlying bedrock over which the rock clasts slide, its mechanical effect on sliding involves replacing τ_B in (31)-(40) by $\tau_B - \tau_f^*$. The magnitude of the frictional drag τ_f^* depends on the abundance, size, shape, and frictional properties of the rock clasts, on the effective pressure $\hat{P}$, and on the basal melting rate m (including regelation). It has been treated theoretically by Hallet (1979, 1981) and Shoemaker (1988), and experimentally by Iverson (1990), but has been incorporated only to a small extent into glacier models (Harbor et al., 1988), even though τ_f^* is probably often a significant fraction of τ_B.

If a discrete layer of unconsolidated, unfrozen rock debris or till intervenes between the ice sole and bedrock, sliding of the ice over the top of the debris layer may be very slow, as argued theoretically by Alley (1989b, p. 120; 1989c, p. 132), and basal motion may instead result from shear deformation within the underlying unfrozen debris. This is called the deforming bed mechanism or "soft bed" mechanism of glacier sliding (sensu lato), in contrast to Section 8's "hard bed" sliding mechanism (which should include the role of rock friction discussed in the last paragraph). Boulton (1986) called the deforming bed mechanism a "new paradigm" of glacier motion and argued that it was the main flow mechanism for large parts of the ice-age North American and Fennoscandian Ice Sheets (Boulton and Jones, 1979; Boulton et al., 1985).

A deforming bed model involves the glacier modeling features already discussed, with the following changes and additions: 1. The basal sliding law (Section 8) is replaced by a u_B-vs.-τ_B relation (call it for short a "soft sliding law") based on the rheological properties and thickness h_B of the bed material and including a dependence on basal effective pressure $\hat{P} = P_1 - P'_w$. 2. Features of the deforming bed that affect the transport of basal water and hence water pressure P_w (Section 10) need to be incorporated. 3. A relation needs to be specified between the basal water pressure P_w and the pore pressure P'_w in the water-saturated basal debris, which is the water-pressure quantity that controls the debris' mechanical properties (item 1); in steady state or in long-term transients, P_w and P'_w are the same. 4. Since the deforming-bed material is in (partial) transport with the ice, it needs its own continuity (mass conservation) condition similar to but separate from the ice continuity condition (42) or (43). 5. A bedrock erosion rate at which fresh rock debris is generated and added to the deforming bed layer needs to be incorporated as the source term in the continuity condition of item 4, analogous to a in (42). 6. As is sometimes also the case with basal sliding, the contribution of bed deformation to ice motion can under some conditions (particularly low $\hat{P}$) be so much larger than the contribution (16) from internal ice deformation that the latter can be neglected. Knowledge of soft sliding laws (item 1) and bedrock erosion laws (item 5) is very limited, and of item 2 practically nonexistent, so that the modeling of glaciers that move by the deforming bed mechanism is necessarily in a primitive state.

Application of the deforming bed mechanism in glacier modeling has come from an unexpected quarter – the Antarctic Ice Streams, discussed already in Sections 8 and 10. The cause of their rapid motion (~500 m a^{-1}, contrasting with ~10 m a^{-1} in the adjacent ice sheet outside the ice streams) was proposed to be the deformation of a meters-thick layer of soft, water-saturated till underlying the ice, and a model of this flow system was generated (Alley et al., 1987a,b). The original (1987b) model included all of the items in the last paragraph, except item 2 and the $\hat{P}$ dependence in item 1, which were added in later versions of the model (Alley, 1989a; Alley et al., 1989). The model assumes a linearly (or nearly linearly) viscous rheology for the till, with a viscosity η_B that is allowed to be spatially variable, similar to the variable sliding coefficient $\hat{S}$ in the Lingle and Brown (1987) aquifer model discussed in Section 10.1. A 300-km length of Ice Stream B (where the basal till layer was discovered) is modeled one-dimensionally, both in steady state and with time dependence. By prescribing the steady-state ice stream configuration as z_B = constant, α = constant, $h(\xi)$ = linear function of ξ, a = constant, hu_B = linear function of ξ, the model creates a prescription for $\tau_B(\xi)$, by (14), and for $u_B(\xi)$, which was set to the observed surface velocity 1.2 m d^{-1} at Up B (at $\xi = 100$ km). This then requires a till viscosity

$$\eta_B(\xi) = \tau_B(\xi)h_B(\xi)/u_B(\xi) \tag{61}$$

to produce the prescribed $u_B(\xi)$, by uniform shear through a till layer of thickness h_B. Model values of η_B can be calculated from (61) once the till thickness $h_B(\xi)$ is known. In the (1989) model, h_B is simply prescribed to be a constant 6 m, while in the (1987b) model a calculation of $h_B(\xi)$ was made from the till continuity equation (item 3 above; steady state assumed) with $h_B(100$ km$) = 6$ m and with the following assumed linear dependence of bedrock erosion rate a_B (item 5) on the till shear strain rate u_B/h_B:

$$a_B = \kappa_B u_B/h_B \tag{62}$$

where κ_B is a constant whose value is not known a priori. By adjusting κ_B it was possible to obtain an $h_B(\xi)$ that was nearly constant (6 m), independent of ξ (Alley et al., 1987c, Fig. 2, bottom panel, curve 4). The resulting $\eta_B(\xi)$ from (61) decreases roughly linearly from about 3×10^{-3} bar a at $\xi = 0$ to about 1×10^{-3} bar a at $\xi = 300$ km (ibid., top panel). The decrease in η_B with ξ is from (61) required by the fact that u_B increases downstream while τ_B decreases (Fig. 26a,b). The model η_B values are somewhat smaller than those obtained from (61) for two other glaciers cited (ibid., Table 2 with a typographical error in line 4), because these other glaciers move more slowly, at higher τ_B, with till thicknesses that are comparable or somewhat smaller. The η_B values are much smaller than typical effective viscosities for ice, $\eta \sim 1$-10 bar a.

The decrease in η_B with distance ξ is attributed to a decrease in $\hat{P}$ as the flotation condition at the grounding line is approached (Alley et al., 1987b, p. 8937). The effect of $\hat{P}$ is incorporated into the (1989) model (Alley, 1989b, p. 123) by introducing a $\hat{P}$ dependence via a till flow law of the type suggested by Boulton and Hindmarsh (1987, p. 9063): written as a "soft sliding" law by multiplying by h_B on the assumption of uniform shear in the till, it is

$$u_B = \omega \hat{P}^{-q} \tau_B^n h_B \tag{63}$$

in which ω, n, and q are till flow-law constants[§]. (63) is of course analogous to the sliding law (32). In addition to (63), Alley et al. (1989, p. 132) introduce a hydraulic model from which they can calculate $\hat{P}(\xi)$; it is found to be a decreasing function of ξ (see below). They then find that with the choice $n = 1$, $q = 2$ they can reproduce the prescribed $u_B(\xi)$ by (63) with a nearly constant value of ω, suggesting an internally consistent model of water-pressure-dependent soft-bed sliding. They tested other values $n = 1.33$ and $q = 1.8, 3, 4, 5$ without finding as nearly constant $\omega(\xi)$. It is interesting for comparison that in Lingle and Brown's (1987) aquifer model of Ice Stream B, discussed in Section 10.1, the sliding coefficient $\hat{S}(\xi)$ would have been nearly constant if $q = 2$ had been used instead of $q = 1$. This consistency ($q = 2$ favored in both models) is curious, because

---

[§]Boulton and Hindmarsh (1987) also suggested an alternative flow law of the form (63) but with τ_B replaced by $\tau_B - \tau_f$, where τ_f is shear strength (discussed later); this law, with $\tau_f \neq 0$, has not been used in modeling, however.

although both models are based on a linear sliding law ($p = 1$), the effective pressure $\hat{P}(\xi)$ is calculated on a very different basis in the two models (see below and Section 10.1).

Crucial to the above treatment is the Alley et al. (1989) hydraulic model, which was mentioned in Section 10.1 as an example of an "incomplete gap conduit" model. It is made up of the following elements (Alley et al., 1989, p. 132). Basal meltwater from (52), giving a water flux density

$$\Phi_w(\xi) = \int_0^\xi m\,d\xi \tag{64}$$

(on the assumption of constant flowband width b), is transported via the incomplete ice-bed gap according to (54), corrected for the areal fraction of the bed occupied by the gap, here designated φ. (Actually, such a correction was not included in the model of Alley et al. [1989, eqn. (1)]. The simplest correction is to multiply the middle and right-hand sides of (54) by φ, which represents a gap system in which all of the gap areas are accessible to and used for water transport.) In this way, the gap thickness $\delta(\xi)$ is specified from (54) and (64). It is then linked to $\varphi(\xi)$ by the following relation

$$\varphi = 0.7 + 0.1\log_{10}\delta \tag{65}$$

(with δ in mm) which is deduced by Alley (1989b, p. 119) on the assumption that the gap area is the same as the area of voids plus "fine" particles in a random slice through the till, "fine" particles being defined as all particles of grain size less than the gap width δ. This could be a valid assumption if all of the "fine" particles from the surface of the till presented to the ice sole were somehow eroded away down to a depth δ, but that has not been demonstrated. Finally, a relation between φ and $\hat{P}$ is adopted (Alley et al., 1989, eqn. (6)):

$$\hat{P} = \beta\tau_B/\varphi \tag{66}$$

with β a "geometric factor" that is treated as a constant (≈ 2) in the model. (This β is not to be confused with the bed slope, for which β is used elsewhere in this review.)

The chain of equations (52), (64), (54) solved for δ, (65), and (66) makes $\hat{P}(\xi)$ a decreasing function of ξ, as needed to explain the decrease in $\eta_B(\xi)$ with ξ and to give near constancy of ω. It does this without any involvement of a pressure constraint based on approach to the grounding line ($\hat{P}{\to}0$), which is seemingly at odds with the original interpretation of the decreasing $\eta_B(\xi)$ mentioned above (Alley et al., 1987c, p. 8937).

The physical basis for (65) and, as the following discussion shows, for (66) seems rather shaky, and it is not clear that they represent the actual hydraulic model, even though the overall model performs satisfactorily as rated by near constancy of ω. It seems at first sight intuitively reasonable that the gap area φ should increase with P_w, as (66) requires, and that it should increase with τ_B if it is generated by a basal cavitation process (at least below point A in Fig. 28a). Such a process is visualized in the derivation of (66) by Alley (1989a, p. 115): the ice sole rests on a bump on a horizontal bed (one bump per "unit" area), and "the bump supports the horizontal shear stress τ_B on its [ice-bed contact area;]... [the] shear stress causes an excess vertical force on the bump, which causes ice flow over the bump..." This is a partial description of the connection in basal sliding between τ_B and the stoss-side pressure on bed protuberances. Alley's model can thus be visualized as shown in Figure 32a. The (fractional) contact area is $s = 1 - \varphi$, and the vertical stress across the contact area is here called $\langle P_c \rangle$ (called P_b by Alley), the brackets indicating that $\langle P_c \rangle$ is a horizontal average over the contact area. Because the detailed shape of the bump or contact area does not come

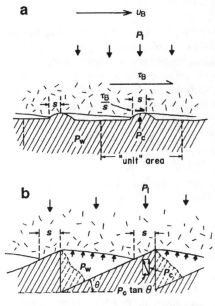

Figure 32. Gap conduit models. (a) is the model of Alley et al. (1989) as I understand it from the description by Alley (1989, p . 132). The ice (patterned), moving to the right, rests on bumps in the bed (surface of the cross-hatched region); the water filled gap conduit (the narrow blank area between the ice and the bed in the figure) is kept open by water pressure P_w in the gap and by a contact pressure P_c in the area of ice-bed contact s. These pressures support the ice overburden pressure P_I. ($<P_c>$ in the text is here abbreviated to simply P_c to save space in the drawing.) The basal shear stress τ_B is supported by a concentrated shear stress τ_B/s across each bump, which is related to P_c and the geometrical shape of the bump. (b) shows a special case of (a) in which the upstream-facing surfaces of the bumps are inclined to the horizontal at a constant angle θ, notation is the same as in (a). The force resolution diagram directly below the contact area shows an unlabeled force vector normal to the inclined, frictionless stoss surface being resolved into a vertical component P_c and a horizontal component $P_c \tan \theta$.

into the derivation, we can for simplicity regard the drawing in Figure 32a as a longitudinal cross section of a two-dimensional bed form, and therefore identify the area s as the longitudinal interval of ice-bed contact. The equilibrium of vertical forces across the ice sole (Alley, 1989a, eqn. (18)) is

$$P_I = s<P_c> + (1 - s)P_w \tag{67}$$

The expected condition for equilibrium of horizontal forces across the sole is not written down by Alley; instead, the following calculation of the "excess vertical [stress] on the bump" is given:

$$<P_c> - P_I = \frac{\beta \tau_B}{s} \tag{68}$$

"where β is the ratio of the excess vertical stress to the shear stress on the areas of the bump" (Alley, 1989a, eqn. (20)). By "excess" is evidently meant "in excess of P_I"; the "shear stress on the area s of the bump" is taken to be τ_B/s because the water in the ice-bed separation gap supports no shear stress. Now (68) is really only a statement of the definition of β just quoted, as is seen by rewriting it in the form

$$\beta = \frac{<P_c> - P_I}{\tau_B/s} \tag{69}$$

It adds nothing physically new, unless we can assume (as is done in the model) that β is a constant, but no justification for this assumption is given. It is incorrect at least in the simple special case where the contact area s is inclined upstream at a constant slope θ, as shown in Figure 32b. In this case the equilibrium of horizontal forces across the ice sole can be written exactly:

$$\tau_B = s<P_c> \tan \theta - sP_w \tan \theta \tag{70}$$

which differs from (68) with constant β. If we set $\beta = (\tan \theta)^{-1}$ in (68) we almost get agreement with (70), but the term $-sP_w \tan \theta$ on the right side of (70) is replaced by $-sP_1 \tan \theta$ in (68). In writing (70) as well as (67) we have assumed that the actual ice-bed contact surface is shear-stress free, as is normally assumed in basal sliding models in the absence of rock friction (Section 8.1); Alley (1989a, p. 115) does not address this point explicitly but implicitly abides by the normal assumption in writing (67). Now if $<P_c>$ is eliminated between (67) and (68) (with $s = 1 - \varphi$), as Alley does, one gets (66), whereas if it is eliminated between (67) and (70) we get

$$P_1 - P_w \equiv \hat{P} = \tau_B/\tan \theta \tag{71}$$

in which the gap area φ does not appear. Since (71) is exact for the special case in Figure 32b, we can calculate β in this case by equating $\hat{P}$ in (66) and (71):

$$\beta = \varphi/\tan \theta \tag{72}$$

this shows that at least in this simple special case β is not a constant, independent of φ, as assumed in the Alley et al. (1989) model.

From (71) and the underlying mechanical equilibrium it follows, somewhat contrary to the "at-first-sight" intuition stated at the beginning of the last paragraph, that for a basal cavitation model like Figure 32b the cavitation area or gap area φ is a function of $\hat{P}$ only in the following somewhat peculiar way: if $\hat{P} > \tau_B/\tan \theta$, then $\varphi = 0$; if $\hat{P} = \tau_B/\tan \theta$, then φ can assume any value from 0 to 1; if $\hat{P} < \tau_B/\tan \theta$, the system cannot be in mechanical equilibrium. These conclusions and the basal sliding model (Fig. 32b) on which they are based were given by Iken (1981, p. 409), who attributes them originally to C.F. Raymond and H. Röthlisberger. They are reinforced by Humphrey's (1987) treatment of the three-dimensional "tesselated bed" model of basal sliding. The appeal to these results by Alley (1989a, p. 115) in defense of the need to set a lower limit "$\beta'\tau_B$" on $\hat{P}$, in order to prevent the unacceptable behavior $\varphi > 1$ that is required by (66) when $\hat{P} < \beta\tau_B$, represents somewhat of a misapplication of the above conclusions, which are not in accord with the type of φ-dependence of P_w given by (66).

In addition to all this, the mental picture that one is likely to have of the Alley et al. (1989) basal hydraulic model, involving a thin gap between ice and soft till with little or no sliding across the gap, is very different from Iken's (1981) and Humphrey's (1987) picture of rapid basal sliding over a hard "tesselated" or "staircase" bed with large lee-side cavities. This makes Alley's (1989a, p. 115) derivation of (66) based on an Iken-like model (Fig. 32a, with its similarity to Fig. 32b) seem physically inappropriate.

As the above discussion suggests, the proper conduit model for subpolar glaciers moving over deformable beds is really up in the air at this time, although efforts to consider the problem have been made (Hughes, 1981, p. 222-231; Shoemaker, 1986; Boulton and Hindmarsh, 1987, p. 907; Alley, 1989a, p. 112; Kamb, 1991, p. 16,589; Fowler and Walder, 1992). Particularly in need of clarification is the possible role of till erosion by moving basal water in development of the conduit system (cutting of channels; piping).

The ice stream model of MacAyeal (1989) is an areal, vertically integrated, two-dimensional model that includes a 400-km length of Ice Stream B, from margin to margin, and extending about 50 km beyond the grounding line; it includes also much of the downstream and ungrounded part of Ice Stream A in order to reach the southern margin of the combined A+B flow system (see Fig. 19). Like the Alley et al. (1987b) model it assumes linearly viscous rheology for the till. However, it lumps the viscosity η_B and till thickness h_B into a single parameter (η_B/h_B), and uses (61) in the form $u_B = (h_B/\eta_B)\tau_B$, which has the form of a sliding law, like (31) with $p = 1$. Thus the model makes no operational distinction between bed deformation and (enhanced) basal sliding as the ice stream mechanism, except to the extent that the stress nonlinearity (p) may distinguish them, which is problematical (see below). (η_B/h_B) is regarded as the "major adjustable parameter" and is allowed to vary spatially, as in the Alley et al. (1987b) model. But no $\overset{?}{P}$ dependence in the sliding law is included and no hydraulic model introduced. Also, the model omits till continuity and bedrock erosion rate, which was possible because of h_B being lumped into the parameter (η_B/h_B). The ice shear flow contribution to the velocity (16) was neglected on the assumption that η_B is much smaller than the effective ice viscosity, as in Alley et al. (1987b).

In relation to the two-dimensional models of Sections 5, 8, and 9, an unusual feature of the MacAyeal (1989) ice stream model is inclusion of longitudinal and lateral stress coupling (Section 6) by using the two-dimensional equivalent of (19):

$$\tau_x = \rho g h \alpha_x + \frac{\partial}{\partial x}\left[\overline{\eta}h(4\frac{\partial u}{\partial x} + 2\frac{\partial v}{\partial y})\right] + \frac{\partial}{\partial y}\left[\overline{\eta}h\left(\frac{\partial u}{\partial y} + \frac{\partial v}{\partial x}\right)\right] \tag{73}$$

$$\tau_y = \rho g h \alpha_y + \frac{\partial}{\partial y}\left[\overline{\eta}h(4\frac{\partial v}{\partial y} + 2\frac{\partial u}{\partial x})\right] + \frac{\partial}{\partial x}\left[\overline{\eta}h\left(\frac{\partial u}{\partial y} + \frac{\partial v}{\partial x}\right)\right] \tag{74}$$

where τ_x and τ_y are the x and y components of the basal shear stress vector τ_B, and $\alpha_x = -\partial z_S/\partial x$, $\alpha_y = -\partial z_S/\partial y$; $\overline{\eta}$ is the vertically averaged effective ice viscosity, from (2); and u and v are the x and y components of ice flow velocity (assumed independent of z). (73) and (74) are based on the stress equilibrium equations (8) in the horizontal (x,y) plane, with vertical integration and the approximation $\tau_{zz} = -\rho(z_S - z)$; a detailed derivation is given by MacAyeal (1989, Appendix A). The stress coupling terms are all terms on the right side of (73) and (74) other than the "driving stress" terms $\rho g h \alpha_x$ and $\rho g h \alpha_y$. These equations, or the simpler (19) in the one-dimensional case, can be used both for an ice sheet, where the stress coupling terms are small and $\tau_B \cong \rho g h \alpha$ as in (14), and in a floating ice shelf, where $\tau_B = 0$ and the stress coupling terms become dominant, of order $\rho g h \alpha$ (but α is of course very small for an ice shelf). (73) and (74) are thus able to describe (i) the free or confined settling/spreading of an ice shelf as a function of ice thickness h and (ii) the transmission through an ice shelf of backpressure from pinning points (Section 9.4.2); the use of (73) and (74) in ice shelf models is described by Huybrechts (1992, p. 97). MacAyeal's (1989) use of (73) and (74) in the ice stream model is predicated on the idea that ice streams, with their abnormally low τ_B (~0.2 bar, compared to ~0.5 to 1 bar for normal glaciers and ice sheets) and their rapid, seemingly little-impeded motions are intermediate in mechanical character between ice shelves and ice sheets.

In the ice stream model (MacAyeal, 1989), τ_x and τ_y in (73) and (74) are replaced by (η_B/h_B)u and (η_B/h_B)v, from the assumed soft sliding law, and $\overline{\eta}$ is calculated from (2) with a constant value $\overline{B}$, vertically averaged for a chosen temperature profile $T(z)$. (Thermal modeling as in Chapter 7 was not included.) In this form (73) and (74) are solved for the velocity field $u(x,y)$, $v(x,y)$ by the finite element method. Observed velocities were placed as boundary conditions on the upstream boundary of the model area, where Ice Streams A and B enter; on the lateral margins the velocity was set to 0. On the downstream boundary a (nonuniform) backpressure condition, the three dimensional equivalent of (51) on $\overline{\tau}'_{\xi\xi}$, was placed on the basis of σ_B values calculated from field measurements (see Section 9.4.2). (The location of these boundaries is shown by the dotted and heavy-dashed lines bounding much of Ice Stream B and the lower part of Ice Stream A in Figure 19.) The parameter (η_B/h_B) was adjusted by trial and error for a best fit between the modeled and observed velocities; in this adjustment, (η_B/h_B) was assumed to be laterally uniform, and was adjusted as a function of longitudinal distance ξ along the length of Ice Stream B.

Some of the model results are shown in Figures 33 and 34. Figure 33a and b give observed and modeled velocity vectors in map view, and Figure 34 shows input and output variables as a function of ξ along the length of Ice Stream B. Figure 34a shows the ice surface and bed configuration,

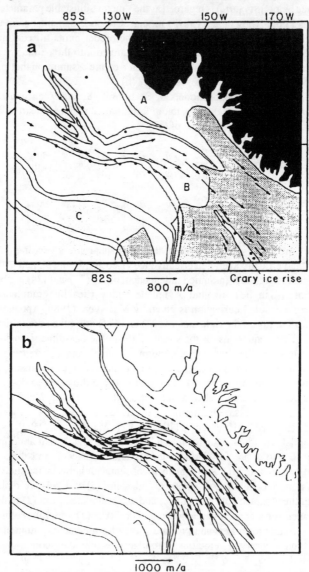

Figure 33. Two-dimensional flow model of Ice Stream B (and the lowest part of A) by MacAyeal (1989a, Figs. 2a, 6). In (a) the ice streams are labeled A, B, C, the black area in the upper right is the Transantarctic Mountains and East Antarctic Ice Sheet, and the gray area in the lower right is the (ungrounded) Ross Ice Shelf. The vectors in (a) are actual flow velocities observed by Whillans et al. (1987) and Bindschadler et al. (1987). (b) shows the model results in terms of calculated flow vectors at the grid points. The velocity vector scale is indicated by the scale arrow (800 m/a and 1000 m/a) at the bottom of each figure.

which, like the Alley et al. (1987b) model, appears rather idealized (straight lines) but it contains a noteworthy feature absent from the (1987b) model, namely, the "ice plain" of very low α downstream from $\xi = 170$ km. This explains the abrupt drop in "driving stress" $\rho g h \alpha$ at $\xi = 170$ km in Figure 34c.

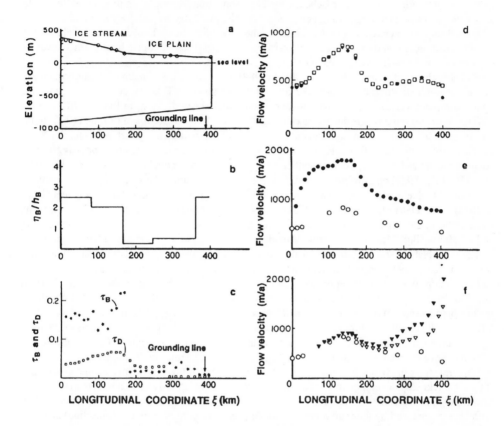

Figure 34. Results of the two-dimensional model of Ice Stream B, West Antarctica, by MacAyeal (1989a, Figs. 3, 5, 10, 7a, 11). (a) shows the input surface and bed profiles along the centerline of the ice stream. (b) shows the model parameter value η_B/h_B as a function of longitudinal coordinate ξ along the ice stream. (c) gives model values of the basal shear stress τ_B (open squares) and the "driving stress" $\tau_D = \rho g h \alpha$ (stars) along the ice stream. (d) compares flow velocities for the best-fit model (open squares) with observed velocities along the ice stream (solid circles). (e) compares the observed velocities (large open circles) with model values obtained by altering the best-fit model by reducing the flow "hardness" B in the marginal shear zones to one tenth its (constant) value in the rest of the model. In (f) the backpressure at the lower boundary of the modeled area, near the grounding line, has been reduced from an average of about 2 bar in the best-fit model to about 1 bar in the model represented by the open triangles; this reduction is based on removing the calculated backpressure due to the Crary Ice Rise (narrow blank area near the lower right corner of Fig. 33a). Further reduction to 0 backpressure gives the solid triangles. The observed velocities are again shown with large open circles.

The (rather coarsely resolved) viscosity parameter values (η_B/h_B) in Figure 34b show a progressive drop from $\xi = 0$ to $\xi \sim 200$ km, as in the (1987b) model, but thereafter a rise, which the (1987b) model did not have. The largest drop in (η_B/h_B), at $\xi \approx 170$ km, is directly associated with the abrupt drop in α on entering the ice plain, and is evidently needed by the model to maintain the high u_B under a large abrupt drop in τ_B. The idea that $\eta_B(\xi)$ decreases downstream because of a decrease in $\hat{P}$ as the grounding line is approached, which had some appeal in relation to the (1987b) model as discussed above, is less appealing here, because of the localized drop at $\xi \approx 170$ km and because of the subsequent increase in (η_B/h_B) closer to the grounding line. However, in the absence of reliable information on till thickness, any variations in (η_B/h_B) can of course be explained away by corresponding reciprocal variations in $h_B(\xi)$.

Probably the most noteworthy result of the (1989) model is its assessment of the importance of lateral/longitudinal stress coupling in the flow of the ice stream. This can be seen in Figure 34c, which compares values of $\rho g h \alpha$ with values of the basal shear stress τ_B in the best-fit model; a difference between the two is due to the effect of the stress coupling terms in (73) and (74). In the ice stream above the ice plain, τ_B is only about $\frac{1}{4}$ to $\frac{1}{2}$ of $\rho g h \alpha$, indicating that in this model, stress gradients (probably lateral) dominate over basal shear stress in mechanically supporting the ice stream against the down-slope component of gravity. The contrary conclusion was reached by Alley et al. (1987a, p. 8922) on a different, less definite basis. The difference between the two assessments is presumably due in effect to a relatively large $\overline{B}$ value assumed for ice in (2) by MacAyeal (1989, p. 4075), i.e. a relative large $\overline{\eta}$ value in (73) and (74). It is curious that in the best model, τ_B is about twice $\rho g h \alpha$ in the upper half of the ice plain, and then suddenly drops to a small fraction of $\rho g h \alpha$ at $\xi = 280$ km (Fig. 34c); no clear cause is evident for these seemingly anomalous flow features.

The role of lateral stress gradients can also be seen in the results of a variant of the model in which $\overline{B}$ near the ice-stream margins was reduced to a tenth the value assumed in the best-fit model: this caused an approximate doubling of flow velocity (Fig. 34e) over the best-fit model (Fig. 34d). A doubling would be expected if basal shear stress and lateral stress gradients contributed equally to the mechanical support of the ice stream in the best-fit model ($\tau_B = \frac{1}{2}\rho g h \alpha$), because reducing $\overline{B}$ by a factor of 10 essentially eliminates mechanical support of the ice stream by the margins, forcing $\tau_B = \rho g h \alpha$ and thus requiring the flow velocity to double, since $u_B = (h_B/\eta_B)\tau_B$. The doubling could of course have been compensated by increasing (η_B/h_B) by a factor 2 in the model.

If lateral stress gradients dominate the mechanical support of the ice streams against gravity and if the ice thickness and temperature are approximately constant laterally as assumed, then transverse profiles of flow velocity should have the fourth-order-parabola form given by integration of (15) with ($z_S - z$) replaced by the transverse coordinate (say, y, measured from the centerline) and with velocity set to 0 at the margins. The model profiles (MacAyeal, 1989, Fig. 7b,c) are qualitatively in accord with this. However, the observed profiles (Bindschadler, 1987, Fig. 7; K. Echelmeyer, pers. com., 1989) are much more like "plug flow," with nearly constant velocity across most of the ice-stream width and a narrow (~4-9 km), sharply defined shear zone of high strain rate at the margins. This may indicate a much increased enhancement factor $\ni$ in the marginal shear zone, as suggested originally by Hughes (1977) and as modeled by reducing $\overline{B}$ in the marginal zones, as noted above. (Transverse profiles of velocity for this model were not given, however.) This suggests that lateral stress gradients do not dominate in the actual ice sheet.

Another noteworthy result of the MacAyeal (1989) model is the influence of backpressure from the ice shelf: this was brought out by calculating a model with σ_b reduced to 1.2 bar, about half its original value, and another with σ_b set to 0. These reductions in σ_b cause a large increase in the ice flow upstream (Fig. 34f); the effect drops off upstream with an e-folding distance of about 50 km and is detectable as far as about the head of the ice plain, some 200 km upstream from the grounding line. This result seems to indicate a much wider ice-shelf/ice-sheet transition zone than indicated by the models discussed in Section 9.4.3. These different indications can be reconciled to some extent by the longitudinal coupling model of Section 6: when τ_B is very small, as in the ice plain where $\tau_B \sim 0.01$ bar, the longitudinal coupling length l from (22) (with $\tau_0 = \tau_B$) becomes long, i.e. ~50 km for the ice plain. This represents a situation approaching that of a free floating ice shelf, where $\tau_B = 0$ and $l \rightarrow \infty$: horizontal forces are transmitted infinitely far. An additional aspect is the effect on l of the lateral stress gradients: if they are dominant over τ_B in (73) and (74), then l becomes scaled to the half width w of the ice stream, that is, h in (22) is replaced by w, and τ_0 is the marginal shear stress, ~1 bar; (22) then indicates $l \sim 150$ km.

As pointed out by MacAyeal (1989, p. 4080), the substantial, long-range stress coupling effects seen in the above model provide a means by which climatically-induced changes at the grounding line can be transmitted upstream instantaneously, instead of only on the long time scale of time-dependent models without longitudinal stress coupling, in which longitudinal interactions occur only via changes in ice thickness, i.e., by kinematic waves (Paterson, 1981, p. 244).

The assumption of linear rheology for the till, which is made in both of the foregoing ice stream models, has been questioned by Kamb (1991) on the following grounds. A priori one would expect till to have mechanical properties like those of soils, as codified in the discipline of soil mechanics: these are "treiboplastic" materials, with a plastic failure strength (shear strength τ_f) that is frictionally based and effective-pressure ($\hat{P}$) controlled[§]. Within the framework of soil mechanics, stress-dependent quasiviscous flow enters as a slight strain-rate dependence of the yield strength τ_f. In terms of the type of rheology described by (1)-(6) this corresponds to an extremely high stress nonlinearity, $n \sim 100$. Core samples of till from the base of Ice Stream B, obtained by drilling to the bottom (Engelhardt et al., 1990), show this type of mechanical behavior in laboratory tests[*]. The $\hat{P}$ dependence of τ_f, when combined with the strain-rate dependence on an empirical basis, translates into a soft sliding law like that in (63) with $q = n$ (Kamb, 1991, p. 16,588).

When incorporated into a deforming-bed model of ice stream motion the highly nonlinear till flow law discussed above makes a mechanical system very different in performance and stability from models based on linear rheology. This has been examined in a one-dimensional model (Kamb, 1991, p. 16,589), which explores the stability of a deforming bed system that includes nonlinear rheology together with the feedback mechanism provided by basal melting and water transport: If a perturbation in such a system causes an increase in the basal water pressure, so that the till pore pressure increases, the till deformation rate will increase; this will increase the basal melting rate, which will in turn cause a further increase in the water pressure, and so on. The magnitude of the further increase can be limited by enlargement of the water-transport conduits in response to the pressure perturbation. For different values of the system parameters, the system can be either stable

---

[§]Jones (1979) suggested that the till immediately below the ice sole could become a slurry, which would have linear rheology; however, the viscosity of slurries is so low that a slurry layer would have to be extremely thin (~10 μm thick) in order to support the ice stream at the observed flow velocity. The bulk of the till layer, estimated to be ~6 m thick, cannot therefore be a slurry and must instead behave mechanically as a soil. If the ice moves by shear of a slurry layer ~10 μm thick, the motion would look macroscopically like basal sliding, quite different from what is visualized in most models of the deforming bed mechanism.

[*]The modified flow law in which τ_B in (63) is replaced by $(\tau_B - \tau_f)$, mentioned above in a footnote in connection with (63), can represent an arbitrarily large flow nonlinearity if $\tau_f \neq 0$, even when $n = 1$, which (with $q = 0$) is the well-known case of the hypothetical Bingham solid/fluid. However, the equivalent nonlinearity n, expressed as $n = \partial \ln u_B / \partial \ln \tau_B$, decreases rapidly with τ_B. There is no observational indication that for till a flow law of this type is preferable to unmodified (63). In soil mechanics it is standard practice to state the strain-rate dependence of the residual strength of clay rich soils in terms of Σ, the percentage change in strength per decade change in strain rate, which for constant Σ corresponds to soft sliding law (63) with $n = (100 \ln 10)/\Sigma$ (for references see Kamb, 1981, p. 16,586).

or unstable under such a perturbation; the stable/unstable behavior can manifest itself almost immediately, before appreciable changes in ice thickness h or surface slope α have had time to develop via (43).

To study this type of behavior the model is formulated as follows. To a nonlinear soft sliding law (63) (with $q = n$) is coupled a hydraulic model consisting of basal melting rate (52) and water transport (54)+(64) plus a simple heuristic relationship between gap-conduit width δ and effective pressure $\hat{P} = P_{\mathrm{I}} - P_{\mathrm{w}}$

$$\delta = c\hat{P}^{-r} \tag{75}$$

(with r and c constants), taking the place of (65)+(66) in the Alley et al. (1989) hydraulic model discussed above. The system is completed with the continuity condition for basal water storage ($\varphi\delta$ per unit area of the bed), analogous to (43) for ice thickness h. (Continuity conditions for ice h and till h_{B} are not included in the model because of the "immediacy" of the response investigated.) A Liapunov stability analysis is carried out by linearizing the foregoing relationships for a small perturbation ΔP_{w} from a steady state (datum state), with an assumed sinusoidal variation of ΔP_{w} as a function of longitudinal coordinate ξ, with wavelength λ. The resulting linearized, time-dependent system has exponential solutions $\Delta P_{\mathrm{w}} \propto \exp(\upsilon t)$, the coefficient υ being

$$\upsilon = t_0^{-1}\left(\frac{n-m}{r} - 1\right) \tag{76}$$

where the time constant t_0 and reference exponent m can be calculated from the system parameters (Kamb, 1991, eqns. (32) and (33)). For parameters appropriate to Ice Stream B, and $\lambda \sim 30$ km, the model results are $t_0 \sim 14$ d and $m \sim 13$. The system responds unstably ($\upsilon > 0$) if $n > m + r$. The "immediacy" of the response is indicated by the short time constant t_0. There is no observational information on the exponent r in (70), but from theoretical models (such as (65)+(66)) it can be estimated to lie in the wide range $1 \lesssim r \lesssim 12$ (Kamb, 1991, p. 16,591). Thus it seems reasonably safe, despite the various parameter uncertainties, to conclude that the system will be unstable for $n \gtrsim 20$.

Since from soil mechanics and geotechnical tests on the till we estimate $n \sim 100$, as discussed earlier, it seems to be a reasonably firm conclusion that the deforming bed mechanism of ice stream motion, at least as represented by the foregoing model, is unstable. The possible significance of this rather drastic conclusion has been discussed briefly (Kamb, 1991, p. 16,592). The further course of development of the system response, going beyond the "immediate" response analyzed in the above model and involving the possibly stabilizing effects of ice continuity and temperature distribution, may perhaps be surmised by analogy with the model of Oerlemans and van der Veen (1984, p.191), which leads to large-scale self-excited oscillations of the Antarctic Ice Sheet as discussed in Section 11.2 above.

Acknowledgements. I thank Bill Budd, James Fastook, Terry Hughes, Craig Lingle, Doug MacAyeal, Uve Radok, Charles Raymond, C.J. van der Veen, and Ed Waddington for providing glacier modeling material and suggestions in the preparation of this review. I am particularly grateful to Dick Jenssen and Kia Mavrakis for preparing and allowing me to use here the Antarctic model results in Figures 8, 14, 16, 17, and 25. I thank Kerry Etheridge for the typesetting and Jan Mayne for the illustration preparation. I am also grateful to David Stone for his editorial patience.

References

Alley, R.B. (1989a) Water-pressure coupling of sliding and bed deformation: I. Water system: *J. Glaciol.* **35**, 108-118.

Alley, R.B. (1989b) Water-pressure coupling of sliding and bed deformation: II. Velocity-depth profiles: *J. Glaciol.* **35**, 119-129.

Alley, R.B. (1989c) Water-pressure coupling of sliding and bed deformation: III. Application to Ice Stream B, Antarctica: *J. Glaciol.* **35**, 130-139.

Alley, R.B. (1990) West Antarctic collapse How likely? *Episodes* **13**, 231-238.

Alley, R.B. (1992) Flow-law hypotheses for ice-sheet modeling: *J. Glaciol.* **38** (129), 245-256.

Alley, R.B., and Whillans, I.M. (1984) Response of the East Antarctic Ice Sheet to sea-level rise: *J. Geophys. Res.* **89**, 6487-6493.

Alley, R.B., Blankenship, D.D., Bentley, C.R., and Rooney, S.T. (1987a) Till beneath Ice Stream B. 3. Till deformation: Evidence and implications: *J. Geophys. Res.* **92**, 8921-8929.

Alley, R.B., Blankenship, D.D., Rooney, S.T., and Bentley, C.R. (1987b) Till beneath Ice Stream B. 4. A coupled ice-till flow model, *J. Geophys. Res.* **92**, 8931-8940.

Alley, R.B., Blankenship, D.D., Rooney, S.T., and Bentley, C.R. (1989) Water-pressure coupling of sliding and bed deformation: II. Velocity-depth profiles: *J. Glaciol.* **35**, 119-129.

Bamber, J.L. (1987) Internal reflecting horizons in Spitsbergen glaciers: *Ann. Glaciol.* **9**, 5-10.

Barnes, P., Tabor, D., and Walker, J.C.W. (1971) Friction and creep of polycrystalline ice: *Proc. Roy. Soc. Lond., Ser. A.* **324**, 127-155.

Bentley, C.R. (1987) Antarctic ice streams: a review: *J. Geophys. Res.* **92**, 8843-8858.

Bindschadler, R. (1983) The importance of pressurized subglacial water in separation and sliding at the glacier bed: *J. Glaciol.* **29**, 3-19.

Bindschadler, R.A., Stephenson, S.N., MacAyeal, D.R., and Shabtaie, S. (1987) Ice dynamics at the mouth of Ice Stream B, Antarctica: *J. Geophys. Res.* **92**, 8885-8894.

Blankenship, D.D., Bentley, C.R., Rooney, S.T., and Alley, R.B. (1987) Till beneath ice stream B. I. Properties derived from seismic travel times: *J. Geophys. Res.* **92**, 8903-8911.

Blatter, H. (1987) On the thermal regime of an arctic valley glacier, a study of White Glacier, Axel Heiberg Island, N.W.T., Canada: *J. Glaciol.* **33** (114), 200-211.

Blatter, H., and Hutter, K. (1991) Polythermal conditions in Arctic glaciers: *J. Glaciol.* **37** (126), 261-269.

Blatter, H., and Klappenberger, G. (1988) Mass balance and thermal regime of the Laika ice cap, Coburg Island, N.W.T., Canada: *J. Glaciol.* **34** (116), 102-110.

Böhmer, W.J., and Herterich, K. (1990) A simplified three-dimensional ice-sheet model including ice shelves: *Ann. Glaciol.* **14**, 17-19.

Boulton, G.S. (1986) A paradigm shift in glaciology? *Nature* **322**, 18.

Boulton, G.S., and Hindmarsh, R.C.A. (1987) Sediment deformation beneath glaciers: rheology and geological consequences, *J. Geophys. Res.* **92**, 9059-9082.

Boulton, G.S., and Jones, A.S. (1979) Stability of temperate ice sheets resting on beds of deformable sediment: *J. Glaciol.* **24**, 29-43.

Boulton, G.S., Smith, G.D., and Morland, L.W. (1984) The reconstruction of former ice sheets and their mass-balance characteristics using a non-linearly viscous flow model: *J. Glaciol.* **30** (105), 140-152.

Boulton, G.S., Smith, G.D., Jones, A.S., and Newsome, J. (1985) Glacial geology and glaciology of the last mid-latitude ice sheets: *J. Geol. Soc. London* **142**, 447-474.

Bowen, O.Q. (1980) Antarctic ice surges and theories of glaciation: *Nature (Lond.)* **283**, 619-620.

Brown, C.S., Meier, M.F., and Post, A. (1982) Calving speeds of Alaskan tidewater glaciers, with application to Columbia Glacier: U.S. Geol. Surv. Prof. Paper 1258C.

Bucher, W.H. (1956) The role of gravity in orogenesis: *Geol. Soc. Amer. Bull.* **67**, 1295-1318.

Budd, W.F. (1975) A first simple model for periodically self-surging glaciers: *J. Glaciol.* **14** (70), 3-21.

Budd, W.F., and Jacka, T.H. (1989) A review of ice rheology for ice sheet modeling: *Cold Regions Sci. and Tech.* **16**, 107-144.

Budd, W.F., and Jenssen, D. (1975) Numerical modelling of glacier systems: *Int. Assoc. Hydrol. Sci., Pub.* **104**, 257-291.

Budd, W.F., and Jenssen, D. (1987) Numerical modeling of basal water flux under the ice sheet in the West Antarctic Ross Ice Shelf basin: in C.J. van der Veen and J. Oerlemans (eds.), *Dynamics of the West Antarctic Ice Sheet*, Reidel, Dordrecht, p. 293-320.

Budd, W.F., and Jenssen, D. (1989) The dynamics of the Antarctic ice sheet: *Ann. Glaciol.* **12**, 16-22.

Budd, W.F., and McInnes, B.J. (1978) The periodically surging Medvezhi Glacier matched with a general ice flow model: U.S.S.R. Academy of Sciences, Soviet Geophysical Committee, Section of Glaciology and Institute of Geography, Data of Glaciological Studies, Publication No. 32, Moscow, 247-260.

Budd, W.F., and McInnes, B.J. (1979) Periodic surging of the Antarctic ice sheet an assessment by modelling: *Hydrol. Sci. Bull.* **24**, 95-104.

Budd, W.F., and Radok, U. (1971) Glaciers and other large ice masses: *Rep. Prog. Phys.* **34**, 1-70.

Budd, W.F., and Smith, I.N. (1982) Large-scale numerical modeling of the Antarctic Ice Sheet: *Ann. Glaciol.* **3**, 42-49.

Budd, W.F., and Smith, I.N. (1987) Conditions for growth and retreat of the Laurentide Ice Sheet: *Géogr. phys. et quatern.* **41**, 279-290.

Budd, W.F., and Young, N.W. (1983a) Techniques for the analysis of temperature-depth profiles in ice sheets: in G. de Q. Robin (ed.), *The Climatic Record in Polar Ice Sheets*, Cambridge Univ. Press, Ch. 5.4, p. 145-150.

Budd, W.F., and Young, N.W. (1983b) Application of modeling techniques to measured profiles of temperatures and isotopes: in ibid., Ch. 5.5, p. 150-177.

Budd, W.F., Jenssen, D., and Radok, U. (1971) Derived physical characteristics of the Antarctic ice sheet: Austral. Natl. Antarctic Res. Exped., Interim Reports, No. 120, Melbourne, 177 pp.; Univ. Melbourne, Meteorol. Dept., Publ. No. 18 (2nd printing, 1984).

Budd, W.F., Keage, P.L., and Blundy, N.A. (1979) Empirical studies of ice sliding: *J. Glaciol.* **23**, 157-170.

Budd, W.F., Jacka, T.H., Jenssen, D., Radok, U., and Young, N.W. (1982) Derived physical characteristics of the Greenland Ice Sheet: Univ. Melbourne, Meteorol. Dept., Publ. No. 23.

Budd, W.F., Jenssen, D., and Smith, I.N. (1984) A three-dimensional time-dependent model of the West Antarctic Ice Sheet: *Ann. Glaciol.* **5**, 29-36.

Budd, W.F., Jenssen, D., and McInnes, B.J. (1985) Numerical modelling of ice stream flow with sliding: ANARE (Australian National Antarctic Research Expeditions), Research Notes 28 (ed. T.H. Jacka), Antarctic Div., Dept. of Science, p. 130-137.

Budd, W.F., Smith, I.N., and Radok, U. (1986a) On the surging potential of polar ice streams. Part III. Sliding and surging analyses for two West Antarctic ice streams: U.S. Dept. of Energy, Report DE/ER/60197-4.

Budd, W.F., Jenssen, D., Leach, J.H.I., Smith, I.N., and Radok, U. (1986b) The north polar ice cap of Mars as a steady-state system: *Polarforschung* **56**, 43-63.

Budd, W.F., McInnes, B.J., Jenssen, D., and Smith, I.N. (1987) Modeling the response of the West Antarctic ice sheet to a climactic warming: in C.J. van der Veen and J. Oerlemans (eds.), *Dynamics of the West Antarctic Ice Sheet*, Reidel, Dordrecht, p. 321-358.

Campbell, W.J., and Rasmussen, L.A. (1969) Three-dimensional surges and recoveries in a numerical glacier model: *Can. J. Earth Sci.* **6**, 979-986.

Campbell, W.J., and Rasmussen, L.A. (1970) A heuristic numerical model for three-dimensional time-dependent glacier flow: *Int. Assoc. Hydrol. Sci. Publ.* **86**, 177-190.

Clarke, G.K.C. (1976) Thermal regulation of glacier surging: *J. Glaciol.* **16**, 231-250.

Clarke, G.K.C., Nitsan, V., and Paterson, W.S.B. (1977) Strain heating and creep instability in glaciers and ice sheets: *Rev. Geophys. Space Phys.* **15**, 235-247.

Clifford, S.M. (1987) Polar basal melting on Mars: *J. Geophys. Res.* **92**, 9135-9152.

Dahl-Jensen, D. (1989) Two-dimensional thermo-mechanical modeling of flow and depth-age profiles near the ice divide in central Greenland: *Ann. Glaciol.* **12**, 31-36.

Dansgaard, W., and Johnsen, S.J. (1969) A flow model and a time scale for the ice core from Camp Century, Greenland: *J. Glaciol.* **8** (53), 215-223.

Deblonde, G., and Peltier, W.R. (1990) A paleoclimatic model of the mid-Pleistocene climate transition: *Ann. Glaciol.* **14**, 47-50.

Durham, W.B., Kirby, S.H., Heard, H.C., Stern, L.A., and Bors, C.O. (1988) Water ice phases II, III, and V: plastic deformation and phase relations: *J. Geophys. Res.* **93**, 10,191-10,208.

Echelmeyer, K. (1983) Response of Blue Glacier to a perturbation in ice thickness theory and observation: Ph.D. thesis, Calif. Inst. of Technol.

Echelmeyer, K., and Kamb, B. (1986) Rheology of ice II and ice III from high-pressure extrusion: *Geophys. Res. Let.* **13**, 693-696.

Echelmeyer, K., and Kamb, B. (1987) Glacier flow in a curving channel: *J. Glaciol.* **33** (115), 1-12.

Echelmeyer, K., and Zhongxiang, W. (1987) Direct observation of basal sliding and deformation of basal drift at sub-freezing temperatures: *J. Glaciol.* **33** (113), 83-98.

Engelhardt, H., Humphrey, N., Kamb, B., and Fahnestock, M. (1990) Physical conditions at the base of a fast moving Antarctic ice stream: *Science* **248**, 57-59.

Engelhardt, H.F., Harrison, W.D., and Kamb, B. (1978) Basal sliding and conditions at the glacier bed as revealed by bore-hole photography: *J. Glaciol.* **20** (84), 469-508.

Fahnestock, M.A. (1991) Hydrologic control of sliding velocity in two Alaskan glaciers: observations and theory: Ph.D. thesis, California Institute of Technology.

Fastook, J.L. (1987a) The finite element method applied to a time-dependent flow band model: in C.J. van der Veen and J. Oerlemans (eds.) *Dynamics of the West Antarctic Ice Sheet*, Reidel, Dordrecht, p. 203-221.

Fastook, J.L. (1987b) Use of a new finite element continuity model to study the transient behavior of Ice Stream C and causes of its present low velocity: *J. Geophys. Res.* **92**, 8941-8949.

Fastook, J.L. (1990) A map-plane finite-element program for ice-sheet reconstruction: in H.U. Brown III (ed.), *Computer Assisted Analysis and Modeling on the IBM 3090*, IBM Corp., Sci. and Technical Computing Dept., White Plains, NY, p. 45-80.

Fastook, J.L., and Chapman, J.E. (1989) A map-plane finite-element model: three modeling experiments: *J. Glaciol.* **35** (119), 48-52.

Firestone, J., Waddington, E., and Cunningham, J. (1990) The potential for basal melting under Summit, Greenland: *J. Glaciol.* **36** (123), 163-168.

Flohn, M. (1974) Background of a geophysical model of the initiation of the next glaciation: *Quat. Res.* **4**, 385-404.

Fowler, A.C. (1986a) A sliding law for glaciers of constant viscosity in the presence of subglacial cavitation: *Proc. Roy. Soc. Lond. Ser. A* **407**, 147-170.

Fowler, A.C. (1984) On the transport of moisture in polythermal glaciers: *J. Geophys. Astrophys. Fluid Dynam.* **28**, 99-140.

Fowler, A.C. (1986b) Sub-temperate basal sliding: *J. Glaciol.* **32** (110), 3-5.

Fowler, A.C. (1987a) Sliding with cavity formation: *J. Glaciol.* **33** (115), 255-267.

Fowler, A.C. (1987b) A theory of glacier surges: *J. Geophys. Res.* **92**, 9111-9120.

Fowler, A.C., and Larson, D.A. (1978) On the flow of polythermal glaciers. I. Model and preliminary analysis: *Proc. Roy. Soc. Lond.* **A363**, 217-242.

Fowler, A.C., and Larson, D.A. (1980) Thermal stability properties of a model of glacier flow: *Geophys. J. R. Astron. Soc.* **63**, 347-359.

Gruell, W. (1992) Hintereisferner, Austria: mass-balance reconstruction and numerical modelling of the historical length variations: *J. Glaciol.* **38** (129), 233-244.

Hallet, B. (1979) A theoretical model of glacial abrasion: *J. Glaciol.* **23** (89), 39-50.

Hallet, B. (1981) Glacial abrasion and sliding: their dependence on the debris concentration in basal ice: *Ann. Glaciol.* **2**, 23-28.

Hanson, B. (1990) Thermal response of a small ice cap to climatic forcing: *J. Glaciol.* **36** (122), 49-56.

Harbor, J.M. (1992) Application of a general sliding law to simulating flow in a glacier cross-section: *J. Glaciol.* **38** (128), 182-190.

Harbor, J.M., Hallet, B., and Raymond, C.F. (1988) A numerical model of landform development by erosion: *Nature (Lond.)* **333**, 347-349.

Herrick, D.L., and Stevenson, D.J. (1990) Extensional and compressional instabilities in icy satellite lithospheres: *Icarus* **85**, 191-204.

Herterich, K. (1987) On the flow within the transition zone between ice sheet and ice shelf: in C.J. van der Veen and J. Oerlemans (eds.) *Dynamics of the West Antarctic Ice Sheet*, Reidel, Dordrecht, p. 185-202

Herterich, K. (1988) A three-dimensional model of the Antarctic ice sheet: *Ann. Glaciol.* **11**, 32-35.

Hills, E.S. (1963) *Elements of Structural Geology*: Wiley, NY.

Hindmarsh, R.C.A., Boulton, G.S., and Hutter, K. (1988) Modes of operation of thermo-mechanically coupled ice sheets: *Ann. Glaciol.* **12**, 57-69.

Hofstadter, M., and Murray, B.C. (1990) Ice sublimation and rheology: implications for the Martian polar layered deposits: *Icarus* **84**, 352-361.

Hooke, R.L. (1981) Flow law for polycrystalline ice in glaciers: comparison of theoretical predictions, laboratory data, and field measurements: *Rev. Geophys. Space Phys.* **19**, 664-672.

Hooke, R.L. (1989) Englacial and subglacial hydrology: a qualitative review: *Arctic and Alpine Res.* **21**, 221-233.

Hooke, R.L., Raymond, C.F., Hotchkiss, R.L., and Gustafson, R.J. (1979) Calculations of velocity and temperature in a polar glacier using the finite-element method: *J. Glaciol.* **24** (90), 131-146.

Hughes, T.J. (1972) Thermal convection in polar ice sheets related to the various empirical flow laws of ice: *Geoph. J. Roy. Astr. Soc.* **27**, 215-229.

Hughes, T.J. (1973) Is the West Antarctic ice sheet disintegrating? *J. Geophys. Res.* **78**, 7884-7910.

Hughes, T.J. (1976) The theory of thermal convection in polar ice sheets: *J. Glaciol.* **16**, 41-71.

Hughes, T.J. (1977) West Antarctic ice streams: *Rev. Geophys. Space Phys.* **15**, 1-46.

Hughes, T. (1981) Numerical reconstruction of paleo-ice sheets: in G.H. Denton and T.J. Hughes (eds.), *The Last Great Ice Sheets*, Wiley, New York, Ch. 5, p. 221-261.

Hughes, T.J. (1985) Thermal convection in ice sheets: we look but do not see: *J. Glaciol.* **31** (107), 39-48.

Hughes, T.J. (1992a) Abrupt climatic change related to unstable ice-sheet dynamics: toward a new paradigm: *Palaeogeogr., Palaeoclimatol., Palaeoecol. (Global and Planetary Change Section)* **97**, 203-234.

Hughes, T.J. (1992b) Theoretical calving rates from glaciers along ice walls grounded in water of variable depths: *J. Glaciol.* **38** (129), 282-294.

501

Hughes, T.J., Denton, G.H., Andersen, B.G., Schilling, D.H., Fastook, J.H., and Lingle, C.S. (1981) The last great ice sheets: a global view: in G.H. Denton and T.J. Hughes (eds.), *The Last Great Ice Sheets*, Wiley, New York, Ch. 5, p. 275-318.

Humphrey, N. (1987) Coupling between water pressure and basal sliding in a linked-cavity hydraulic system: in E.D. Waddington and J.S. Walder (eds.), *The Physical Basis of Ice Sheet Modelling*: Int. Assoc. Hydrol. Sci. Publ. No. 170, p. 105-119.

Hutter, K. (1982) A mathematical model of polythermal glaciers and ice sheets: *J. Geophys. Astrophys. Fluid Dynam.* **21**, 201-224.

Hutter, K. (1983) *Theoretical Glaciology*: Reidel, Dordrecht, 510 p.

Hutter, K., Blatter, H., and Funk, M. (1988) A model computation of moisture content in polythermal glaciers: *J. Geophys. Res.* **93**, 12,205-12,214.

Huybrechts, P. (1990) A 3-D model for the Antarctic ice sheet: a sensitivity study on the glacial-interglacial contrast: *Climate Dynamics* **5**, 79-92.

Huybrechts, P. (1992) The Antarctic ice sheet and environmental change: a three-dimensional modelling study: *Ber. Polarforsch.* **99**, 1-241.

Huybrechts, P., and Oerlemans, J. (1988) Evolution of the east Antarctic ice sheet: a numerical study of thermo-mechanical response patterns with changing climate: *Ann. Glaciol.* **11**, 52-59.

Huybrechts, Ph., deNooze, P., and Decleir, H. (1989) Numerical modelling of Glacier d'Argentière and its historic front variations: in J. Oerlemans (ed.) *Glacier Fluctuations and Climatic Change*, Reidel, Dordrecht, p. 373-389.

Hyde, W.T., and Peltier, W.R. (1987) Sensitivity experiments with a model of the ice age cycle: the response to Milankovich forcing: *J. Atm. Sci.* **44**, 1351-1374.

Iken, A. (1981) The effect of subglacial water pressure on the sliding velocity of a glacier in an idealized numerical model: *J. Glaciol.* **27**, 407-421.

Iken, A., Echelmeyer, K., Harrison, W.D., and Funk, M. (1992) Mechanisms of fast flow in Jakobshavns Isbrae, Greenland. Part I: Measurements of temperature and water level in deep boreholes: *J. Glaciol.*, in press.

Iverson, N. (1990) Laboratory simulations of glacial abrasion: comparison with theory: *J. Glaciol.* **36** (124), 304-314.

Jenssen, D. (1977) A three-dimensional polar ice-sheet model: *J. Glaciol.* **18**, 373-389.

Jenssen, D., Budd, W.F., Smith, I.N., and Radok, U. (1985) On the surging potential of polar ice streams. Part II. Ice streams and physical characteristics of the Ross Sea drainage basin, West Antarctica: Report DE/ER/60197-3, Meteorol. Dept. Univ. Melbourne and C.I.R.E.S., Univ. Colo., Boulder, 36 pp. + 3 appendices.

Jenssen, D., Mavrakis, E., and Budd, W.F. (1989) Modeling Antarctic ice sheet characteristics for planning of deep core drilling: *Abstracts of Conference and Workshop on Antarctic Weather and Climate*, Flinders University, South Australia, 5-7 July 1989.

Jones, A.S. (1979) The flow of ice over a till bed: *J. Glaciol.* **22** (87), 393-395.

Kamb, B. (1970) Sliding motion of glaciers: theory and observation: *Rev. Geophys. Space Phys.* **8**, 673-728.

Kamb, B. (1972) Experimental recrystallization of ice under stress: *Am. Geophys. Union, Geophys. Mono.* **16**, 211-241.

Kamb, B. (1986) Stress-gradient coupling in glacier flow: III. Exact longitudinal equilibrium equation: *J. Glaciol.* **32** (112), 335-341.

Kamb, B. (1987) Glacier surge mechanism based on linked cavity configuration of the basal water conduit system: *J. Geophys. Res.* **92**, 9083-9100.

Kamb, B. (1991) Rheological nonlinearity and flow instability in the deforming bed mechanism of ice stream motion: *J. Geophys. Res.* **96**, 16,585-16,595.

502

Kamb, B., and Echelmeyer, K.A. (1986) Stress-gradient coupling in glacier flow: I. Longitudinal averaging of the influence of ice thickness and surface slope: *J. Glaciol.* **32** (111), 267-284.

Kamb, B., and Engelhardt, H.F. (1987) Waves of accelerated motion in a glacier approaching surge: The mini-surges of Variegated Glacier, Alaska: *J. Glaciol.* **33**, 2746.

Kamb, B., and Engelhardt, H. (1991) Antarctic Ice Stream B: conditions controlling its motion and interactions with the climatic system: *Int. Assoc. Hydrol. Sci. Publ.* **208**, 145-154.

Kamb, B., and LaChapelle, E.R. (1964) Direct observation of the mechanism of glacier sliding over bedrock: *J. Glaciol.* **5**, 159-172.

Kamb, B., Pollard, D., and Johnson, C.B. (1976) Rock-frictional resistance to glacier sliding: *EOS Trans. Am. Geophys. Union* **57**, 325 (abst.).

Kamb, B., Raymond, C.F., Harrison, W.D., Engelhardt, H., Echelmeyer, K.A., Humphrey, N., Brugman, M.M., and Pfeller, T. (1985) Glacier surge mechanism: the 1982-3 surge of Variegated Glacier, Alaska: *Science* **227**, 469479.

Kamb, B., Meier, M.F., Engelhardt, H., Fahnestock, M.A., Humphrey, N., and Stone, D. (1992b) Mechanical and Hydrologic Basis for the Rapid Motion of a Large Tidewater Glacier: Part II. Interpretation: *J. Geophys. Res.*, ms. submitted for publication.

Kirk, R.L., and Stevenson, D.J. (1987) Thermal evolution of a differentiated Ganymede and implications for surface features: *Icarus* **69**, 91-135.

Kotlyakov, V.M., and Macheret, Yu.Yu. (1987) Radio-echo sounding of sub-polar glaciers in Svalbard: some problems and results of Soviet studies: *Ann.Glaciol.* **9**, 151-159.

Krimmel, R.M. (1992) Photogrammetric determination of surface altitude, velocity, and calving rate of Columbia Glacier, Alaska, 1983-91: *U.S. Geol. Surv., Open File Report* **92-104**.

Kruss, P. (1984) Terminus response of Lewis Glacier, Mt. Kenya, to sinusoidal net balance forcing: *J. Glaciol.* **30**, 212-217.

Lange, M.A., and MacAyeal, D.R. (1988) Numerical models of steady-state thickness and basal ice configurations of the central Ronne ice shelf, Antarctica: *Ann. Glaciol.* **11**, 64-70.

Letréguilly, A., Huybrechts, P., and Reeh, N. (1991) Steady-state characteristics of the Greenland ice sheet under different climates: *J. Glaciol.* **37** (125), 149-157.

Lile, R.C. (1978) The effect of anisotropy on the creep of polycrystalline ice: *J. Glaciol.* **21** (85), 475-483.

Lindstrom, D.R., and MacAyeal, D.R. (1987) Environmental constraints on West Antarctic Ice-Sheet formation: *J. Glaciol.* **33** (115), 1-11.

Lingle, C.S. (1984) A numerical model of interactions between a polar ice stream and the ocean: applications to Ice Stream E, West Antarctica: *J. Geophys. Res.* **89**, 3523-3549.

Lingle, C.S., and Brown, T.J. (1987) A subglacial aquifer bed model and water-pressure-dependent basal sliding relationship for a West Antarctic ice stream: in C.J. van der Veen and J. Oerlemans (eds.), *The Dynamics of the West Antarctic Ice Sheet*, Reidel, Dordrecht, p. 249-285.

Lingle, C.S., and Clark, J.A. (1985) A numerical model of interactions btween a marine ice sheet and the solid earth: application to a west Antarctic ice stream: *J. Geophys. Res.* **90**, 1100-1114.

Lliboutry, L. (1959) Une théorie du frottement du glacier sur son lit: *Ann. Géophys.* **15**, 250.

Lliboutry, L.A. (1964) Sub-glacial "supercavitation" as a cause of the rapid advances of glaciers: *Nature* **202**, 77.

Lliboutry, L. (1968) General theory of subglacial cavitation and sliding of temperate glaciers: *J. Glaciol.* **7**, 21-58.

Lliboutry, L. (1979a) Local friction laws for glaciers: a critical review and new openings: *J. Glaciol.* **23**, 67-95.

Lliboutry, L. (1979b) A critical review of analytical approximate solutions for steady state velocities and temperatures in cold ice-sheets: *Z. Gletscherkunde u. Glazialgeol.* **15**, 135-148.

Lliboutry, L. (1983) Viscosité selon une loi puissance anisotrope: calcul theoretique approprié à certaines glaces polaires: *Comptes rendus Acad. Sci. Paris* **297**-II, 787-790.

Lliboutry, L.A. (1987a) *Very Slow Flows of Solids: basics of modeling in geodynamics and glaciology*: M. Nijhoff, Dordrecht, 510 pp.

Lliboutry, L. (1987b) Realistic, yet simple boundary conditions for glaciers and ice sheets: *J. Geophys. Res.* **92**, 9101-9109.

Lliboutry, L., and Duval, P. (1985) Various isotropic and anisotropic ices found in glaciers and polar ice caps and their corresponding rheologies: *Ann. Geophys.* **3**, 207-224.

MacAyeal, D.R. (1987) Ice-shelf backpressure: form drag vs. dynamic drag: in C.J. van der Veen and J. Oerlemans (eds.), *Dynamics of the West Antarctic Ice Sheet*, Reidel, Dordrecht, p. 141-160.

MacAyeal, D.R. (1989a) Large-scale ice flow over a viscous basal sediment: theory and application to Ice Stream B, Antarctica: *J. Geophys. Res.* **94**, 4071-4087.

MacAyeal, D.R. (1989b) Ice-shelf response to ice-stream discharge fluctuations: III. The effects of ice-stream imbalance on the Ross Ice Shelf, Antarctica: *J. Glaciol.* **35** (119), 38-42.

MacAyeal, D.R. (1992) Irregular oscillations of the West Antarctic Ice Sheet: *Nature* **359**, 29-32.

MacAyeal, D.R., and Lange, M.A. (1988) Ice-shelf response to ice-stream discharge fluctuations: II. Ideal rectangular ice shelf: *J. Glaciol.* **31** (116), 128-134.

MacAyeal, D.R., and Thomas, R.H. (1982) Numerical modeling of ice-shelf motion: *Ann. Glaciol.* **3**, 189-194.

MacAyeal, D.R., and Thomas, R.H. (1986) The effects of basal melting on the present flow of the Ross Ice Shelf, Antarctica: *J. Glaciol.* **32** (110), 72-86.

MacAyeal, D.R., Shabtaie, S., Bentley, C.R., and King, S.D. (1986) Formulation of ice shelf dynamic boundary conditions in terms of a Coulomb rheology: *J. Geophys. Res.* **91**, 8177-8191.

MacAyeal, D.R., Bindschadler, R.A., Shabtaie, S., Stephenson, S., and Bentley, C.R. (1987) Force, mass, and energy budgets of the Crary Ice Rise complex, Antarctica: *J. Glaciol.* **33** (114), 218-230.

MacAyeal, D.R., Bindschadler, R.A., Stephenson, S., Shabtaie, S., and Bentley, C.R. (1989) Correction to: force, mass, and energy budgets of the Crary Ice Rise complex, Antarctica: *J. Glaciol.* **35** (119), 151-152.

Mahaffy, M.W. (1976) A three-dimensional numerical model of ice sheets: tests on the Barnes Ice Cap, Northwest Territories: *J. Geophys. Res.* **81**, 1059-1066.

McInnes, B.J., and Budd, W.F. (1984) A cross-sectional model for West Antarctica: *Ann. Glaciol.* **5**, 95-99.

McInnes, B., Radok, U., Budd, W.F., and Smith, I.N. (1985) On the surging potential of polar ice streams. Part I. Sliding and surging of large ice masses a review: Report DE/ER/60197-2, C.I.R.E.S., Univ. Colo., Boulder, and Meterol. Dept. Univ. Melbourne, 53 pp.

McInnes, B.J., Budd, W.F., Smith, I.N., and Radok, U. (1986) On the surging potential of polar ice streams. Part III. Sliding and surging analysis for two West Antarctic ice streams: Report DE/ER/60197-4, Meteorol. Dept. Univ. Melbourne and C.I.R.E.S. Univ. Colo., Boulder, 55 pp. + 3 appendices.

McMeeking, R.M., and Johnson, R.E. (1986) On the mechanics of surging glaciers: *J. Glaciol.* **32** (110), 120-132.

Meier, M.F., and Post, A. (1969) What are glacier surges? *Can. J. Earth Sci.* **6**, 807-817.

Meier, M.F., and Post, A. (1987) Fast tidewater glaciers: *J. Geophys. Res.*, **92**, 9051-9058.

Meier, M.F., Rasmussen, L.A., Post, A., Brown, C.S., Sikonia, W.G., Bindschadler, R.A., Mayo, L.R., and Trabant, D.C. (1980) Predicted timing of the disintegration of the lower reach of Columbia Glacier, Alaska: *U.S. Geol. Surv. Open File Rep.*, **80-582**.

Metcalf, R.C. (1979) Energy dissipation during subglacial abrasion at Nisqually Glacier, Washington, U.S.A.: *J. Glaciol.* **23** (89), 233-245.

Morland, L.W., Smith, G.D., and Boulton, G.S. (1984) Basal sliding relations deduced from ice sheet data: *J. Glaciol.* **30** (105), 131-139.

Nye, J.F. (1960) The response of glaciers and ice sheets to seasonal and climatic changes: *Proc. Roy. Soc. Lond.* **A239**, 559-584.

Nye, J.F. (1963) The response of a glacier to changes in the rate of nourishment and wastage: *Proc. Roy. Soc.* **A275**, 87-112.

Nye, J.F. (1965a) A numerical model of inferring the budget history of a glacier from its advance and retreat: *J. Glaciol.* **5** (41), 589-607.

Nye, J.F. (1965b) The flow of a glacier in a channel of rectangular, elliptic, or parabolic cross-section: *J. Glaciol.* **5** (41), 661-690.

Oerlemans, J. (1982) A model of the Antarctic Ice Sheet: *Nature* **297**, 550-553.

Oerlemans, J. (1983) A numerical study on cyclic behavior of polar ice sheets: *Tellus* **35A**, 81-87.

Oerlemans, J. (1986) An attempt to simulate historic front variations of Nigardsbreen, Norway: *Theor. and Appl. Climatol.* **37**, 126-135.

Oerlemans, J. (1989) On the response of valley glaciers to climatic change: in J. Oerlemans (ed.) *Glacier Fluctuations and Climatic Change*, Kluwer, Dordrecht, p. 353-372.

Oerlemans, J. (1992) Climate sensitivity of glaciers in southern Norway: aplication of an energy-balance model to Nigardsbreen, Hellstugubreen, and Alfotbreen: *J. Glaciol.* **38** (129), 223-232.

Oerlemans, J., and van der Veen, C.J. (1984) *Ice Sheets and Climate*: Reidel, Dordrecht, 217 pp.

Oeschger, H., and Langway, C.C. (eds.) (1989) *The Environmental Record in Glaciers and Ice Sheets*: Wiley, New York, 401 pp.

Oswald, G.K.A., and Robin, G. de Q. (1973) Lakes beneath the Antarctic ice sheet: *Nature* **245**, 251-254.

Paterson, W.S.B. (1980) Ice sheets and ice shelves: in S.C. Colbeck (ed.) *Dynamics of Snow and Ice Masses*, Academic Press, New York, Ch. 1, p. 3-79.

Paterson, W.S.B. (1981) *The Physics of Glaciers*: Pergamon, Oxford, 380 pp.

Paterson, W.S.B. (1991) Why ice-age ice is sometimes "soft": *Cold Regions Sci. and Technol.* **20**, 75-98.

Peltier, W.R., and Hyde, W.T. (1987) Glacial isostasy and the ice-age cycle: Int. Assoc. Hydrol. Sci., Publ. **170**, 247-260.

Philberth, K., and Federer, B. (1971) On the temperature profile and the age profile in the central part of cold ice sheets: *J. Glaciol.* **10**, 3-14.

Poirier, J.-P. (1982) Rheology of ices: a key to the tectonics of the icy moons of Jupiter and Saturn: *Nature* **299**, 638-640.

Post, A. (1975) Preliminary hydrography and historic terminal changes of Columbia Glacier, Alaska: *U.S. Geol. Surv. Hydrol. Invest. Atlas*, 559.

Radok, U. Barry, R.G., Jenssen, D., Keen, R.A., Kiladis, G.N. and McInnes, B. (1982) *Climatic and Physical Characteristics of the Greenland Ice Sheet*: Univ. Colorado, Coop. Inst. for Res. in Environm. Sci., Boulder, CO.

Radok, U., Jenssen, D., and McInnes, B. (1987) On the surging potential of polar ice streams [Part V.]. Antarctic surges a clear and present danger? U.S. Dept. of Energy, Report DOE/ER/60197-H1, 62 pp.

Radok, U., McInnes, B.J., Jenssen, D., and Budd, W.F. (1989) Model studies on ice-stream surging: *Ann. Glac.* **12**, 132-137.

Rasmussen, L.A., and Campbell, W.J. (1973) Comparison of three contemporary flow laws in a three-dimensional, time-dependent glacier model: *J. Glaciol.* **12** (66), 361-373.

Raymond, C.F. (1971) Flow in a transverse section of Athabasca Glacier, Alberta, Canada: *J. Glaciol.* **10** (58), 55-84.

Raymond, C.F. (1983) Deformation in the vicinity of ice divides: *J. Glaciol.* **29** (103), 357-373.

Raymond, C.F. (1987) How do glaciers surge? A review: *J. Geophys. Res.* **92**, 9121-9134.

Raymond, C.F. Unpublished. Numerical calculation of glacier flow by finite element methods. Final technical report for National Science Foundation Grant No. DPP74-19075.

Raymond, C.F., and Harrison, W.D. (1987) Fit of ice motion models to observations from Variegated Glacier, Alaska: Int. Assoc. Hydrol. Sci., Publ. 170, 153-166.

Reeh, N. (1968) On the calving of ice from floating glaciers and ice shelves: *J. Glaciol.* **7** (50), 215-232.

Reeh, N. (1989) Dating by flow modeling: a useful tool or an exercise in applied mathematics?: in H. Oeschger and C.C. Langway (eds.), *The Environmental Record in Glaciers and Ice Sheets*, Wiley, NY, p. 141-159.

Reynaud, L. (1973) Flow of a valley glacier with a solid friction law: *J. Glaciol.* **12** (65), 251-258.

Ritz, C. (1987) Time-dependent boundary conditions for calculation of temperature fields in ice sheets: in E.D. Waddington and J.S. Walder (eds.), *The Physical Basis of Ice Sheet Modelling*: Int. Assoc. Hydrol. Sci. Publ. No. 170, p. 207-216.

Ritz, C. (1989) Interpretation of the temperature profile measured at Vostok, East Antarctica: *Ann. Glaciol.* **12**, 138-144.

Robin, G. de Q. (1955) Ice movement and temperature distribution in glaciers and ice sheets: *J. Glaciol.* **2**, 523-532.

Robin, G. de Q. (1983) *The Climatic Record in Polar Ice Sheets*: Cambridge, 212 pp.

Robin, G. de Q., Drewry, D.J., and Meldrun, D.T. (1977) International studies of ice sheet and bedrock: *Phil. Trans. Roy. Soc. Lond.* **B279**, 185-196.

Röthlisberger, H. (1972) Water pressure in intra- and sub-glacial channels: *J. Glaciol.* **11**, 177-203.

Röthlisberger, H. (1980) Les glaciers, force naturelle: in Swiss Tourist Office (ed.) *La Suisse et ses Glaciers*: Kümmerly u. Frey, Bern, p. 128-164.

Röthlisberger, H., and Lang, H. (1987) Glacial hydrology: in A.M. Gurnell and M.J. Clark (eds.) *Glacio-fluvial Sediment Transfer*, Wiley, New York, p. 207-284.

Shoemaker, E.M. (1986) Subglacial hydrology for an ice sheet resting on a deformable aquifer: *J. Glaciol.* **32** (110), 20-30.

Shoemaker, E.M. (1988) On the formulation of basal debris drag for the case of sparse debris: *J. Glaciol.* **34** (118), 259-264.

Schøtt, C., Waddington, E.O., and Raymond, C.F. (1991) Predicted time scales for GISP2 and GRIP boreholes at Summit, Greenland: *J. Glaciol.* **38** (128), 162-168.

Shoji, H., and Langway, C.C. (1985) Mechanical properties of fresh ice core from Dye 3, Greenland: in C.C. Langway, Jr., H. Oeschger, and W. Dansgaard (eds.) *Greenland Ice Core: Geophysics, Geochemistry, and the Environment*, Am. Geophys. Union, Geophys, Monogr. 33, Washington, p. 39-48.

Shreve, R.L. (1972) Movement of water in glaciers: *J. Glaciol.* **32**, 20-30.

Shreve, R.L. (1984) Glacier sliding at subfreezing temperatures: *J. Glaciol.* **30** (106), 341-347.

Spring, U., and Hutter, K. (1981) Numerical studies of jökulhlaups: *Cold Reg. Sci. Technol.* **4**, 227-244.

Stevenson, D.J., and Lunine, J.I. (1986) Mobilization of cryogenic ice in outer solar system satellites: *Nature* **323**, 46-48.

Stroeven, A., van de Wal, R., and Oerlemans, J. (1989) Historic front variations of the Rhone Glacier: simulation with an ice flow model: in J. Oerlemans (ed.) *Glacier Fluctuations and Climatic Change*, Reidel, Dordrecht, p. 391-405.

Stuiver, M., Denton, G.H., Hughes, T.J., and Fastook, J.L. (1981) History of the marine ice sheet in West Antarctica during the last glaciation: a working hypothesis: in G.H. Denton and T.J. Hughes (eds.) *The Last Great Ice Sheets*, Wiley, New York, p. 319-439.

Thomas, R.H. (1973a) The creep of ice shelves: theory: *J. Glaciol.* **12** (64), 45-54.

Thomas, R.H. (1973b) The creep of ice shelves: interpretation of observed behavior: *J. Glaciol.* **12** (64), 55-70.

Thomas, R.H. (1979a) The dynamics of marine ice sheets: *J. Glaciol.* **24** (90), 167-177.

Thomas, R.H. (1979b) Ice shelves: a review: *J. Glaciol.* **24** (90), 273-286.

Thomas, R.H., and Bentley, C.R. (1978) A model for Holocene retreat of the West Antarctic ice sheet: *Quat. Res.* **10**, 150-170.

Thomas, R.H., and MacAyeal, D.R. (1982) Derived characteristics of the Ross Ice Shelf, Antarctica: *J. Glaciol.* **28** (100), 397-412.

Thomas, R.H., Sanderson, T.J.O., and Rose, K.E. (1979) Effect of climatic warming on the West Antarctic Ice Sheet: *Nature* **277**, 355-358.

van der Veen, C.J. (1985) Response of a marine ice sheet to changes at the grounding line: *Quat. Res.* **24**, 257-267.

van der Veen, C.J. (1986) Numerical modelling of ice shelves and ice tongues: *Ann. Geophysicae* **4B**, 45-54.

van der Veen, C.J. (1987) Longitudinal stresses and basal sliding: a comparative study: in C.J. van der Veen and J. Oerlemans (eds.) *Dynamics of the West Antarctic Ice Sheet*, Reidel, Dordrecht, p. 223-248.

van der Veen, C.J. (1991) Numerical modeling of glaciers and ice sheets: unpublished ms.

van der Veen, C.J. (1992) Land ice and climate: in K. Trenberth (ed.) *Climate Systems Modeling*, Chap. 13, Cambridge University Press, in press.

van der Veen, C.J., and Whillans, I.M. (1989) Force budget: I. Theory and numerical methods: *J. Glaciol.* **35**, 53-60.

van der Veen, C.J., and Whillans, I.M. (1990) Flow laws for glacier ice: comparison of numerical methods and field measurements: *J. Glaciol.* **36** (124), 324-339.

Vivian, R. (1966) La catastrophe du glacier Allalin: *Rev. Géogr. Alpine* **54** (1), 97-112.

Waddington, E.D., and Walder, J.S. (eds.) (1987) *The Physical Basis of Ice Sheet Modelling*: Int. Assoc. Hydrol. Sci. Publ. No. 170, I.A.H.S. Press, Wallingford, Oxfordshire, 390 pp.

Walder, J.S. (1982) Stability of sheet flow of water beneath temperate glaciers and implications for glacier surging: *J. Glaciol.* **28** (99), 273-294.

Weertman, J. (1957a) On the sliding of glaciers: *J. Glaciol.* **3**, 33-38.

Weertman, J. (1957b) Deformation of floating ice shelves: *J. Glaciol.* **3**, 38-42.

Weertman, J. (1962) Catastrophic glacier advances: *Int. Assoc. Hydrol. Sci. Publ.* **58**, 31-39.

Weertman, J. (1964) The theory of glacier sliding: *J. Glaciol.* **5**, 287-303.

Weertman, J. (1966) Effect of a basal water layer on the dimensions of ice sheets: *J. Glaciol.* **6**, 191-207.

Weertman, J. (1969) Water lubrication mechanism of glacier surges: *Can. J. Earth Sci.* **6**, 929-942.

Weertman, J. (1972) General theory of water flow at the base of a glacier or ice sheet: *Rev. Geophys. Space Phys.* **10**, 287-333.

Weertman, J. (1979) The unsolved general glacier sliding problem: *J. Glaciol.* **23** (89), 97-116.

Weertman, J., and Birchfield, G.E. (1982) Subglacial water flow under ice streams and West-Antarctic ice-sheet stability: *Ann. Glaciol.* **3**, 316-320.

Whillans, I.M., Bolzan, J., and Shabtaie, S. (1987) Velocity of ice streams B and C, Antarctica: *J. Geophys. Res.* **92**, 8895-8902.

Wilson, A.T. (1964) Origin of ice ages: an ice shelf theory for Pleistocene glaciation: *Nature (Lond.)* **201**, 147-149.